Cancer Screening

BASIC AND CLINICAL ONCOLOGY

Editor

Bruce D. Cheson, M.D.

National Cancer Institute
National Institutes of Health
Bethesda, Maryland

1. Chronic Lymphocytic Leukemia: Scientific Advances and Clinical Developments, *edited by Bruce D. Cheson*
2. Therapeutic Applications of Interleukin-2, *edited by Michael B. Atkins and James W. Mier*
3. Cancer of the Prostate, *edited by Sakti Das and E. David Crawford*
4. Retinoids in Oncology, *edited by Waun Ki Hong and Reuben Lotan*
5. Filgrastim (r-metHuG-CSF) in Clinical Practice, *edited by George Morstyn and T. Michael Dexter*
6. Cancer Prevention and Control, *edited by Peter Greenwald, Barnett S. Kramer, and Douglas L. Weed*
7. Handbook of Supportive Care in Cancer, *edited by Jean Klastersky, Stephen C. Schimpff, and Hans-Jörg Senn*
8. Paclitaxel in Cancer Treatment, *edited by William P. McGuire and Eric K. Rowinsky*
9. Principles of Antineoplastic Drug Development and Pharmacology, *edited by Richard L. Schilsky, Gérard A. Milano, and Mark J. Ratain*
10. Gene Therapy in Cancer, *edited by Malcolm K. Brenner and Robert C. Moen*
11. Expert Consultations in Gynecological Cancers, *edited by Maurie Markman and Jerome L. Belinson*
12. Nucleoside Analogs in Cancer Therapy, *edited by Bruce D. Cheson, Michael J. Keating, and William Plunkett*
13. Drug Resistance in Oncology, *edited by Samuel D. Bernal*
14. Medical Management of Hematological Malignant Diseases, *edited by Emil J Freireich and Hagop M. Kantarjian*
15. Monoclonal Antibody-Based Therapy of Cancer, *edited by Michael L. Grossbard*
16. Medical Management of Chronic Myelogenous Leukemia, *edited by Moshe Talpaz and Hagop M. Kantarjian*
17. Expert Consultations in Breast Cancer: Critical Pathways and Clinical Decision Making, *edited by William N. Hait, David A. August, and Bruce G. Haffty*

18. Cancer Screening: Theory and Practice, *edited by Barnett S. Kramer, John K. Gohagan, and Philip C. Prorok*

ADDITIONAL VOLUMES IN PREPARATION

Supportive Care in Cancer: A Handbook for Oncologists: Second Edition, Revised and Expanded, *edited by Jean Klastersky, Stephen C. Schimpff, and Hans-Jörg Senn*

Integrated Cancer Management: Surgery, Medical Oncology, and Radiation Therapy, *edited by Michael H. Torosian*

AIDS-Related Cancers and Their Treatment, *edited by Ellen Feigal, Alexandra Levine, and Robert Biggar*

The Graft Versus Leukemia Effect in Allogeneic Stem Cell Transplantation, *edited by John Barrett and Yin-Zheng Jiang*

Cancer Screening
Theory and Practice

edited by
Barnett S. Kramer
National Cancer Institute and
Uniformed Services University
of the Health Sciences
Bethesda, Maryland

John K. Gohagan
Philip C. Prorok
National Cancer Institute
Bethesda, Maryland

MARCEL DEKKER

MARCEL DEKKER, INC. NEW YORK • BASEL

ISBN: 0-8247-0200-X

This book is printed on acid-free paper.

Headquarters
Marcel Dekker, Inc.
270 Madison Avenue, New York, NY 10016
tel: 212-696-9000; fax: 212-685-4540

Eastern Hemisphere Distribution
Marcel Dekker AG
Hutgasse 4, Postfach 812, CH-4001 Basel, Switzerland
tel: 41-61-261-8482; fax: 41-61-261-8896

World Wide Web
http://www.dekker.com

The publisher offers discounts on this book when ordered in bulk quantities. For more information, write to Special Sales/Professional Marketing at the headquarters address above.

Current printing (last digit):
10 9 8 7 6 5 4 3 2 1

PRINTED IN THE UNITED STATES OF AMERICA

Series Introduction

The current volume, *Cancer Screening*, is Volume 18 in the Basic and Clinical Oncology series. Many of the advances in oncology have resulted from close interaction between the basic scientist and the clinical researcher. The current volume follows, expands on, and illustrates the success of this relationship as demonstrated by new and promising areas for scientific research.

As editor of the series, my goal has been to recruit volume editors who not only have established reputations based on their outstanding contributions to oncology, but also have an appreciation for the dynamic interface between the laboratory and the clinic. To date, the series has consisted of monographs on topics such as chronic lymphocytic leukemia, nucleoside analogs in cancer therapy, therapeutic applications of interleukin-2, retinoids in oncology, gene therapy of cancer, and principles of antineoplastic drug development and pharmacology. *Cancer Screening* is certainly a most important addition to the series.

Volumes in progress include works on AIDS-related malignancies, secondary malignancies, chronic lymphoid leukemias, and controversies in gynecologic oncology. I anticipate that these volumes will provide a valuable contribution to the oncology literature.

Bruce D. Cheson, M.D.

Preface

Unlike diagnostic testing in the workup of a suspected medical problem, cancer screening is applied to asymptomatic individuals with no a priori suspicion of disease. As such, different principles guide the practice of diagnostic testing versus screening. In the former situation, a patient is seeking help from a health care professional in order to investigate a problem and alleviate suffering. In the latter situation, the health care system actively encourages groups of individuals who are thought to be healthy to undergo tests they would not ordinarily seek. This process may result in the transformation of asymptomatic people into patients, triggering a series of invasive tests and therapies that sometimes have morbid physical, emotional, and social consequences. To our knowledge, no medical technology is without risk. Therefore, it is incumbent upon those who would actively encourage individuals to be screened to have solid knowledge of the risks and benefits of the screening process and subsequent workup of abnormal results, as well as of the therapeutic outcomes. It is also important to communicate what is known about these issues to the screening candidate so that informed choice is possible. These complexities have their counterpart in formulating public health policy and in measuring the utility of screening versus other health care strategies.

Determining the efficacy and advisability of screening is a complicated process. Rarely does one size fit all. Early adverse outcomes in some people must be weighed against possible delayed benefits. Strength of evidence should be explicitly measured and incorporated into decisions at both the individual and the societal level. The aim of this book is to describe how one might translate what is known about a particular screening test into action and how to involve the person being considered for screening in the decision. Varying levels of evidence exist regarding the efficacy of screening tools that are currently available, some already in widespread use. We hope to equip the reader with sufficient information to judge the utility of such tests already in use, and to provide the basis for the reader to judge future screening technologies, such as genetic testing. The book also provides a practical approach to screening implementation for a number of cancer sites. The potential audiences of the book are medical students, physicians, and health care professionals, who must give guidance and advice to the public regarding cancer screening.

If the reader comes away from the book with a sense that the principles of cancer screening are more complex than is often conveyed in the proliferating list of screening guidelines, media reports, public campaigns, and testimonials, then we have accomplished our goal. Informed decision-making in medicine and public health requires some appreciation of its complexity. On the other hand, a firm grasp of the underlying concepts is not difficult to achieve, and many of the issues are applicable to a very broad range of medical screening and interventions. Disease pathogenesis is like an iceberg: Lurking beneath the surface of clinical overt disease is an immense reservoir of asymptomatic states in the pathogenic pathways. The more we seek, the more we will find. It pays to anticipate the consequences, both good and bad, as we approach the icebergs.

We are indebted to all the contributors, whose hard work is evident in their chapters. We are also grateful to Ms. Cindy Rooney, who served as managing editor, keeping the project on track, on time, and up to her high standard of quality.

Barnett S. Kramer
John K. Gohagan
Philip C. Prorok

Contents

Contributors

Mark L. Bernstein, M.D., F.R.C.P.C. Clinical Director, Hematology Service, Montreal Children's Hospital, and Associate Professor, Department of Pediatrics and Oncology, McGill University, Montreal, Quebec, Canada

William C. Black, M.D. Associate Professor of Radiology and Community and Family Medicine, Department of Radiology, Dartmouth-Hitchcock Medical Center, Lebanon, and Dartmouth Medical School, Hanover, New Hampshire

Robert S. Brown, Jr., M.D., M.P.H. Chief of Clinical Hepatology, Medical Director, Center for Liver Disease and Transplantation, Columbia Presbyterian Medical Center, New York, New York

Jack Cuzick, Ph.D. Head, Prevention and Control Section, Imperial Cancer Research Fund, London, England

Geoffrey E. Dougherty, M.D.C.M., M.Sc., F.R.C.P.C. Division Head, Intensive Ambulatory Care Service, Montreal Children's Hospital, and Assistant Professor, Pediatrics and Epidemiology and Biostatistics, McGill University, Montreal, Quebec, Canada

Alan C. Geller, R.N., M.P.H. Assistant Professor, Department of Dermatology, Cancer Prevention and Control Center, Boston University School of Medicine, and Department of Epidemiology and Biostatistics, Boston University School of Public Health, Boston, Massachusetts

Jack L. Gluckman, M.D Professor and Chairman, Department of Head and Neck Surgery, University of Cincinnati Medical Center, Cincinnati, Ohio

John K. Gohagan, Ph.D., F.A.C.E. Chief, Early Detection Branch, Division of Cancer Prevention, National Cancer Institute, Bethesda, Maryland

Edward S. Golub, Ph.D. Pacific Center for Ethics and Applied Biology, Solana Beach, California

Steven I. Gutman, M.D., M.D.A. Director, Division of Clinical Laboratory Devices, Center for Devices and Radiological Health, Food and Drug Administration, Rockville, Maryland

Emily L. Harris, Ph.D., M.P.H. Senior Investigator, Kaiser Permanente Center for Health Research, Portland, Oregon

Russell Harris, M.D., M.P.H. Associate Professor of Medicine, University of North Carolina School of Medicine; Co-Director, Program on Health Promotion and Disease Prevention; and Member, Cancer Control Program, UNC—Lineberger Comprehensive Cancer Center, Chapel Hill, North Carolina

Gerry B. Hill, M.B. B.Ch., Ph.D., F.R.C.P. Adjunct Professor, Department of Epidemiology and Community Medicine, University of Ottawa, Ottawa, Ontario, Canada

Ralph H. Hruban, M.D. Associate Professor, Department of Pathology and Oncology, The Johns Hopkins Medical Institutions, Baltimore, Maryland

Scott E. Kern, M.D. Associate Professor, Department of Oncology, The Johns Hopkins Medical Institutions, Baltimore, Maryland

Howard K. Koh, M.D., M.P.H., F.A.C.P. Commissioner, Massachusetts Department of Public Health; Department of Dermatology, Cancer Prevention and Control Center, Boston University School of Medicine; and Department of Epidemiology and Biostatistics, Boston University School of Public Health, Boston, Massachusetts

Barnett S. Kramer, M.D., M.P.H. Deputy Director, Division of Cancer Prevention, National Cancer Institute, and Clinical Professor of Medicine, Uniformed Services University of the Health Sciences, Bethesda, Maryland

Isra G. Levy, M.B. B.Ch., M.Sc., F.R.C.P.C. Director, Health Programs, Canadian Medical Association and University of Ottawa, Ottawa, Ontario, Canada

Robert A. Lew, Ph.D. Associate Professor, Department of Dermatology, Cancer Prevention and Control Center, Boston University School of Medicine, and Biometry Group, Multipurpose Arthritis Center, Brigham and Women's Hospital, Boston, Massachusetts

Scott M. Lippman, M.D. Professor of Medicine, Department of Clinical Cancer Prevention, University of Texas M.D. Anderson Cancer Center, Houston, Texas

Peter E. Maxim, Ph.D. Chief, Immunology Branch, Division of Clinical Laboratory Devices, Center for Devices and Radiological Health, Food and Drug Administration, Rockville, Maryland

Anthony B. Miller, M.D., F.R.C.P. Professor Emeritus, Department of Public Health Sciences, University of Toronto, Toronto, Ontario, Canada

Richard Moriarty, M.D. Professor and Chairman, Department of Pathology, Eastern Virginia Medical School, Norfolk, Virginia

S. M. Moss, Ph.D. Acting Head, Cancer Screening Evaluation Unit, Section of Epidemiology, Institute of Cancer Research, Sutton, Surrey, England

Evelyn Rosas Nuñez, M.D. Assistant Professor, Division of Gynecologic Oncology, Department of Obstetrics and Gynecology, Tufts University School of Medicine, and Tufts New England Medical Center, Boston, Massachusetts

D. Maxwell Parkin, M.D. Chief, Unit of Descriptive Epidemiology, International Agency for Research on Cancer, Lyon, France

Paola Pisani, Ph.D. Scientist, Unit of Descriptive Epidemiology, International Agency for Research on Cancer, Lyon, France

Philip C. Prorok, Ph.D. Acting Chief, Biometry Branch, Division of Cancer Prevention, National Cancer Institute, Bethesda, Maryland

Susan C. Rossi, Ph.D., M.P.H. Senior Program Analyst, Office of Science Policy, National Cancer Institute, Bethesda, Maryland

Anita L. Sabichi, M.D. Assistant Professor of Medicine, Department of Clinical Cancer Prevention, University of Texas M.D. Anderson Cancer Center, Houston, Texas

Bruce F. Scharschmidt, M.D. Professor of Medicine, University of California, San Francisco, San Francisco, California

Paul F. Schellhammer, M.D., F.A.C.S. Professor and Chairman, Department of Urology, Eastern Virginia Medical School, Norfolk, Virginia

Fritz H. Schröder, M.D., Ph.D. Professor and Chairman, Department of Urology, Erasmus University and Academic Hospital, Rotterdam, The Netherlands

Vicki L. Seltzer, M.D. Chairman, Department of Obstetrics and Gynecology, Long Island Jewish Medical Center, New Hyde Park, and Professor, Obstetrics and Gynecology and Women's Health, Albert Einstein College of Medicine, Bronx, New York

Sam Shapiro, B.S., D.H.L. (Hon.) Professor Emeritus, Department of Health Policy and Management, School of Hygiene and Public Health, The Johns Hopkins University, Baltimore, Maryland

Sudhir Srivastava, Ph.D., M.P.H. Program Director, Early Detection Branch, Division of Cancer Prevention, National Cancer Institute, Bethesda, Maryland

Joan L. Walker, M.D. Associate Professor, Chief of Gynecologic Oncology, Department of Obstetrics and Gynecology, University of Oklahoma Health Sciences Center, Oklahoma City, Oklahoma

Douglas L. Weed, M.D., Ph.D. Senior Research Fellow, Kennedy Institute of Ethics, Georgetown University, Washington, D.C., and Chief, Preventive Oncology Branch, National Cancer Institute, Bethesda, Maryland

William G. Woods, M.D. Director, South Carolina Cancer Center, Palmetto Richland Hospital, and University of South Carolina, Columbia, South Carolina

Charles J. Yeo, M.D., F.A.C.S. Professor, Department of Surgery and Oncology, The Johns Hopkins Medical Institutions, Baltimore, Maryland

I

BACKGROUND

1

History

Sam Shapiro
School of Hygiene and Public Health, The Johns Hopkins University, Baltimore, Maryland

I. BACKGROUND

One of the important developments in medical science and community action during the past quarter century has been the emergence of screening for breast and colorectal cancers, with the certainty of saving lives from the target conditions.

We are currently in the early stages of testing the efficacy of screening at ages 55–74 years for cancer of the prostate, lung cancer, and ovarian cancer and determining the value of flexible sigmoidoscopy for colorectal cancer screening (1). Screening is conducted at an initial examination and annually thereafter for 3 years, except for sigmoidoscopy, which is scheduled for one examination at 3 years after the initial examination.

These screening protocols are being addressed through clinical interventions. Basically, the question is whether an earlier detection through screening than through usual medical care makes a difference in the outcome of the disease. Modification of environmental conditions and exposure may play an important role in the prevention of the condition, as it does for lung cancer. But, the issue of interest concerns the extent to which medical care can alter the course of the cancer, or a precursor, through earlier detection.

Even though screening has been used for a long time, early studies were faulted because of failure to have an appropriate control group, and no firm conclusions about efficacy of the procedure could be drawn. In 1968, a publication of the World Health Organization formalized the requirements for screening and clarified ways in which the procedure could be an important resource for answering key questions about screening (2). A key point made was that the efforts of the past and those still needed to produce necessary data were aimed at the following precepts for screening application.

1. The condition screened for is an important cause of morbidity, disability, or mortality.
2. The natural history of the disease is sufficiently well known, treatments are sufficiently effective, risks resulting from the screening are low enough, and efficacy in reducing morbidity, disability, or mortality is high enough to conclude that detection of the disease through screening will lead to a net benefit to the target population and that the resources required to administer the tests under screening conditions are justified in terms of the net benefits.
3. The tests must have high levels of sensitivity and specificity, where sensitivity is the proportion of persons screened who have the disease and are positive to the test, and specificity is the proportion of persons screened who do not have the disease and are negative to the test.
4. The tests must be acceptable to the target population and their physicians, and appropriate follow-up of positive findings must be assured.

There are no absolute standards to determine when these conditions have been met or how to combine them to reach a decision. All are stated in relative terms and a specific disease may rate extremely high in importance as a cause of mortality but only moderately high on the basis of knowledge about its natural history. Also, it is frequently necessary to trade off reduced specificity for increased sensitivity, if a small increase in sensitivity leads to a substantial decrease in specifity in a low-incidence condition that is accompanied by high anxiety. The point is that value judgments play a large role in weighing the relative importance of the aforementioned conditions. However, this does not obviate the need for information from a carefully designed study; policy decisions regarding use of screening protocols in large segments of the population will depend on data from such sources.

II. SCREENING FOR BREAST AND COLON CANCER

A. Breast Cancer

In the past, efforts to establish the utility of screening for breast and colorectal cancers demonstrated the uncertainty that prevailed about techniques for useful screening. Present discussion of breast cancer screening, however, no longer concentrates on the value of periodic mammography as a general proposition. Remaining critically important issues include how to organize and conduct a program that reaches a large majority of the target group of women. The dominant question in the United States is whether routine screening should begin at 40 or 50 years of age. The National Cancer Institute, the American Cancer Society, and many other professional groups advocate the younger age, but a recent consensus development conference concluded that the baseline age for starting to screen on a broad basis is 50 years (3,4).

Early efforts were made to explore the utility of X-ray equipment in studies of breast cancer. In 1915, Salomon, a German surgeon, used roentgenography to study gross mastectomy specimens (5); Warren, an American, reported in 1930 on clinical use of roentgenography of the breast (6). The possibility of early detection and the value of mammography in prolonging life was advanced, amid expressions of fatalism about the possibility of changing the course of the disease. Despite misgivings, Ingleby and Gershon-Cohen compiled results of mammography during the 1950s (7), encouraging the outlook for the procedure. In 1956, R. L. Egan at the M. D. Anderson Hospital and Tumor Institute, University of Texas, Houston, TX, after extensive experimentation with various X-ray factors to produce radiographs of the breast, developed a high-milliamperage–low-kilovoltage, low-exposure time technique with industrial film (Kodak M) to produce a simple and reproducible mammogram with a great advance in image detail. Egan's publication in 1960 (8), more than any other event, thrust mammography into the field as a procedure that had to be tested under rigorous conditions for screening on a population basis.

During this period, Michael Shimkin of the National Cancer Institute sought sites to determine the value of screening with mammography. A similar interest in assessing the utility of mammography was held by Philip Strax and Sam Shapiro at the Health Insurance Plan of Greater New York (HIP). Funding from the federal agency for the research at HIP was assured after a trial period at this program. In December 1963, the first prospective study with an appropriate control group to determine the value of screening got underway (9). This marked the dawn of the randomized controlled trial (RCT) in cancer screening. Mammography and clinical examinations of the breast, each carried out independently, were provided to women aged 40–64 years. Clinical examinations were included because of the uncertainty of mammography. The end point was reduction in mortality rates from breast cancer, and the rationale was that all measures that could become valuable had to be provided. Prior studies in which women had been screened with clinical examination of the breast had been carried out and showed promising results, but no conclusion could be reached because they did not have a control group.

During the 1960s and early 1970s, the HIP trial was carefully watched; soon, other studies were undertaken to test the utility of mammography. Many of these studies were case-control studies in the Netherlands and Italy (10,11), but, even though they generated great interest, they did not convincingly answer many of the questions that remained about screening.

Randomized controlled trials were started in Sweden and Edinburgh to establish whether screening was valuable and to determine whether the HIP results were replicable under various conditions (12,13). A point of special issue was the attempt to determine the utility of screening in women younger than 50 years of age. The new trials varied in their designs. Some started at age 40 years, others at age 45; no commonality existed in the upper age limit (14). Mammography techniques had improved since the start of the HIP study and, in a few instances, the new trials used a single view for each breast; others continued to apply double views at least in the initial (prevalence) screening. Other differences existed, but the important point was that testing had begun from which it could be determined whether screening should be introduced on a population basis.

There are differences among countries (in fact, within countries) in the content of screening programs, but screening with mammography is advocated in 22 countries (15). Whether mammography will be improved or replaced by another procedure is an open question. The important point is that attention within many countries is being given to broadening the use of mammography under screening conditions and determining the extent to which it contributes to the decline in breast cancer mortality rate.

B. Colorectal Cancer

Another case in which there is a history of determined effort to find a viable, successful intervention through screening is colorectal cancer (16). This cancer has consistently been the top priority in developing a screening technique that would result in a reduction in mortality rate from cancer. Colorectal cancer has long been among the more common causes of death from cancer. In 1997, colorectal cancer was responsible for about 54,900 deaths in the United States, ranking it second among cancers as a cause of death (17). The condition is more common among males than females, but the cancer is prominent in both sexes. A high incidence of colorectal cancer is found among persons with hereditary conditions, such as familial polyposis, familial nonpolyposis, a cancer family syndrome (autosomal dominant), and hereditary site-specific colon cancer (18,19). However, these conditions account for a small proportion of colorectal cancer cases (5%–10%). There are more common conditions that carry an increased risk, such as a family history of colorectal cancer, but these account for only about 25% of all colorectal cancers, and most screening has been aimed at the total population at middle ages when the incidence of the disease begins to increase.

Application of rigid sigmoidoscopy in seeking methods to reduce mortality rates from colorectal cancer started about 50 years ago. Early studies of this procedure suggested that mortality rates from colorectal cancer could be reduced, but the studies did not have a rigorous design in which controls have a prominent role. Since then, two case-control studies based on rigid sigmoidoscopy (20) and rigid and flexible sigmoidoscopy (21) have been viewed as promising for screening; both showed a marked reduction in colorectal cancer mortality rates among those who were screened. These studies have led to a firmer recommendation to "offer screening sigmoidoscopy with a flexible sigmoidoscope every 5 years" than previously existed (22).

Major efforts have been underway in the United States and Europe to test the value of the fecal occult blood test (FOBT), a procedure that generated interest because the specimen could be obtained at home and did not cause the discomfort of sigmoidoscopy. The Minnesota Colon Cancer Study, which started in 1975–1977 among persons 50–80 years of age, has produced information on the subject (23). Thirteen years of follow-up from the entry date to the study showed a reduction of about 33% in the death rate from colorectal cancer in the group designated for annual screening compared with the death rate of the control group. Although it has been speculated that these results were influenced by a high rate of false-positive findings that led to screening colonoscopy, further analysis has indicated that screening with FOBT was responsible for a large part of the gain.

Five controlled trials that test the effectiveness of FOBT screening have been completed or are in progress. Randomized controlled trials in Funen, Denmark (24), and Nottingham, United Kingdom (25), have recently shown a mortality rate reduction using biennial FOBT. All trials have experienced a consistent improvement in the stage of disease in the screened population and colorectal cancers diagnosed in the screened group were smaller than in the control group.

III. SCREENING FOR CANCER OF THE PROSTATE, LUNG, AND OVARY

A. Prostate Cancer

Screening approaches have been advanced in the control of several other cancer sites. The most common is the cancer of the prostate, the most frequently diagnosed cancer (excluding skin) among men in the United States. In 1997, an estimated 210,000 new cases and 40,400 prostate-related deaths occurred in the United States (17). Incidence rates have increased sharply in recent years and the death rate from this condition may have begun to decline slightly. Clinical prostate cancer is rarely seen in men younger than 50 years; it increases rapidly thereafter. Potential risks are related to family and genetic predisposition, dietary fat, alcohol and vitamin consumption, and plasma lipid profiles (26–28). Whereas attention continues to be directed at understanding how these risk factors, among others, may function, the literature has been preoccupied in recent years with the issue of early detection through screening.

There is a sharp division of opinion on the readiness of general screening, with the American Cancer Society and various professional groups favoring it (29) and others (e.g., United States Preventive Services Task Force) being neutral (30). After considering these views, the physician–data query (PDQ) of the National Cancer Institute concluded that ''there is insufficient evidence to establish whether a decrease in mortality from prostate cancer occurs with screening by digital rectal examination, transrectal ultrasound, or serum markers including prostate specific antigen'' (31). Rectal examination is inexpensive, noninvasive, and nonmorbid, but no study has determined the extent to which this procedure leads to a reduction in prostate cancer mortality rate; best estimates predict only a small proportion of prostate cancer cases would be affected by this procedure. Ultrasound has been found to have low specificity and its invasiveness and costs make it unlikely that it will be used for screening purposes.

Measurement of prostate specific antigen (PSA) is the most promising test developed. It has been tested in observational studies for initial diagnosis of prostate cancer and as a tool to monitor for recurrence (32). Its appeal rests on simplicity, objectivity, lack of invasiveness, and relatively low cost compared with ultrasound. Efforts are underway to improve the sensitivity, specificity, and positive predictive value of PSA. A great deal has been learned about these values, but there is no completed RCT of the efficacy of this procedure. In effect, there are two camps: one holds that enough has been learned to warrant population application of this procedure on a screening level, and the other holds that the value of the PSA is not known and that a trial is needed to answer many questions about its utility as a routine screening tool.

One of the troubling problems in prostatic cancer is that some of the cases diagnosed early may otherwise never be detected; autopsies have suggested that a large proportion of men have cancer of the prostate that does not become known clinically. Furthermore, in many cases, screening and subsequent treatment could have substantial risks and morbidity that include urinary incontinence, urethral strictures, sexual impotence, and rectal injury (33). The inclusion of prostatic cancer in the trials that are being carried out by the National Cancer Institute is designed to provide definitive answers to questions about the efficacy of screening. The European Randomized Study of Prostatic Cancer (ERSPC) includes collaborating countries using similar but not identical RCT designs (34).

B. Lung Cancer

There is still no evidence that screening for lung cancer leads to decreased mortality rates, but interest is high in assessing the value of screening for the condition. This situation still holds after many years of studies in the United States and abroad that followed recognition that cigarette smoking was responsible for the rapid increase in the death rate from lung cancer. In 1997, there were an estimated 178,100 cases diagnosed and 160,400 deaths from lung cancer, reflecting both the high occurrence of the disease and the high fatality rate among these cases (17).

The earliest studies were conducted more than 40 years ago and were not randomized. These were replaced by carefully designed trials in the 1970s in the United States by the Mayo Lung Project (35), the Johns Hopkins Hospital (36), and the Memorial-Sloan Kettering Cancer Center (37). The Mayo Lung Project was started in 1971 for men aged 45 years or older who were heavy smokers. Subjects found to be free of lung cancer on initial screening were randomized to a group offered cytologic and radiologic screening every 4 months or to a control group who received advice to be screened annually with both modalities. Although cases in the study group were diagnosed in earlier stages than the control group, there was no statistically significant reduction in lung cancer deaths in the study group. In the Hopkins and Memorial-Sloan Kettering studies, where sputum cytology for the study group was added every 4 months to annual radiologic examinations performed on both the study and control groups, no significant lung cancer mortality difference was found between the two groups.

A reservation about these trials was the large effect sought through screening, that is, 50% reduction in mortality rate from lung cancer. Reappraisal of the results of these trials and reaching for a far more modest impact of screening, 15%–18% decrease in lung cancer mortality, has resulted in the inclusion of X-ray screening in the study initiated by the National Cancer Institute (1). Both men and women are included because lung cancer is a major cause of cancer death in both sexes.

C. Ovarian Cancer

Cancer of the ovary has an incidence of about 26,800 cases per year with more than 40% of patients dying within 5 years (17). Most ovarian cancers are first identified in late stage disease, and the expectation of improvement was not promising for a long time. With the advent of two new techniques—CA 125 screening, a more specific marker for earlier detection of ovarian cancer than had existed, and the transvaginal ultrasound, which images the ovaries in about 5–10 minutes—a new outlook has been generated for this condition (38). These techniques are added to the pelvic examination, which was the primary method for detecting ovarian cancer.

A heavy reduction in mortality rate because of ovarian cancer screening is projected. If this outlook is too optimistic, the sample size will either be increased or a meta-analysis drawing on other relevant studies is speculated as a possibility to increase the power of detecting a 25% reduction in mortality rate.

IV. CONCLUSION

Success in reducing breast cancer deaths through screening and the follow-up of suspicious cases created an environment favorable for screening for other cancer sites. There are issues that need to be resolved in breast cancer screening, for example, at what age should screening begin. But, there is a clear indication that screening can change the course of the condition for a substantial proportion of breast cancer occurrences. A favorable result from screening with FOBT has been found in colorectal cancer, but there is a question concerning the contribution that could be made by use of sigmoidoscopy.

In the United States, the new efforts to test the efficacy of screening are aimed at major cancer sites, with confidence in the production of a favorable result varying among the sites. Some commentators believe that RCTs for the flexible sigmoidoscopy in colorectal cancer and the PSA in prostate cancer (39) may not be relevant at this time because of their conclusion that benefit will occur. In the case of prostate cancer, some critics believe there is considerable probability of net harm associated with screening and ''ethical justification has as yet to be publicly heard'' (40). In view of the large costs and uncertainty that exist concerning the actual effect of these procedures in a population-wide screening environment, the decision to move ahead with trials appears to be sound.

The goal for improved lung cancer mortality rates is relatively modest, less than 20%, but given the magnitude of the existing rate, this is a consequential decrease. Furthermore, a substantial proportion of the population, about 25%, still smoke cigarettes, and, even optimistically, a sizable decrease in the death rate from lung cancer cannot be expected for many years. The condition has a high 5-year case fatality rate, almost 80%, and the low reduction in the death rate that has been targeted would be medically important.

Ovarian cancer has been included in the trials of the U.S. National Cancer Institute, providing a balance between the genders in the objectives of the new studies. There is little experience with the cancer detection methods available under screening conditions for this site and the chances of producing a favorable result in the current study are subject to question. But, this is the point behind a clinical trial.

The outlook for reducing mortality rates from the cancer sites being tested under screening conditions seems favorable, and, if the procedures being used are efficacious, many issues concerning participation of the target groups will need to be resolved at high costs. But, this is the price of progress through screening in cancer mortality control and few would set aside the procedure because of this.

REFERENCES

1. Concept for Prostate, Lung, Colorectal and Ovarian Cancer Screening Trial, January 31, 1991, National Cancer Institute, Working document.
2. Wilson JMG, Jungren G. Principles and practice of screening for disease. Public Health Paper (No. 34). Geneva, World Health Organization, 1968:26.
3. Mammography Recommendations for Women Ages 40–49. Bethesda, National Cancer Institute, National Cancer Advisory Board (NCAB), 1997.
4. Breast Cancer Screening for Women Ages 40–49. NIH Consensus Statement 1997, January 21–23. Natl Cancer Inst 1997; 15(1):1–35.
5. Salomon A. Beitrage zur pathologic und klinik des mammahaarzinoms. Arch f Klin Chir 1913; 101:573.
6. Warren SL. Roentgenologic study of the breast. Am J Roentgenol, 1930; 24–113.
7. Ingleby H, Gershon-Cohen J. Comparative Anatomy, Pathology, and Roentgenology of the Breast. Philadelphia: University of Pennsylvania Press, 1960.
8. Egan RL. Experience with mammography in a tumor institution. Evaluation of 1,000 studies. Radiology 1960; 75:894–900.
9. Shapiro S, Venet W, Strax P, et al. Periodic Screening for Breast Cancer: The Health Insurance Plan Project and Its Sequelae, 1963–1986. Baltimore: The Johns Hopkins University Press, 1988.
10. Collette HJA, Day NE, Rombach JJ, de Waard F. Evaluation of screening for breast cancer in a non-randomized study (the Dom project) by means of a case-control study. Lancet 1984; 1:1224–1226.
11. Palli D, del Turco R, Buiatti E, Ciatto S, Crocetti E, Paci E. A case-control study of the efficacy of a non-randomized breast cancer screening program in Florence (Italy). Int J Cancer 1986, 38:501–504.
12. Nystrom L, Rutqvist LE, Wall S, Lingren A, Lindqvist M, Ryden S, et al. Breast cancer screening with mammography, an overview of the Swedish randomized trials. Lancet 1993; 341:973–978.
13. Roberts MM, Alexander FE, Anderson TJ, Chetty U, Donnan PT, Forrest P, et al. Edinburgh trial of screening for breast cancer mortality at seven years. Lancet 1990; 335:241–246.
14. Hurley SF, Kaldor JM. The benefits and risks of mammographic screening for breast cancer. Epidemiol Rev 1992; 14:101–129.
15. Shapiro S, Coleman EA, Broeder M, et al. Breast Cancer Screening Programs in 22 Countries: Current Policies, Administration, and Guidelines. Int J Epid.
16. Winawer SJ, Fletcher RH, Miller L, et al. Colorectal cancer screening: clinical guidelines and rationale. Am Gastroenterol 1997; 112:594–642.
17. Parker SL, Tong T, Bolden S, et al. Cancer statistics, 1997. Ca-A Cancer J Clin, 1997; 47(1):5–27.
18. Ransohoff DF, Lang CA. Screening for colorectal cancer. N Engl J Med 1991; 325(1):37–41.
19. Fuchs CS, Giovannuccci EL, Colditz GA, et al. A prospective study of family history and the risk of colorectal cancer. N Engl J Med 1994; 331(25):1669–1674.
20. Selby JV, Friedman GD, Quesenberry CP Jr, et al. A case-control study of screening

sigmoidoscopy and mortality from colorectal cancer. N Engl J Med 1992; 326:653–657.
21. Newcomb PA, Norfleet RG, Storer BE, et al. Screening sigmoidoscopy and colorectal cancer mortality. J Natl Cancer Inst 1992; 84:1572–1575.
22. Fuchs CS, Giovannuccci EL, Colditz GA, et al. A prospective study of family history and the risk of colorectal cancer. N Engl J Med 1994; 331(25):597.
23. Mandel JS, Bond JH, Church TR, et al. Reducing mortality from colorectal cancer by screening for fecal occult blood. N Engl J Med 1993; 328(19):1365–1370.
24. Kronborg O, Fenger C, Olsen J, et al. Randomised study of screening for colorectal cancer with faecal-occult-blood test. Lancet 1996; 348(9040):1467–1471.
25. Hardcastle JD, Chamberlain JO, Robinson MHE, et al. Randomised controlled trial of faecal-occult-blood screening for colorectal cancer. Lancet 1996, 348(9040): 1472–1477.
26. Hayes RB, Brown LM, Schoenberg JB, et al. Alcohol use and prostate cancer risk in US blacks and whites. Am J Epidemiol 1996; 143(7):692–697.
27. Eichholzer M, Stahelin HB, Gey KF, et al. Prediction of male cancer mortality by plasma levels of interacting vitamins: 17-year follow-up of the prospective basel study. Int J Cancer 1996; 66(2):145–150.
28. Gann PH, Hennekens CH, Sacks FM, et al. Prospective study of plasma fatty acids and risk of prostate cancer. J Nat Cancer Inst 1994; 86(4):281–286.
29. American Cancer Society. Guidelines for the cancer-related checkup: recommendations and rationale. Ca-A Cancer J Clin 1980; 30(4):4–50.
30. U. S. Preventive Services Task Force. Guide to Clinical Preventive Services: An assessment of the Effectiveness of 169 Interventions. Report of the U.S. Preventive Services Task Force. Baltimore: Williams & Wilkins, 1989.
31. Screening for Prostate Cancer. Bethesda, PDQ, National Cancer Institute, 1996.
32. Mettlin C, Murphy GP, Lee F, et al. Characteristics of prostate cancers detected in a multimodality early detection program. Cancer 1993; 72(5):1701–1708.
33. Kramer BS, Brown ML, Prorok PC, et al. Prostate cancer screening: what we know and what we need to know. Ann Intern Med 1993; 119:914–923.
34. Schroder FH, Bangma CH. The European Randomized Study of Screening for Prostate Cancer (ERSPC). Br J Urol 1997; 79(suppl I):68–71.
35. Fontana RS. Screening for lung cancer. Recent experience in the United States. In: Hansen HH, ed. Lung Cancer: Basic and Clinical Aspects. Boston: Martinus Nijhoff, 1986: 91–111.
36. Tockman MS, Frost JK, Stitik FP, et al. Screening and detection of lung cancer. In: Aisner J, ed. Lung Cancer. New York: Churchill Livingstone, 1985: 25–38.
37. Melamed MR, Flehinger BJ, Zaman MB, et al. Screening for lung cancer. Results of the Memorial Sloan-Kettering study in New York. Chest 1984; 86:44–53.
38. Jacobs I, Bast RC Jr. The CA 125 tumor-associated antigen: a review of the literature. Hum Reprod 1989; 174:1783–1788.
39. Walsh PC. Using prostate-specific antigen to diagnose prostate cancer: sailing in uncharted waters. Ann Intern Med 1993; 119:948–949.
40. Adami HO, Baron JA, Rothman KJ. Ethics of a prostate cancer screening trial. Lancet 1994; 343:958–960.

2

Heroic Myths and Probability: Confronting the Limits of Screening

Edward S. Golub
Pacific Center for Ethics and Applied Biology, Solana Beach, California

> In some ways disease does not exist until we have agreed that it does, by perceiving, naming and responding to it.
>
> Everyone who is born holds dual citizenship, in the kingdom of the well and in the kingdom of the sick . . . sooner or later each of us is obliged, at least for a spell, to identify ourselves as citizens of that other place.

I. INTRODUCTION: FRAMING DISEASE AND DIAGNOSTIC "TESTS"

Health and disease are not terms that describe absolute states, rather they describe conditions that societies frame differently at different times; how we frame them determines which passport each of us will carry. In the present era, health and disease are framed technologically—health is the absence of disease, and disease is defined in terms of scientific tests. Individuals who feel ill present themselves to physicians with the expectation that there will be "tests" that tell them which passport they will be using. If the tests are negative, they are relieved to know that they are not sick, even though they may still experience the same sensations that prompted them to be seen by the doctor. It is not overdramatic to say that the entire nature of the future life of a patient can depend on the result of those tests, what the person can and cannot do, how much time the person has to do it in, what the person's self-perception is, and how others think of and behave toward the person.

Consider then, the potential impact of laboratory tests on people who do not feel ill and have not described themselves as patients, but who have submitted to laboratory tests as part of a screening program. For them, disease does not exist until their doctor tells them that their "tests came back positive," and, with no warning, they find that they must use their other passport because they have suddenly entered the kingdom of the sick. If positive tests allow treatment and the restoration of health, then the screening process is a triumph of science. If they allow management of a condition by medication or life choices such as diet and exercise, then they are a valuable adjunct to disease management or public health. We even see value in positive tests that identify a condition for which there is no treatment because the information at least allows the individual to make important life decisions that affect family and friends.

But, we are beginning to deal with the realities of screening tests that frame the probability of disease; it is in this context that myths about the triumphs of technology in medicine begin to break down and a new area of concerns emerges. This chapter provides a framework in which to better evaluate the lines along which we want technology to move us. The alternative is that we are moved by the technology, perhaps in directions that are not in the best interests of the public that has placed their faith in the clinician's ability to make proper choices.

II. HISTORICAL CONTEXT

A. Addressing Heroic Myths

To draw some generalities about the role and value of screening, it is important to understand the historical context in which reliance on the laboratory arose. We have been taught a heroic version of the history of science and medicine, and heroic histories are equally dangerous whether they are of nations, politicians, or professions. To understand how the science of the laboratory replaced the art of the bedside it is necessary to understand the history of the framing of disease and the emergence of the concept of the cure. We are pleased to see ourselves in the heroic mode as the latest inheritors of the tradition of Hippocrates, Galen, Harvey, and Pasteur, part of a long unbroken line, marching toward the clear, unambiguous goal of health. But, the change has not been linear, and the goals have not been clear.

We are on the verge of an era of technological change in which the ability to do the tests can determine whether we do them, which ones we do, and on whom they are done; it is crucial to understand the value and limits of the tests, and how they fit into the goals of medicine. We are in danger of behaving as if all technological change is progress, and of confusing being swept along on the wave of change with responsible exercise of the authority that society has given to clinicians in that which matters most to people, their health.

B. The Constancy of Death and the Concept of Disease

1. *The Constant Presence of Death*

From as far back as there are reliable records until the beginning of this century, the most pressing reality for humans was the constant presence of death. Through all of that time, infant and childhood mortality rates were about 25%. Mean life expectancy was about 30 years; until well into the 1900s, no more than 2% or 3% of the population was 65 years of age or older. At the end of the 20th century, in the industrialized nations, for the first time ever, death is associated with old age, and the aged are the fastest growing population group.

In the heroic version of history, scientific advances in medicine were responsible for the dramatic change. So, it is almost demeaning to think that, even though we are descended from Hippocrates, Galen, and Pasteur, the major credit for changing this situation must go to the introduction of sewage disposal systems in 1840 for purely economic reasons. The first sanitation laws in England were conceived and executed when diseases were thought to be due to ''miasmas and mephitic vapors'' arising from dirt in which the preindustrial world was awash. In a time before disease was associated with specific cause, Edwin Chadwick convinced the Parliament to allocate £1,000,000 to build sewers so that the workers in the new mills in industrializing England would be healthy enough to work. The argument was essentially that, if the mill owners spent money to maintain their spinning jennies, they should also spend money keeping labor, the other essential commodity of the Industrial Revolution, running.

As a result of the sanitation laws, the decline in infant mortality rate could already be measured in England years before the microbial concept of infectious disease (the germ theory of Pasteur) was introduced. Ironically, science and medicine receive the credit for the most profound change in health in human history. Throughout France, streets are named after Pasteur, but Edwin Chadwick's name is all but unknown. Once modern scientific understanding of the nature of disease was acquired, public health and sanitation were more rapidly incorporated into governmental and personal lives. But it is wise to keep the past in perspective when thinking about the present and the future.

2. *The Constancy of the Concept of Disease*

Because we live in an era in which science and technology have been raised to the level of a secular religion, we may think that the seeds of this belief should be readily seen in history. But, in this case, the change is difficult to see because, for most of recorded history, nothing changed. The concept of health and disease remained constant from the time of Hippocrates and Galen to the time of Pasteur, and this constancy of concept resulted in constancy in therapy and expectation of outcome. For those 2 millennia, disease was conceived of as the humors being out of balance, and therapy was the reestablishment of the proper balance.

Humoral medicine originated in the Hippocratic school of physicians on the island of Cos in the 5th century BC, and was refined by the school of Galen, a Greek who lived in Rome during the 1st century AD. Medicine reflects its times, and those times were dominated by the Aristotelian description of the world, a world in which everything is composed of four elements (fire, air, earth, and water) and has four qualities (hot, cold, dry, and wet). The elements are represented in human bodies as four humors (blood, phlegm, yellow bile, and black bile), all of which are in balance in the healthy individual. An imbalance in the qualities, whether by a change in climate, diet, emotion, or miasmas, can lead to an imbalance in the humors, resulting in disease. Therapy, therefore, consisted of putting the humors back into proper balance. We look back on the bloodletting, sweating, diuretics, emetics, herbs, incantations, and amulets that were used to reestablish this balance with either horror or amusement, but the patients of the day looked on these therapies with the same hope that diabetics looked at the introduction of insulin in 1920 and that the general public looked at the introduction of penicillin in the 1950s. Those therapies that are so wrong-headed to us because they are founded on such a bizarre idea made as much sense to ancient peoples as our ideas do to us. We have had modern scientific explanations of health and disease for less than 200 years; Galen's explanations lasted 2000 years.

We realize how deeply humoral medicine was ingrained in western civilization when we see how it survived with only slight refinements until almost the beginning of the 20th century, even in the face of the growth of modern science. In 1628, at the dawn of the Scientific Revolution, William Harvey, who is traditionally hailed as one of the founders of modern medicine, carried out his brilliant experiments showing that the blood circulated through the body. The impact of these experiments on the concept of disease or on therapy was nil, because the experiment fit into the humoral paradigm and was seen how blood, the most important of the humors, got around in the body. One of the central figures in the Scientific Revolution, the chemist Robert Boyle, used his understanding of the nature of gases (Boyle's Law) to explain how amulets placed at the skin could alter the humors in the body by drawing gases dissolved in blood to the surface of the body where they could be more easily dissipated. In the 19th century, François Magendie, the teacher of Claude Bernard and himself one of the founders of modern experimental physiology, differed with his colleagues about when it was appropriate to bleed a patient, not whether it was appropriate. And, just before the dawn of the 20th century, the most renowned professor of hygiene in the world, Max von Pettenkofer, was convinced that cholera was caused by changing levels of ground water that emitted miasmas and mephitic vapors. If microbes were involved in cholera, he was sure that they acquired their ability to cause disease because of contact with these emanations.

During this long period of time, many changes took place—ideas about democratic government and personal liberty developed, scientific developments flowered, making technology of the Industrial Revolution possible and allowing the rise of the middle classes, sanitation and electrification changed the way people lived and worked, and culture and the arts underwent enormous changes. Yet, disease continued to be conceived of as the humors being out of balance and the logical therapy was to put them back into balance.

C. Changing Nature of Medical Practice and Disease

1. *From "Bedside" to "Hospital" Medicine*

Even though the concept of disease remained constant, the way medicine was practiced did not. Medicine, or more correctly, healing, was practiced for centuries by a wide range of people. Physicians were the elite group; descended from the school of Hippocrates, they took his oath, were gentlemen who had attended university and so knew some Latin and a little Greek. As gentry, they treated gentry but never touched the patient, a practice that was below their station. Because their patients were social and intellectual peers who knew a great deal about medical practice, they had distinct ideas about what was wrong with them and what their treatment should be, the physician probably spent as much time listening to the patient's ideas about therapy as asking about their symptoms. In contrast, surgeons were descended from barbers because of their agility with cutting instruments. They were called the ''poor man's physician'' and were of the same social class as most of their patients. They had no qualms about touching patients, and they performed bleedings, lanced boils, cleaned wounds, set bones, and, for the less socially exalted, made diagnoses and prescribed drugs. It was the surgeons who were called in by the physician to bleed the gentry, lance their boils, and set their broken bones. Finally, apothecaries prescribed herbs and diets, and village wise women, ladies of the manor, and practically anyone who cared to took part in the healing of the sick.

Ultimately, it mattered little that healing was carried out by such a motley group, because they were all doing essentially the same thing: deciding which humors were out of balance and then reestablishing balance by the usual means. It was a matter of art to determine whether bleeding was called for (it usually was), or sweating (not as often), or diuretics (a great favorite), or which combination of herbs were to be used as teas or poultices. Furthermore, even when the concept changed and people knew that the really serious diseases were caused by living infectious agents, there was still nothing else to be done and these practices continued into the early years of the 20th century.

It was not science, but the French Revolution in 1794 that brought this long era of what historians call ''bedside medicine'' to an end. In the aftermath of the Reign of Terror, the National Convention suppressed all corporations and faculties as part of its plan to abolish all the hated bastions of privilege. This included both the faculty of medicine and the hospitals, medicine because physicians treated the elite and hospitals because they were rightly seen as repositories for the sick and dying poor. A 1788 report on the conditions of the hospitals of Paris described up to six people in one bed (some of them dead), and narrow, windowless rooms with the air so foul that it was impossible to breathe. It took only a short time to realize that the grand gesture of abolishing training and hospitals was bad policy. The law was revised so that the faculty of medicine was opened (in theory at least) to people of all classes. Surgeons and physicians would be trained side by side, and the hospitals would be cleaned up and opened to the poor as places where they could receive the best medical care. The best medical care was still balancing the humors, but at least a poor patient would receive food and clean shelter in the reformed hospitals during their therapy.

The move was no doubt good social policy, but its unforeseen effect was to change the way medicine was practiced. In the late 18th and early 19th centuries, there was an enormous interest in medical statistics, and for the first time large numbers of patients were together in the same place for treatment; the statisticians then were able to compare the efficacy of different treatments for similar symptoms. Furthermore, morbid anatomy was becoming a science, and, after a patient died in the hospital (the mortality rate was extremely high), the state provided the relatives with 60 francs for burial. If the body was not claimed within 24 hours, the physician had the right to perform an autopsy on the deceased patient before the state provided burial, so the correlation between symptoms and pathology for the first time was direct. This systematic approach to diagnosis, treatment, and pathology was the beginning of a more scientific medicine, but it also had the profound effect of converting patients with disease into subjects of disease. The practice of medicine had entered what historians call the period of ''hospital medicine'' in which the health of the patient began to be seen in terms of deviation from statistical norms.

2. *The Rise of Laboratory Medicine and the Idea of the Cure*

For all practical purposes, modern medicine began in 1854 when the French chemist Louis Pasteur, newly appointed dean of the Faculté des Sciences at Lille University, was approached by the owner of a local sugar beet distillery who wanted scientific help with a problem at his factory. The fermentation of beet sugar to alcohol was proceeding normally for part of the process, but then the alcohol became vinegar. In agreeing to try to solve this problem, Pasteur the chemist became Pasteur the microbiologist. At the time, the yeasts associated with the fermentation of sugar to alcohol were not considered the cause of the alcoholic fermentation, but were thought to be nonliving by-products of the process. Pasteur noted that on microscopic examination of the contents of the vats in which the ethanol was turning into acetic acid, there were small rod-shaped things among the larger, round yeast bodies. The great intellectual leap that changed medicine was the realization that the yeasts were living entities that caused the fermentation of sugar to alcohol and that the smaller things (that he called ''microbes'') were also living entities that converted the alcohol to acetic acid. Pasteur's great insight was that living entities not only carry out useful chemical transformations but also cause damage.

To Justin von Liebig and the other great chemists of the day, this was not only nonsense, it was dangerous nonsense. They were fighting the concept of vitalism, the idea that living things had a mysterious ''vital'' quality that distinguished them from nonliving entities, and even though Pasteur the chemist was one of their own kind, Pasteur the microbiologist was to them seriously in error because the idea of living entities carrying out chemical processes smacked of vitalism. Liebig claimed that the idea that fermentations of any kind could be carried out by ''microscopic animalcula . . . may be compared to that of a child who would explain the rapidity of the Rhine current by attributing it to the violent movement of the many millwheels at Mainz.'' But the elegant scientific experiments Pasteur carried out quickly began to convince his professional skeptics, and when he went on to show that the periodic diseases of silkworm that crippled the French silk industry were due to a similar ''infection'' with microbes, the idea that microbes could carry out useful biological processes and that they could cause disease took hold in the popular mind. If microbes could cause ''disease'' of alcohol fermentation and silk worms, why not cause disease of humans?

The idea was so powerful that, by the last quarter of the 19th century, enormous energy was put into identifying the bacterial causes of the scourges of humankind. It is at this point that history begins for most persons trained in science and medicine. In heroic history, Pasteur's monumental intellectual leap is portrayed as a logical next step in the long march of medical progress since the time of Hippocrates. The purpose of this short historical review is to dispel that notion. To overturn in two decades an idea that had flourished for two millennia is not the orderly march of history, and the change in world view that resulted from the germ theory was as important as the change brought about by Copernicus, and it had a much more rapid impact on the world. Before Pasteur's discovery, disease was seen as nonspecific; afterward, it was not only seen as specific, it was, in theory, treatable by eliminating the cause. With the realization that tuberculosis, cholera, pneumonia, dysentery, child bed fever, scarlet fever, diphtheria—the diseases that were the scourge of humankind—were caused by specific microbes, came the realization that they could be unequivocally diagnosed, if not yet treated.

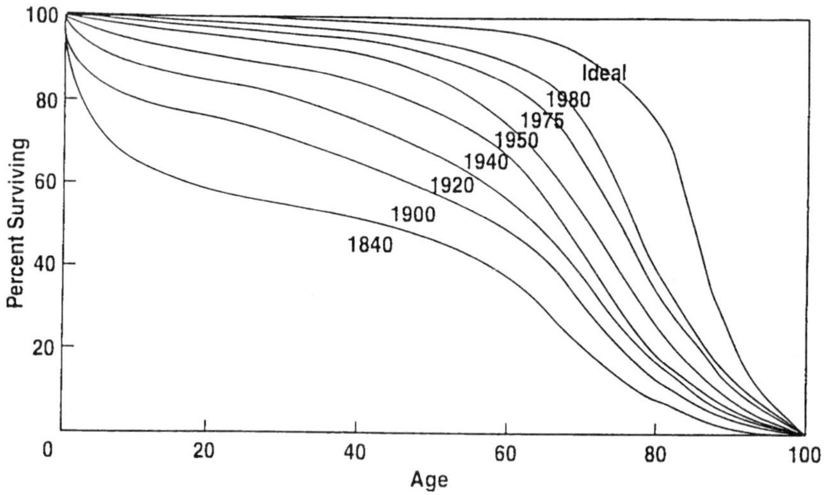

Figure 1 Progressive elimination of early death. The family of mortality curves between 1840 and 1980 show a progressive decline of early death. Note that while the mean life expectancy has increased steadily, there has been no increase in life span, which is a biological limit. The ideal curve is theoretical and reflects a population in which the vast majority live out their biological life span. Redrawn from JF Fries and LM Crapo, Vitality and Aging: Implications of the Rectangular Curve, New York, WH Freeman, 1981. With permission.

For 2000 years, nothing had changed in either the concept or treatment of disease; then, in less than a century, the goals as well as the practice of medicine had changed twice, first to "hospital medicine," in which patients became the object of disease, and then to "laboratory medicine," in which the determinants of disease became removed from the patient. Diagnosis required only that a specimen be collected and brought to the hospital laboratory. For the first time, health or disease was determined on the basis of the laboratory report, the "tests" that the physician runs. Until the middle of the 20th century, the most important laboratory diagnosis was for the presence of bacteria responsible for infectious disease, and the meaning of the "tests" in the mind of both the public and professionals was unequivocal. But, as infectious diseases were brought under control and the population became older, the nature of disease changed, and the meaning of the "tests" became more equivocal.

3. *The Changing Nature of Disease*

The combination of better sanitation, better nutrition, and better housing in the 19th century and of vaccination and antibiotics in the 20th resulted in one of the most startling changes in human history: for the first time, illness and death became associated with old age. This is obvious with even a cursory examination of the survival curve in Figure 1. Between 1840 (the year of the first sanitation laws in England) and 1980 there was a steady increase in mean life expectancy, most of which can be accounted for by the steady decline in infant and childhood mortality rates. Two crucial conclusions come from these data, one is that the increase is not in life span but in life expectancy. True, the percentage of the oldest old is increasing rapidly, but the age to which a much smaller percentage of people always lived is not being extended. The second conclusion is that cures are changing, a phenomenon that has been called their ''rectangularization.'' Whereas some deaths are inevitable through disease and accident, the philosophical underpinning of the idea of the rectangular curve is that, in modern industrial society, a combination of social and scientific factors can allow the survival of the overwhelming majority of people to live out their biologically allotted time.

Whether couched in these terms or not, extension of life has been the primary end of medicine in this century, as is reflected in the fact that efficacies are still reported primarily as reduction in mortality rates, even when the group being discussed is 85 years old. But, as the populations of the industrialized nations have aged, the primary causes of mortality and morbidity have changed from acute and infectious to chronic and noninfectious diseases. This dramatic shift is seen in Figure 2. Influenza, tuberculosis, diphtheria, gastritis, and other infectious diseases that accounted for a significant proportion of deaths in 1900, by 1960 had been replaced by heart disease and cancer, which together accounted for some 60% of deaths. Cancer, cerebrovascular, and chronic obstructive pulmonary diseases are the leading causes of death in Americans older than 45 years of age. In the age group from 1 to 24 years, accidents, homicide, and suicide are the leading causes of death, followed by cancer and heart disease. In the 24- to 44-year-old group, accidents and complications of infection with human immunodeficiency virus are the leading causes of death, followed by cancer, heart disease, and homicide. The only other infectious disease that makes it into the top five causes of death is pneumonia, ''the old person's friend,'' which is number 5 in the 65- to 74-year-old group and is number 4 in other persons 85 years old and older.

But even though chronic diseases are, for the most part, complex diseases of metabolism, biomedical scientists and the public approach them as if they were as straightforward as infectious disease. The military metaphor of ''search and destroy'' using the ''armamentarium'' of the body's own defense turned against ''the enemy'' or ''magic bullets'' that selectively kill the ''invader'' were perhaps appropriate for diseases caused by a single, identifiable, foreign entity. And because the idea of a cure by eliminating the cause is a 20th century phenomenon, the metaphor rang especially true because most of the century was

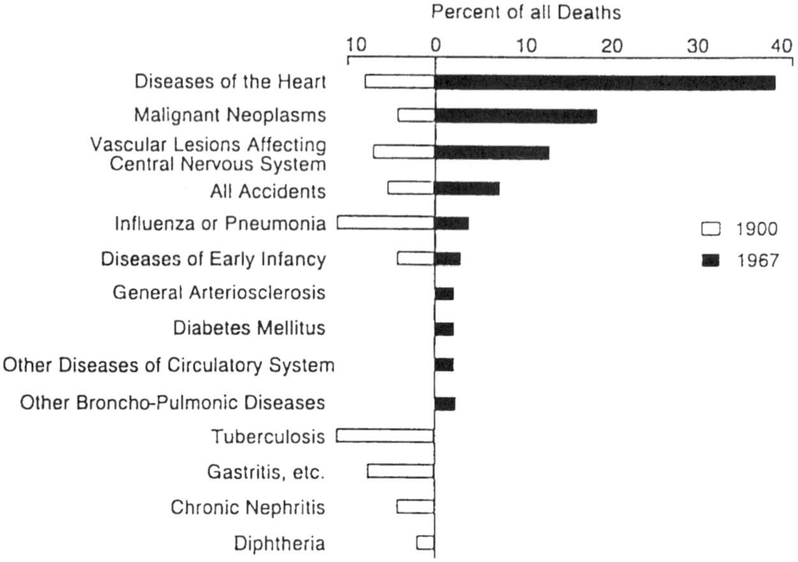

Figure 2 Changing pattern of mortality. The figure shows the dramatic change in the causes of death between the early and middle twentieth century in which chronic diseases replaced infectious diseases as the major causes of death. Redrawn from JF Fries and LM Crapo, Vitality and Aging: Implications of the Rectangular Curve, New York, W.H. Freeman, 1981. With permission.

spent fighting either hot or cold wars. ''Tests'' that are performed when cancer or heart disease is suspected are internalized by many patients and their physicians in the same manner that a throat swab or sputum culture was in the past. The phrase ''cancer is cancer'' removes indeterminacy and probability from the discourse, and reaches its zenith when prominent surgeons claim they do not need data from clinical trials to know when surgery is the preferred action in prostate cancer. These surgeons, according to a report in *The New York Times*, claim that ''they can make inferences and study the case histories of their own patients to conclude that they are doing the right thing.''

The problem is that, with the chronic diseases, the very definition of disease changes from being a laboratory diagnosed absolute to an interpretation of the clinical meaning of laboratory data. Is a PSA level that is slightly over the age-adjusted norm an indication of prostate cancer? Does a mutation in the BRCA-1 gene call for clinical action? Both the definition of the diseases and the significance of the laboratory ''tests'' associated with them have become more indeterminate and require the development of a new mind set.

III. SCREENING FOR CANCER AS AN EXERCISE IN PROBABILITY, SELF-DEFINITION, AND RESPONSIBILITY

A. Probability and Power

The ''search and destroy'' metaphor of infectious diseases underlies the hopes and promises for cures to complex chronic diseases that permeate American society. It is seen in appeals for funds from specific disease groups, announcements from drug companies, testimony by scientists before the Congress at budget time, and in the way the media covers news of science and medicine. It feeds and fuels the hope for a quick technological fix and thereby undermines efforts at prevention that require personal discipline such as cessation of smoking and weight reduction, and undermines the ability to gain scientific information about incidence of disease and outcomes of therapeutic modalities. Ironically, it also undermines the position of authority of the surgeons and the scientific biomedical establishment because, when the evaluation of the clinical value of ''watchful waiting'' in prostate cancer or the recommended age for mammograms are seen to be based on the intuition of an individual with special powers of insight, the whole basis for rational medicine is threatened. Science and medicine hold privileged positions in society because their practitioners are seen to have special knowledge that is gained through the scientific method. Most of the public cannot define what that method is, but they have placed their trust in it, and when the practitioners themselves are seen to deny its validity, the foundation of the system is threatened.

There is little in scientific training to prepare clinicians for the fact that, as disease has become more biologically complex, it has also become more socially complex. A stool specimen positive for *Vibrios cholerae* or *Mycobacterium tuberculosis* in sputum gives a reasonably definitive diagnosis with a reasonably definitive set of therapeutic and public health responses, but a ''positive test'' for a BRCA mutation or an elevated prostate specific antigen (PSA) do not. For the first time, laboratory results have an element of probability about them, whether the specimens are obtained from people who feel symptoms and are self declared ''patients,'' or from self-declared ''healthy'' people who are part of a screening program. Patients, physicians, and scientists need to be educated in this new dimension of the ''tests'' they develop, advocate, and rely upon. There must be an increased awareness of the profoundly complex effects that these ''tests'' have on the person who is hearing the result.

For example, more that 100 variants of the BRCA1 and BRCA2 genes have been identified, but only a few have been associated with increased risk of breast cancer, and these are in a well-defined population of women with a family history of early breast cancer. These ''tests,'' unlike bacteriological analysis, are for susceptibility; women from families with a strong history of early breast cancer for whom the test is ''positive'' have as high as an 85% probability of development of breast cancer and as high as 45 percent chance of development of ovarian cancer. But women with a similar family history for whom the test is ''negative,'' that is, they do not have a mutation in these genes, still have the same probability of breast cancer development later in life as do women without a family history of early breast cancer. This notion is abstruse to the lay person experiencing the anxiety of the situation, but the medical profession and the press are not always sensitive to this fact. The report in the media that a particular mutation of BRCA2 had been found in a high percentage of Ashkenazi Jewish women, a group that includes more than 90% of Jews living in the United States, understandably caused a great deal of consternation. Media accounts metamorphosed from reports of increased susceptibility of these women to breast cancer to increased incidence of cancer. Only the rare reporter asked whether there was in fact a markedly higher incidence of breast cancer in Jewish women and what the role of an Ashkenazi Jewish woman was.

There is heated debate about the value of annual mammograms for women younger that 50 years old. There are many anecdotal accounts from women who had small tumors detected by mammograph in their younger years and who underwent successful surgery. The counter argument is that there is no statistical advantage to this population undergoing mass mammography, an argument that, in contrast, seems uncaring. In a similar manner, the advent of PSA measurement has made discussion of prostrate cancer very difficult; in advertisements and interviews with celebrities, men are told that there is an epidemic of the disease. The fact that disease-specific mortality rates have remained reasonably flat and that a large proportion of older men who die of other causes and were asymptomatic even though they have histological evidence of tumors at autopsy seems to be no match for the personal accounts of survivors.

B. "Public Privacy" and "Private Privacy"

The problem of facing the probabilistic nature of many ''tests'' carries an important set of social ramifications dealing with privacy. One is the right not to have private information about oneself made available without the expressed consent of the individual; I call this public privacy to indicate that it is the protection of an individual from the use of information about them by others. From its inception, the Human Genome Project was aware of this kind of problem and set aside a small percentage of its budget for the ethical, legal, and social implications (ELSI) of the project. The technology of the project was such that genetic information about individuals might become available to others and could be used to exclude them from employment, insurance coverage, and so on. Even though little has come from this program in terms of actual recommendations, its existence alone has probably served as a persistent reminder to legislators, civil libertarians, and informed individuals of the importance of safeguarding personal genetic information. There are times, however, when government has a legitimate interest in the private genetic information about citizens. For example, narcolepsy (or a predisposition to it, if such a test existed) need not necessarily disqualify an individual from being an airline pilot, but it is important that the physical condition be known so that it can be regularly monitored and appropriate safeguards instituted when applicable. On the other hand, neither the state nor anyone else has an obvious interest in knowing the vast majority of medical conditions or predispositions of individuals.

In the end, public privacy about genetic predisposition or other medical information is really not very different from the great array of other information about our credit and employment, driving record, and buying habits that are kept in government and industrial data bases.

Public privacy is an important issue, but it is one that is negotiated through legislative and regulatory processes. Given the public awareness of the problem, medical information will eventually be as protected as all other information (for better or for worse), and some forms of legislation preventing unfair employment or insurance practices will surely be enacted.

The second kind of privacy is more profound and more directly related to screening. I call this private privacy because it goes to the heart of the question of what medical ''tests'' mean to individuals. Because it deals not with what someone else knows, but with what a person knows about himself or herself, private privacy cannot be legislated, and it is here that the probabilistic and indeterminate nature of so many of the ''tests'' has its most serious implications for the individual.

In this era of increasing genetic determinism (some would even say rebirth of eugenics), private privacy concerns the existential meaning of what a person sees as his or her genetically encoded destiny. Genetic counselors have dealt with this problem for a limited number of conditions, but biotechnology and increased interest in screening presents us with the specter of having what we have been told is our own hidden biological fate routinely revealed. Technology is forcing people to confront the terror of their predetermined, innermost selves, to constantly redefine the meaning of normal, and to constantly reshape the knowledge of what scientists have taught them is destiny. It is both ironic and frightening that the term ''normal,'' which is at the heart of probabilistic meaning, is in danger of being given the ultimate nonprobabilistic meaning.

IV. CONCLUSION

Our mythic understanding of the path that brought us to the present juncture and our lack of understanding of the nonscientific influence of science and medicine on society make it difficult to have the complex nature of chronic disease and the probabilistic nature of the result of screening incorporated into the thinking and psyches of the public. A nation that spends billions of dollars on lotteries and casino gaming does not seem ready for the niceties of probability analysis, and the constant barrage of scientific absolutism that the public is immersed in makes the task of conveying complexity even more daunting. Every week, the media reports the discovery of another gene for a disease or a behavioral trait, fueling the idea that destiny is encoded in genes, and, with equal regularity, there are claims of breakthrough discoveries that are certain to lead to cures of the full range of diseases. It is impossible for the public to give up the idea of single cause and single remedy, so why should they be expected to give up the idea of absolute diagnostic or prognostic "tests?"

It is comfortable to blame the media for this situation, but the blame must be layed on the institution of science because it has willfully confused indoctrination with education, measuring the success of public education by the number of facts learned and by a positive attitude toward technology. This is qualitatively the same as equating the compliant patient with the informed patient, and, in the process, creating a population that relies on the authority of science and the physician rather than on informed judgment. Clearly, if the population is to understand the indeterminate nature of diagnosis and screening in chronic disease, this attitude must change, but, there is not universal understanding of the problem within the profession, so change will be a long time in coming. There is little chance that the existential meaning of the new kinds of personal information (private privacy) that goes hand in hand with the technical problem of probability of disease will be addressed, but, in the short term, those who are involved in dealing with health professionals and the public in areas of screening can consider rhetorical devices that dampen the absoluteness of the dominant message being transmitted.

II

BASIC PRINCIPLES OF SCREENING

3

Screening Theory and Study Design: The Basics

Philip C. Prorok, Barnett S. Kramer,* and John K. Gohagan
Division of Cancer Prevention, National Cancer Institute, Bethesda, Maryland

I. INTRODUCTION

In contrast to diagnostic testing of an individual with symptoms of a cancer, cancer screening involves active encouragement by some sector of the health care system directed at asymptomatic individuals to undergo tests that they would not ordinarily seek. This process results in the transformation of some individuals from a healthy state to that of patient, possibly triggering a series of invasive procedures and therapy that may have attendant morbidity. It is therefore incumbent upon those who would actively promote screening to have solid knowledge of the risks and benefits of screening and to communicate these to potential participants.

Determining the benefit and advisability of cancer screening is a difficult task. Early adverse outcomes in some people must be weighed against possible delayed benefits. Evidence should be rigorously obtained and the strength of the evidence explicitly measured and incorporated into decision-making. The strength of evidence should also be explained to the individuals considering screening. There are several levels of evidence that can be used to support a given position on cancer screening. The strongest evidence is that obtained from a well-designed and carefully implemented randomized controlled trial (RCT). Because it is not always possible or practical to conduct such a trial for every question that arises in cancer screening, it is sometimes necessary to base determination of benefit or recommendations on information not arising from an RCT. Consequently, various levels of evidence have been articulated for estab-

**Also:* Uniformed Services University of the Health Sciences, Bethesda, Maryland lishing screening benefit and associated recommendations or guidelines. In order of strength of evidence, the levels are as follows:

1. Evidence obtained from at least one RCT
2. Evidence obtained from controlled trials that use allocation methods other than randomization (e.g., by birthdate or hospital chart number)
3. Evidence obtained from cohort or case-control analytic studies, preferably from more than one center or research group
4. Evidence obtained from multiple time series with or without intervention (quasiexperimental designs)
5. Opinions of respected authorities based on clinical experience, descriptive studies, or reports of expert committees (1)

This chapter focuses on the issues, biases, and study designs that apply to these levels of evidence and that should be addressed in an appropriate assessment of cancer screening.

II. DEFINITION AND CRITERIA

Cancer screening is herein defined as the testing of apparently healthy volunteers from the general population for the purpose of separating them into groups with high and low probabilities of having a given cancer. Screening is typically initiated by those who offer the tests, thereby involving an implicit promise of benefit to those who are screened. Hence, those individuals having a positive screening test should be followed up to exact diagnosis and receive treatment of proven efficacy if they are determined to have cancer (2). From these considerations, there are two fundamental components to an effective screening program: 1. The screening test must detect cancer early, in a recognizable preclinical state, and 2. Therapy initiated as a result of a cancer being detected by screening must be more effective than therapy undertaken at the usual time of diagnosis. There can be no benefit from screening if either component fails. Thus, it is important to determine how each component contributes or fails to contribute in the assessment of a screening intervention.

In general, there are several key criteria that should be considered before screening is adopted.

1. The disease should be a serious health problem and the cause of substantial mortality and morbidity.
2. The target population should be clearly defined and have a reasonable disease prevalence.
3. The target population should be accessible, with a reasonable expectation of response to a screening invitation.

4. The screening test should have acceptable performance characteristics (e.g., sensitivity, specificity) and be acceptable to those being screened.
5. An effective treatment for the target cancer should be available.
6. There should be a reasonable expectation that individuals having positive screening tests will undergo diagnostic workup and that individuals diagnosed with disease and their physicians will comply with appropriate therapeutic practices.
7. Sufficient personnel and facilities should be available both for screening and for subsequent diagnosis and therapy.
8. There should be an agreed policy for early recall of individuals with suspicious findings and on the frequency of routine recall for those with negative findings.
9. A quality control procedure should be implemented to maintain the sensitivity and specificity of the test and high compliance with diagnostic and therapeutic follow-up (3,4).

Further perspective on screening assessment follows from considering the possible advantages and disadvantages of such testing. The possible outcomes of screening can be summarized into four categories (Figure 1):

1. True-positive results are given for individuals who have the cancer of interest and for whom the screening test is positive.
2. True negative results are given for persons who do not have the cancer and have a negative test.
3. False-positive results are given for persons who do not have the cancer but have a positive test.
4. False-negative results are given for individuals who have the cancer but test negative.

Only individuals with true-positive results can potentially benefit in any substantial way from screening. Some of these individuals may undergo less radical treatment or have prolonged life expectancy as a result of having their cancer diagnosed early. Alternatively, screening may result in longer morbidity for individuals whose life expectancy is unchanged despite having their cancer detected early by screening or for individuals who receive overtreatment of borderline abnormalities. Persons with true-negative results may gain a degree of reassurance that they are not likely to have the cancer, but, for low prevalence diseases such as cancer, most people can be reassured that they do not have cancer whether or not they undergo screening. One would be wrong only a few times per thousand. In contrast, screening results in net harm to persons with false-positive and false-negative results. Both incur the costs of screening and are exposed to any hazards of the testing process. False-negative results give

Disease Status

Test Result	Disease +	Disease -
+	True Positives (TP)	False Positives (FP)
-	False Negatives (FN)	True Negatives (TN)

$$\text{Sensitivity} = \frac{TP}{TP+FN}$$

$$\text{Specificity} = \frac{TN}{TN+FP}$$

$$\text{Positive Predictive Value} = \frac{TP}{TP+FP}$$

Figure 1 Cross classification of individuals by disease status and screening test result.

false reassurance that may lead to a delay in diagnosis, whereas false-positive results incur anxiety as well as any morbidity associated with the diagnostic procedures that follow a positive screen. Given these circumstances, it is imperative that both the properties of a screening test and the impact of screening on disease outcome be carefully evaluated.

III. SCREENING TEST PROPERTIES

Test performance is often summarized by three quantities: sensitivity (Se), specificity (Sp), and positive predictive value (PPV). Sensitivity is the proportion of persons designated positive by the screening test among all individuals who have the disease, whereas Sp is the proportion designated negative by the test among all those who do not have the disease. The PPV is the proportion of individuals with a positive screening test who have the disease. This quantity is a function of sensitivity, specificity, and disease prevalence (5–7). These quantities are typically estimated in small studies before the implementation of a screening trial or program, but accurate population estimates are usually only obtainable from actual population-based screening programs.

Traditionally, sensitivity is estimated using data from a 2 × 2 classification of individuals as having or not having cancer, and as having or not having a positive test result (see Figure 1). Sensitivity is then the proportion of all those with cancer who have a positive test. However, this approach requires the assumption of some arbitrary time interval during which cancers are diagnosed that could have been detected by the screening test. An alternative method that takes into account the time evolution of the screening process is to define sensitivity for each successive time period of a screening program, for example, yearly, which represents the ability of the test to detect cancers that would have surfaced increasingly further in the future after the test. An overall estimate of sensitivity can then be calculated as an average of these quantities (8).

The mathematical relationship between the screening test properties is

$$\text{PPV} = (\text{P})\,(\text{Se}) \,/\, \{(\text{P})\,(\text{Se}) + (1 - \text{P})(1 - \text{Sp})\}$$

where P is the prevalence, or probability, of having disease. From this formula, the implications of various parameter values can be explored. Prevalence values of 5 per 1000 and 1 per 10,000 are used for illustration. These values span the range of most cancers. The results are shown in Table 1. In addition to prevalence, specificity is a key parameter in assessing a screening test. Specificity must reach 99% or greater to avoid large numbers of false-positive results. Most available screening tests for cancer have considerably lower specificity, however, some below 90%.

The sensitivity, specificity, and predictive value requirements of a test differ in different settings. For diagnosis or monitoring following therapy, high sen-

Table 1 Positive Predictive Value (%) as a Function of Sensitivity, Specificity, and Prevalence

		Sensitivity		
Prevalence	Specificity	.8	.9	.95
.005	.95	7	8	9
	.99	29	31	32
	.999	80	82	83
.0001	.95	.2	.2	2
	.99	.8	.9	.9
	.999	.7	8	9

sitivity is important but high specificity is not, whereas for screening, high specificity is required while moderate sensitivity is often acceptable. The importance of specificity in the screening context is that low specificity leads to an undesirably large number of false-positive cases. This can overwhelm diagnostic facilities and result in prohibitive costs, and it exposes individuals to the risks of unnecessary diagnostic procedures. Thus, the test properties should be estimated as accurately as possible in well-designed studies before a test is used in practice. However, screening test properties do not provide information about the impact of screening on cancer mortality rate or life expectancy. Instead, once a screening test is known to have acceptable properties, it can be incorporated into a screening program that must then be evaluated in terms of any effect on cancer mortality.

IV. NATURAL HISTORY, SCREENING BIASES, AND STUDY ENDPOINTS

In attempting to conceptualize the disease process, the simplest model is the progressive disease model comprising three states: the disease free state, in which a person has no cancer discernible by any existing procedure; a preclinical disease state, in which the person has cancer which has not led to signs or symptoms, the person is unaware of the disease, but the cancer can be detected if a screening test is performed; and a clinical disease state, in which the cancer has become sufficiently consequential to produce symptoms leading to clinical diagnosis and possibly morbidity or mortality. For screening, the most relevant component of this natural history is the time from when the cancer is screen detected during the preclinical period until death. Cancer screening is predicated upon the assumption that there exists some manifestation of a cancer in a preclinical state that does not generate noticeable symptoms but which can be identified by a screening test, during which intervention can lead to improved outcome for the individual (i.e., decreased risk of dying of the cancer).

This natural history is usually not known or well understood, and potential early endpoints such as stage shift or case survival, which depend on knowledge of this natural history for their validity, cannot provide a definitive assessment of screening. The only outcome variable known to be valid for evaluating cancer screening is the population cancer mortality rate. This is the number of cancer deaths per unit time per unit population at risk (5,7,9–11). Mortality rate provides an overall evaluation of both components of screening program effectiveness by giving a combined assessment of early detection plus therapy. Demonstration of reduction in cancer mortality rate for a population offered screening is the gold standard by which the human and economic costs of a screening program can be justified and the promise of net benefit to those who elect to participate in screening can be fulfilled.

Careful study design and long-term follow-up of large populations are required to obtain an accurate estimate of mortality rate reduction. This is typically a costly undertaking. Consequently, intermediate or surrogate outcome measures have been sought that could provide answers in a shorter time. The most frequently mentioned are shift to earlier stage at diagnosis and improvement in case survival. However, there are critical shortcomings associated with these endpoints (7,9–11). The shortcomings result from biases inherent in screening programs: lead time bias, length bias, and overdiagnosis bias.

Individuals who participate in a screening program may harbor cancer and the cancer may be detected earlier than it would have been in the absence of screening. The amount of time by which the diagnosis is advanced as a result of early detection by screening is termed the lead time (Figure 2). Because of the lead time phenomenon, the time or age of diagnosis is advanced. Therefore, survival duration, which is typically measured from diagnosis, is automatically lengthened for cancers detected by screening, even if the individual does not live longer. This is lead time bias, and this bias renders the case survival endpoint invalid (7,9–13).

Length bias is the phenomenon that cancers detected by screening are not a random sample from the population of cases under consideration. As it is assumed that each cancer has some preclinical duration, all cancers in a population taken together comprise a distribution of preclinical durations, or sojourn times. However, the sojourn times of the cancers detected by screening are not a random sample for this distribution. Instead, cancers with longer duration preclinical disease are overrepresented among the detected cases (7,9,10,14,15) (Figure 3). If, as seems reasonable, cancers with long preclinical sojourn times

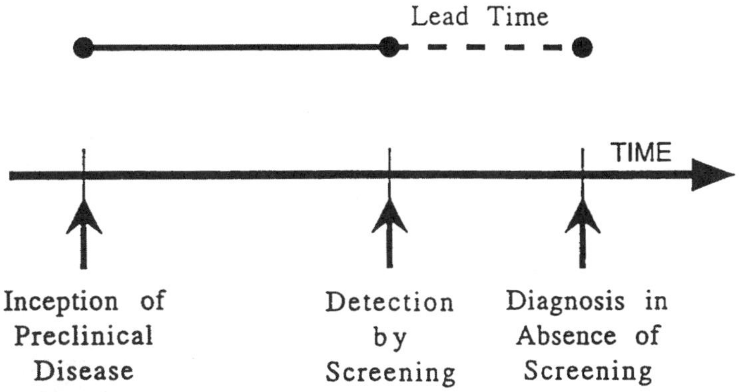

Figure 2 Relationship between duration of preclinical disease, screening point, and lead time.

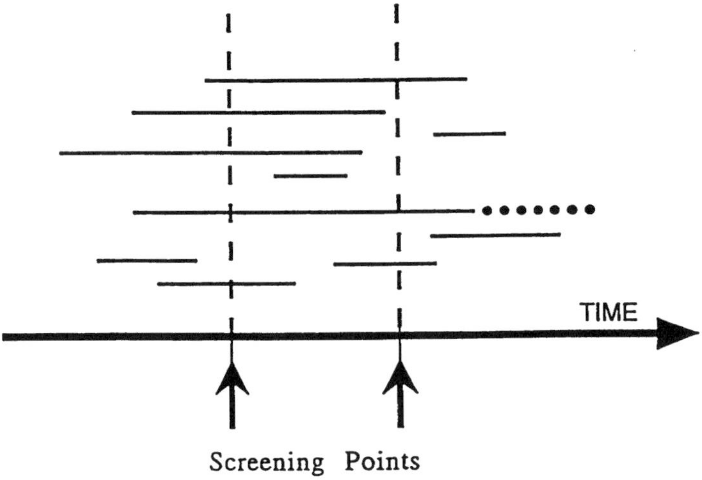

Figure 3 Length biased sampling in repeated screening. Solid horizontal lines represent a distribution of preclinical durations. The line ending in dots represents nonprogressive preclinical disease.

represent slow-growing preclinical disease which then progresses to slow-growing clinical disease, cancers with more favorable progression rates are the ones more likely to be detected by screening. In other words, more indolent tumors remain in a preclinical ''dormant'' state, during which they can be detected in a screening program, than faster moving, virulent cancers. Thus, screen-detected cancers tend to have good prognostic characteristics, such as lack of regional lymph node involvement, and tend to have a more favorable outcome, even in the absence of screening. Of key importance is that the inherent favorable prognostic characteristics of the disease increase the likelihood of these cancers being detected at screening, rather than the favorable characteristics being beneficial effects of screening resulting from detection at an earlier time. The latter may also happen, but this is a separate effect that cannot be established simply by the fact that cancers detected at screening have a high survival rate or a low incidence of nodal involvement (16).

Overdiagnosis bias is related to both lead time bias and length bias. It is possible that a subset of cancers in the preclinical state (or even in the clinical state) are nonprogressive or regressive in nature. These cancers are detectable by the screening test, but they would not progress to clinical disease during the person's lifetime; hence, they would never be diagnosed in the absence of screening. This is a problem of particular concern in screening for prostate cancer and neuroblastoma (17,18). One way to visualize the situation is that many cancers may be viewed as a pyramid-shaped iceberg of disease. Clinical or symptomatic disease is the tip of the iceberg, which appears above the waterline. However, there is a huge base of preclinical disease below the waterline, which lies hidden or undiscovered in the absence of screening. The pyramid expands rapidly below the surface, so screening tests, particularly those with increased sensitivity, detect increasingly large numbers of cancers, many of which may never reach the surface (19). Certainly the detection of such cancers is of no benefit to the individuals. However, such cases likely remain preclinical for a long time and are therefore more likely to be detected in a program with repeated screenings (See Figure 3). One consequence is that, in a screening trial, for example, the screened arm contains a higher proportion of early stage cases even if there is no mortality effect from screening because the counterparts to the nonprogressive cancers never surface in the control group. Overdiagnosis, or detection of disease that would never have caused morbidity if undetected in the normal life span, is a realistic consideration for many cancers because malignancy is generally a disease of aging.

Endpoints most frequently proposed as alternatives to the mortality rate are the case-finding rate or yield, stage of disease, and case-survival rate. The case-finding rate, or the incidence rate (because similar arguments apply to both), is of interest as an early clue as to whether screening might have an effect because more cases should be detected in the presence of than in the absence of screening. However, this rate provides no information about the effect of screening on disease outcome. The case-finding rate would be expected to increase in a screened population, especially in the early years of a screening program, relative to an unscreened population because of lead time. This can occur whether or not the screening results in a mortality effect. Furthermore, care must be exercised in determining what to count as discovered cancer in an early detection program. Many borderline lesions found by newer screening modalities may not be progressive disease, resulting in overdiagnosis bias. If this occurs, individuals undergo treatment unnecessarily, and they are exposed to other possible risks of screening. Thus, an increased cancer rate in a screening program is only an indication of increased cost, both human and economic, which may or may not be compensated for by benefit realized from the screening.

Another intermediate criterion that has been suggested for the evaluation of screening is a shift to a more favorable stage distribution of the cancers in a screened group. This can be an early indicator that screening might be effective, but it can also be misleading and is unsatisfactory as a final endpoint. The classification of cancers into stages involves some subjectivity and its validity requires strict guidelines and tight control over pathology, which may be difficult to implement in practice. Furthermore, the relationship of stage to survival or mortality has not been generally established in the screening setting. The quantitative relationship between extent of shift in stage distribution of cancers as a result of screening and size of mortality rate reduction is not well understood. The impact of the detection of in situ or borderline cases on stage distribution and subsequent mortality is also not clear.

The most noticeable impact may be for stage I or localized cases, particularly in a screening trial that has a cut-off point after which new cases are not accrued. Lead time and length bias can then result in the detection of slow-growing or nonprogressive stage I cancers in the screened arm to a greater extent than in the control arm. The control group counterparts of these stage I cancers may not surface by the cut-off point, if ever, so the screened arm contains a higher proportion of stage I cases even if screening has no effect on mortality rate. If there is a mortality effect, the magnitude estimated on the basis of a stage shift could be exaggerated. Therefore, observation of a stage shift does provide an indication of early detection, but it is insufficient evidence of an effect on disease outcome.

The case-survival rate is another proposed endpoint. The *N*-year survival rate is the number of cases alive after *N* years of observation divided by the number of cases diagnosed at the beginning of the time period. This measure is ordinarily calculated using life table methods to account for losses to follow-up. The case-survival rate refers only to diagnosed cancer cases, in contrast to mortality rate, which is a population measure. The case survival is attractive on the surface because it does address disease outcome and gives suggestive evidence of screening effectiveness. However, an improvement in survival of cancer cases in a screened group may not accurately imply a reduction in population mortality rate for reasons that center on lead time, length, and overdiagnosis biases and the lack of methodology to completely adjust for these biases.

The objective of screening is to detect and treat cancer earlier than usual and thereby prevent or postpone death from the cancer. If this occurs, there should be an increased survival rate as well as a mortality rate reduction. However, because some cancers are diagnosed earlier by screening, any observed improvement in survival measured from time of diagnosis is, at least partly, a reflection of lead time. Lead time cannot be directly observed for ethical reasons. Therefore, for any particular cancer detected by screening, it is impossible to distinguish between a true increase in survival time and an artificial increase resulting from added lead time. Because there is no universally accepted method to estimate lead time or to adjust survival for lead time, case survival cannot be considered a valid measure of screening effectiveness.

Even if one could compensate for lead time bias, survival comparisons could still be confounded by length bias. For example, length bias is likely to create differences between two groups of cancers diagnosed in different ways, such as by screening versus no screening, or by different screening modalities. The two groups likely have different natural history distributions. Thus, even if adjustment for lead time were possible, any remaining survival difference could be a real benefit, or it could simply result from the differing disease natural histories between the two groups, or both. There is no generally applicable method to either estimate length bias or adjust survival for length bias. A method has been proposed for separating the effects of treatment, lead time, and length bias in certain circumstances (20).

An alternative approach to analysis of survival data in screening studies is to consider survival time measured from entry into a study instead of from time of diagnosis. Then, the survival time of all cancer cases starts from the same time origin, thereby eliminating lead time bias. Even though survival calculated in this way is not comparable to that typically calculated in treatment studies or registries, its use can lead to a valid comparison between cancers in an intervention group and the cancers in a control group within a study. Even though this procedure avoids lead time bias, length bias and overdiagnosis are still concerns. For a valid analysis, it must still be established that the two groups of cancers whose survival is being compared are comparable with respect to their disease natural histories (21).

A final measure that has received attention as an endpoint for screening studies is the population incidence rate of advanced stage cancer (22–26). As discussed, the overall incidence rate or the rate of early stage cancer should increase with screening. These measures are invalid endpoints because each could be artificially inflated because of length bias or overdiagnosis bias. However, if the rate of advanced cancer, cancer which has metastasized or is likely to lead to death, is reduced by screening, then it might be reasonably expected that the cancer mortality rate will also decline. The validity of this endpoint must be established for each cancer separately by defining advanced stage cancer and then determining the relationship between the advanced cancer and mortality rates in properly designed studies.

V. STUDY DESIGNS

Experimental studies and observational studies are the two basic types that can be used to evaluate cancer screening (5,7,9). The most common type of experimental study is the RCT, and it is the method of choice because its inherent structure is designed to produce an unbiased assessment of effect. Observational or quasiexperimental designs are used when an RCT is not possible. These may be similar in many respects to experimental studies, but because their comparison groups are not established by randomization, they are generally difficult to analyze and interpret. Nonetheless, cohort and case-control observational studies have been used to evaluate cancer screening, and the design and interpretation of these studies are the subject of research (7,10,27–29).

In cancer screening evaluation, both the effectiveness of offering screening and the efficacy of screening participation are important measures. The effectiveness of offering screening is the ratio of the cancer mortality rate of those offered screening to what the cancer mortality rate of these same individuals would have been had they not been offered screening. Effectiveness is of particular importance for establishing public health policy. The efficacy of screening is the ratio of the cancer mortality rate for those actually screened to what the cancer mortality rate of these same individuals would have been had they not been screened. Efficacy is applicable to determining the value of screening for those who accept the screening test, but it is not a direct measure of the effect of a screening program under consideration for a population.

Effectiveness can be estimated directly by comparing the mortality rates between the randomized groups in an RCT. An indirect estimate of efficacy can also be obtained (30). If compliance with the screening maneuver is high, effectiveness and efficacy are nearly identical. On the other hand, observational study designs compare those who are actually screened with those who are not, and therefore provide estimates of efficacy. For accurate estimation of either quantity, the design features typically found in an RCT should be met by an observational study just as for an experimental study. These features include explicit inclusion criteria, a well-defined target population, a carefully articulated intervention protocol with quality control provisions, and a clear definition of the study endpoint, including careful (possibly blind) ascertainment. It takes just as long to measure mortality endpoints when screening is instituted using the weaker designs as it does using the stronger RCT design.

A. Selection Bias and Observational Studies

Observational studies are deficient in that they lack a control group constructed by a chance mechanism. The purposes and advantages of randomization are well known (31). Interpretation of a study that uses a control group chosen by any method other than randomization requires the assumption either that the control and intervention groups are identical in all important variables except the intervention under study or that one can correct for all relevant differences. To accomplish the latter, it must be further assumed that all factors affecting the course of the disease in question are identified and measured. It is difficult to conceive of circumstances in which these necessary assumptions are satisfied or can be verified for any known cancer.

The paramount problem with observational studies of screening is the potential noncomparability of the groups being compared. There are severe limitations on the ability to draw valid inferences from studies of patient groups or observations of changes in community rates resulting from the lack of detailed knowledge of disease natural history and the inadequate understanding of reasons for changes over time in incidence, survival, and mortality rates. In screening, a strong selection bias may render the characteristics of individuals who agree to participate in screening very different from those of individuals who refuse to participate. This has been observed in the Health Insurance Plan (HIP) breast cancer screening trial (32). This trial compared breast cancer mortality rates between a control group and a total study group to evaluate screening with clinical examination plus mammography. The study group included women who were screened and women who refused screening. There were substantial differences in disease characteristics between the compliers and the refusers. Among study group women who were screened, over a 5-year period, the rate of case detection was about 2.3 per 1000. However, the incidence rate among study group women who refused screening was 1.59 per 1000 per year, whereas the control group rate was 1.95 per 1000 per year. This suggests that women with a higher risk of breast cancer selected themselves for screening. Furthermore, the death rates from all causes excluding breast cancer per 10,000 per year after 5 years of follow-up were 1. total study group, 55.1; 2. study screened women, 42.4; 3. study women who refused screening, 81.0; 4. control group, 56.4. Thus, among the women who attended screening, general mortality rate excluding breast cancer was far lower. Construction of an appropriate comparison group for the women who elected to be screened was not possible in view of this self-selection bias (32,33).

B. Observational Study Designs

If a screening test is in wide use, precluding the availability of an unscreened comparison group, or if the cancer under consideration is very rare, thus requiring a prohibitively large sample size for a trial, an attractive observational study design that can be used for evaluation involves a comparison of cancer incidence and mortality in a defined population before and after the introduction of a screening program. Alternatively, one can compare geographic regions that are as comparable as possible with respect to disease mortality rate, examining time trends in incidence and mortality between intensively screened areas versus nonscreened areas. Rapid introduction of the screening program and virtually full coverage of the population at risk are required for both approaches. Reliable incidence and mortality data for at least a 10-year period before the start of screening should be available for use in predicting trends. The capability for accurate, long-term follow-up of the entire population at risk is also required. Under some circumstances, this type of observational study may actually require larger populations than an RCT and may eventually prove more costly than the more rigorous and definitive study design. Apart from the potential biases involved, this design may be used to estimate either effectiveness or efficacy, depending on the intervention population chosen. Cervical cancer screening has been evaluated using this approach (34).

The case-control design is another observational approach that can be used for screening evaluation (5,27–29). Such studies can be used to evaluate the efficacy of screening in the prevention of death, or of invasive disease and consequently death in situations in which the screening test detects precursor lesions. A number of case-control studies have been reported, primarily for evaluation of cervical and breast cancer screening (30,35–39). Methodological issues that arise in the screening case-control studies have also received attention (7,27–30,40–44).

Appropriate definition of cases and controls is required for a proper case-control design. Because the primary cancer screening endpoint is cause-specific mortality, eligible cases are deaths from the cancer of interest in the population under study, regardless of the means of diagnosis. Eligible controls include all living individuals in the population from which the cases were derived, including individuals with the diagnosis of cancer. The analysis then centers on determining whether exposure to screening is associated with a reduction in the risk of death from the cancer. Because screening histories of cases and controls are required, the approach is only applicable for screening tests that have been in use for several years.

A major concern with both the cohort and case-control designs is the potential for selection bias among individuals who choose to be screened versus those who do not, and the impact of this bias on interpretation of the study (30,43,45,46). The underlying assumption in all cohort and case-control study analyses is that the screened and unscreened cohorts, or the cases and controls, must be drawn from the same population. Selection biases in the choice to be screened or not make the basic assumption questionable in most such studies. Both designs investigate whether screening reduces the mortality rate of those individuals who elect to be screened relative to those who do not. This comparison does not estimate efficacy, in that the self-selected comparison group does not, in general, provide an estimate of the mortality rate of the screenees if they had not been screened. Both designs can provide a valid efficacy comparison only if those refusing to be screened have the same cancer mortality rate as those accepting screening would have if there was no screening, that is, if there is no self-selection bias (30).

C. Randomized Screening Trials

1. *Design Issues*

A randomized controlled screening trial involves the testing of an intervention population and long-term follow-up of intervention and control populations according to a protocol. Several key features must be considered. The target cancers, the screening tests, and the diagnostic and therapeutic regimens must be determined, although diagnostic and therapeutic procedures may not be under the control of those administering the trial. The appropriate outcome variable must be chosen and the sampling unit (individual versus group or cluster) selected. Because cluster allocation introduces correlation among trial participants, larger sample sizes are generally required and statistical adjustment must be performed in the analysis. However, the approach is sometimes attractive in that it may reduce contamination in the control arm. Admission and exclusion criteria must be established and a randomization procedure adopted. The intervention and control groups should be followed to ascertain predetermined endpoints with equal intensity over the same time period, and the endpoint measured in a blind fashion, if possible. Adherence to the study protocol should be maximized. Finally, in the analysis, all individuals in the control arm should be compared to all individuals in the intervention arm, including individuals who comply with screening and those who do not (i.e., an ''intent to treat'' rather than a ''treatment received'' analysis).

The number of screening examinations or screening rounds and the interval between examinations must be chosen. Choosing the number of screens involves a tradeoff between a number thought sufficient to realize an effect, if there is one, and the cost of the screening procedures. These issues have been addressed by several modeling efforts (47–50). Another concern in design is the relationship between study duration, sample size, and the expected timing of any effect. Study duration and sample size are inversely related, in that the number of endpoints (cancer deaths) is a major factor in determining the trial size sufficient to answer hypotheses. If only these two parameters were involved, the relationship between follow-up costs on the one hand and recruitment and screening costs on the other would determine the design. The time at which a reduction in mortality rate may occur must also be considered, however. For completed cancer screening trials in which a reduction in mortality rate occurred, a separation between the mortality rates in the screened and control arms typically emerged 4 to 5 years or more after randomization (51,52). In addition, the difference may increase with time, even after screening stops (51). Thus follow-up may have to continue for many years to observe the full effect of the screening, even with a very large sample size trial. Whereas a follow-up period of at least 10 years is usually appropriate, if the screening effect is manifested primarily among a subset of patients with slowly growing cancer, a longer period may be necessary.

Determination of the appropriate endpoint and the required sample size in a cancer screening study are intimately related. In the hypothesis testing framework, the sample size can be calculated from an appropriate formula if one knows the mortality rate, effect size, and statistical procedure, all of which depend on the choice of endpoint for the study. In cancer screening trials, this is the cancer mortality rate. Because the analysis involves a comparison of the numbers or rates of deaths, methods for Poisson-distributed data can be used

(53,54). Other factors that must be taken into consideration are noncompliance in the screened and control groups (i.e., contamination rates), randomized groups of different sizes, and lower than expected event rates among the individuals who participate in the study (''healthy volunteer effect''). Several approaches have been formulated to address these problems (7,55,56).

2. *Trial Designs*

a. Classic Two-Arm Trial to Answer a Single Question. In this trial design, a designated population is randomly allocated to two groups, one offered screening according to a protocol and the other a control group not offered screening. The mortality rates in the two groups are compared at the end of the follow-up period (56). The first large randomized cancer screening trial, the Health Insurance Plan of Greater New York (HIP) trial of breast cancer screening, used this design (51).

b. Designs to Answer More Than One Question in the Same Trial. Extensions of the classic two-arm design have been considered for cancer screening trials to address more than one question at the same time (57). As an example, the National Breast Cancer Screening Study in Canada includes two different study populations within the same administrative structure. The questions being addressed are

1. In women aged 40 to 49 years at entry, does annual screening by mammography and physical examination, when used as an adjunct to the highest standard of care in the Canadian health care system, reduce the mortality rate associated with breast cancer?
2. In women aged 50 to 59 years at entry, what is the additional contribution of routine annual mammographic screening to screening by physical examination alone in reducing the breast cancer mortality rate?

This is accomplished by separately randomizing individual women in the two age groups (58). In some circumstances, it is possible to address several related questions by including additional randomized groups. The colon cancer screening trial conducted at the University of Minnesota is an example (59). The two questions being asked are whether screening can reduce the mortality rate and whether there is a different effect when different screening frequencies are used. The design involves three randomized groups: a control group, a group offered annual screening with a test to detect occult blood in the stool, and a group offered the occult blood test every 2 years.

Another extension of the basic design is the all-versus-none design. This is typically a two-group trial in which the intervention group includes multiple interventions (57). One version of this design uses several interventions, with each intervention being an early detection maneuver for a different type of cancer. Two assumptions are required for this design to be valid: first, the test for any particular cancer must not affect the detection or mortality rate of any other cancer, and, second, disease-specific mortality rate is independent among the cancers under study. An example of this design is the Prostate, Lung, Colorectal and Ovarian Cancer (PLCO) Screening Trial sponsored by the National Cancer Institute (60). The objectives of this trial are to determine the following: 1. in females and males, whether screening with 60-cm flexible sigmoidoscopy can reduce the mortality rate owing to colorectal cancer and whether screening with chest radiograph can reduce the mortality rate from lung cancer; 2. in males, whether screening with digital rectal examination plus serum PSA can reduce the mortality rate from prostate cancer; and 3. in females, whether screening with pelvic examination plus serum CA 125 and transvaginal ultrasound can reduce the mortality rate associated with ovarian cancer. The reciprocal control design is yet another design option to answer more than one question at a time (57). The participants in each arm of the trial receive an intervention, and they also serve simultaneously as controls for an intervention in another arm of the trial. The assumption is required that the intervention aimed at a particular cancer does not affect any of the other cancers under investigation.

There are options for the relationship between screening and follow-up within each of the aforementioned designs (61). The basic design involves randomizing individuals to either an intervention or a control group, with intervention arm participants receiving periodic screening throughout the trial. Individuals in the control arm are not offered screening; they are simply expected to follow their usual medical care practices. This has been termed the continuous-screen design. The NCI Cooperative Lung Cancer Screening RCT essentially followed this design some 25 years ago (62).

A potentially serious drawback of the continuous-screen design is that all intervention group participants are screened for the duration of the trial, and the cost of this level of intervention may be prohibitive. An alternative is the stop-screen design. In this method, screening is offered to intervention participants for a limited time and both arms are followed up for disease incidence and mortality rate until the conclusion of the trial. This design is preferred when it is expected that a reduction in mortality rate will emerge only after a long follow-up period and when it is too expensive or unfeasible to continue the periodic screening for the entire duration of the trial. The HIP trial used this design (51). A considerable saving in cost can be realized from using the stop-screen design. However, analysis and interpretation can be more complex than in the continuous-screen design, because any difference in disease-specific mortality rate between the two arms may be diluted by deaths from cancers that develop in the intervention group after screening stops.

A variant of the stop-screen design in which the screening test is also offered to all participants in the control arm at the time the last screen is offered to those in the intervention arm is termed the split-screen design. The breast cancer screening trial conducted in the 1980s in Stockholm is an example of this design (63). An advantage of this design is that there is a greater possibility of identifying comparable sets of cancer cases for the analysis.

A variant of the continuous-screen design is the delayed-screen design in which periodic screening is offered to the control arm starting at some time point during the study and continuing until the end of the trial. This design is appropriate for estimation of the marginal effect of introducing screening at some standard time or age, relative to starting the screening at a later time or age. This design is being used to evaluate the effect of beginning screening before the age of 50 years in the UK Breast Cancer Screening Trial of women younger than age 50 years (64).

3. *Trial Analysis*

Special issues often arise in the analysis of the screening trials described, for both primary and secondary analyses. Primary analyses focus on evaluating whether there is a statistically significant difference in disease-specific mortality rate between the control and intervention arms. Secondary analyses focus on ascertaining the magnitude of the mortality rate difference, and with developing a deeper understanding of the underlying mechanisms (61).

a. Primary Analysis. The statistical methodology used for primary analysis typically involves a Poisson test statistic comparing the observed death rates (65), a Fisher exact test comparing the observed proportions of cancer deaths (66), or a logrank test comparing disease-specific death rates over time in the two trial arms (66,67). For illustration, the Poisson test statistic for comparing cumulative mortality rates is

$$Z_T = (PY_S\,D_C - PY_C\,D_S) / \sqrt{(PY_C\,PY_S\,(D_C + D_S))}$$

where D_C = the number of deaths from the cancer of interest in the control group through the time of analysis, D_S = the corresponding number of deaths in the screened group, PY_C = the number of person-years at risk of death from the cancer of interest in the control group through the time of analysis, and PY_S = the corresponding number of person-years in the screened group (53).

The analysis is termed an overall mortality analysis if the disease-specific mortality experience of all randomized participants is the basis for the logrank test. If the test is based on the mortality experience of comparable groups of cancer cases in the two arms of the trial, the term the limited mortality analysis is used (67).

The advantage of the overall mortality analysis is comparability of comparison groups formed by randomization, because the total populations allocated to the two trial arms are compared. However, the logrank test is optimal under proportionality of the disease-specific mortality hazards in the two arms of a trial, whereas, in cancer screening trials, there is generally a delay from the start of the screening intervention until any effect on cancer mortality rate is observed. Furthermore, the magnitude of any effect may vary over time. Thus, logrank tests comparing disease-specific population mortality rate can be relatively inefficient. In stop-screen designs, there is the additional complication that cancers continue to accrue in both trial arms after screening stops. The screening effect can be diluted by the cancer deaths in the intervention arm that are due to cancers developing after screening stops. Thus, the mortality hazard ratio may decrease with time after some point in the trial. To gain efficiency, a weighted logrank statistic may be used instead of the usual (unweighted) logrank statistic if the specific form of departure from proportional mortality hazards is known (68). Several statistical procedures have recently been proposed to improve the analysis when departures from proportional hazards is a likely possibility (69–71).

The limited mortality analysis is restricted to examination of comparable sets of cancers, one set consisting of cancers that arise in the intervention arm and that are diagnosed during a designated time interval from the start of the study and the other set comprising their counterparts in the control arm diagnosed during the same interval. This limited mortality analysis is typically only applied to split-screen or stop-screen designs. The split-screen design naturally generates two presumably comparable sets of cancer cases as those cancers diagnosed up to and including the final screen offered to both arms. Determination of comparable cancer sets is not so straightforward, however, in the stop-screen design. The key is in choosing the time interval during which cancers to be used in the analysis should be ascertained. The endpoint of the case ascertainment interval should not be too long after the cessation of screening, because further accrual of clinically detected cancers in both arms could dilute any observed screening effect, as discussed previously. However, another form of dilution can arise if all cancers diagnosed after screening ends are excluded. Some cancers diagnosed in the control arm after screening stops may correspond to screen-detected cancers in the intervention arm that were diagnosed earlier than they would have been without screening. If this cancer subset benefits from the earlier diagnosis resulting from screening, then the exclusion of their control group counterparts also causes a dilution of the screening effect (68).

When comparable cancer sets are chosen, the true comparability of the sets should be carefully investigated. An evaluation should be undertaken of the cases in the selected sets as well as the cases that are diagnosed after the end of the time interval used to define the selected sets. Methods have been proposed to assess comparability of cancer sets in a stop-screen design (67). The numerical as well as the biological comparability of the cancer sets are considered. Numerical comparability relates to the numbers of cancers in the sets. Biological or qualitative comparability focuses on the natural history of the cancer sets, especially their survival characteristics in the absence of screening. Qualitative comparability of candidate sets can be assessed by examining covariates associated with the cancer cases defined at the time of randomization.

It is often difficult to identify comparable cancer sets. It may be impossible to identify comparable sets in a stop-screen design if screening is available outside of the trial and if the uptake of the outside screening differs between the two arms after the cessation of the screening that is part of the trial protocol. Furthermore, it is unlikely that numerical equalization of cancer sets will ever occur in a continuous-screen design. In this case, the only valid option is an appropriately weighted overall mortality analysis.

b. Secondary Analysis. Secondary analyses of cancer screening trials typically focus on two types of information, that related to outcome of cancer cases captured in survival data and that related to early diagnosis, captured by estimates of lead time, sensitivity, and the extent of any shift to earlier stage of disease at diagnosis among cancers in the screened arm.

Analyses of potential survival improvement are based on the postdiagnosis survival curves in the two cancer sets. However, the postdiagnosis survival of screen-detected cancers includes lead time, and lead time bias can be avoided only by removing the lead time fraction of the survival time. Approaches were initially developed in the analysis of data from the HIP trial assuming a fixed lead time of 1 year and assuming the survival at a given year, k, from diagnosis of control group cancers and interval cancers (cancers diagnosed in the intervals between screens because of signs or symptoms) as equivalent to the $k + 1$-year survival from diagnosis of screen-detected cancers (65). More recently, Walter and Stitt (72) allowed lead time to be a random variable with a known distribution, and this approach was extended nonparametrically by Xu and Prorok (73).

Knowledge of the lead time probability distribution is required to adjust survival for lead time. Estimates of average lead time based on comparing disease incidence in the control and intervention groups were developed by Shapiro et al. (74) and Morrison (7). Kafadar and Prorok (75) used differences in survival rate from entry and from diagnosis between screened and control group cancers to estimate mean benefit time and mean lead time assuming two comparable case sets as in a limited mortality analysis. Statistical modeling methods for estimating lead time have been developed by Zelen and Feinleib (12) and Walter and Day (76).

Caution must be exercised in interpreting information on shift in the distribution of cancer stage at diagnosis because shifts may be due to overdiagnosis bias or length bias, and therefore need not imply any disease-specific mortality benefit. A stage-shift model has been developed that allows estimation of the amount of shift owing to screening both between and within stages, as well as estimation of associated mortality benefits. The model requires comparable sets of cancer cases (77).

VI. CONCLUSION

The evaluation of a cancer screening program is a very complicated undertaking. The act of screening converts some ostensibly healthy individuals into patients who are then subject to medical intervention at some level of economic and human cost, with the hope of improved disease outcome or increased life expectancy. This is quite different from intervening to relieve the suffering of someone known to have cancer. Therefore, when practical, it is best to achieve the highest level of evidence possible in determining whether the screening intervention is effective. This implies use of the RCT design whenever possible. Other designs exist and have been used on occasion, but the results can be misleading, and can even give the false impression of benefit when screening would result in a net harm to the population.

REFERENCES

1. Physicians Data Query. Cancer screening. Bethesda, MD: National Cancer Institute, 1997.
2. Sackett DL, Holland WW. Controversy in the detection of disease. Lancet 1975; 23:357–359.
3. Wilson JMG, Jungner G. Principles and practice of screening for disease. Public Health Paper No. 34. Geneva: World Health Organization, 1968.
4. Summary and general recommendations. In: Miller AB, ed. Screening in Cancer. UICC Technical Report Series, Vol. 40. Geneva: International Union Against Cancer, 1978: 334–338.
5. Cole P, Morrison AS. Basic issues in population screening for cancer. J Natl Cancer Inst 1980; 64:1263–1272.
6. Thorner RM, Remein QR. Principles and procedures in the evaluation of screening for disease. Public Health Monograph No. 67. Washington, DC: U.S. Public Health Service, 1961.
7. Morrison AS. Screening in Chronic Disease. New York: Oxford University Press, 1985.
8. Day NE. The theoretical basis for cancer screening. In: Miller AB, ed. Advances in Cancer Screening. Boston: Kluwer Academic Publishers, 1996:9–24.
9. Prorok PC. Evaluation of screening programs for the early detection of cancer. In: Cornell RG, ed. Statistical Methods for Cancer Studies. New York: Marcel Dekker, 1984:267–328.
10. Prorok PC, Connor RJ. Screening for the early detection of cancer. Cancer Investigation 1986; 4:225–238.
11. Sackett DL. Periodic examination of patients at risk. In: Schottenfeld D, ed. Cancer Epidemiology and Prevention, Current Concepts. Springfield, IL: Charles C. Thomas 1975:437–454.
12. Zelen M, Feinleib M. On the theory of screening for chronic diseases. Biometrika 1969; 56:601–614.

13. Hutchison GB, Shapiro S. Lead time gained by diagnostic screening for breast cancer. J Natl Cancer Inst 1968; 41:665–681.
14. Albert A, Gertman PM, Louis T. Screening for the early detection of cancer—I: the temporal natural history of a progressive disease state. Math Biosci 1978; 40: 1–59.
15. Zelen M. Theory of early detection of breast cancer in the general population. In: Heuson JC, Mattheiem WH, Rozenzweig M, eds. Breast Cancer: Trends in Research and Treatment. New York: Raven Press, 1976:287–300.
16. Bailar JC. Mammography: a contrary view. Ann Intern Med 1976; 84:77–84.
17. Kramer BS, Brown ML, Prorok PC, Potosky AL, Gohagan JK. Prostate cancer screening: what we know and what we need to know. Ann Intern Med 1993; 119: 914–923.
18. Bessho F. Effects of mass screening on age-specific incidence of neuroblastoma. Int J Cancer 1996; 67:520–522.
19. Kramer BS, Gohagan JK, Prorok PC. Is screening for prostate cancer the current gold standard?—''No.'' Euro J Cancer 1997; 33:348–353.
20. Morrison AS. The effects of early treatment, lead time and length bias on the mortality experienced by cases detected by screening. Int J Epidemiol 1982; 11: 261–267.
21. Aron JL, Prorok PC. An analysis of the mortality effect in a breast cancer screening study. Int J Epidemiol 1986; 15:36–43.
22. Chamberlain J. Planning of screening programs for evaluation and non-randomized approaches to evaluation. In: Prorok PC, Miller AB, eds. Screening for Cancer. I—General Principles on Evaluation of Screening for Cancer and Screening for Lung, Bladder and Oral Cancer. UICC Technical Report Series, Volume 78. Geneva: International Union Against Cancer, 1984:5–17.
23. Tabar L, Fagerberg CJG, Gad A, Baldetorp L, Holmberg LH, Grontoft O, Ljungquist U, Lundstrom B, Manson JC, Eklund G, Day NE, Pettersson F. Reduction in mortality from breast cancer after mass screening with mammography. Lancet 1985; i:829–832.
24. Roberts MM, Alexander FE, Anderson TJ, Chetty U, Donnan PT, Forrest P, Hepburn W, Huggins A, Kirkpatrick AE, Lamb J, Muir BR, Prescott RJ. Edinburgh trial of screening for breast cancer: mortality at seven years. Lancet 1990; 335: 241–246.
25. Day NE, Williams DRR, Khaw KT. Breast cancer screening programs: the development of a monitoring and evaluation system. Br J Cancer 1989; 59:954–958.
26. Paci E, Ciatto S, Buiatti E, Cecchini S, Palli D, Rosselli del Turco M. Early indicators of efficacy of breast cancer screening programs. Results of the Florence district program. Int J Cancer 1990; 46:198–202.
27. Morrison AS. Case definition in case-control studies of the efficacy of screening. Am J Epidemiol 1982; 115:6–8.
28. Sasco AJ, Day NE, Walter SD. Case-control studies for the evaluation of screening. J Chron Dis 1986; 39:399–405.
29. Weiss NS. Control definition in case-control studies of the efficacy of screening and diagnostic testing. Am J Epidemiol 1983; 118:457–460.

30. Connor RJ, Prorok PC, Weed DL. The case-control design and the assessment of the efficacy of cancer screening. J Clin Epidemiol 1991; 44:1215–1221.
31. Byar DP, Simon RM, Friedewald WT, Schlesselman JJ, DeMets DL, Ellenberg JH, Gail MH, Ware JH. Randomized clinical trials: perspectives on some recent ideas. N Engl J Med 1976; 295:74–80.
32. Shapiro S, Venet W, Strax P, Venet L, Roeser R. Selection, follow-up, and analysis in the Health Insurance Plan study: a randomized trial with breast cancer screening. Natl Cancer Inst Monogr 1985; 67:65–74.
33. Bearhs OH, Shapiro S, Smart C. Report of the working group to review the National Cancer Institute/American Cancer Society breast cancer detection demonstration projects. J Natl Cancer Inst 1979; 62:639–709.
34. Hakama M, Miller AB, Day NE. Screening for Cancer of the Uterine Cervix. IARC Scientific Publications No. 76. Lyon: International Agency for Research on Cancer, 1986.
35. Ebeling K, Nischan P. Screening for lung cancer—results from a case-control study. Int J Cancer 1987; 40:141–144.
36. Friedman GD, Hiatt RA, Quesenberry CP, Selby JV. Case-control study of screening for prostatic cancer by digital rectal examinations. Lancet 1991; 337:1526–1529.
37. Oshima A, Hirata N, Ubukata T, Urneda K, Fujimato I. Evaluation of a mass screening program for stomach cancer with a case-control design. Int J Cancer 1986; 38:829–833.
38. Selby JV, Friedman GD, Quesenberry CP, Weiss NS. A case-control study of screening sigmoidoscopy and mortality from colorectal cancer. N Engl J Med 1992; 326:653–657.
39. Sobue T, Suzuki T, Naruke T, The Japanese Lung Cancer Screening Research Group. A case-control study for evaluating lung cancer screening in Japan. Int J Cancer 1992; 50:230–237.
40. Berrino F, Gatta G, D'Alto M, Crosignani P, Riboli E. Use of case-control studies in evaluation of screening programmes. In: Prorok PC, Miller AB, eds. Screening for Cancer I—General Principles on Evaluation of Screening for Cancer and Screening for Lung, Bladder and Oral Cancer. UICC Technical Report Series, Volume 78. Geneva: International Union Against Cancer, 1984: 29–43.
41. Dubin N, Friedman DR, Toniola PG, Pasternack BS. Breast cancer detection centers and case-control studies of the efficacy of screening. J Chron Dis 1987; 40:1041–1050.
42. Hosek RS, Flanders WD, Sasco AJ. Bias in case-control studies of screening effectiveness. Am J Epidemiol 1996; 143:193–201.
43. Moss SM. Case-control studies of screening. Int J Cancer 1991; 20:1–6.
44. Weiss NS, McKnight B, Stevens NG. Approaches to the analysis of case-control studies of the efficacy of screening for cancer. Am J Epidemiol 1992; 135:817–823.
45. Friedman DR, Dubin N. Case-control evaluation of breast cancer screening efficacy. Am J Epidemiol 1991; 33:974–984.
46. Janzon L, Andersson I. The Malmo mammographic screening trial. In: Miller AB,

Chamberlain J, Day NE, Hakama M, Prorok PC, eds. Cancer Screening. Cambridge: Cambridge University Press, 1991:37–44.
47. Eddy DM. Screening for Cancer—Theory, Analysis and Design. Englewood Cliffs, NJ: Prentice-Hall, 1980.
48. Habbema JDF, Lubbe JTN, Van der Maas PJ, VanOortmarssen GJ. A computer simulation approach to the evaluation of mass screening. In: Van Bemmel JH, Ball MJ, Wigertz O, eds. MEDINFO 83. Amsterdam: North-Holland, 1983:1222–1225.
49. Kirch RLA, Klein M. Surveillance schedules for medical examinations. Management Science 1974; 20:1403–1409.
50. Knox EG. A simulation system for screening procedures. In: McLachlan G, ed. The Future and Present Indicatives, Problems and Progress in Medical Care. Ninth Series, Nuffield Provincial Hospitals Trust. London: Oxford University Press, 1973: 17–55.
51. Shapiro S, Venet W, Strax P, Venet L. Periodic Screening for Breast Cancer. The Health Insurance Plan Project and Its Sequelae, 1963–1986. Baltimore: The Johns Hopkins University Press, 1988.
52. Tabar L, Fagerberg G, Duffy SW, Day NE, Gad A, Grontoft O. Update of the Swedish two-county program of mammographic screening for breast cancer. Radiologic Clinics of North America 1992; 30: 187–210.
53. Shiue WK, Bain LJ. Experiment size and power comparisons for two-sample Poisson tests. Appl Statist 1982; 31:130–134.
54. Taylor WF, Fontana RS. Biometric design of the Mayo lung project for early detection and localization of bronchogenic carcinoma. Cancer 1972; 30:1344–1347.
55. Moss S, Draper GJ, Hardcastle JD, Chamberlain J. Calculation of sample size in trials of screening for early diagnosis of disease. Int J Epidemiol 1987; 16:104–110.
56. Prorok PC. Screening studies. In: Greenwald P, Kramer BS, Weed DL, eds. Cancer Prevention and Control. New York: Marcel Dekker, 1995:225–242.
57. Freedman LS, Green SB. Statistical designs for investigating several interventions in the same study: methods for cancer prevention trials. J Natl Cancer Inst 1990; 82:910–914.
58. Miller AB, Howe GR, Wall C. The national study of breast cancer screening. Clin Invest Med 1981; 4:227–258.
59. Gilbertsen VA, Church TR, Grewe FA, Mandel JS, McHugh RM, Schuman LM, Williams SE. The design of a study to assess occult-blood screening for colon cancer. J Chron Dis 1980; 33:107–114.
60. Gohagan JK, Prorok PC, Kramer BS, Cornett JE. Prostate cancer screening in the Prostate, Lung, Colorectal and Ovarian Cancer Screening Trial of the National Cancer Institute. Urol 1994; 152:1905–1909.
61. Etzioni RD, Connor RJ, Prorok PC, Self SG. Design and analysis of cancer screening trials. Statistical Methods in Medical Research 1995; 4:3–17.
62. Fontana RS. Screening for lung cancer: recent experience in the United States. In: Hansen HH, ed. Lung Cancer: Basic and Clinical Aspects. Boston: Martinus Nijhoff Publishers, 1986:91–111.
63. Frisell J, Eklund G, Hellstrom L, Glas U, Somell A. The Stockholm Breast Cancer

Screening Trial—5-year results and stage at discovery. Breast Cancer Research and Treatment 1989; 13:79–87.
64. Moss S. Personal communication, 1994.
65. Shapiro S, Venet W, Strax P, Venet L, Roeser R. Ten- to fourteen-year effect of screening on breast cancer mortality. J Natl Cancer Inst 1982; 69:349–355.
66. Chu KC, Smart CR, Tarone RE. Analysis of breast cancer mortality and stage distribution by age for the Health Insurance Plan clinical trial. J Natl Cancer Inst 1988; 80:1125–1132.
67. Connor RJ, Prorok PC. Issues in the mortality analysis of randomized controlled trials of cancer screening. Controlled Clinical Trials 1994; 5:81–99.
68. Etzioni R, Self SG. On the catch-up time method for analyzing cancer screening trials. Biometrics 1995; 51:31–43.
69. Zucker DM, Lakatos E. Weighted logrank-type statistics for comparing survival curves when there is a time lag in the effectiveness of treatment. Biometrika 1990; 77:853–864.
70. Self SG. An adaptive weighted logrank test with application to cancer prevention and screening trials. Biometrics 1991; 47:975–986.
71. Self SG, Etzioni R. A likelihood ratio test for cancer screening trials. Biometrics 1995; 51:44–50.
72. Walter SD, Stitt LW. Evaluating the survival of cancer cases detected by screening. Statistics in Medicine 1987; 6:885–900.
73. Xu JL, Prorok PC. Non-parametric estimation of the post-lead time survival distribution of screen detected cancer cases. Statistics in Medicine 1995; 14:2715–2725.
74. Shapiro S, Goldberg JD, Hutchison GB. Lead time in breast cancer detection and implications for periodicity of screening. Am J Epidemiol 1974; 100:357–366.
75. Kafadar K, Prorok PC. A data-analytic approach for estimating lead time and screening benefit based on survival curves in randomized cancer screening trials. Statistics in Medicine 1994; 13:569–586.
76. Walter SD, Day NE. Estimation of the duration of a pre-clinical disease state using screening data. Am J Epidemiol 1983; 118:865–886.
77. Connor RJ, Chu KC, Smart CR. Stage-shift cancer screening model. J Clin Epidemiol 1989; 42:1083–1095.

4

Decision-Making About Screening: Individual and Policy Levels

Russell Harris
University of North Carolina School of Medicine; Program on Health Promotion and Disease Prevention; and Cancer Control Program, UNC—Lineberger Comprehensive Cancer Center, Chapel Hill, North Carolina

I. RATIONAL DECISION-MAKING

Screening is an intuitively appealing activity. Unexamined, the basic idea of finding disease earlier seems to be desirable in itself, especially when the disease is feared as much as cancer is (1). In fact, screening is more complex and there are more than benefits. Screening is a "cascade" that can lead to either good or harm. Sometimes screening leads to more good than harm, and sometimes more harm than good.

Thus, there is a decision to be made: to screen or not? A process is needed for determining in each situation whether screening is appropriate. A purely rational approach to deciding about screening would include decision-makers (whether policy makers for groups, or physicians and patients for individuals) examining the evidence for three issues:

1. The magnitude of risk of a disease and its associated adverse health effects (such as death and disability) for a population or an individual person
2. The degree to which a screening activity can and does lead to treatment that reduces the risk of these adverse consequences of the disease
3. The magnitude of the health harms (both physical and psychological) and financial costs associated with the reduction in the risk of adverse health effects by screening

The rational decision-maker, fully informed, then weighs the benefits from screening (from numbers 1 and 2) against the harms and costs of screening (from number 3) and decides, compared to other options, whether the screening activity is a "good buy." In performing this weighing of benefits and harms, the decision-maker implicitly or explicitly carries out two tasks:

1. Evaluates a number of factors to develop a "perception" of the magnitude of the benefits, harms, and costs associated with each option
2. Places a value on the perceived outcomes associated with the various options

Thus, rational decisions about screening, whether for populations or for individuals, necessarily involve both a scientific process (developing and evaluating the evidence as discussed) and a weighing of benefits vs. harms and costs based on informed people's values and preferences (2). For decisions about screening to be rational, both the scientific process and the informed weighing must be valid.

This chapter focuses on problems and possibilities for rational decision-making about screening. The chapter begins with a discussion of two ways of assessing benefit: by scientific approach and by human perceptions. There are problems and possibilities with each. In a similar way, assessments of harms and costs, and the different ways of weighing benefits versus harms and costs are also examined. In practice, the process is not always so tidy. Some policy and individual decisions about screening are rational, and some are not. This chapter also gives some ideas of how to make the actual process more rational.

II. BENEFITS: REAL AND IMAGINED

A. Scientific Assessment of Benefit

The scientific idea of benefit from screening is reduction of risk. In the case of screening for cancer, the risk could be either the risk of cancer development (screening for premalignant conditions such as colorectal polyps, cervical dysplasia, or genetic susceptibility to breast cancer) or the risk of suffering the adverse health consequences of cancer (screening for early stages of actual diseases such as breast cancer or prostate cancer). Even in the case of screening for premalignant conditions, however, the bottom line risk to reduce is the risk of worsened health status caused by the disease. That is to say, in the end, it is incumbent upon those advocating screening to show evidence of benefit in terms of helping people live better or longer.

Scientifically, risk is measured in epidemiological studies. Thus, a population or group of people at risk (the denominator) of development of some health problem is followed up over time to determine the number (the numerator) in whom the health problem of interest actually develops. The result is a proportion (either a measure of incidence or mortality, depending on the health outcome) that provides an estimate of the ''risk'' over the specified time period for that health problem. In practice, the risk of developing cancer in the United States is often provided by the Surveillance, Epidemiology, and End Results (SEER) Program of the National Cancer Institute, and the risk of dying of cancer is usually provided by the National Center for Health Statistics (NCHS). The SEER program collects cancer incidence data from nine different areas of the United States (covering about 9.5% of the US population) and reports regularly on incidence from various types of cancer from these areas. The NCHS collects mortality data for the entire country.

Even though the SEER and NCHS data are useful for estimating the risk for large groups of people in the United States, specific targeted research is needed to determine how an intervention (e.g., screening mammography) affects the risk of dying of a cancer (in this case, breast cancer). As discussed in Chapter 3, the best evidence for risk reduction comes from randomized controlled trials (RCTs). Other research designs can also be useful, but are open to important biases that can obscure the true answer.

Randomized controlled trials are useful because they provide risk estimates for two groups identical in every way except that one received the intervention (screening test) of interest and the other did not. One may then compare these risk estimates in a number of different ways:

1. Subtract the risk of the group receiving the intervention from that of the control group not receiving it (''absolute risk reduction'')
2. Form a ratio of the absolute risk reduction to the total risk in the control group (''relative risk reduction'')
3. Invert the absolute risk reduction to find the ''number needed to screen'' (i.e., the number of people who must be screened to benefit one person)
4. Divide the total absolute risk reduction for an entire population of people by the number of people, thus calculating the risk reduction for each individual within a population, as if each benefited equally
5. Multiply the relative risk reduction by the number of people expected to die of the cancer over a specified time period, thus calculating the total number of people in a population expected to benefit
6. Extrapolate the risk of developing or dying of cancer over a lifetime, beginning either at birth or some other age, thus calculating the lifetime risk of cancer in each group.

All of these ways (and others) of expressing risk are scientifically correct, but their impact on people may vary considerably.

There are a number of methodological issues in determining risk, including such problems as ''competing risks,'' time factors, and generalizability. It is, for example, not always clear what is meant by ''risk'' of a single condition when people in the population being observed die of other health problems not being studied, so called competing risks. In a sense, then, the risk of Disease X is a hypothetical idea that assumes that one does not die of Disease Y.

Time is also crucial in determining risk. Essentially, the risk can be made small or large depending on the time frame considered. The time frame must make sense and be useful to real people. If too short a time frame is considered, the occurrence of Disease X may be missed because it takes longer to declare itself and be diagnosed. The shorter the time frame, the lower the incidence and thus the risk. If too long a time frame is considered, the risk may seem enormous, yet be of lesser salience to people who are less concerned with risks far into the future.

Finally, generalizability is always a concern with studies of risk. The risk that is measured in epidemiological studies is an average risk over a number of people, each with different risk factors. Thus, some people in any population have higher risk and others have lower risk, depending on their risk factor status. To accurately determine risk for a specific individual, however, it is preferred to have studies of people whose risk factor status matches the individual in question exactly. This is rarely the case.

There are also questions about the entire notion of applying population risk calculations to individuals. It is often argued that, in the end, an individual has a risk of either one or zero: the individual either gets the disease or does not. Even though this is correct, it is also true that, before an individual has a disease, that individual may be considered as being a member of a group (defined as precisely as possible) that does have a risk. The entire enterprise of screening is one of successively reassigning people into different groups so as to more clearly assign risk. From the risk of the entire group before screening, one is reassigned to either the positive screen group (with a higher risk) or a negative screen group (with a lower risk).

B. Nonmortality Benefits of Screening

Besides extended life, there are other potential benefits to ''quality of life'' from screening. Several of these nonmortality benefits of screening have been proposed. However, these benefits have not been measured in the large RCTs of screening efficacy (such as have been conducted for breast cancer and colorectal cancer screening). Thus, there is primarily anecdotal evidence, sometimes mixed with strongly held belief, about the magnitude and even the existence of these benefits.

Nonmortality benefits generally fall into two categories: 1. less aggressive treatment from finding the cancer at an earlier stage, and 2. psychological effects. The idea that screening results in less aggressive treatment seems intuitively plausible, yet it is not clear whether this does occur. First, screening sometimes unearths disease that is so ''early'' that it is not clear that it would ever have progressed at all. This is true of prostate cancer screening and is likely true for ductal carcinoma in situ, a pathological breast lesion that only sometimes progresses to invasive disease. Because it is often difficult to distinguish lesions that will progress from those that will not, all of these patients are considered for treatment. In this way, screening may actually increase the number of people receiving treatment. There is still much to learn about how much treatment is needed for some early lesions that screening has detected. Some very aggressive treatment has been proposed for early stage cancers, given in the hope of complete cure. Thus, screening may sometimes, but not always, reduce the intensity of treatment. This is a potential benefit that must be documented for each cancer.

In terms of psychological benefits of screening, ''peace of mind'' is often mentioned. This theoretically is a benefit for people who are concerned that they may have a cancer, and then have a negative screening test. There would theoretically be little peace of mind benefit for people who are not greatly concerned about having the disease in question.

Even for those people who do have great concern about having a particular cancer, there remains uncertainty about this benefit. In general, screening is performed on people who have a low (but not zero) risk of a disease, seeking to reclassify them into either a higher risk group (positive test) or an even lower risk group (negative test). Logically, the peace of mind derived from a negative screening test should be equal to the difference between the person's risk before being screened and the risk after a negative screen. Because the risk is initially low, and because the risk after a negative screen is almost never zero (i.e., there are always some false-negative screens), the actual difference between these risks is usually small. For example, for breast cancer screening for women in their 40s, the risk difference is about 0.1%, or about 1/1000. When people report peace of mind out of proportion to these calculations, either the initial risk perception or the perception of benefit from the screening test was probably inflated. In these situations, it may be asked whether simple education about correct risk perceptions would provide peace of mind as well as the screening test. Again, peace of mind is a potential benefit of screening that requires further documentation.

Some people have suggested that there may be a psychological benefit for a person with a new cancer diagnosis resulting from a screening test. The idea is that people find benefit in simply knowing what they have, independent of whether there is effective treatment. Again, there is no research evidence about this issue, and we are left to wonder whether this indeed makes sense.

C. The Inherent Uncertainty of the Scientific Assessment of Benefit

In examining the scientific evidence of benefit of screening, it does not take long to appreciate that there are aspects of current scientific evidence that create problems for decision-makers.

First, there are many screening situations that have not been adequately studied. That is, studies either have not been done or studies have provided only indirect evidence of the possibility of benefit. Studies that determine the sensitivity, specificity, predictive value of screening tests, or the stage of disease detected by the test do not answer the question that decision-makers need to know: Will routine screening help people to live longer or better? Studies like these tempt decision-makers to extrapolate from detection of disease to improved outcomes (i.e., benefit), but it is clear that such a leap is usually not warranted.

Second, most studies have used a limited notion of benefit and no notion of harm at all. Large studies of screening usually examine only the effect of the screening program on disease-specific mortality rate, ignoring such issues as whether the program provided additional benefit in less aggressive therapy (with potentially fewer acute complications and less long-term negative impact) or in greater peace of mind, or instead produced psychological harm in worry or regret. These "quality of life" issues are important to wise decision-makers and should be included in future studies of screening.

Third, even in situations that have been well researched, the nature of science is not to "prove" that a screening program is beneficial both "now and for always." Rather, science provides evidence, sometimes seemingly contradictory evidence, for successive approximations of the truth, always with some degree of uncertainty. The uncertainty comes from several factors:

1. There is always the possibility of unrecognized bias within the design and implementation of a study that could explain the positive or negative results.
2. Each study provides an "answer" that is bounded by a statistical confidence interval, giving a range of "answers" consistent with the data.
3. The subjects and conditions of the studies may or may not be directly translatable into the "real life" situation, leaving some uncertainty about the effect of the screening program in actual practice.
4. Because scientific studies take time (and the best ones often take a long time), screening techniques and treatment regimens used in the study often become technically outdated.
5. Even unbiased studies with narrow confidence intervals of the most up-to-date screening and treatment protocols still only provide a probabi-

listic answer that does not tell an individual whether he or she will actually benefit from being screened.

Thus, even though decision-makers would like to know (indeed, need to know) *now* about the exact nature and magnitude of the various benefits that the present-day screening test will provide, the scientific evidence may be completely silent, may provide partial answers, and, at best, can only provide an approximation of the probability of mortality benefit from screening tests and treatments in use during the time of the study. Even though there are ways to improve the information from scientific studies, there will likely always be some degree of mismatch between scientific evidence and the needs of decision-makers. Sometimes the case for or against screening is strong enough that this ''built-in'' uncertainty is relatively small and can be ignored. In situations in which neither benefits nor harms or costs are clearly dominant (e.g., screening for breast cancer among women in their 40s or screening for prostate cancer), however, decision-makers must accept some degree of uncertainty and act accordingly.

D. Perceptions of Benefit

As discussed earlier, rational decision-making about screening (or other medical interventions) involves both 1. a scientific assessment of the potential outcomes: benefits, harms, and costs; and 2. a weighing of benefits against harms and costs. In assessing the outcomes, decision-makers do not simply take the scientific assessments of benefits, harms, and costs at face value, but rather interpret them, forming perceptions of their magnitude and personal salience (3). In this section, perceptions of benefit, and the possibilities and problems this creates for rational decision-making are discussed.

There are conceptually two different issues that together make up a person's ''perception of benefit'' (or lack of benefit), whether it be a policy maker for groups or physicians and patients deciding for individuals. The first is a person's perception of the risk of the disease in question. The second is the perceived degree to which the screening activity reduces this risk. Theoretically, higher perceived risk and greater perceived reduction of risk would lead to greater recommended use of screening. Although these two issues are discussed separately, it is not clear to what extent real decision-makers actually think this way. Instead, people may integrate these two issues (and others) into an overall perception, almost a visceral feeling, of benefit. Still, it is likely that these two issues are important to the final perception of benefit, however different components are integrated.

1. *Perceived Risk of Disease*

A person's perceptions of risk often differ from the objective, statistical assessments of risk discussed previously. Even though these preferences may appear irrational and even random at times, research in cognitive psychology has shown that risk perceptions fall into predictable patterns and are, in fact, understandable and even ''make sense'' (3,4). Risk perceptions can be thought of as incorporating two factors: 1. perceptions of probabilities and 2. perceptions of outcomes. This may explain why studies have shown that experts, guided by probability tables alone, order the risk of such concerns as global warming, x-rays, food preservatives, and nuclear power very differently from the lay public (3). Most people think of risk as not only the probability of an activity leading to a health condition but also consider their perceptions of the activity and the health problem. Activities or their potential health outcomes that have certain characteristics are more feared and considered more ''risky.'' Some of these characteristics are as follows (3):

The activity or health condition is perceived to be involuntary, not in the individual's control.
The activity or health condition is unfairly shared.
The activity is natural (not man-made).
The activity or health condition is not controllable.
The activity or health condition is potentially catastrophic, such as a dread disease.
Experts disagree about the activity or health condition.
There is a lot of news media exposure.
There are visible advocacy groups associated with the health condition.

In the United States, most people's perceptions of the risk of cancer are very high. Part of this perception may be due to the idea of cancer as a ''dread disease''; the way no one wants to die. Thus, the high perception of risk of cancer is driven by the evaluation of the health condition in addition to the probability of having (or being later found to have) the condition. As a result, risk perceptions for more common conditions such as heart disease and diabetes are lower than for cancer (5).

The public's perceptions of risk also vary among cancers. Risk perceptions for breast and prostate cancers, with mortality rates comparable to that of colorectal cancer and lower than that of lung cancer, are higher than perceptions of other cancers. Research has shown that women overestimate their risk of breast cancer several fold (5,6). This overestimation is greater among women younger than age 50 years, and, among younger women, it is greater in those with higher educational achievement. Likewise, men overestimate their risk of prostate cancer (5). This may be due to the larger amount of publicity given to these two cancers, the existence of advocacy groups for each, or even to their identification with different genders.

Advocacy groups can have a large impact on public perceptions. One way is by choice of the statistical expression of risk. As discussed, there are a number of scientifically correct statements about risk that can be made to help people understand their risk. Although all of these may be statistically correct, the use of some expressions clearly has a different effect from others. For example, the American Cancer Society and other groups have long expressed the risk of breast cancer as ''one in 10'' (more recently, one in seven or eight). This is a well-known statistic, and its use and effect has been telling. Many speakers about breast cancer begin their presentations to women's groups by predicting that ''one in 10'' women in the room will be diagnosed one day with breast cancer. This commonly cited number is the lifetime probability at birth of ever being diagnosed with breast cancer. There is usually no comparison to other diseases; for example, the lifetime probability at birth of heart disease for women is greater than four in 10, and for diabetes nearly three in 10. Other expressions of the risk of breast cancer could have different effects on women. For example, the risk of diagnosing breast cancer in the next year in a 35-year-old woman is about one in 3000, and at age 55 is about one in 400 (7). Because screening for breast cancer presumably has little impact on the probability of ever being given the diagnosis of breast cancer but rather seeks to reduce the risk of ever dying of the disease, the risk of dying of breast cancer is a more appropriate statistic to discuss with women. The lifetime probability at birth of dying of breast cancer is about three in 100; the risk of dying of breast cancer in the next year at age 35 is about one in 21,000 and at age 55 is about one in 1800 (7). These figures sound very different from one in 10, yet they are much less well known.

Another possible reason for people overestimating the risk of breast and prostate cancers is that many people currently found to have these diseases publicly acknowledge their diagnosis. Thus, nearly everyone knows someone who has breast or prostate cancer, while people diagnosed with other conditions are less publicly visible. In addition, at least some people diagnosed with prostate cancer and some diagnosed with the breast lesion ''carcinoma in situ'' (often thought of by patients as breast cancer) have conditions that would never have developed into clinical disease and would never have been discovered had there not been screening (8,9). Both of these factors lead to exaggerated public perceptions of the frequency of clinically important breast and prostate cancer.

Thus a person's assessment of the ''risk'' of a disease is some combination of the perceived probability of being given the diagnosis and perceptions of the disease itself. Given the current body of knowledge of the diseases and the attention that the media and advocacy groups give to them, it is not surprising that risk perceptions of breast and prostate cancer are so high.

2. *Perceived Reduction of Risk by Screening*

The perceived risk of cancer is a major factor influencing decision-makers' perceptions of benefit from screening, but it is not the only one. At least conceptually, another factor is the perceived effectiveness of screening in reducing risk. This perception occurs with the background of the evolution of medicine over the past 100 years toward a bias in favor of technology as the primary means of dealing with disease (10). For example, a recent study showed that television portrayal of the effectiveness of cardiopulmonary resuscitation (CPR) is much greater than the actual effectiveness demonstrated in research studies (11).

As discussed, screening is an intuitively appealing activity. It seems apparent that detecting disease at an earlier stage is preferable to finding it later. Yet, it is clear from such studies as those evaluating screening for lung cancer (12) that it is possible to find disease earlier and still have no impact on the length of a person's life. In addition, there are harms and costs to screening that must be considered along with any benefits. Thus, although it is counterintuitive, it is necessary to consider the evidence that any screening test (and subsequent treatment) actually improves the quantity or quality of people's lives. It is not sufficient to determine only that a screening test can detect earlier disease. This distinction between benefit and disease detection is fundamental to evaluations of screening programs.

This critical issue, however, is little understood among the public. For most people (including many physicians), evidence that a test can detect earlier disease is sufficient to warrant its widespread use (13). This has been seen with evaluation of both mammography as a screening test for breast cancer and prostatic specific antigen (PSA) as a screening test for prostate cancer. Research has shown that people generally overestimate the effectiveness of mammography and PSA screening in reducing the risk of dying of breast and prostate cancers (5). On the other hand, people know little about the effectiveness of sigmoidoscopy as a screening test for colorectal cancer (14), despite the scientific evidence that sigmoidoscopy may be more effective in reducing deaths from colorectal cancer than either mammography for breast cancer or PSA for prostate cancer.

This overestimate of the effectiveness of mammography and PSA may be partly due to the nature of breast and prostate cancer, and the public visibility of those people diagnosed with those conditions. Neither breast nor prostate cancer has ever been universally fatal. That is, there is variation in the natural history of these cancers and some grow so slowly that they can be adequately treated even if they are found only after symptoms develop. This means that many people diagnosed with breast and prostate cancer will do well because of the natural history of the type of cancer they have, regardless of how it is detected. With the advent of widespread screening, many more breast and prostate cancers are detected by screening. Thus, many ''survivors'' of breast and prostate cancer attribute their survival to early detection when, in fact, it is due to having a biologically slow-growing, less aggressive cancer. Because some of these people are in the public eye, the public as a whole also is impressed with the power of screening to improve survival. On the other hand, because of the low utilization rate of screening sigmoidoscopy, few people have had their colorectal cancers detected by this test, and the public has little evidence of its effectiveness. People may also partly evaluate a screening test by its perceived discomfort, and there is an intuitive sense among many that sigmoidoscopy is very uncomfortable, although some research indicates that people who have had it done find it less uncomfortable than expected.

Finally, another reason that people may overestimate the effectiveness of some screening tests is the tendency of people to think of tests and treatments as either all good or all bad (15). It seems difficult for people to understand that most tests and treatments are partially effective, if at all.

Thus, like perceptions of the risk of development of cancer and dying of cancer, perceptions of the effectiveness of a screening test in reducing that risk are heavily influenced by everyday experiences and ingrained ways of thinking, sometimes leading to a biased view compared with scientific evidence.

III. HARMS AND COSTS: THE "DOWN SIDE" OF SCREENING

A. Scientific Assessment of Harms and Costs

Although it is not always obvious, there is always a ''down side'' to medical interventions, including such intuitively beneficial activities as screening. Harms usually fall into one of two categories: 1. physical harm from the screening test, the workup of positive screening tests, or unnecessary treatment; and 2. psychological harm. Costs are also important to consider, and usually fall into four categories: screening, workup, treatment, and long-term surveillance.

As discussed, most large-scale RCTs of screening have not gathered evidence about harms or costs, focusing instead on benefit. Some smaller studies have examined psychological harms, but this issue has, in general, not been well researched. Thus, much of the evidence about harms of screening is anecdotal.

One RCT that did examine a specific harm of screening was the Minnesota Colon Cancer Control Study of fecal occult blood screening (16). This study not only collected data on reduction in the risk of dying of colon cancer but also found that, of 12,246 colonoscopies performed at the university (75% of all positive fecal occult blood tests), four resulted in colonic perforation requiring surgery and 11 resulted in serious bleeding, three of which required surgery. The rates of perforation, bleeding, and surgery for colonoscopy performed in the community setting are not available. Because the number of people dying of colorectal cancer over the 13-year study in the screened group was about 39 less than the control group out of the approximately 15,500 participants in each group, it can quickly be appreciated that, if the rate of colonic perforation and serious bleeding were higher, as could occur in some community settings, the net benefit from screening would be much reduced. The harms of screening must be considered in addition to the benefits.

Another study provided information on the number of follow-up procedures per cancer diagnosed for breast cancer screening (17). In an academic screening mammography program in California, there were more than 60 total procedures and nearly seven biopsies per cancer diagnosed among women in their 30s, about 22 procedures and more than three biopsies per cancer diagnosed among women in their 50s, and about 10 procedures and two biopsies per cancer diagnosed among women in their 70s. These numbers could be even higher in some community settings. Information about resulting patient harm (e.g., pain, scarring, infection, anxiety) as a result of these procedures is not available, but, even if the rates of these problems were low, the net benefit is reduced by these harms. In the rush to determine benefits, harms can be easily overlooked.

One of the most influential studies of psychological harm from screening was by Lerman et al. (18) These investigators conducted telephone interviews with women 3 months after screening mammography, comparing three groups: those who had had a completely normal mammogram, those who had had a ''low suspicion'' abnormality, and those who had had a ''high suspicion'' abnormality. All women with suspicious mammograms had completed a diagnostic workup that had excluded cancer. All women who had had breast cancer diagnosed were excluded. At follow-up, 24% of women with normal mammography and 47% of women with high suspicion abnormal mammogram reported anxiety about having mammography. About 40% of women with either low suspicion or high suspicion mammography reported worries about breast cancer, compared with 28% of women with a normal mammogram. These worries affected the functioning of 17% of women who had had a high suspicion mammogram compared with about 5% of women with either a low suspicion mammogram or normal mammography. The worries affected the mood of 26% of women who had had a high suspicion mammogram, 20% of women with a low suspicion mammogram, and 9% of women with a normal mammogram. When women were asked whether the mammogram changed their breast cancer concerns, 56% of women with a high suspicion mammogram reported that their concerns had been increased, compared with 37% of women with a low suspicion mammogram and 23% of women with a normal mammogram. It seems clear that false-positive mammograms, and even normal mammograms, can sometimes increase breast cancer anxiety, and that, for a sizable number of women, this anxiety results in difficulty functioning or worsened mood even 3 months after the mammogram. Another study suggests that, over a decade of regular screening mammography, about 30% of all women will have at least one false-positive mammogram (19).

Other potential psychological harms of screening have been little studied. These may include regret from false-negative screening tests or in people who were not screened at all but in whom cancer developed. In addition, people who have aggressive, untreatable cancer diagnosed by screening do not live longer, but carry for a longer time the burden of knowledge that they have a fatal disease. Conceivably, this could lead to increased suffering. Finally, for many people, the experience itself of having a screening test raises anxiety about a dread disease.

A few studies have examined the cost and cost effectiveness of screening. In some cases, the costs of screening do not come from the screening test itself, but occur later in either working up abnormal results or in surveillance of people who have had a previous abnormal screening test (e.g., people who have had a previous benign polyp on sigmoidoscopy often undergo repeated colonoscopy in future years) (20). It is also easy to underestimate the costs of screening when the time involved in the screening is not considered. For example, breast self-examination (BSE) was thought to be an inexpensive screening test for breast cancer, until it was realized that women who have not been taught to do BSE do it poorly (21), and it is very expensive to teach all women to do BSE well. Because of the enormous amount of time needed from health care professionals to teach BSE, it may well be one of the most expensive screening tests ever proposed (22). Thus, it is important to consider all aspects of the costs of screening.

Interestingly, there is no consensus about what to do with cost-effectiveness studies of screening. Some people believe that considerations of cost should not be used in making decisions for individuals; others minimize the use of these studies in policy making for groups. However, lack of use of cost-effectiveness evidence could result in another potential harm of screening: the diversion of scarce health care resources from an area where it might do more good to one where it might do less good (23). In this way, screening could also lead, overall, to more harm than good (i.e., a net loss of lives).

B. Decision-Makers' Perceptions of Harms and Costs

Although the evidence is minimal, it appears that the public perceives little harm from screening and generally believes that screening saves money by finding disease earlier. (Note: Very few if any screening tests actually save money.) Thus, a common rationale in "close call" situations is that "if it might help, then we should do it." This implies that there is no "down side" in terms of harms and costs to consider, that the decision is entirely driven by the benefit side of the equation.

When psychological effects of screening are raised as an issue in discussions, many argue that these effects are exaggerated, a perception that may come from the fact that few people who suffer such effects discuss them publicly. Others argue that such effects should be easily resolvable by simple counseling.

Yet one of the origins of anxiety from false-positive tests may come from widespread social perceptions of risk that may prove difficult to change.

Costs play a peculiar role in these decisions. In most discussions of screening, decision-makers appear to act as if this issue does not exist. Many reject costs as a factor in decisions outright, arguing that we should pay any price to extend the life of even one person. Yet, costs play a major role in decisions about medical care in virtually every other part of the health care system. It is difficult to see how screening can be exempt from these concerns.

IV. WEIGHING BENEFITS VERSUS HARMS AND COSTS: THE PROBLEM OF RATIONAL DECISION-MAKING

According to the model of rational decision-making developed at the beginning of this chapter, decision-makers should first understand the scientific evidence for benefits, harms, and costs of screening in the specific situation of interest. Then, using appropriate values and perceptions, they should weigh the benefits against the harms and costs to arrive at a decision of whether screening, compared with other possible uses of the resources, is a "good buy." For decision-making to be "rational," both the scientific assessment of benefits, harms, and costs and the informed weighing of these outcomes must be valid.

Evidence is sometimes of uncertain validity when making a decision about screening. Improving the evidence about the outcomes of screening would be an important step toward more rational decision-making about screening.

The public often perceives risk, risk reduction, harms, and costs differently from scientific evidence. However, this finding of disagreement by itself does not lead to a conclusion that the "weighing" step is invalid. Risk perceptions, for example, may be valid (and thus "rational"), even if they depart from statistical statements of risk. The fact that people perceive some outcomes differently from others that seem similar (i.e., two different fatal diseases) should not be surprising, and, in itself, does not make a risk perception invalid or irrational. This inclusion of evaluation of the outcome is one of the characteristics that separates statistical risk from risk perceptions, and is to be expected.

The validity of the weighing process depends importantly on the degree to which the decision-makers are informed. To be able to take into account benefits, harms, and costs, the decision-maker must have at least a "ball park" understanding of the evidence concerning these outcomes. For decisions about individuals, how the evidence is perceived is an individual matter, and reasonable individuals sometimes differ. For policy decisions about groups, decision-makers need to factor in the perceptions and values of the people most affected by the policy. These perceptions and values are not always the same as those of groups who are most vocal about an issue. How to understand true public perceptions and values so that they can be factored into such decisions is an underdeveloped area of the policy development process.

But, assuming that the decision-makers are accurately (within a reasonable range) informed about the scientific evidence, and assuming that true perceptions and values of those people affected by the decision are used, then it should be assumed that the decision is valid and reasonable. Many decisions made about screening, on both individual and policy levels, are not rational, often because the decision-makers are ill informed or because the actual perceptions and values of those affected were not used in making the decision.

These criticisms certainly apply to recent decisions of physicians, patients, and policy makers about prostate cancer screening and breast cancer screening for women in their 40s. In the situation of screening for prostate cancer, it is unclear whether there is benefit from screening, whereas harms are well documented (24). Yet, the American Cancer Society recently published a recommendation for PSA screening (25) despite the uncertainty about whether such a policy will result in more good than harm. Thus, policy making has not followed our rational decision-making model in this case, given the lack of information about important outcomes, and no attempt to understand the values of men who will be affected by the policy. In contrast, the American College of Physicians' policy (26) is more conservative, in better agreement with the lack of evidence about benefits and harms.

In the situation of breast cancer screening for women in their 40s, policy making has been even more confusing. In this case, there is general agreement about the benefits, and some information about harms and costs, yet the perceptions and values that have predominated in decision-making have been those of some advocacy groups (27,28). Many women are ill informed about the actual risks of breast cancer and the actual benefits of mammography (5,6), and it is unclear what perceptions about screening women would hold if they were better informed. Yet, under pressure from Congress and advocacy groups, in April 1997, the National Cancer Institute changed its policy statement on breast cancer screening for women in their 40s from a "Statement of the Evidence" without a specific recommendation for or against screening to a specific recommendation for screening (29). Interestingly, this came after a National Institutes of Health Consensus Conference in early 1997 that declined to make a general recommendation in favor of screening, but suggested instead that physicians inform women of their actual risks and help them to make their own individual decisions (30). This approach has been called "shared decision-making."

In both of these situations, decision-makers may not have been completely informed, and the predominant perceptions and values used were not those of the group of people most affected by the decision. It is important to emphasize that the problem with some decisions about screening in these two situations is not that there was influence from the public. Indeed, the model of rational decision-making requires input from those most affected by the decision. And, given people's perceptions of benefit and harm, these decisions would be expected to sometimes differ from the decisions of risk experts. But the decision becomes problematic when people are not presented accurate information on which to base their perceptions. The public has often not had accurate information available about screening. As more is learned about genetic markers of cancer and genetic susceptibility to cancer, these difficult decisions about cancer screening will likely continue to arise.

In addition to lack of information and lack of involvement of appropriate people in the decision, another factor is a major potential threat to rational decision-making. When thinking about this and other complex issues, many people do not use a rational decision-making process at all, but instead take cognitive short cuts (''heuristics'') to simplify the task (4). People may focus on one particularly salient outcome without considering the others; or they may be so swayed by social factors, the values and perceptions of others, that they do not consider for themselves whether their own values are more consistent with being screened or not being screened. These problems in rational decision-making lead to the question of what might be done to improve decision-making about screening.

V. CONCLUSION: TOWARD RATIONAL DECISION-MAKING ABOUT SCREENING

Screening is not a single test but rather a cascade of events that can lead to benefit or harm for individuals or for groups. There is a decision that needs to be made about the appropriateness of screening in each situation. This chapter has presented a model of rational decision-making, and explored the problems and possibilities of such a model in deciding about screening. There are many current examples, however, of decision-making not being rational, including some decisions in the ongoing debates about screening for prostate cancer among men and for breast cancer among women ages 40–49 years. There are several problems that make decision-making irrational, including lack of informed decision-makers, lack of input from those most affected by the decision, and people not following the logical sequence of rational decision-making. In this final section, several approaches are suggested for improving decision-making about screening. The suggestions are discussed in categories for each of the three major types of problems.

A. Improved Information About the Outcomes of Screening

A major problem in making decisions about screening is that the decision-makers are not fully informed about the benefits, harms, and costs of screening. One reason for this is that there is a misperception of screening among many people. Some of these misperceptions include the idea that detection of disease is equal to benefit (rather than requiring evidence of improved quantity or quality of life for those screened), that there are no harms from screening, and that screening saves money and thus a person does not need to consider alternative uses of resources. Other misperceptions are specific to the cancer and the screening test of interest. One is that screening mammography for younger women and PSA for men are very effective in reducing risk from breast and prostate cancer, as opposed to, at best, being partially effective. Many of these misperceptions are reinforced by advocacy groups with heightened interest in promoting screening for their disease as opposed to rational decision-making. Indeed, advocacy groups are one of the major ways the American public receives information about screening.

The public and individuals need alternative sources of information about screening. One of the best-placed groups to provide such information is physicians and physician organizations. Physicians are seen as credible sources of health-related information, are in a good position to have accurate information, and are in a good position to provide that information to the public. For physicians and physician groups to speak out more forcefully on screening might well change the nature of the public discussion.

B. More Widespread Involvement in Decision-Making by Those Affected by the Decision

Perceptions and values for decisions about policy come primarily from advocacy groups, and it is often assumed by physicians that individual patients share these perceptions and values. Along with informing people about the outcomes of screening, policy makers need to find ways to systematically sample informed public perceptions about screening. Helping people participate in making decisions about how scarce health care resources are spent may be a useful way of approaching the issue.

For individuals, an encouraging approach has been termed ''shared decision-making'' (31). This is the approach suggested by the Consensus Development Panel in its report on breast cancer screening for women ages 40–49 (30). Suggested especially for decisions that are ''close calls,'' shared decision-making begins with the physician engaging the patient in a discussion about the outcomes of screening, making sure that the patient's information is accurate. Then the discussion turns to defining and clarifying the perceptions and values of the patient. Finally, the physician and patient discuss the decision to be made and which option would be most consistent with the patient's informed weighing of benefits versus harms and costs. The underlying assumption from the start is that the decision rightly belongs to the patient, with the physician seeking to provide whatever assistance is necessary. Some patients may not wish to participate, and the physician must use his or her understanding of the patient's values to determine the optimal strategy for the individual.

Criticism of shared decision-making has followed four somewhat contradictory lines. First, it is claimed that physicians are already sharing decisions with patients. Research is clear, however, that there is variation in the amount of patient participation allowed by physicians (32,33), that a participatory style is not the predominant pattern of practice (32,33), and that there are advantages both to the patient and the physician when participation is maximized (34,35).

Second, many people wonder whether it is feasible for physicians to spend the time required to properly inform patients and help them clarify values to make their decision. The fact that some physicians are finding time to encourage patient participation (32,33) is one answer to this criticism, but another is the development of new, more efficient approaches to informing patients and assisting them with decision-making. Several studies have evaluated videos (36), tailored newsletters (37,38), and other decision aids for patients. Further development is proceeding. This process could well become much more efficient than it has been in the past.

Third, some people say that physicians are not well trained to carry out shared decision-making with patients. Although this may be true for some physicians, it is certainly not true for all. The shared decision-making process is not so difficult to adopt as to require advanced training in an esoteric discipline. Standardizing the process within medical practice and involving nonphysicians also helps answer this criticism. Finally, if some physicians lack the training to carry out shared decision-making with patients and if this is a better model for practice, then medical education should incorporate such training.

Fourth, some people claim that patients are not interested in sharing these decisions but would prefer to be told what to do. Although research is clear that generally, patients prefer to receive more information than they are presently receiving, there is contradictory evidence concerning patient interest in participating in decisions (39–42). Probably this stems from differences in the decisions themselves, and perhaps also from patients' lack of experience with sharing decisions with their physicians (43). There is evidence that, in situations like screening, patients would respond if encouraged to participate (44,45).

In summary, although shared decision-making is not fully evaluated, there are promising indications that it is feasible and has positive effects on patients and physicians. It is an approach that should be further explored.

There is also a need to develop approaches to systematically understanding the informed perceptions and values of the public at large. This remains a challenge. The first step in solving this problem is for decision makers to understand that the most vocal members of society do not necessarily speak for the entire population of potentially affected people.

C. Increased Use of the Process of Rational Decision-Making

It is clear that many people do not take the time and trouble to go through the rational process of informing themselves about outcomes, clarifying values and perceptions, and carefully weighing the pros and cons of screening to make a choice. It is still unclear, however, whether simplified ways can be found to assist people in going through enough of the process to improve decision-making. Shared decision-making in the physician's office is one such process and would be eased by improved public education about the issues involved in deciding about screening. Even though some patients may have their minds made up and are not interested in considering all options, many patients, with encouragement and assistance, may be able to and may be interested in going through a simplified decision-making process.

In conclusion, the intuitive appeal of screening sometimes makes decision-making difficult. If we are to maximize the benefit and minimize the harms and costs of screening, we must first inform ourselves about these outcomes and then factor in the informed perceptions and values of the people most affected by the screening decision. This process is termed ''rational decision-making.'' Some people may call it common sense. It is also a necessary and attainable goal.

REFERENCES

1. Sontag S. Illness as Metaphor. Boston: Peter Smith Publisher, 1995.
2. Eddy DM. Anatomy of a decision. In: Clinical Decision Making: From Theory to Practice. Sudbury, MA: Jones and Bartlett Publishers, 1996:10–17.
3. Slovic P. Perception of risk. Science 1987; 236:280–285.
4. Tversky A, Kahneman D. Judgment under uncertainty: heuristics and biases. Science 1974; 185:1124–1131.
5. Harris R, Kinsinger L, Qaqish B. Misperception in breast and prostate screening: patients' estimates of risk and benefit (abstr). J Gen Int Med 1996; 11(suppl):124
6. Black WC, Nease RF, Tosteson ANA. Perceptions of breast cancer risk and screening effectiveness in women younger than 50 years of age. J Natl Cancer Inst 1995; 87:720–731.
7. Miller BA, Ries LA, Hankey BF, Kosary CL, Edwards BK, eds. Cancer Statistics Review: 1973–1989. NIH Publication No. 92-2789. Bethesda, MD: National Cancer Institute, 1992.
8. Kramer BS, Brown ML, Prorok PC, Potosky AL, Gohagan JK. Prostate cancer screening: what we know and what we need to know. Ann Intern Med 1993; 119: 914–923.
9. Ernster VL, Barclay J, Kerlikowske K, Grady D, Henderson IC. Incidence and treatment for ductal carcinoma in situ of the breast. JAMA 1996; 275:913–918.

10. Reiser SJ. Medicine and the Reign of Technology. Cambridge: Cambridge University Press, 1978.
11. Diem SJ, Lantos JD, Tulsky JA. Cardiopulmonary resuscitation on television—miracles and misinformation. 1996; 334:1578–1582.
12. Brett GZ. The value of lung cancer detection by six-monthly chest radiographs. Thorax 1968; 23:414–420.
13. Black WC, Welch HG. Advances in diagnostic imaging and overestimations of disease prevalence and the benefits of therapy. N Engl J Med 1993; 328:1237–1243.
14. Harris R. Unpublished data, 1995–1996, Making Prevention Work project, funded by the National Cancer Institute.
15. Redelmeier DA, Rozin P, Kahneman D. Understanding patients' decisions: cognitive and emotional perspectives. JAMA 1993; 270:72–76.
16. Mandel JS, Bond JH, Church TR, Snover DC, Bradley GM, Schuman LM, Ederer F. Reducing mortality from colorectal cancer by screening for fecal occult blood. N Engl J Med 1993; 328:1365–1371.
17. Kerlikowske K, Grady D, Barclay J, Sickles EA, Eaton A, Ernster V. Positive predictive value of screening mammography by age and family history of breast cancer. JAMA 1993; 270:2444–2450.
18. Lerman C, Trock B, Rimer BK, Boyce A, Jepson C, Engstrom P. Psychological and behavioral implications of abnormal mammograms. Ann Intern Med 1991; 114:657–661.
19. Elmore JG, Barton MB, Moceri VM, Polk S, Arena PJ, Fletcher SW. Ten-year risk of false positive screening mammograms and clinical breast exams. N Engl J Med 1998; 338:1089–1096.
20. Ransohoff DR, Lang CA, Kuo S. Colonoscopic surveillance after polypectomy: considerations of cost-effectiveness. Ann Intern Med 1991; 114:177–182.
21. Fletcher SW, O'Malley MS, Earp JL, Morgan TM, Lin S, Degnan D. How best to teach women breast self-examination: a randomized controlled trial. Ann Intern Med 1990; 112:772-779.
22. O'Malley MS. Cost-effectiveness of two nurse-led programs to teach breast self-examination. Am J Prevent Med 1993; 9:139–145.
23. Eddy DM. Cost-effectiveness analysis: a conversation with my father. In: Clinical Decision Making: From Theory to Practice. Sudbury, MA, Jones and Bartlett Publishers, 1996:156–170.
24. Woolf SH. Current concepts: Screening for prostate cancer with prostate-specific antigen—an examination of the evidence. N Engl J Med 1995; 333:1401–1405.
25. Mettlin C, Jones G, Averette H, Gusberg SB, Murphy GP. Defining and updating the American Cancer Society guidelines for the cancer-related checkup: prostate and endometrial cancers. CA Cancer J Clin 1993; 43:42–46.
26. American College of Physicians. Screening for prostate cancer. Ann Intern Med 1997; 126:480–484.
27. Ransohoff DR, Harris R. Lessons from the mammography screening controversy: can we improve the debate? Ann Intern Med 1997; 127:1029–1034.
28. Fletcher SW. Whither scientific deliberation in health policy recommendations?

Alice in the Wonderland of breast cancer screening. N Engl J Med 1997; 336: 1180–1183.
29. News article JAMA re NCI guideline
30. National Institutes of Health, Consensus Development Panel: National Institutes of Health Consensus Development Conference Statement: Breast Cancer Screening for Women Ages 40–49, January 21–23, 1997. Monogr Natl Cancer Inst 1997; 22:vii–xviii.
31. Kasper JF, Mulley AG, Wennberg JE. Developing shared decision-making programs to improve the quality of health care. Quality Review Bulletin 1992:183–190.
32. Roter DL, Stewart M, Putnam SM, Lipkin M, Stiles W, Inui TS. Communication patterns of primary care physicians. JAMA 1997; 277:350-356.
33. Braddock CH, Fihn SD, Levinson W, Jonsen AR, Pearlman RA. How doctors and patients discuss routine clinical decisions: informed decision making in the outpatient setting. J Gen Intern Med 1997; 12:339–345.
34. Kaplan SH, Greenfield S, Ware JE. Assessing the effect of physician-patient interactions on the outcomes of chronic disease. Med Care 1989; 27:S110–S127.
35. Kaplan SH, Greenfield S, Rogers WH, Ware JE. Characteristics of physicians with participatory decision-making styles. Ann Intern Med 1996; 124:497–504.
36. Flood AB, Wennberg JE, Nease RF, Fowler FJ, Ding J, Hynes LM. The importance of patient preference in the decision to screen for prostate cancer. J Gen Intern Med 1996; 11:342–349.
37. Skinner CS, Strecher VJ, Hospers H. Physicians' recommendations for mammography: do tailored messages make a difference? Am J Public Health 1994; 84:43–49.
38. Kreuter MW, Strecher VJ. Changing inaccurate perceptions of health risk: results from a randomized trial. Health Psychology 1995; 14:56–63.
39. Mazur DJ, Hickam DH. Patients' preferences for risk disclosure and role in decision making for invasive medical procedures. J Gen Intern Med 1997; 12:114–117.
40. Degner LF, Kristjanson LJ, Bowman D, Sloan JA, Carriere KC, O'Neill J, Bilodeau B, Watson P, Mueller B. Information needs and decisional preferences in women with breast cancer. JAMA 1997; 277:1485–1492.
41. Ende J, Kazis L, Ash A, Moskowitz MA. Measuring patients' desire for autonomy: decision making and information-seeking preferences among medical patients. J Gen Intern Med 1989; 4:23–30.
42. Strull W, Lo B, Charles G. Do patients want to participate in medical decision making? JAMA 1984; 252:2990–2994.
43. Deber RB. Shared decision making in the real world. J Gen Intern Med 1996; 11: 377–378.
44. Kinsinger L, Harris R, Karnitschnig J. Interest in decision-making about breast cancer screening in younger women (abstract). J Gen Intern Med 1998; 13:98.
45. Harris R, Kinsinger L, Karnitschnig J. Men want to participate in the PSA decision (abstract). J Gen Intern Med 1998; 13:98.

5

Regulation of Medical Devices by the Food and Drug Administration* with Emphasis on In Vitro Diagnostic Devices

Peter E. Maxim and Steven I. Gutman
Center for Devices and Radiological Health, Food and Drug Administration, Rockville, Maryland

I. HISTORICAL PERSPECTIVES

Historically, civilizations have been concerned about the safety and purity of their food and medicines. Given the opportunity, some merchants would short-weight bread; mix expensive spices, coffees, or teas with fillers; or dilute wines all in the interest of increasing profit. Governments usually responded by setting laws or standards in order to control the adulteration of important products intended for human consumption. Trade associations or guilds were also formed to provide voluntary standards and some oversight to their members. Punishments were as variable as the rules and standards, ranging from public humiliation to death.

In the United States, most of the states had enacted some form of safe food laws by the end of the 19th century. In 1880, Peter Collier in the Department of Agriculture recommended passage of a National Food and Drug Law with no success. In 1883, Dr. Harvey Wiley became Chief Chemist of the Department of Agriculture and conducted research into the degree of product adulteration in the United States. His research showed that problems associated with the

*The views expressed are those of the authors and do not represent official FDA policy.

nation's food and drug supply were not merely economic but had fundamental adverse impacts on public health. Starting in the 1890s, a series of bills to introduce federal regulation were proposed in Congress but were consistently and successfully opposed by industry (1–3).

Wiley began a grass roots campaign to bring public opinion to bear on this matter. His efforts resulted in the passage of the Pure Food and Drugs Act of 1906 and the formation of the Food and Drug Administration (FDA). The act provided definitions of misbranding and adulteration and allowed for seizure of those products. The basis for proof was at first limited to a chemical or microbiological analysis of the products. In 1912, the Sherley Amendment was passed, which declared that false and fraudulent labeling could also be used to establish misbranding.

In 1933, a bill was introduced that would allow for a modernization of the 1906 act. As was the case in the past, the bill drew an enormous amount of resistance. In the late 1930s, Elixir Sulfanilamide containing the toxic solvent diethylene glycol was commercialized and resulted in the deaths of more than 100 persons. The Secretary of Agriculture was able to remove the product from the market but explained to Congress that this was only because the elixir did not contain any alcohol as required by the United States Pharmacopeia definition. Largely in response to this public health disaster, in 1938 Congress passed the Food, Drug and Cosmetics Act. This act greatly expanded the authority of the FDA and allowed for preclearance determination of safety for new drugs.

Additional amendments to the 1938 act were added in 1945 and again in 1962 when, as a result of a tragedy involving thalidomide, drugs had to be shown to be safe and effective before marketing. It was not until 1976 that the Medical Device Amendments legislation went into effect to regulate medical devices. The 1976 amendments, in part, provided for the classification of all medical devices, required assurances that the device was safe and effective, provided for premarket notification and premarket approval as routes to the market, and provided for participation of outside experts in device evaluation. The Safe Medical Devices Act of 1990 served as a codification of existing FDA practice and expanded the FDA's authority in premarket and postmarketing activities (2).

More recently the Mammography Quality Standards Act of 1992 was enacted to ensure that all mammography facilities in the United States are adequately and consistently evaluated for compliance with quality standards for mammography and to establish quality standards for mammography on a nationwide level (4).

II. CLASSIFICATION OF DEVICES

Following the passage of the Medical Device Amendments of 1976, the FDA formed advisory panels of outside experts to classify all existing medical devices based on their knowledge of device performance and the risk associated with the use of the device into one of three classes. The devices were to be placed into the lowest class reasonable to ensure the safe and effective use of the device. The advisory panels recommended the appropriate class for the devices, then the sessions were held as open public hearings to obtain additional input from other interested parties. The three classes established were:

1. Class I—Devices whose safety and effectiveness can be assured by the application of general regulatory controls that are applicable to all medical devices such as labeling and good manufacturing requirements.
2. Class II—Devices for which the general controls are not sufficient to provide reasonable assurances of safety and effectiveness and for which sufficient information exists to establish special controls. Examples of special controls include guidelines, guidance documents, voluntary standards, and post market surveillance.
3. Class III—Devices for which the general and special controls are not adequate to reasonably assure safety and effectiveness. Studies must be performed to obtain valid scientific evidence that the use of the device will provide clinically scientific results and that the probable benefit to health outweighs any probable risk of injury or illness.

At the time that the classification meetings were conducted, there were certain devices such as tumor-associated antigen assays that were being regulated as investigational new drugs. As a result of device classification, these were termed transitional devices and were placed into class III. Devices that come to have new uses other than that which was originally regulated may be placed into a different (higher) class based on the new intended use.

As more information is obtained regarding the safety and effectiveness of particular devices, the FDA of its own volition, or interested parties through a petition can act to reclassify these devices to either a higher or lower level that may even result in exemption from many of the regulatory controls.

On September 19, 1996, the FDA responded to a citizens' petition and reclassified tumor-associated antigen assays intended to monitor a patient for recurrence of disease or for response to therapy from class III to class II. Based on several factors, including the FDA's own experience in reviewing these assays and clinical knowledge accumulated through many years of use of tumor markers for monitoring, it was decided that assays for these particular intended uses could be regulated at a lower level than class III.

III. PATHWAYS TO THE MARKET

The Medical Device Amendments of 1976 allowed for several mechanisms to put medical devices into commercial distribution. Those devices that were already marketed were grandfathered as preamendment devices and formed the basis for the classification panels. For manufacturers that wanted to introduce a new product with the same intended use as a preamendment class (predicate) device, a premarket notification (510[k]) could be filed with the FDA providing information that demonstrated the equivalence of the new product and notifying the FDA that the device would be marketed in 90 days. The FDA would review the information and ''clear'' the new product for commercial distribution as being ''substantially equivalent'' in its safety and effectiveness as the predicate device.

The FDA has accommodated changes in device design such as new technology, new analytes for the same intended use, and the incorporation of monoclonal antibodies and nucleic acid probes into tests by using the most recently cleared device as the predicate rather than depending on preamendment products. In many cases, data obtained from clinical studies were used to support the product substantial equivalence claim. The new device would then be classified at the same level as the predicate device as long as there were no new issues of safety and effectiveness. This procedure was codified with the passage of the Safe Medical Device Act of 1990. The administrative requirements for a premarket notification application are described in Title 21 of the Code of Federal Regulations (CFR) Part 807 Subpart E (4). These references are usually stated as 21 CFR 807 Subpart E or, in some cases, simply as 21 CFR 807.

Devices that are determined to be ''not substantially equivalent'' to those in commercial distribution before the 1976 amendments or to a legally marketed predicate device are then determined to be class III and needed premarket approval before entering into commercial distribution. When a device is classified as class III, a premarket approval application (PMA) must be submitted to the FDA. This application must provide valid scientific information to support the safety and effectiveness of the product for the proposed intended use. This generally includes data obtained from controlled clinical trials conducted at several sites. The application also contains the documentation supporting the manufacturing process and describing the manufacturing facility as specified in 21 CFR 820 in addition to administrative requirements described in 21 CFR 814. The application is reviewed by FDA scientific, statistical, and medical staff and may be referred to an advisory panel where the submission is discussed in an open public meeting and the panel of experts makes a recommendation to the agency.

When clinical studies are needed to support an approval by the FDA and there is a significant risk of injury associated with the use of the device in patients, an Investigational Device Exemption (IDE) may have to be filed with the agency and reviewed before the start of the study. The criteria established to determine when an IDE is necessary and the regulations governing an IDE study are found in 21 CFR 812. In general, clinical studies for in vitro diagnostic devices (IVD) are rarely performed under the IDE regulations unless there is no confirmatory test available and the results of the assay are being used in patient management. Clinical studies to evaluate the effectiveness of an IVD should be performed with a defined protocol in place, Institutional Review Board approval, and, in some cases, informed patient consent. The FDA strongly suggests that the manufacturer meet with the agency and have the clinical study protocol reviewed before starting the study. Additionally, there are requirements that the kits and the product inserts for the assays be labeled ''For Investigational Use Only.''

IV. ANALYTICAL DATA REQUIREMENTS FOR SUBMISSIONS TO THE FDA

Based on the many variables involved, it is possible to state data requirements only in general terms. Much depends on the class of the device and whether a 510 (k) or PMA submission is being prepared for the FDA. The submission, however, needs to contain sufficient analytical performance data and clinical data to either establish that the device is substantially equivalent to the predicate device or to establish that it is safe and effective for its intended use.

For assays, the analytical information contains descriptions of the assay components including the characterization of the antigens and antibodies. Whether the antigen and antibodies are produced internally or purchased from a vendor, it is important to ensure that an adequate supply of these critical reagents is available to avoid frequent material changes to new lots. Antigen characterization by any of a variety of biochemical, immunochemical, or immunological techniques is necessary to demonstrate purity, the nature and possibly quantity of potentially cross-reacting antigens, and the presence of all essential reactants in the antigen preparation. If the antigen is prepared from cell culture or by recombinant procedures, appropriate documentation should be available. The antibodies should be tested by the most sensitive procedure available to determine the specificity and whether there are any cross-reactions occurring. A popular technique to establish antibody specificity is Western immunoblotting using a crude antigen mixture. In some cases, additional information such as calculation of binding constants may be needed. If a monoclonal antibody is used, there should be adequate documentation of the procedures used to produce the hybridoma and to select and expand the monoclonal antibody. If these reagents are purchased, sufficient characterization may have already been done and the needed documentation can be supplied by the vendor.

The assay developer can use these materials after they have passed established acceptance criteria.

Calibrators are materials with established values that are used to standardize the instrument or assay method at established intervals. The calibrators set the working range of the assay and should encompass the critical determination points on the operating curve. For example, if the key medical decision point for the assay is 2.0 ng/mL, there should be a calibrator set close to that value.

Controls are materials that are run along with patient samples to ascertain the reliability of the assay. Rather than a preset value, controls have a range of values that should be consistent with the variability seen in patient samples. Generally, positive and negative controls are used and the positive control should contain levels of analyte that realistically controls the assay at the decision point. Generally, real-time stability studies are performed to establish the expiration dating for the entire kit and individual components under recommended storage conditions and this is provided on the label.

Warnings and precautions may need to be provided for some reagents. Materials containing metallic azides have to be identified and proper disposal conditions described. Special warnings are required for any kit that contains radioactive materials. Those that contain reagents made with human source materials must contain a statement that the source materials were tested for the presence of human immunodeficiency virus (HIV) and hepatitis antigens. Additional warnings and handling recommendations may be necessary depending on the kit contents.

Information should also be provided to support the use of various specimen types in the assay. Serum is the most commonly recommended specimen for immunoassay. It should be clearly stated if hemolysis, lipemia, cloudiness, or any other sample conditions are to be avoided. When plasma is included as a specimen, the recommended anticoagulants should also be listed because some may interfere in the assay. Comparison studies with paired samples should be performed to show that the same results are obtained with each sample type. Similar studies are needed to show that whole blood, urine, or another specimen type is adequate. It is incumbent on the assay developer or manufacturer to describe the conditions whereby a sample or specimen can be held and for how long. There should be data generated to support that the sample remains stable under those conditions.

The materials and procedures necessary to ensure proper and accurate performance of the test should be determined. Included are the control materials and levels for positive and negative controls, any additional control parameters that need to be addressed, directions for performing quality control (QC), interpretation of QC, and recommendations for action on discrepant QC results. If the device contains internal or procedural controls, these should be described, including what exactly is being controlled. Depending on the complexity of the test, training requirements can range from a detailed step-by-step procedure that is provided as a part of the product labeling, to an in-depth user manual, to a formal on-site training session with practice specimens.

The analytical performance chacteristics of the assay also need to be established and usually include accuracy, precision (interassay, intra-assay, lot to lot), linearity, specificity, analytical sensitivity (minimal detectable level), recovery of analyte, and method comparison to another commercially available kit or to a reference method. The FDA often relies on voluntary standards promulgated by organizations such as the National Committee for Clinical Laboratory Standards (NCCLS) to serve as guidance for these studies (5–7).

The effect of common interfering substances that are routinely tested during assay development include hemoglobin, lipids, and bilirubin (8). Additional potential interferences are highly dependent on the assay being developed and the proposed test population. For example, the interfering substances evaluated for a tumor marker assay proposed for monitoring previously treated cancer patients would be quite different from those considered for a screening population. For a screening application, a wide variety of over-the-counter and prescription medications expected to be used by the test population would be tested as potentially interfering substances. The interfering effect of drugs such as anti-inflammatory agents, multivitamins, antiarthritic medications, and antihypertensives may need to be evaluated in a screening population. For a follow-up to therapy indication, chemotherapeutic drugs and other medications directly related to the management of a cancer patient would be evaluated also.

Finally, the optimization of the assay would include establishing normal ranges of the test analyte using the proposed device, along with the determination of expected values in the test populations. Also at this time, the cut-offs for the device or the key medical decision points need to be established so that they may be challenged and validated in the clinical studies (9).

V. CLINICAL DATA REQUIREMENTS FOR SUBMISSIONS TO THE FDA

As was the case for analytical data, the requirements for clinical data also are highly dependent upon the class of the device, the type of submission being prepared, and the proposed intended use of the device. For class I and II devices that can be cleared by a routine 510 (k), the data requirements may simply be a sufficient number of patient samples to assure an appropriately high degree of correlation between the test and predicate devices. In these cases, the device can be determined to have a relative sensitivity and relative specificity when compared to the predicate device. For a class I or II device being proposed for a new clinical indication, a clinical study may have to be conducted to establish device performance in the population intended for use.

With class III devices, there is an increased burden in that the safety and effectiveness of the device has to be established in the proposed test population, and this almost always means that the device has to be tested in controlled clinical trials to establish safety and effectiveness. In many cases, a randomized, prospective clinical trial may be necessary but the intended use of the device may allow for retrospective studies or perhaps historical case-controlled studies. In contrast to drug trials, clinical trials for medical devices are rarely performed in three phases. Although, at times, a strong case can be made for a pilot or feasibility study before initiating the comprehensive clinical trial. There may even be times when this is appropriate for an in vitro device (10).

For medical device evaluations to determine safety and effectiveness, a device is considered safe when the probable benefits to health outweigh any probable risks when the device is used for its intended purpose and is accompanied by adequate directions and warnings against unsafe use. A device is considered effective when, in a significant portion of the test population, the use of the device for its intended uses and conditions of use yields clinically significant results.

In addition to safety and effectiveness, the clinical usefulness as well as the performance of the test may also be evaluated. The clinical usefulness of a device is generally determined as part of the agency review but may require input from experts on the advisory panels. Not only does a device need to function in accordance with its design, but it also is expected to provide information that measurably contributes to a patient's diagnosis or management. Performance evaluation is simply that the device does what it is supposed to do. In some cases, this may be evaluated by bench or animal testing but may be a part of the clinical trial. For in vitro devices, the key element of performance that is being evaluated during trials is the challenge of the cut-offs or medical decision points of the device.

Generally, the FDA views that a device can deliver clinically useful information without knowledge of the eventual clinical outcome. For example, it was sufficient to demonstrate that an elevated prostate specific antigen level in a man older than age 50 years, when considered along with the results of the digital rectal examination, was effective in aiding in the early detection of more prostate cancers. However, it was not necessary to show that the eventual outcome of the patient was improved as a result of the earlier detection. This question will be answered by ongoing randomized clinical trials designed to assess the outcome of patients with ''early'' prostate cancer detected with PSA. The medical community will then have an opportunity to reassess ''clinical usefulness'' with respect to the long-term outcomes measure.

Before the start of the study, the clinically relevant patient populations must be defined, appropriate inclusion and exclusion criteria set, and the patients entered with all of the critical medical documentation. In identifying patient populations, care should be taken to ensure that demographics are appropriate and that gender and ethnic differences have been considered.

The study should be conducted in a masked fashion so that test results of the in vitro device are not available for patient management decisions. If this is not possible and the results of the investigational device are going to be used for patient management, an IDE may have to be filed with the FDA. The sponsor is expected to have a well-defined protocol in place at all of the test sites, to monitor all of the sites for compliance to the protocol, and to ensure data integrity. When the study is terminated, data analysis is conducted to evaluate the performance of the device against preset standards. Tests performed during studies should conform with good laboratory practices.

Although there are many ways to describe the clinical performance of the device, the FDA generally likes to see several specific parameters described:

1. Sensitivity—The ability to detect a positive in the affected population.
2. Specificity—The percent of patients that tested negative in the unaffected population.
3. Predictive value of a positive—Varies with the prevalence of the disease and is the likelihood that a person with a positive test result has the disease.
4. Predictive value of a negative—also is dependent upon prevalence and is the likelihood that a person with a negative test result does not have the disease.

The basis for cut-off points chosen for clinical decision-making should be carefully considered and, whenever possible, receiver operator characteristics (ROC) plots generated to help clarify device performance (11). Some of the advantages of using ROC displays is that all information provided regarding the sensitivity and specificity a device can be viewed at all cut-off points. Cut-offs other than the one proposed can be evaluated and two devices with different cut-offs can be compared. It allows for an evaluation of the performance of the device with regard to clinical decision-making and can allow for the comparison of two devices that may be on different sides.

The performance parameters of particular interest in the review process can differ depending on the intended use of the device (12). Endpoints used to define clinical performance and the relative degree of performance expected may differ substantially based on whether a device is intended for diagnosis, screening, monitoring for recurrence, predicting the probability of recurrence, or monitoring for response to therapy. A screening test, for example, might be expected to have high sensitivity with a relative compromise in the degree of specificity; a confirmatory test leading to radical therapy might conversely be expected to have high specificity with a relative compromise in the degree of sensitivity. Many screening tests for a low prevalence disease would perform badly (too many false-positive results) if specificities decrease much. This would trigger large numbers of diagnostic workups if applied to the whole population. The relationship and interdependencies among various performance measures of in vitro diagnostic tests have been described in detail by Galen and Gambino (13) and more recently by Einstein et al. (14).

VI. LABELING OF IN VITRO DEVICES

In vitro diagnostic devices are unique in the area of device regulation in that they have their own labeling regulations, which are presented in 21 CFR 809.10. These regulations clearly indicate the information required to support device labeling and submissions. The labeling regulation is divided into the following 15 components:

The proprietary and established name of the device
Intended use
Summary and explanation of the test
A description or the principle of the procedure
Information on reagents
Information on instruments
Information on specimen collection and preparation
Procedure of operation
Results
Limitations of the procedures
Expected values
Specific performance characteristics
Bibliography
Name and place of business
Date of the package insert

Of these various elements, the most important is evaluation of the intended use and the related indications for use. The intended use and indications for use of a product determine the type of review, the questions likely to be raised, and the data requirements that will be required in the course of review.

VII. FDA INTERACTIONS WITH INDUSTRY

The FDA recognizes that it has a dual responsibility: to keep bad or incompletely characterized products off of the commercial market while ensuring that good and well-characterized products do reach the medical market in a timely fashion. As a result, the agency works with manufacturers in all phases of product development, schedules meetings between company scientists and appropriate FDA review personnel before, during, and after the design and conduct of device studies, and works in any way possible to assist in the development of high-quality submissions to allow rapid approval of effective medical diagnostic products. The agency is continually looking for ways to modify and increase the efficiency of the preapproval review process in order to meet its goals and to keep abreast of the rapid changes in the medical device arena.

Similarly, the public and the regulated industry have opportunities to aid the agency in modifying the review process. For example, a manufacturer of tumor marker assays recently filed a petition with the agency providing information and data that serum tumor marker assays, when used to monitor patients for recurrence of disease or response to therapy, should be regulated as class II rather than class III devices. After obtaining panel input and comments during a public meeting, the FDA concurred and approved the downward classification of those assays for the specified intended uses. The process was facilitated because the petitioner also provided a draft copy of a guidance document that described the types of data that would be needed to review the devices in this class. After some agency modification, the document was made available for additional public comment and serves as a special control for these class II devices.

VIII. CONCLUSION

This Chapter is an overview of FDA processes involved in the regulation of medical devices. An excellent source of additional information is the FDA Internet homepage at www.fda.gov, where the reader can access homepages for each of the centers (Food, Drugs, Biologics, Devices) as well as that of the Division of Small Manufacturers Assistance to obtain more detailed information on how to interact with the FDA.

REFERENCES

1. Young JH. Pure Food. Securing the Federal Food and Drug Act of 1906. Princeton: Princeton University Press, 1989.
2. Hutt PB. A history of government regulation of adulteration and misbranding of medical devices. In: Estrin NF, ed. The Medical Device Industry. New York: Marcel Dekker, 1990.
3. Estrin NF. The Medical Device Industry. New York: Marcel Dekker, 1990.
4. Mammography Quality Standards Act. Federal Register 1993; 58:67562.
5. National Committee for Clinical Laboratory Standards Evaluation for Precision Performance of Clinical Chemistry Devices: Tentative Guideline. 2d ed. NCCLS document EP5-T2. Wayne, PA: National Committee for Clinical Laboratory Standards, 1992.
6. National Committee for Clinical Laboratory Standards. Evaluation of the Linearity of Quantitative Analytical Methods: Proposed Guideline. NCCLS document EP6-P. Wayne, PA: National Committee for Clinical Laboratory Standards, 1986.

7. National Committee for Clinical Laboratory Standards. Method Comparison and Bias Estimation Using Patient Samples: Approved Guideline. NCCLS document EP9-A. Wayne, PA: National Committee for Clinical Laboratory Standards, 1995.
8. National Committee for Clinical Laboratory Standards. Interference Testing in Clinical Chemistry: Proposed Guideline. NCCLS document EP7-P. Wayne, PA: National Committee for Clinical Laboratory Standards 1986.
9. National Committee for Clinical Laboratory Standards. Preliminary Evaluation of Quantitative Clinical Laboratory Methods. 2d ed. NCCLS document EP10-T2. Wayne, PA: National Committee for Clinical Laboratory Standards 1993.
10. McCormack RT. Resolving IVD submissions: a role for feasibility studies. Applied Clinical Trials 1995; 4:14–16.
11. National Committee for Clinical Laboratory Standards. Assessment of the Clinical Accuracy of Laboratory Tests Using Receiver Operating Characteristic (ROC) Plots Approved Guideline. NCCLS document number GP10-A: Wayne, PA: National Committee for Clinical Laboratory Standards 1995.
12. Gail MH. Some statistical methods for evaluating immunodiagnostic tests. In: Herberman R, McIntire R, eds. Immunodiagnosis of Cancer. New York: Marcel Dekker, 1979:30–37.
13. Galen RS, Gambino SR. Beyond Normality: The Predictive Value and Efficiency of Medical Diagnoses. New York: John Wiley and Sons, 1975.
14. Einstein AJ, Bodian CA, Gil J. The relationship among performance measures in the selection of diagnostic tests. Arch Pathol Lab Med 1997; 121:110–117.

6

Ethics and Consent

Douglas L. Weed
Georgetown University, Washington, D.C., and National Cancer Institute, Bethesda, Maryland

I. INTRODUCTION

There is much to discuss about ethics and cancer screening. Screening tests are interventions, performed *by* health care professionals *on* individuals (usually healthy, symptom-free individuals). Ideally, these tests are designed to find disease early so that it can be treated effectively and with a net gain in life expectancy. In practice, however, cancer screening tests may pose potential health risks as well as potential benefits to those tested. Information provided about the likelihood of these potential risks and benefits helps individuals to decide whether they wish to be screened. Screening is performed in different cultural settings and may involve populations with special morally relevant designations such as ethnicity and socioeconomic status. Furthermore, screening is often recommended—even advocated—by public health agencies, private health organizations, and by health care professionals. Finally, screening involves economic and other resources, often in considerable amounts.

Inasmuch as ethics is the study of what is *right* and *good* and it includes the practice of *acceptable* and *appropriate* behavior, the fact that screening involves interventions on healthy people, the behavior of health care professionals, informed consent, health education and promotion, benefits and risks to the health of diverse populations, and extensive resources makes clear that ethics and cancer screening are inextricably entwined.

Cancer screening interventions may be part of research studies, part of public health programs, or provided upon request in clinics and hospitals as a part of preventive medical practice. Although the emphasis in this chapter is on the first two categories—research and public health—many of the ethical issues are general enough to be relevant to screening in medical practice. It is not always possible or desirable to refrain from considering all three categories; as discussed later in the chapter, the controversy regarding prostate-specific antigen (PSA) as a potential screening test for prostate cancer involves ethical issues that span research, preventive medical practice, and public health.

In general, the central ethical issues of cancer screening—informed consent, balancing benefits and risks, just allocation of resources, and so on—are applicable to all screening test–cancer pairs, although any real-life answers to ethical questions will be influenced by the particular circumstances found in specific situations. There are many different specific screening technologies and therefore organ-specific cancer sites to consider: Papanicolaou smears for cervical cancer, mammography for breast cancer, sigmoidoscopy for colorectal cancer, and PSA testing for prostate cancer are just a few examples. Each of these pairs has particularistic features; that is, each cancer has different incidence rates, mortality rates, and survival characteristics, and different treatment regimens with side effects that vary in severity. Similarly, each screening test has its specific features: measures of sensitivity, specificity, positive predictive value, and side effects.

The first task, however, is not to provide more detail on the specifics of screening test-cancer pairs; much of that information can be found in other chapters of this book. Rather, I begin with a foundation for discussing ethics itself. What is needed is a framework for ethics, emphasizing practical rather than theoretical concerns, one that rests securely upon a foundation of accepted bioethical theory. Ideally, such a framework would be a beacon to guide us to justifiable answers to questions that emerge within the many ethical issues of cancer screening. It may also enlighten us on what ethical positions are held by participants in the turbulent controversies that so often characterize this topic. It can be thought of as a locus of moral consensus.

II. A LOCUS OF MORAL CONSENSUS

Across most groups of health care professionals (and in the public at large) there are points of agreement—i.e., norms or rules—representing a common morality (1). Even though it is possible that these norms may have changed over the centuries and may have somewhat different interpretations and meanings in different contemporary cultures, they are (for now) near-universal rules of behavior. The following list is a reasonable (if incomplete) set of rules that both the public *and* contemporary health care professionals would agree are morally correct in nearly all circumstances. Expressed in the language of obligations, these rules should be accepted by health care professionals in their interactions with individuals and communities. In no particular order, these rules are

Truth telling
Respect for others
Obtaining consent before touching or invading a person's body
Preventing harm to others
Not causing pain or suffering
Helping others
Just and fair allocation of resources
Confidentiality of private information

Similarly, there are proscriptive rules that are clearly immoral (i.e., inappropriate and unacceptable) in nearly all circumstances:

Lying
Coercion
Disrespect for an individual
Knowingly and purposefully harming another
Discriminating unfairly on the basis of gender, race, ethnicity, education
Conflicts of interest
Unjust allocation of resources

The first list can be expanded to include other rules, such as ''do not kill,'' and ''rescue those in danger,'' but these have limited applicability for cancer screening. The second list can also be expanded, such as to include unfair discrimination on the basis of genetic test results. For the purposes of this chapter, though, these lists provide a set of necessary moral rules to follow in nearly all circumstances involving research on cancer screening technologies as well as those circumstances involving the application of screening in preventive medical practice and in public health programs.

To say that rules are ''followed'' brings to the fore an important practical problem. As stated, these rules are broad in scope. They provide neither specific conditions nor constraints on their application. In addition, when these rules—even universally accepted rules—are applied to real-life situations, they may conflict. An obvious example in cancer screening is the conflict between helping others and preventing harm to come to them. Cancer screening nearly always involves a conflict between these two widely accepted rules of conduct; screening can be beneficial and it can do harm. How then can rules be followed, that is, applied, if they are rather general and may also conflict? Solutions to these (related) problems are likely to have important implications for dealing with the inevitable ethical dilemmas of cancer screening wherein a dilemma is defined as a decision made in untidy circumstances. First is the problem of conflicting rules, because within its solution lies the solution to the accompanying problem of the nonspecific nature of the rules found in this locus of moral consensus.

Although the aforementioned rules are reasonable, none are absolute. According to the bioethicist Beauchamp, any rule may be overriden in specific situations depending upon the particular context and the individuals or social groups involved (1). Even so, it is unlikely that one rule will always or completely ''trump'' another. Rather, it is more likely that some balancing of rules will occur. Balancing requires conscious deliberation (a judgment) regarding what, in the end, is the right action to take. The methodological issue, and one vigorously debated in professional bioethics, is how to balance rules when they conflict.

An important methodological strategy in such situations is called *specification*. In this technique, rules are made more specific for the particular situation such that their implications can be examined. Specification, therefore, links the general rules with the specifics of the decision at hand, often described in a ''case study,'' and may also allow the decision maker to choose between alternative rules that, at a more general level, appear to conflict. Other strategies, which are often used in combination with specification, include obtaining more information (e.g., scientific information) that may facilitate decision-making, invoking codes of personal or professional conduct, providing definitional clarity, and using examples and counterexamples. In addition, it may be helpful to carefully analyze the arguments of decision makers. Such an analysis may reveal errors in understanding or in applying the underlying rules. It may also reveal the more general ethical precepts, such as principles or ethical theories, implicitly held by the proponents and opponents of diverging positions.

Indeed, the resolution of specific ethical dilemmas may require making reference to general ethical theories, which, in turn, provide a foundation for the aforementioned rules. It is uncommon, however, for health care professionals directly involved in decision-making to disclose their theoretical commitments (2), perhaps because they are unaware of them, lack the means to express them explicitly, or believe that theories are too far removed from ''real life'' to be useful. Nevertheless, there *are* several such theories and health care professionals may benefit from even a cursory look at their respective structures and implications (see references 3–6).

Some ethical theories are familiar, such as the principle-based approach of the Belmont Report, and are discussed at length in Beauchamp and Childress' classic bioethics textbook (4). The four principles include autonomy (or respect for persons), beneficence (doing good, preventing harm, and balancing good against harm), nonmaleficence (not doing harm), and justice (fairness and equity). These principles are only somewhat more general than the rules discussed previously. Even more general theories include those called utilitarian (or consequentialist) and deontological (or Kantian). Another is the theory of virtue, which is important to all morally relevant situations because it emphasizes the character of the individual decision maker. Virtue or character ethics have special relevance for the issue of scientific misconduct (5). Feminist ethics and the ethics of care are other examples of prominent ethical theories. Finally, there is casuistry, which is more methodology than theory, but is often included under the heading of the theoretical foundations of ethics perhaps because, like virtue theory, it has its origins in some of the earliest writings on ethics in Western philosophy (6).

Casuistry emphasizes the particularistic nature of practical ethical decision-making, that is, the details of the situation that make each decision unique. It is a technique that helps the ''user'' generate decisions, given the specific circumstances, using moral maxims, which may be the rules described previously, specific moral intuitions, or more general notions such as the familiar bioethical principles. In turn, these maxims may require specification if the decision maker carefully examines their implications in particular circumstances. Casuistry and the method of specifying principles and rules, therefore, can be considered two sides of the same coin. A sharp distinction between them, at least in decisions that emerge within public health practice, may not be possible (7). As a method, casuistry begins with a careful description of the particular situation and employs, when necessary, progressively more general moral maxims (including rules and principles). Specified principlism, on the other hand, begins with the principles or rules and specifies them for the particular situation at hand. Either way, a balancing of rules may be required and judgment plays an important role. Illustrations are given in the text that follows.

III. CANCER SCREENING: GENERAL ISSUES AND GENERAL RULES

Several recent journal articles and textbook chapters have described ethical issues that may arise in cancer screening (8–12). Other sources are papers discussing ethical issues in public health research and practice (13–19). In addition, the recently published *Encyclopedia of Bioethics* (20) provides a useful compendium of current thought on many general ethical issues discussed in this chapter. Some of the most common concerns in this literature are the benefits and harms of screening as well as issues of informed consent, advocacy, and other forms of health promotion, that is, issues involving the flow of information between those who screen and those who are screened. Less attention has been paid to the ethics of resource allocation and issues specifically relating to the ethics of randomized screening trials; rare are discussions of scientific misconduct.

To provide a comprehensive yet reasonable discussion of ethics and cancer screening, several general ethical issues are described in this section, mirroring many of those mentioned previously. Issues of just allocation of resources and discrimination on the basis of race, gender, or genetic factors are not examined in detail.

For each issue, background, definitions, and general rules appropriate for making ethically correct decisions in situations of moral uncertainty are provided. A distinction is made between ethical issues in cancer screening research and those involving cancer screening programs. Some overlap is inevitable, although population-based screening programs are not labeled ''experiments'' as recently suggested (14). There may also be overlap in the normative rules applied to these issues. Conflicts between rules are resolved in the sections of this chapter in which particular screening test-cancer pairs are examined in terms of current controversies.

IV. ETHICAL ISSUES IN RESEARCH ON CANCER SCREENING

Five general ethical issues are examined, each with direct relevance to research on cancer screening tests: scientific misconduct, informed consent, a duty to not harm study participants, the balance of benefits and risks weighed before and during research, and equipoise in randomized clinical trials of screening. Following the discussion of these general issues (and their accompanying rules), attention is paid to the considerable controversy surrounding the ethics of screening for prostate cancer.

A. Scientific Misconduct

As an ethical issue, scientific misconduct is not usually associated with a specific type of biomedical research, such as cancer screening research. Nevertheless, the importance of the topic and its potential impact on the public's perception of biomedical science and on the professional health of scientists engaged in any form of biomedical research (including research on cancer screening) cannot be underestimated nor wished away. No one knows with certainty how much scientific misconduct has been committed, although several recent studies have begun to answer this question (21–26). A recent estimate, probably no more than an educated guess, is that for every 1000 research studies undertaken, one to four have some fraudulent aspect (27). The definition of misconduct is extremely important in efforts to estimate its occurrence. Although the vigorous debates on the definition of misconduct are not discussed in depth in this chapter, some concerns include the distinction between intentional error and the unintentional error that is essential in science, and the distinction between so-called ''serious'' misconduct—fabrication, falsification, and plagiarism—and other lesser forms, including misuse of research funds, conflicts of interest, noncompliance with research regulations, authorship issues, and deceptive scientific practices such as the failure to explain weaknesses in data or selective reporting of results (28–38).

Fabrication, falsification, and plagiarism are always unacceptable. Each, in its own way, contradicts what science claims as its primary goal: to seek the truth. Although scientists are often uncertain about what constitutes truth, as witnessed by the much more common terms of ''evidence'' and ''hypothesis'' and the relatively infrequent use of terms such as ''proven'' or ''disproven,'' scientists nevertheless explicitly and unanimously accept the idea that falsification and fabrication are direct affronts to their search for the truth. Plagiarism, on the other hand, is truth's clown-like mimic.

To better see the connection between the general rules found in the locus of moral consensus described previously and rules more clearly guiding our behaviors involved in scientific (mis)conduct, some specification is required. From the general obligation of truth telling and respect for others' ideas can be specified clear and unequivocable statements:

Never fabricate scientific data.
Do not falsify scientific information.
Do not copy others' scientific work verbatim without acknowledging the source.

It is reasonable, nevertheless, to ask why scientific misconduct occurs in a social environment (i.e., the practice of science in the community of scientists) within which these rules exist and nearly everyone agrees with them. To answer this question, it is necessary to examine what *causes* scientific misconduct, or if strong evidence for causality does not exist, then what factors are hypothesized as responsible for the occurrence of scientific misconduct. There are many opinions on the topic of what (and who) is responsible for scientific misconduct, including mental illness (39,40), personal factors such as ego, financial gain, and vanity (40–42), and factors such as publication pressure, intense scientific competition, and the failure of senior scientists to be effective mentors for their junior colleagues (43–49). Aside from the phenomenon of psychiatric disease, for which it is reasonable to argue that a disturbed individual may be aware of rules prohibiting misconduct yet nevertheless openly defy them, the remaining (often overlapping) factors suggest that there is something at a deeper level that must be described and explained, that is, some fundamental failing in ethical standards at a personal level. Misconduct may be perpetrated by scientists who succumb to pressure and put their personal interests ahead of the interests of society (including the society of scientists). The ethics of character, or virtue ethics, is likely the best way to understand the motivations of such mentally competent scientists who are aware of the rules proscribing misconduct and yet do not follow them. Virtue ethics involve those attributes of character that motivate people to act in the right manner, including their propensity to follow accepted rules of conduct (5). Scientific misconduct, therefore, can be understood as a failing in professional character, that is, as a lack of integrity or honesty or both. Teaching or learning the virtues relevant to appropriate scientific conduct (and therefore relevant to the prevention of scientific misconduct) are discussed in references (5 and 50–52).

B. Voluntary Informed Consent

Voluntary informed consent is a relatively recent phenomenon, emerging in general codes of research ethics in the 1950s. In research practice, however, it has an even shorter history, dating from the mid-1970s (53). Since then, consent has become a central and necessary part of the practice of medical and public health research worldwide. Its moral justification derives primarily from the principle of autonomy or respect for persons. Indeed, the original description in the Nuremberg Code used the word ''voluntary'' rather than ''informed'' consent, thereby emphasizing that the key element has been and continues to be the voluntary nature of participation by individual study subjects. Voluntary informed consent means that each individual voluntarily agrees to participate after receiving the information outlining the research. The person may also discontinue participation at any time without penalty or loss of benefits (54). Additional but insufficient moral justification can arise from several other arguments: that consent helps to protect people from harm, that consent provides some measure of social benefit by protecting the research endeavor itself from suspicion, and that consent represents an effort on the part of researchers to disclose the truth about their intended scientific research to others.

Obtaining consent is important enough to warrant its appearance as a distinct item on the list of morally acceptable rules of behavior (i.e., the locus of moral consensus). Specifying this rule, however, to make clear its precise requirements is not straightforward. Some attention must be paid to the dual nature of informed consent and to the various elements or criteria that make up the process of consent. The issue of informed consent is reexamined in a later section of this chapter for its role in preventive medical practice and especially in cancer screening programs offered to populations once efficacy has been established.

1. *Dual Nature of Informed Consent*

Informed consent has two different though related meanings. One concerns the individual research participant and the other concerns the institution sponsoring (or carrying out) the research project. For the *individual*, to give informed consent means that participation in research has been autonomously authorized. Several elements are involved in authorization: the person must be competent to understand and decide, the person must provide a decision voluntarily, the information must have been disclosed to the person, the person must have understood the information and what it means in terms of a plan of action, and the person must be in agreement with the plan, thereby providing authorization for it to occur (55). For the *institution*, informed consent is a legally effective approval by the research participant. It is typically provided on a document that, according to federal regulations (56), must have the following components:

A statement that the study involves research
An explanation of the purposes of the research
The expected duration of the subject's participation
Description of the procedures to be followed
Which procedures are considered experimental
A description of reasonably foreseeable risks or discomforts
A description of benefits to the subjects or to others
Appropriate alternative procedures, if any
The extent to which confidentiality of records will be maintained
Compensation and the availability of medical treatments if injury occurs
Contacts for answers to questions about research or injuries
A statement that participation is voluntary and that no penalty or loss of benefits will occur if the subject refuses to participate or discontinues participation at any time

The same federal regulations state that the following additional elements of informed consent may be provided, when appropriate:

A statement that the procedure may involve risks that are currently unforeseeable
The circumstances under which the investigator may decide to terminate the experiment without the subject's consent
Additional costs to the subject that may result from participation in the research
The consequences of withdrawing from the research
A statement that significant new findings will be made available to the subject that may affect the subject's participation
The approximate number of subjects involved in the study

2. *Balance in Informed Consent*

In practice, some balance must be struck between the two meanings of informed consent. As Beauchamp and Faden (55) argue, it may in practice be an overdemanding criterion to require *full* disclosure and *complete* understanding for every research participant. Some consideration must also be paid to the efficiency and effectiveness of the research process, although not to the extent that the document itself becomes the primary consideration in determining whether or not consent has been obtained. In other words, it is an underdemanding criterion to require only that the consent form have the appropriate components and be signed by all participants. Some commentators believe that informed consent should be seen as something more than a single event, that is, as a process of discussion and negotiation between research institution and research participant, in which the primary purpose of the process is to foster self-determination (i.e., autonomy) and to empower prospective participants to protect themselves. The primary purpose of the document itself, however, remains to protect the institution from legal liability (54). A prototype of the informed consent document for the National Cancer Institute's Prostate, Lung, Colorectal, and Ovarian Cancer Screening Trial (PLCO Trial) is the Appendix to this chapter.

C. Not Causing Harm to Research Participants

In medical practice, an obligation to not cause harm is a central ethical precept steeped in the Hippocratic tradition. In biomedical research, an obligation to not cause harm to volunteer participants is also vital and is historically linked to an obligation to obtain voluntary consent. Both ideas were prominently featured in the Nuremberg Code published in 1949. However, as Beecher's landmark paper in a 1966 issue (57) of *The New England Journal of Medicine* made clear, progress in codifying the ethics of medical research (in the wake of the Nazi criminal trials) did not have an immediate or wide-ranging impact on research practice in the United States. Beecher's paper revealed a litany of research studies that occurred after World War II, all of which revealed significant ethical lapses: failure on the part of investigators to obtain consent and some involving deliberate harmful interventions. Two infamous examples in the latter category are the feeding of live hepatitis virus to mentally retarded residents of the Willowbrook Institution and the injection of live cancer cells into elderly and senile hospitalized patients at the Brooklyn Jewish Chronic Disease Hospital (58). More recently, examples of unethical research that occurred primarily in the 1950s and 1960s have been uncovered. These involved exposure to radioactive materials by research participants who either did not give consent or gave a superficial consent because investigators avoided careful and reasonable explanations of potential harms (59). Although these examples do not involve cancer screening tests, they nevertheless provide important reminders of the idea that not causing harm to research participants is a central ethical precept closely linked in practice to the equally important rule that potential research participants be informed of the research and its potential harms (60).

D. Benefits and Risks of Research

Research on cancer screening tests may involve potential harms to study participants, including but not limited to characteristics of the test itself, such as the minimal risk of slight discomfort and bruising of venous blood collection, low doses of X-irradiation, and minimally invasive procedures such as fiberoptic devices that are inserted into body orifices. In addition, screening in a research setting may identify malignancies that require aggressive treatment with its attendant side effects and which may not provide benefits. There may also be harmful psychological effects (e.g., anxiety) associated with participation in research studies. A more complete rendering of the benefits and risks of screening is provided in a later section.

It is important, therefore, to consider what benefits can be weighed against such risks, inasmuch as the efficacy of the research intervention (i.e., the cancer screening test) has not been established. It is not uncommon for research studies to be justified in part on the benefit to society, a benefit described in terms of the scientific knowledge gained once the research has been completed and in terms of the potential benefit to the participants of the study itself, at least to those who are randomly assigned to the intervention group. Decisions about the ethical acceptability of the research project in terms of the balance of risks to individuals, benefits to society, and potential benefits to participants, must be made carefully and judiciously and in an open forum, typically in the proceedings of an institutional review board (IRB).

In the discussion of benefits and risks of research, the term ''balance'' is used but not carefully defined. The implicit definition is that the potential benefits will at least equal the potential harms, although there are many additional issues, including the fact that benefits and harms may be measured in different units. The next section, discusses the extent to which such a definition—emphasizing the need for equivalence between benefits and harms—is warranted for research on healthy individuals, that is, those who are the most frequent participants in randomized screening trials.

E. Ethical Justifications for Randomized Screening Trials: Equipoise

Decisions about the balance of beneifts and risks is an important part of determining whether any randomized clinical trial is ethically justified, whether therapeutic or preventive. However, the ethical justification for undertaking screening trials may nevertheless be guided by somewhat different considerations than for cancer treatment trials. The size, cost, and duration of cancer screening trials, which are typically several orders of magnitude larger than most cancer treatment trials, are not further considered here. A more troublesome concern is the difference between the populations enrolled in such studies. Treatment trials enroll patients (i.e., those who are ill); screening trials enroll ''healthy individuals'' (i.e., nonpatients who, by virtue of their participation in the trial, may become patients if a cancer is detected). These healthy volunteers are also exposed to other risks of research.

Given this important difference in the type of population studied, some commentators (61) have recently argued that, before a randomized trial of screening can be ethically justified, there must be a ''stronger promise of benefit for screening healthy people than for treating ill people'' (62–65). Although it is intuitively appealing, this claim has not received any scrutiny regarding its own ethical justification and practical implications. For example, does such a requirement refer to the amount and type of evidence (short of randomized trial evidence) supporting the efficacy of the screening test or does it refer to the point at which potential benefits match or exceed potential risks, that is, the relative balance of benefits versus risks? Perhaps it refers to both considerations. To examine these questions more carefully, I turn to a discussion in the clinical research and bioethics literature regarding the need for randomized trials on the basis of what is known or believed to be known about the potential benefits and potential harms of therapeutic interventions. In that literature, the issue is referred to as *equipoise* (66–70).

There are two basic types of equipoise: ''theoretical'' and ''clinical.'' These are used to describe the cognitive state of knowledge on the part of individual investigators and on the part of the scientific community, respectively, before the onset of a randomized trial. Inasmuch as any screening test almost surely identifies new patients, what is the meaning and relevance of equipoise in randomized screening trials? Furthermore, how can the issue of equipoise in randomized studies of patients contribute to the understanding of the conceptual and ethical difficulties faced by prevention scientists studying healthy volunteers?

1. *Equipoise in Randomized Screening Trials*

In therapeutic research, equipoise refers to the state of genuine uncertainty on the part of an individual physician-investigator about the relative merits of the treatments to be tested in each arm of the trial. This individual or ''theoretical'' equipoise implies that the investigator has no rational treatment preference whatsoever (67). Such stark uncertainty is necessary from an ethical perspective because, if an investigator believed one treatment to be superior to the other, then the investigator would be ethically obliged to offer that treatment to patients and, as a result, could not in good faith (i.e., could not ethically) enter these individuals on trial. Freedman (66), citing ''nearly insuperable obstacles'' of individual or theoretical equipoise, suggested an alternative but closely related concept: ''clinical equipoise.'' Clinical equipoise—sometimes called ''community equipoise''—is based on the idea that the expert medical community rather than the individual investigator must exhibit genuine uncertainty about the relative merits of the two (or more) treatments. Thus, the medical community provides the determination (a judgment) that the two treatments appear to be equally effective on the basis of current evidence available before the onset of the trial. Clinical equipoise allows the investigator who believes that there are differences in treatment options to suspend that belief for the purposes of entering patients on a trial, based upon a consensus within the medical community that the differences are genuinely unknown. Some controversy exists over who best represents the medical community and the extent to which patients or patient-advocates should be involved (68).

The ethical problem of equipoise in therapeutic research arises in part because of the professional role of the physician-investigator. These research-oriented clinicians are faced with a difficult situation: they must (by virtue of their professional obligation to patients) suggest one treatment or another. But, at the same time they are also committed to the need for research whose goal is to test whether one treatment is better than another.

Research on screening reveals some cogent differences from the situation in therapeutics: different professional roles of investigators, very different study participants (i.e., healthy rather than ill), and therefore different relationships between investigator and participant, where ''different'' means something other than the classic physician–patient relationship.

Although physicians may often be investigators in screening trials, they may or may not accept a general obligation to provide prevention messages to their patients. Clearly, physicians have an obligation to *treat* those study participants who are identified with cancers. An ethically acceptable screening trial, in turn, must involve access to the best possible therapeutic care for those individuals identified with cancer. But whether these same physicians have as a part of their regular medical practice an obligation to provide preventive recommendations or procedures is a somewhat different question. Furthermore, screening trials are not undertaken only by physicians. They may also involve health care professionals and researchers who have no (or perhaps little) direct influence on the health care (therapeutic or preventive) of the individuals or communities participating in the research study, much less a professional obligation to provide screening tests. Clearly, if a health care professional *does not* accept an obligation to recommend preventive interventions, then the difficult dilemma of choosing between being offered a preferred intervention and being offered an opportunity to participate in a clinical (screening) trial may not so strongly emerge as it does in the case of therapeutic research. On the other hand, if a general obligation to provide preventive messages and interventions exists for medical and public health professionals, then determining whether or not to encourage participation in a screening trials emerges as a difficult dilemma analogous to that faced by physicians contemplating putting ill patients on treatment protocols. Whereas a more detailed analysis of the public health (i.e., preventive) obligations of physicians and other public health professionals is beyond the scope of the chapter, it remains true that screening studies may not proceed (from an ethical perspective) without a careful consideration of the potential benefits and risks and that there may be important differences between individual investigators and the scientific community regarding where in the balance the benefits and risks lie.

Thus, there is a distinction between ''theoretical equipoise'' and ''clinical equipoise'' for randomized screening trials inasmuch as there can be prominent differences between the state of uncertainty on the part of individual investigators undertaking a screening trial and the state of uncertainty on the part of the general scientific community regarding the relative merits (and harms) of the screening test. In screening it may be more often the case that there are vocal and influential members of the medical community who believe that a screening trial is unnecessary because in their judgment sufficient evidence exists to warrant general screening in the absence of randomized trial evidence.

We can, therefore, define clinical equipoise in the screening situation as genuine uncertainty on the part of the scientific community regarding whether being offered the screening test is better than not offering it. This definition, however, does not take into account that screening study participants are healthy asymptomatic individuals and therefore may require something more than an assurance that offering the screening test may be no better than not offering it (whether in terms of the type of evidence or in terms of the balance of benefits and risks). For the latter concern, clinical equipoise in screening research may require a judgment made by the scientific community that the potential benefits of the test considerably outweigh its potential harms.

For therapeutic research (67), it is naive to think of equipoise as a decision solely about benefit, that is, as a decision solely about whether or not an intervention is better. There are also the potential harms, that is, side effects of treatment, and the point of equipoise may not be ''no effect'' but rather the point at which the potential benefits are big enough to compensate for the potential harms. Applying this approach to screening, however, means that this point of compensation is reached when the potential benefits are bigger still—that is, bigger than what is judged ''balanced'' for comparable therapeutic situations—thereby more than compensating for potential harms. To better understand this issue, making a distinction between ''absolute'' and ''effective'' equipoise is helpful (67).

Absolute equipoise, a form of clinical equipoise, is used to describe the situation of ''no effect'' of intervention in the expert judgment of the scientific community. ''Effective'' equipoise, on the other hand, describes the situation in which the potential benefits clearly compensate for the perceived risks of intervention, where these risks are described in terms of side effects. Decisions regarding the point of ''effective'' equipoise likely involve the patient's preference for treatment inasmuch as the patient may value the balance of potential benefits and potential risks somewhat differently than the investigator or, for that matter, differently than the scientific community (67).

It follows, then, that for randomized screening trials, there may be a need to reach a point somewhat beyond the point of ''effective'' equipoise, inasmuch as the benefits will therefore (at least in the value system of the study participant) not only outweigh the potential risks, but do so by a large enough margin to compensate for the fact that the participants are healthy volunteers. This point is likely somewhat different than that reached by a study participant contemplating a randomized treatment trial, if the arguments of some commentators are to be accepted. We might call it a point of ''extra-effective'' equipoise.

Given this analysis, it is not unusual that so much controversy can develop about whether a screening trial should take place. If the potential benefits must outweigh the potential harms by a considerable margin (to compensate for the fact that the participants are healthy volunteers), then there cannot be much evidentiary room between such a judgment and a judgment that enough evidence exists to forego implementing a trial at all. Resolution of such quandaries is difficult unless a randomized trial is considered a necessary test (i.e., an absolute criterion or standard) for judging the effectiveness of screening.

In sum, the proposed ethical justification for randomized screening trials—that the investigators and potential study participants believe that the benefits likely outweigh the risks and by a larger margin than that acceptable to a group of comparable physician-investigators and patients in a comparable therapeutic situation—seems intuitively acceptable on the surface, yet rather difficult to defend in theory and in practice. The reasons for these difficulties seem straightforward: there is a lack of clarity about what constitutes a ''comparable therapeutic situation,'' there is a lack of clarity in what comprises the professional obligations of investigators with regard to preventive messages, and, finally, there is a lack of precision regarding how benefits and risks are ''balanced'' in theory and in practice.

How specific judgments are made about the ethical appropriateness of randomized screening trials cannot be laid out (at least at this time) in any more detail. Examples are necessary to illustrate the problem of equipoise as well as the other ethical issues discussed in this first section of the chapter: problems of informed consent, not harming research participants, and the balance of benefits and risks. Prostate cancer screening is an informative example.

V. EXAMPLE: PROSTATE CANCER SCREENING WITH PROSTATE-SPECIFIC ANTIGEN

Currently, at least two randomized trials of screening tests for prostate cancer are underway (71,72) and several randomized trials of treatment of localized lesions have also begun (73). Definitive results from the screening trials are not expected much before the year 2008. At least two ethical issues have received attention, one directly related to the ethical acceptability of prostate cancer screening trials and another regarding whether screening should be recommended in the absence of results from randomized screening trials.

A. Are Prostate Cancer Screening Trials Ethical?

In a 1994 paper published in *Lancet*, Adami, Baron, and Rothman (61) argue that, in their judgment, the prostate cancer screening trials initiated in the United

States and in Europe have not been adequately ethically justified. They proposed that these trials may be ethically inappropriate because there exists the likelihood that the overall (i.e., final) trial results may reveal more harm done than good. In such a situation, Adami et al. argued that the screening trials were in direct conflict with the generally accepted rule of research that studies should never be designed to demonstrate harm and, furthermore, that informed consent was insufficient to provide a safeguard against research participants being exposed to studies with unfavorable risk-to-benefit outcomes. In addition, Adami et al. argued that ''there should be widespread perception that the prospect of net gain should surpass the prospect of harm before an experiment on healthy people is considered ethical,'' an appeal to the notion of ''extra-effective'' equipoise previously discussed.

A response to this paper appeared as a letter to the editor in *Lancet*. Written by Schroder, the principal investigator of the European prostate cancer screening trial (74), it deserves careful consideration. Key to the response was the fact that Adami et al. failed to consider the possibility that the available ethical justifications for both screening trials had already been provided by the ethics committees that reviewed the trial protocols. Seven considerations were provided regarding the decision to undertake the European trial:

1. That prostate cancer has a potentially curable (early) stage
2. That the natural history of prostate cancer is understood well enough to support a screening trial
3. That available data support testing the hypothesis that screening could result in a 20% mortality reduction
4. That screening tests are sufficiently reliable and are acceptable for healthy men
5. That treatment trials involving delayed treatment and radical prostatectomy involve unacceptable ethical problems
6. That the proportion of clinically unimportant cancers detected is estimated to be less than 10%
7. That randomized screening studies are ethically justified inasmuch as they are the only feasible way of resolving the dilemma of screening and early treatment

There was no mention in Schroder's letter of the balance of potential benefits versus potential harms, although such a determination may have been made in the deliberations of the ethics committees.

Summing up, it is possible that the final result of the screening trial could reveal an overall harm to the study participants if screening is not effective in reducing mortality rate (75). But, that possibility is clearly not what these studies were designed to demonstrate. Indeed, any randomized trial may not produce an overall benefit inasmuch as there is at least some (and sometimes considerable) uncertainty regarding the outcome of the trial. In my view, adequate consideration was given to the ethical acceptability of randomized screening trials of prostate cancer.

B. Is Prostate Cancer Screening Appropriate in the Absence of Clinical Trial Evidence of a Mortality Reduction?

While randomized screening trials recruit, screen, and follow study participants (and while the world waits for their results), a controversy of considerable intensity has emerged regarding the appropriateness of prostate cancer screening in practice (76). At least three organizations espouse screening asymptomatic men in medical practice situations: the American Urological Association (AUA), the American Cancer Society (ACS), and the American College of Radiology (ACR). Several other organizations recommend against PSA screening unless the individuals are at high personal risk: the United States Preventive Services Task Force (USPSTF), the American College of Physicians (ACP), the National Health Service of the United Kingdom (NHS), and the Canadian Task Force on the Period Health Examination (CTFPHE). At this time, none of these organizations espouse general population-based public health screening programs for prostate cancer (77–79).

Commentators from many different medical specialties—primarily urologists, family practitioners, general internists, and oncologists—defend what can only be considered opposite positions. Some examples of contrary claims appearing in the 5-year period from 1993 through 1997 include the following:

1. Schroder (a Dutch urologist and principal investigator of the European screening trial) who writes in a 1993 *British Medical Journal* editorial (80) that ". . . screening should not be recommended . . . until clear benefit in terms of reduced mortality . . . can be shown in prospective studies . . ."

2. Catalona (an American urologist) writes in a 1993 editorial in the journal *Urology* (81) that ". . . it is unrealistic to expect clinicians to refrain from PSA testing . . ."

3. Voss (an American general internist) writes in a 1994 editorial in the *Journal of General Internal Medicine* (82) that ". . . annual PSA (tests) for asymptomatic men are not warranted . . ."

4. Middleton (an American urologist) writes in a 1997 editorial in the *Annals of Internal Medicine* (83) that, in his practice, ". . . I subscribe to the ACS recommendation that a man older than 50 years should have an annual digital rectal exam and a PSA . . ."

It seems reasonable to consider why such starkly different screening recommendations have emerged. One approach is to carefully examine the ethical justifications provided by these and other commentators, that is, the arguments used to buttress the claims. Once these justifications are identified, it may be possible to determine which are the strongest (i.e., best reasoned) in terms of the ethical rules and methods of ethical reasoning described previously. In addition, it may be possible to to gain a better understanding of how ethical reasoning works in daily research and practice settings (as reflected in the medical literature). On this second point, it is interesting that in no published paper on prostate cancer screening is there mention made of any general ethical theory or method of moral reasoning, such as casuistry or specified principlism.

A paper by a urologist (84) who questions the value of screening for prostate cancer in asymptomatic men, however, does mention the Hippocratic Oath, in particular its admonition to ''be of benefit/do no harm.'' Indeed, the controversy appears to be focused on the different opinions investigators and physicans hold about the relative balance of benefits and risks. Chodak (84) and others who do not support general screening recommendations in preventive medical practice (75,80,82,85,86) cite the lack of scientific evidence for a mortality benefit to outweigh the known harms of screening and treatment, where ''scientific'' refers to a randomized screening trial. Those who support screening in asymptomatic men have a number of arguments for their position (87); an interesting and representative insight is provided by a urologist writing in the *Western Journal of Medicine* in 1995 (88). He notes that prostate cancer affects nearly 250,000 men each year and is responsible for 40,000 deaths, thus arguing (as is a popular argument in this controversy) that the disease is an important public health phenomenon. He goes on to argue that

> it is understandable—perhaps even justifiable—that the public and their concerned physicians should react and that this reaction should include requests and proposals to actively screen men for prostate cancer in the *hope* [my emphasis] that earlier detection might improve survival. If we (as physicians) feel it is best to screen men for prostate cancer today—and as physicians and human beings we must at times do what we ''feel'' is best when our science cannot yet provide every answer, then let us proceed as rationally as possible, based on those facts we do know.

The problem with the justification is that it is weak. After all, both opponents and proponents of PSA screening ''hope'' that, in the end, the risks of prostate cancer screening will be outweighed by the benefits. Put another way, ''hope'' is used in the quote as something more than an abstract feeling. It is used as an ethical justification for the writer's belief that waiting for clinical trial results is unnecessary and that it is ethically justified to ''do all that one can'' as a practicing physician. Even with such strong feelings, this same screening proponent does not recommend mass screening in public health programs. Although it is unclear how many other practicing physicians—primarily urologists and family practitioners—agree with a justification based on ''feeling,'' it is clear that PSA screening has broad support in the clinical practice community (89).

Critics of prostate cancer screening, on the other hand, believe that the risks may outweigh the benefits and therefore believe that it is inappropriate to expose individuals, whether in private medical practice or in the public at large, to (known) risks without known benefits (86).

The situation, therefore, is one in which the correctness of one's ethical position is either a matter of ''feeling,'' a matter of allegiance to the screening recommendations of a professional organization (e.g., the AUA or the ACS), or it is a matter of an absolute reliance upon clinical trial evidence, evidence that will not be available for years. Clearly, it is a matter of competing beliefs and values (90,91). And it is a matter of practical significance, that is, real-life decision-making, because the following ethical problem remains: what should be done for asymptomatic individuals who arrive at a physician's office and ask (or not) for a test to screen for prostate cancer?

Several choices are possible (92–95): proponents of screening may simply tell patients that they should be screened (whether they ask or not); opponents of screening may tell such patients that they should not be screened (if asked) and may ignore the issue when patients do not ask. Neither choice seems appropriate because inadequate attention is given to the moral rule that the truth should be told even when that ''truth'' implies a lack of consensus amongst scientists and physicians. Thus, several authors argue for an ''informed consent'' model of preventive medical practice: regardless of whether the patient asks or not, the physician should explain the uncertainties, the lack of consensus, and, in some detail, the possible risks and benefits of the procedures. Then, the patient should decide for himself (including or not the opinion of the physician) and should sign a consent document or provide verbal consent if he chooses to be screened. The ethical justification for this third choice stems primarily from its attention to several ethical rules found in the locus of moral consensus described in this chapter: primarily, respect for persons, truth-telling, and obtaining consent, with due consideration toward the uncertainty surrounding benefits and risks. Interestingly, in a recent randomized trial of the impact of an informed consent process on asymptomatic men 50 years of age or older seen at primary care practices, those receiving informed consent were significantly less interested in undergoing PSA screening (96).

Prostate cancer screening continues to be one of the most controversial problems facing contemporary medical science and preventive practice (97–105).

VI. ETHICAL ISSUES IN CANCER SCREENING PROGRAMS

General ethical issues relevant to population-based (mass) screening programs include: the balance of benefits and harms, information issues including advocacy, health promotion, health education and informed consent, and justice in resource allocation.

A. Benefits and Risks of Screening Programs

Accounts of the ethics of population screening programs often ask the question, ''When is a public mass screening program justified?'' A key component of an ethically acceptable answer to this question is that the screening test must be efficacious, that is, the test must have been demonstrated to have a clear benefit. Benefit is most often defined in terms of improved overall mortality rates in comparisons made between those offered screening and those not offered screening in well-designed and scrupulously performed randomized clinical trials. Considerable controversy may exist about the extent to which such evidence is available for specific screening tests in specific populations; mammography in women aged 40–49 is a current example as is the issue of prostate cancer screening by PSA discussed earlier. The importance of the evaluation of a screening test in a rigorous fashion and with a specific endpoint (i.e., death) cannot be underestimated, inasmuch as screening evaluated in nonrandomized studies can appear to provide a benefit in terms of increased disease-yields, early stage at diagnosis, or even improved survival among those screened. But these are not necessarily benefits to those screened. A screening test almost always advances the diagnosis of the cancer and therefore appears to improve survival when, in fact, the individual or group being screened may simply have had their date of diagnosis occur earlier than would have occurred once symptoms appeared.

In other words, sometimes a fundamental doctrine of preventive medicine is wrong: early intervention is not always the right thing to do (106). Put succinctly, screening is beguiling (9). Careful and rigorous demonstration of efficacy is required. Nevertheless, even when a clear benefit has been demonstrated in a randomized clinical trial, there remain risks that must be taken into account. These include the physical harms that may emerge from using any imperfect cancer screening test that will produce some (hopefully few) false-positive and false-negative results and the psychological harms that may also occur from participation in screening programs. In addition, there may be overdiagnosis and treatment of lesions that would never have caused harm if not detected.

As described previously, cancer screening tests have measurable rates of identifying false-positive and false-negative results. A highly sensitive test, for example, rarely misses finding a cancer, but does so at the risk of identifying individuals with positive tests but no cancer, that is, false-positive results. These individuals undergo further testing, often more invasive testing such as biopsies, and often experience psychological distress (17,107). A highly specific cancer screening test, on the other hand, rarely produces false-positive results, but does so at the risk of missing cancers. The harms include the false reassurance provided to those with cancer who would otherwise be benefitted by having their cancers found early and treated successfully (17,107).

B. Ethics and Information in Cancer Screening Programs

An important area in the ethics of cancer screening involves, at its center, decisions about the flow of information from the institutions that design, sponsor, and promote mass public screening programs to the populations (and the individuals who make up those populations) who will either choose or not choose to participate. The challenges arise even in the ''best case'' scenario in which randomized trial evidence exists demonstrating a clear mortality rate reduction in those populations screened. Although evidence of a beneficial effect may be considered by some to be necessary—that is, an ethical imperative—to undertake a screening program (108), there are several competing and sometimes conflicting concerns that arise when institutions undertake to promote (using various information and media systems) mass screening programs and when these same institutions provide information to those who choose to participate.

To illustrate the situation, the locus of moral consensus described earlier is used. At least four different rules are involved in the ethics of information flow for cancer screening programs:

1. Helping others. An effective screening test in an efficacious program will produce benefits and thus provide an opportunity for public health professionals to realize a central goal: to help populations (communities) by improving their health status through reduced cancer mortality rates.

2. Preventing harm to others. An effective screening test will almost surely cause some harm in the same population it will benefit. Indeed, harm may come to individuals who do not benefit at all. As described, there may be relatively minor harms associated with the physical characteristics of the test itself, and there are the physical, emotional, and economic harms associated with false-positive and false-negative results. Harm from therapy may be much more substantial.

3. Respect for persons. Potential participants in population-based mass screening programs should decide for themselves whether or not they wish to participate.

4. Truth-telling. Public health professionals should tell potential participants the truth about the screening program. There are, as described, many different and pertinent aspects of such a program about which the truth must be told: the scientific evidence supporting the claim that the test is efficacious, the projected benefits of the program, the known risks (harms), and attendant costs to the individual and to the community.

To show how these rules interconnect, it is relevant to point out what is not at stake. Screening programs do not mandate participation as do some other forms of public health interventions such as seat belt use, immunization of children, and reporting of communicable diseases. Thus, in cancer screening, there is no extreme threat to personal autonomy inasmuch as individuals (and communities) ideally have an opportunity to agree to participate or not based on the information provided. Nevertheless, there remains a potential infringement on autonomy if the information provided is inaccurate, misleading, or incomplete regarding known benefits, known risks, and their attendant uncertainties. At the heart of the balancing of these rules, therefore, is the problem of *how much* information is provided, at what point in the process it appears, and its accuracy.

VII. EXAMPLE: PROMOTING BREAST CANCER SCREENING

As an example, consider a hypothetical public health poster attached to the wall of a subway station. The poster shows a well-dressed woman, well into middle age. Splashed across the surface is a simple message: "Mammograms save lives; Get one today!"

The ethical concerns can be succinctly framed in the following question: Is the poster a coercive message produced by a paternalistic public health institution, or is it an effort on the part of a beneficent public (or private) institution to briefly inform the public about an important opportunity for them to obtain more information about (and then choose to participate in) a program that may reduce their chance of dying from a cancer and thereby improve their health? As in most real-life situations, the answer is neither black nor white but lies somewhere in between.

Even though the basic message of the poster is true, there is more truth to tell. But is the wall of a public building the best place to inform potential screening program participants that mammograms in women older than 50 years of age reduce mortality rate, that all screening tests have a false-positive rate, and that false-positive results mean that unnecessary additional testing will be necessary? Nor does the poster explain the notion of false-negative rates and the fact that, if the mammogram identifies a suspicious lesion, then further tests, biopsies, and treatment (and time off from work, copayments to insurance companies, and money for child care) may be required. In sum, the poster (as well as 30-second television public interest spots and many magazine advertisements) provides a brief message designed primarily to convince the reader (or viewer) that she should *do* something, namely, have a mammogram. Implicit in the message—or so we hope—is the idea that before undergoing the test, the participant will first be informed about the nature of the test and the implications of participating.

Concerns about indirect pressure to participate in screening programs (8) and about efforts to "hard sell" public health products (14,15) have been expressed. O'Hagan (10), for example, notes that there is a subtle coercion involved when a public health message says that mammography reduces a woman's chance of dying of breast cancer by 30% inasmuch as the mammogram either improves her survival or it does not. Furthermore, as the paradox of prevention dictates, many women do not benefit at all from their participation beyond being informed that they did not test positively. The cynic could argue that these instances represent situations in which the women would have been better off to have stayed home.

All public health interventions involve some effort to convince the public to change their behavior and these persuasive techniques are likely grounded in a beneficent paternalism. While such a statement may sound insensitive or politically incorrect, the truth is that someone other than the person reading the poster has decided that the screening intervention is beneficial. Indeed, it is generally agreed that all admonitions to participate in a public intervention program should be based on solid evidence of efficacy (17) and that all persons considering screening in a public health program should be given the information about the sensitivity and specificity of the tests, the consequences of positive and negative findings (16) so that they can make a reasoned choice about participation. In practice, however, there are some situations in which it is not so clear that the evidence can be interpreted as demonstrating efficacy, and there are many constraints on our ability to fully inform the public. These constraints include the fact that potential screening program participants come from many different ethnic backgrounds with different educational and socioeconomic experiences. The definitions, mathematics, and individual implications of false-positive rates, false-negative rates, risks of dying, and other often used medical terms may seem straightforward to the public health practitioner or to the college-educated professional business woman, but to the 55-year-old mother with an 8th grade education received in a developing country, it may seem to be a complex source of confusion. Nevertheless, despite the difficulties imposed on public health institutions by a public with diverse educational and cultural backgrounds and therefore diverse capacities to understand relatively complex technical knowledge, the ethics of the situation remain clear: all individuals should understand what they are agreeing to do, what *will* happen to them, and what *may* happen to them or not, as a result of their participation. It is possible, in other words, to balance the four moral rules involved when information flows from screening programs to potential screening participants. It follows that such a balance may be more easily achieved in those situations in which public input to developing the public health messages is an integral part of the process. How this balance is achieved may be different in different populations (109).

VIII. JUSTICE IN THE ALLOCATION OF SCREENING PROGRAMS

In the locus of moral consensus described previously, a rule to justly allocate resources was included. For screening programs as part of public health initiatives, this rule implies that all populations, regardless of cultural identity, ethnicity, or socioeconomic status, should be given the opportunity to receive screening tests judged to be effective and which reduce mortality rate without unacceptable side effects. How a just allocation of resources can be achieved, and, indeed, whether it has been achieved, requires attention to specific locations where screening programs have been made available to populations, that is, countries or other geopolitical designations. In addition, some attention must be paid to the ways in which resources across all public health practice categories are allocated. In some countries, for example, general health care systems that essentially guarantee some level of health care for all citizens. In other countries, such as the United States, health care is provided both publicly and privately. It is not necessarily true that one system is more or less likely to be unjust in resource allocation.

IX. ISSUES OF DISCRIMINATION AND CONFIDENTIALITY

Some of the most interesting issues in contemporary medical ethics involve the problems associated with the collection of biological information on cancer predisposition. Genes and other biomarkers are increasingly available as a direct result of the rapidly advancing technologies associated with molecular biology and genetics. Currently, such measurements are not used as screening tests per se. Thus, it is not necessary to discuss the ethics of biomarkers and predisposition genes in a chapter on ethics and consent in cancer screening. On the other hand, the possibility of using genetic markers as screening tools has been discussed. The confidentiality of this information and the potential for discrimination (for health insurance, health care, and employment) are at the forefront of the concerns.

X. CONCLUSION

This chapter has discussed several central ethical issues that inhabit the everyday world of cancer screening researchers and those who practice screening in their medical practices or in community programs. The ways in which ethical reasoning can be used and the importance of justifying an ethical position with sound reasoning have been emphasized. Issues of benefits, risks, and consent are key concerns in cancer screening.

There are many other issues and perspectives that could have been discussed but were given either brief mention or no mention at all. Ethical considerations in different randomization schemes for clinical trials were not described. Many contemporary controversies were not discussed, including mammography in women younger than 50 years of age and colorectal cancer screening. Largely ignored are issues of professional codes of ethics and conflicts of interest. Conflicts of interest, however, are clearly possible inasmuch as practicing physicians may benefit financially from the extensive workups and therapeutic interventions that may be necessary after screening test results are positive. Screening in children and in incompetent adults was not addressed. Nor did I discuss in much detail the ways in which institutional review boards and other ethics committees deliberate over research proposals and community intervention programs.

Nevertheless, in all these various situations, the norms (e.g., rules, theories, methodologies) of contemporary bioethics, indeed of contemporary Western culture, are useful in coming to justifiable decisions about specific cases. There remain the important concerns about the extent to which researchers and medical practitioners have learned these basic tenets of contemporary bioethics and the extent to which some of these norms are culturally influenced (110). I did not examine how ethics can be taught (or learned) in educational institutions, training programs, and in everyday practice. Nor did I discuss current trends in the professional philosophical literature beyond a brief introduction to various types of ethical theories and methodologies, some of which I did not discuss with the kind of attention they deserve. Indeed, there are many interesting problems (and controversies) involving not only ethics but also its relationship to contemporary philosophy of science (111). These I leave for another day.

REFERENCES

1. Beauchamp TL. Moral foundations. In: Coughlin SS, Beauchamp TL, eds. Ethics and Epidemiology. New York: Oxford University Press, 1996:24–52.
2. Nijhuis HGJ, Van der Maesen LJG. The philosophical foundations of public health: an invitation to debate (editorial). J Epidemiol Commun Health 1994; 48:1–3.
3. Peach L. An introduction to ethical theory. In: Penslar RL, ed. Research ethics: cases and materials. Bloomington, IN: Indiana University Press, 1995:13–26.
4. Beauchamp TL, Childress JF. Principles of biomedical ethics. 4th ed. New York: Oxford University Press, 1994.
5. Pellegrino ED, Thomasma DC. The virtues in medical practice. New York: Oxford University Press, 1993.
6. Jonsen AR, Toulmin S. The abuse of casuistry: a history of moral reasoning. Berkeley; CA: University of California, 1988.
7. Weed DL. Epistemology and ethics in epidemiology. In: Coughlin SS, Beauchamp TL, eds. Ethics in epidemiology. New York: Oxford University Press, 1996:76–94.

8. Riis P. Mass screening procedures and programmes. In: Doxiadis S, ed. Ethical Dilemmas in Health Promotion. New York: John Wiley and Sons, 1987: 171–182.
9. Mant D, Fowler G. Mass screening: theory and ethics. BMJ 1990; 300:916–918.
10. O'Hagan J. The ethics of informed consent in relation to prevention screening programmes. N Z Med J 1991; 104:121–123.
11. Edwards PJ, Hall DMB. Screening, ethics, and the law. BMJ 1992; 305:267–268.
12. Holland WW. Screening: reasons to be cautious. BMJ 1993; 306:1222–1223.
13. Blaney R. Why prevent disease? In: Doxiadis S, ed. Ethical dilemmas in health promotion. New York: John Wiley and Sons, 1987:47–56.
14. Skrabanek P. Why is preventive medicine exempted from ethical constraints? J Med Ethics 1990; 16:187–190.
15. Gillon R. Ethics in health promotion and prevention of disease. J Med Ethics 1990; 16:171–172.
16. Horner JS. Medical ethics and the public health. Public Health 1992; 106:185–192.
17. Cole P. The moral bases for public health interventions. Epidemiology 1994; 6(1): 78–83.
18. Last JM. Ethics and public health policy. In: Last JM, Wallace RB, eds. Public Health and Preventive Medicine. Norwalk, CT: 1992:1187–1196.
19. Beauchamp DE. Philosophy of public health. In: Reich WT, ed. Encyclopedia of Bioethics. New York: Simon & Schuster MacMillan 1995; 4:2161-2166.
20. Reich WT, ed. Encyclopedia of Bioethics. New York: Simon & Schuster MacMillan, 1995.
21. deLacey G, Record C, Wade J. How accurate are quotations and references in medical journals? BMJ 1985; 291:884–886.
22. Sekas G, Hutson WR. Misrepresentation of academic accomplishments by applicants for gastroenterology fellowships. Ann Int Med 1995; 123:38–41.
23. Gurudevan SV, Mower WR. Misrepresentation of research publications among emergency medicine residency applicants. Ann Emerg Med 1996; 27:327–330.
24. Anderson C. Survey tracks misconduct, to an extent. Science 1993; 262:1203–1204.
25. Shapiro MF, Charrow RP. Scientific misconduct in investigational drug trials. N Engl J Med 1985; 312:731–736.
26. Shapiro MF, Charrow RP. The role of data audits in detecting scientific misconduct: results of the FDA program. JAMA 1989; 261:2505–2511.
27. Dealing with deception (editorial). Lancet 1996; 347:843.
28. Gunsalus CK. Institutional structure to ensure research integrity. Acad Med 1993; 68:S33–S38.
29. Schachman HK. What is misconduct in science? Science 1993; 261:148–149.
30. Buzzelli DE. The definition of misconduct in science: a view from NSF. Science 1993; 259:584–585, 647–648.
31. Klein DF. Should the government assure scientific integrity? Acad Med 1993; 68: S56–S59.
32. Rennie D, Gunsalus CK. Scientific misconduct: new definition, procedures and office—perhaps a new leaf. JAMA 1993; 269:915–917.
33. US Department of Health and Human Services. Integrity and misconduct in re-

search: report of the Commission on Research Integrity [PHS 1996-746-425]. Washington, DC: US Government Printing Office, 1995.
34. Ryan KJ. Commission on Research Integrity report sparks debate on science and ethics. Professional Ethics Report 1996; 9:1, 6–7.
35. Kaiser J. Scientific misconduct: panels look for common ground. Science 1996; 272:476.
36. Bailar JC. The real threats to the integrity of science. Chronicle of Higher Education 1995; April 21:B1.
37. Chalmers I. Underreporting research is scientific misconduct. JAMA 1990; 263: 1405–1408.
38. Huth EJ. Irresponsible authorship and wasteful publication. Ann Intern Med 1986; 104:257–259.
39. Woolf P. Fraud in science: how much, how serious? Hastings Cent Rep 1981; 11:9–14.
40. Fraud and Misconduct in Medical Research: Causes, Investigation, and Prevention. A report of the Royal College of Physicians. London: Royal College of Physicians, 1991.
41. Kassirer JP. The frustrations of scientific misconduct. N Engl J Med 1993; 328: 1634–1636.
42. Dingell JD. Shattuck lecture: misconduct in medical research. N Engl J Med 1993; 328:1610–1615.
43. Lock S. Misconduct in medical research: does it exist in Britain? BMJ 1988; 297: 1531–1535.
44. Angell M. Publish or perish: a proposal. Ann Intern Med 1986; 261–262.
45. Woolf PK. Pressure to publish and fraud in science. Ann Intern Med 1986; 104: 254–256.
46. Andersen D, Attrup L, Axelsen N, Riis P. Scientific dishonesty and good scientific practice. Copenhagen: Danish Medical Research Council, 1992.
47. Danforth WH, Schoenhoff DM. Fostering integrity in scientific research. Acad Med 1992; 67:351–356.
48. Petersdorf RG. The pathogenesis of fraud in medical science. Ann Intern Med 1986; 104:252–254.
49. Alberts B, Shine K. Scientists and the integrity of research. Science 1994; 266: 1660–1661.
50. Weed DL, McKeown RE. Epidemiology and virtue ethics. Int J Epidemiol 1998; 27:343–348.
51. Weed DL. Preventing scientific misconduct. Am J Public Health. 1998; 88:125–129.
52. Pellegrino ED. Character and the ethical conduct of research. Accountability in Research 1992; 2:1–11.
53. Beauchamp TL, Faden RR. History of Informed Consent. In: Reich WT, ed. Encyclopedia of Bioethics. New York: Simon & Schuster MacMillan, 1995; 3:1232–1238.
54. Levine RJ III. Consent issues in human research.. In: Reich WT, ed. Encyclopedia of Bioethics. New York: Simon & Schuster MacMillan, 1995; 3:1241–1250.

55. Beauchamp TL, Faden RR II. Meaning and elements of informed consent. In: Reich WT, ed. Encyclopedia of Bioethics. New York: Simon & Schuster MacMillan, 1995: 3:1238–1241.
56. Protection of Human Subjects. OPRR Reports. (NIH, PHS, HHS) Title 45 Code of Fed Reps.
57. Beecher HK. Ethics and clinical research. N Engl J Med 1966; 274:1354–1360.
58. Rothman DJ. Human research: historical aspects. In: Reich WT, ed. Encyclopedia of Bioethics. New York: Simon & Schuster MacMillan 1995; 4:2248–2258.
59. Freedman DJ. Unethical research. In: Reich WT, ed. Encyclopedia of Bioethics. New York: Simon & Schuster MacMillan. 1995: 4:2258–2261.
60. Rothman DJ. Ethics and human experimentation: Henry Beecher revisited. N Engl J Med 1987; 317:1195–1199.
61. Adami HO, Baron JA, Rothman KJ. Ethics of a prostate cancer screening trial. Lancet 1994; 343:958–960.
62. Morrison AS. Screening in chronic disease. New York:Oxford, 1992.
63. Wilson JMG, Jungner G. Principles and practice of screening for disease. Geneva: WHO, 1968.
64. Wald N. Ethical issues in randomized prevention trials. BMJ 1993; 306:563–565.
65. Cochrane AL, Holland WW. Validity of screening procedures. Br Med Bull 1971; 27:3–8.
66. Freedman B. Equipoise and the ethics of clinical research. N Engl J Med 1987; 317:141–145.
67. Lilford RJ, Jackson J. Equipoise and the ethics of randomization. J R Soc Med 1995; 88:552–559.
68. Kopelman LM. Research methodology: controlled clinical trials. In: The Encyclopedia of Bioethics. Reich WT, ed. 1995; 4:2278–2285.
69. Lantos J. Ethics, randomization and technology assessment. Cancer 1994; 74(suppl):2653–2656.
70. Alderson P. Equipoise as a means of managing uncertainty: personal, communal and proxy. J Med Ethics 1996; 22:135–139.
71. Gohagan JK, Prorok PC, Kramer BS, Cornett JE. Prostate cancer screening in the Prostate, Lung, Colorectal and Ovarian Cancer Screening Trial of the National Cancer Institute. J Urol 1994; 152:1905–1909.
72. Schroder FH, Bangma CH. The European Randomized Study of Screening for Prostate Cancer (ERSPC). Br J Urol 1997; 79(suppl 1):68–71.
73. Wilt TJ, Brawer MK. The prostate cancer intervention versus observation trial: a randomized trial comparing radical prostatectomy versus expectant management for the treatment of clinically localized prostate cancer. J Urol 1994; 152:1910–1914.
74. Schroder FH. Screening for prostate cancer (letter). Lancet 1994; 343:1438–1439.
75. Kramer BS, Brown ML, Prorok PC, Potosky AL, Gohagan JK. Prostate cancer screening: what we know and what we need to know. Ann Intern Med 1993; 119: 914–923.
76. Lange PH. The next era for prostate cancer: controlled clinical trials. JAMA 1993; 269:95–96.

77. Garnick MB. Prostate cancer: screening diagnosis, and management. Ann Intern Med 1993; 118:804–818.
78. Mettlin C, Jones G, Averette H, Gusberg SB, Murphy GP. Defining and updating the American Cancer Society guidelines for the cancer-related checkup: prostate and endometrial cancers. CA Cancer J Clinicians 1993; 43:42–46.
79. U.S. Preventive Services Task Force. Screening for prostate cancer: commentary on the recommendations of the Canadian Task Force for the Periodic Health Examination. Am J Prev Med 1994; 10:187–193.
80. Schroder FH. Prostate cancer: to screen or not to screen? BMJ 1993; 306:407–408.
81. Catalona WJ. Screening for prostate cancer: enthusiasm. Urology 1993; 42:113–115.
82. Voss JD. Prostate cancer: screening and prostate-specific antigen: promise or peril? J Gen Intern Med 1994; 9:468–474.
83. Middleton RG. Prostate cancer: are we screening and treating too much? Ann Intern Med 1997; 126:465–467.
84. Chodak GW. Questioning the value of screening for prostate cancer in asymptomatic men. Urology 1993; 42:116–118.
85. Sox HC. Preventive health services in adults. N Engl J Med 1994; 330:1589–1595.
86. Woolf SH. Screening for prostate cancer with prostate-specific antigen. N Engl J Med 1995; 333:1401–1405.
87. Cher ML, Carroll PR. Screening for prostate cancer. West J Med 1995; 162:235–242.
88. Kabalin JN. Screening for prostate cancer today. West J Med 1995; 162:272–273.
89. Thompson IM, Zeidman EJ. Current urological practice: routine urological examination and early detection of carcinoma of the prostate. J Urol 1992; 148:326–330.
90. Aronowitz R. To screen or not to screen: what is the question? J Gen Intern Med 1995; 10:295–297.
91. Wolf AMD, Becker DM. Cancer screening and informed patient discussions. Arch Intern Med 1996; 156:1069–1072.
92. Lee JM. Screening and informed consent. N Engl J Med 1993; 328:438–440.
93. Marshall KG. Screening for prostate cancer. Can Fam Physician 1993; 39:2385–2390.
94. Hahn DL, Roberts RG. PSA screening for asymptomatic prostate cancer: truth in advertising. J Fam Practice 1993; 37:432–434.
95. Brett AS. The mammography and prostate-specific antigen controversies: implications for patient-physician encounters and public policy. J Gen Intern Med 1995; 10:266–270.
96. Wolf AMD, Nasser JF, Wolf A, Schorling JB. The impact of informed consent on patient interest in prostate-specific antigen screening. Arch Intern Med 1996; 156: 1333–1336.
97. Morris K. UK experts advise against prostate cancer screening. Lancet 1997; 349: 477.
98. Anonymous. The prostate question, unanswered still (editorial). Lancet 1997; 349: 443.
99. Coley CM, Barry MJ, Fleming C, Fahs MC, Mulley AG. Clinical Guidelines: Part

II. Early detection of prostate cancer. Estimating the risks, benefits, and costs. Ann Intern Med 1997; 126:468–479.
100. American College of Physicians. Clinical Guideline: Part III. Screening for prostate cancer. Ann Intern Med 1997; 126:480–484.
101. Woolf SH. Should we screen for prostate cancer? (editorial). BMJ 1997; 314:989–990.
102. Raghavan D. Prostate cancer, populations, politicians and power (editorial). Eur J CA 1997; 33:329–330.
103. Kramer BS, Gohagan JK, Prorok PC. Is screening for prostate cancer the current gold standard?—''No.'' Eur J Ca 1997; 33:348–353.
104. Lange PH. Is screening for prostate cancer the current gold standard?—''Yes.'' Eur J Ca 1997; 33:354–356.
105. Mettlin C. Public health impact of prostate cancer early detection (editorial). Prostate 1997; 31:71–73.
106. Welch HG. Questions about the value of early intervention. N Engl J Med 1996; 334:1472–1473.
107. Marteau TM. Screening in practice: reducing the psychological costs. BMJ 1990; 301:26–28.
108. Wald N, Law M. Screening, ethics, and the law. BMJ 1992; 305:892.
109. Pasick RJ, D'Onofrio CN, Hiatt RA, eds. Promoting cancer screening in ethnically diverse and underserved communities. Health Educ Q 1996; 23(suppl).
110. Glick SM. Unlimited human autonomy—a cultural bias? N Engl J Med 1997; 336:954–956.
111. Veatch RM, ed. Bioethics and contemporary philosophy of science. J Med Philos 1995; 20.

Appendix

Informed Consent Form (Prototype) for the Prostate, Lung, Colorectal, and Ovarian Screening Trial

(2) PROTOCOL

Precis

The Division of Cancer Prevention and Control (DCPC), under extramural contracts to 10 U.S. clinical centers, is evaluating the effectiveness of screening for prostate. lung, colorectal and ovarian cancer (The PLCO Trial). The NCI Executive Committee approved the expansion of the PLCO Trial to collect additional materials and to conduct additional studies. About, 74,000 men and 74,000 women, aged 55–74 years, are being randomized on a 50/50 basis into screening or usual care arms. Additional blood and urine will be collected from screened subjects and saliva from control subjects. Pathologic tissues will be obtained for cases that develop cancer or selected related diseases (e.g. colon polyps, benign prostatic hyperplasia). Additional questionnaire-based risk information will also collected. Genetic, biochemical and questionnaire-based risk information will be related to the development of cancer and other diseases in this population. Volunteers who provide samples for these studies will not receive their individual results from the Additional Investigations. This protocol Review Application is for the collaboration in the Additional Studies of intramural scientists in the Division of Cancer Epidemiology and Genetics (DCEG).

Introduction

The Division of Cancer Prevention and Control (DCPC), under extramural contracts to 10 U.S. clinical centers, is evaluating the effectiveness of screening for prostate, lung, colorectal and ovarian cancer (The PLCO Trial). The Trial includes a biorepository (blood samples from screened subjects) and accompanying data collections on risk factors. Informed consent documents approved by the respective Screening Center IRBs and a summary of their current review status are attached.

In December, 1996 the NCI Executive Committee approved a collaborative proposal from the Division of Cancer Epidemiology and Genetics and DCPC to substantially expand questionnaire and biologic sample collections from PLCO Trial screened and non-screened study subjects. The PLCO Trial is now seen as a key resource for early marker and etiologic studies of cancer in which NCI intramural investigators will play a major role. Because these investigations will involve genetic studies, we feel that all present and future study subjects should be asked to provide an informed consent which includes permission for genetic testing. Also, data security must be assured. Thus, we present for your review the informed consent documents for Additional Studies and a plan for data security. It is intended that all current participants and new enrollees will receive an Informed Consent document for Additional Studies as a supplement to the routine PLCO Trial Informed Consent.

In the PLCO Trial, 74,000 men and 74,000 women, aged 55–74 years, will be randomized on a 50/50 basis into screening or usual care arms. The study goal is to include approximately equal numbers of men and women and special efforts are made to enhance minority representation. Screening centers with minority populations are encouraged and additional funds are used for developing approaches to minority recruitment. At present, about 45% of the study group are women and about 15% are members of minority ethnic/racial groups.

Objectives

The overall objective of the additional biologic and risk factor data collections in the PLCO Trial is to provide a resource for early marker and etiologic studies of cancer and related diseases. Initial objectives are to characterize early markers, and genetic and environmental risk factors for benign and malignant conditions of the prostate, colorectum, lung and breast. As the study progresses, investigations of less frequently occurring tumors and other diseases which occur in this age group will be carried out. Specific objectives of the procedures being proposed are to:

a. provide an opportunity for PLCO Trial participants to voluntarily participate in additional studies of the development and carses of cancer and other diseases.
b. establish a procedure to maintain the secrity and confidentiality of the collected data.

Study Design and Methods

PLCO participants are recruited at 10 U.S. clinical centers for the PLCO Trial, by a variety of methods including direct mailings, clinic population recruitment, solicitations to various organizations, and by general advertising. Men and women, age 55–74 who do not have a prior history of PLCO tumors and have not undergone routine screening are eligible. Subjects complete an informed consent document and are randomized to the screening or control arms. Screened subjects follow a 3-year protocol (Initial plus 3 yearly screens) for the early detection of prostate, lung, colorectal, and ovarian cancer and are followed for incidence of tumors and all causes mortality for at least 10 more years. Blood samples are drawn from screening arm participants at each of the four screening visits (PSA testing for men, CA-125 testing for women). Women without ovaries do not receive CA-125 testing, but may still choose to provide blood samples for this additional medical research. The expected number of cancer cases in the PLCO Trial are shown in Appendix B.

Current Collections

The current collections for etiology and early marker studies are:

For screened and control subjects

Risk factor questionnaire at entry (T0)

For screened subjects:

Blood collections

45 ml (aliquoted to serum. plasma buffy coat. rbc) at the entry (T0) screening visit

10 ml (serum) at Year 1 (second) screening visit (T1)

10 ml (serum) at Year 2 (third) screening visit (T2)

Dietary questionnaire at entry (T0) screening visit

Proposed Collections

Based upon the 1996 Executive Committee review, the proposed expansion of collections for etiology and early marker studies are:

For screened and control subjects:

Supplemental risk factor questionnaires every 3–4 years during follow-up.

Pathology samples, for selected cancer cases (slides prepared from paraffin blocks).

For screened subjects:

50–100 ml urine collection aliquoted as a cellular pellet and as supernatant

30 ml blood (aliquoted to plasma, buffy coat, rbc), T3 (final) screening visit
For control subjects
Self-administered buccal cell collection, once
Dietary questionnaire, once

Monitoring and Criteria for Withdrawal of Subjects from Study

This Protocol Review Application is for an observational component within a randomized control trial. PLCO Trial participants may volunteer for these Additional Studies and may withdraw from them at any time, without impact on their status in the Trial.

Analysis of Study

The biologic samples and questionnaire data collected in the PLCO Trial provide a major resource for investigations into cancer etiology and prevention. Over the course of the PLCO Trial, proposals for early marker and etiologic investigations will be reviewed by a scientific oversight committee for relevance and efficient use of biorepository resources. The studies on biologic samples will include determination of biochemical values (e.g. protein markers of early disease, hormones, serum nutrients) and the characterization of genetic traits. NCI staff with expertise in genetic, nutritional, early-detection/screening and disease-specific epidemiology will be closely involved in intramural and collaborative extramural investigations. When specific uses of the biorepostory materials are proposed, appropriate amendments to the protocol will be made and NCI IRB approval will be obtained before testing is carried out.

Human Subject Protections

Informed Consent Documents

Because the original consent did not specifically address genetic studies, PLCO participants already enrolled in the PLCO Trial will be reconsented for collection and use of biorepository specimens explicitly for these types of studies. New enrollees into the PLCO Trial will receive two informed consent documents: the original document (Screening Center Informed Consents for the PLCO Trial) and a document concerning additional medical research. The Informed Consent Documents for Additional Medical Research are:

Document A: for new subjects entering the Trial
Document B: for screening arm subjects in active screening (language about control arm procedures removed)

Document C: for screening arm subjects who have already completed the T3 final visit.
(language about control arm and about urine collection removed)
Document D: for subjects already in the control arm
(language about screening arm procedures removed)

Voluntary Participation

PLCO Trial participants may volunteer for these Additional Studies. They may continue in the PLCO Trial for the evaluation of early detection methods regardless of their consent decision on genetic studies.

Benefits

No direct benefits to the participants are expected from the early marker and etiologic components of this study and the subjects receive no compensation for their participation. They will not receive notice that specific assays are being carried out nor will they receive any individual results from this additional research.

Harm

Minimal physical harm is expected from the drawing of blood. No physical harm is expected from the collection of a urine sample from screening arm participants. Buccal cells are collected by the participants using a small cotton swab or cytobrush to brush the tongue and buccal mucosa followed by expectoration into a collecting device and placement of the sample onto a storage card. No physical harm is expected from this procedure. Should adverse events occur from the blood draw, clinic staff are trained and prepared to handle such situations. These adverse events are reported to the NCI through the study data management system and are reviewed on a regular basis.

Confidentiality

To ensure confidentiality of the data, the following procedures will be adhered to:

a) data including personal identifiers are kept securely at the 10 Screening Centers,
b) only data and biologic materials without personal identifiers are sent to the NCI, using an identifier code.
c) the NCI and its scientific collaborators carry out analyses only on materials that have no personal identifiers and do not link the results of these tests to personal identifier information held at the screening center,
d) no individual results will be presented in publications or other reports, and

e) it will be possible to link test results with past and future test results, through the individual code identifier.

We plan to maintain the confidentiality of the results, to the full extent permitted by law (A DHHS Certificate of Confidentiality will be applied for if the NCI IRB makes such a recommendation). We plan to inform participants, on a yearly basis, of important scientific results from these studies which could potentially impact upon their health, but this reporting will be on a group basis only, without information on individual results.

PROTOTYPE DOCUMENT A:FULL INFORMED CONSENT

Prostate, Lung, Colorectal and Ovarian Cancer Screening Trial

RESEARCH ON THE DEVELOPMENT AND CAUSES OF CANCER AND OTHER DISEASES
CONSENT FOR PARTICIPATION
NAME OF CLINICAL CENTER

DESCRIPTION OF ADDITIONAL STUDIES ON THE DEVELOPMENT AND CAUSES OF CANCER AND OTHER DISEASES

The purpose of the Prostate, Lung, Colorectal and Ovarian (PLCO) Cancer Screening Trial is to determine if certain screening tests can lead to early diagnosis of prostate, lung, colorectal and ovarian cancer and so reduce deaths caused by these cancers.

In addition to evaluating screening tests, the PLCO Trial also seeks to study factors which may cause these cancers to develop and progress. Additional research on cancer and other diseases which occur in your age group will be carried out among PLCO Trial participants who volunteer for these additional studies. **We are requesting your consent to participate in these additional studies of cancer and other diseases that occur in your age group**.

ADDITIONAL STUDY PROCEDURES

If you are in the PLCO Trial screened group and you volunteer for these additional studies, blood not used for the prostate cancer screening test (PSA) or the ovarian cancer screening test (CA-125) test will be stored and used in future medical research. (If you are a women without ovaries and will not receive the CA-125 blood screening test, you may still choose to participate in this additional medical research by signing this document.) As a participant in this additional medical research, you will also provide a urine sample during one of your screening visits to be stored for future medical research. These future studies will include investigations to determine if genetic (inherited) factors and chemicals in blood and urine, such as dietary nutrients and hormones, are related to the risk of developing cancer and other diseases that occur in your age group.

If you are in the PLCO Trial control group and you volunteer for these additional studies, you will provide a sample of saliva which will be stored and used in future medical research. Such studies will include investigations to determine if genetic (inherited) factors are related to the risk of developing cancer and other diseases that occur in your age group. Briefly, the collection of saliva requires that, at home, you brush the inside of your mouth with a soft bristle brush and then spit the saliva into a small cup. You will be provided a kit and instructions for the collection and pre-paid return mailing of this sample to the researchers.

If you volunteer for these additional studies and you later have surgery for diagnosis or treatment of cancer or a related condition, the study investigators may then ask your doctors and the hospitals where you were treated to provide a small sample of the surgical material. This material may be used by the study investigators only for medical research about genetic (inherited) factors and chemical changes that lead to the development of cancer and other diseases that occur in your age group.

These materials will be stored at a National Cancer Institute research storage facility for up to 25 years and used to help scientists learn what causes cancer and how to prevent its progression. It is believed that cancer may be caused by both environmental and genetic factors. Therefore, the samples which you contribute may be used in biochemical and genetic studies to identify these causes.

BENEFITS OF ADDITIONAL STUDIES

The additional studies will not provide direct benefit to you other than the satisfaction of participating in this research for the possible benefit of future generations. However, your participation in these additional studies will help answer questions related to the health and longevity of persons in your age group and will help establish a scientific understanding of the factors which influence the development and progression of cancer.

RISKS OF ADDITIONAL STUDIES

There are certain risks and discomforts that might be associated with the additional procedures.

- When blood is drawn, there may be local bruising or bleeding at the puncture site. There may also be uneasiness associated with needles.
- When saliva is collected, (by brushing sites in the mouth with a small soft brush) there may be a feeling of pressure on the inside of the cheeks. This may include a small amount of bleeding.

NOTIFICATION AND COSTS

The additional biologic samples are for medical research only and the research results are not suitable for use as clinical tests for your medical care. The scientific studies require only looking at all lab results together. Therefore, the results of these <u>additional studies</u> will not be available to you.

The samples of blood, urine or saliva will only be used for research and will not be sold. If any new scientific or medical testing products are developed using the results of the research done on your samples, you will not be paid, even if these medical products are sold.

COMPENSATION FOR RESEARCH-RELATED INJURIES

If you are in the screened group: In the unlikely event of physical injury from drawing a blood sample, you will be provided with immediate medical treatment. No payment of medical treatment is available from the National Cancer Institute for any such injury.

If you are in the control group: No physical injury is likely from the collection of a saliva sample. No payment of medical treatment is available from the National Cancer Institute for any such injury.

INFORMATION ON NEW FINDINGS

As a participant in the PLCO Trial, you will be contributing to an important scientific investigation and will be notified about important scientific findings of these studies. This notification will be of results for all participants together. No individual results will be provided from these additional studies.

CONFIDENTIALITY

The information concerning your participation in the study will be kept confidential and used only for scientific purposes, in accordance with applicable state and federal laws.

As genetic research may discover genetic information about you, this information will be kept confidential to the full extent permitted by law. Your blood, urine, or saliva samples will not be labeled with your name and no one who works with these samples will have access to that information. Your name and any identifying information will not be used in any reports.

RIGHT TO WITHDRAW

Your participation in the additional medical research is voluntary and you may refuse to participate and/or withdraw your consent and discontinue participation at any time without penalty or loss of benefits to which you are otherwise entitled.

You may participate in the screening part of the PLCO Trial and yet decline to have biologic samples (blood and urine, or saliva) stored for research purposes. Further, if you initially decide to have your biologic samples stored for research purposes ,but later change your mind by written notification of (principal investigator) at (the screening center), whatever remains of your biologic samples will then be destroyed. Your decision will not affect your care.

CERTIFICATION

I have read this form or it has been read to me and I understand its contents. Any questions concerning the research or the rights of the participants involved have been and will be answered by NAMES, TITLES, PHONE NUMBERS. A copy of this consent form has been given to me. My signature(s) below means that I freely agree to participate in this additional medical research. **My consent to participate in the screening part of the PLCO Trial has been provided in a separate document**

By signing this consent, I agree to have biologic samples (some of my blood, and urine or saliva , and possibly surgical tissue) stored for **future research on cancer**. I understand that it may involve genetic and biochemical studies.

PARTICIPANT'S NAME (PRINT) DATE

PARTICIPANT'S SIGNATURE WITNESS' SIGNATURE

By signing this consent, I agree to have biologic samples (some of my blood, and urine or saliva , and possibly surgical tissue) stored for future research **other diseases that effect my age group**. I understand that it may involve genetic and biochemical studies.

PARTICIPANT'S NAME (PRINT) DATE

PARTICIPANT'S SIGNATURE WITNESS' SIGNATURE

If you have questions about this document or about the research and your right's as a participant in this research, please call:

Screening Center Representative
Telephone number

PROTOTYPE DOCUMENT B:INFORMED CONSENT FOR ACTIVE SCREENING ARM PARTICIPANTS

Prostate, Lung, Colorectal and Ovarian Cancer Screening Trial

RESEARCH ON THE DEVELOPMENT AND CAUSES OF CANCER AND OTHER DISEASES
CONSENT FOR PARTICIPATION
NAME OF CLINICAL CENTER

DESCRIPTION OF ADDITIONAL STUDIES ON THE DEVELOPMENT AND CAUSES OF CANCER AND OTHER DISEASES

The purpose of the Prostate, Lung, Colorectal and Ovarian (PLCO) Cancer Screening Trial is to determine if certain screening tests can lead to early diagnosis of prostate, lung, colorectal and ovarian cancer and so reduce deaths caused by these cancers.

In addition to evaluating screening tests, the PLCO Trial also seeks to study factors which may cause these cancers to develop and progress. Additional research on cancer and other diseases which occur in your age group will be carried out among PLCO Trial participants who volunteer for these additional studies. **We are requesting your consent to participate in these additional studies of cancer and other diseases that occur in your age group**.

ADDITIONAL STUDY PROCEDURES

If you volunteer for these additional studies, blood not used for the prostate cancer screening test (PSA) or the ovarian cancer screening test (CA-125) test will be stored and used in future medical research. (If you are a women without ovaries and will not receive the CA-125 blood screening test, you may still choose to participate in this additional medical research by signing this document.) As a participant in this additional medical research, you will also provide a urine sample during one of your screening visits to be stored for future medical research. These future studies will include investigations to determine if genetic (inherited) factors and chemicals in blood and urine, such as dietary nutrients and hormones, are related to the risk of developing cancer and other diseases that occur in your age group.

If you volunteer for these additional studies and you later have surgery for diagnosis or treatment of cancer or a related condition, the study investigators may then ask your doctors and the hospitals where you were treated to provide a small sample of the surgical material. This material may be used by the study investigators for medical research about genetic (inherited) factors and chemical changes that lead to the development of cancer and other diseases that occur in your age group.

These materials will be stored at a National Cancer Institute research storage facility for up to 25 years and used to help scientists learn what causes cancer and how to prevent its progression. It is believed that cancer may be caused by both environmental and genetic factors. Therefore, the samples which you contribute may be used in biochemical and genetic studies to identify these causes.

BENEFITS OF ADDITIONAL STUDIES

The additional studies will not provide direct benefit to you other than the satisfaction of participating in this research for the possible benefit of future generations. However, your participation in these additional studies will help answer questions related to the health and longevity of persons in your age group and will help establish a scientific understanding of the factors which influence the development and progression of cancer.

RISKS OF ADDITIONAL STUDIES

There are certain risks and discomforts that might be associated with the additional procedures.

- When blood is drawn, there may be local bruising or bleeding at the puncture site. There may also be uneasiness associated with needles.

NOTIFICATION AND COSTS

The additional biologic samples are for medical research only and the research results are not suitable for use as clinical tests for your medical care. The scientific studies require only looking at all lab results together. Therefore, the results of these additional studies will not be available to you.

The samples of blood, and urine will only be used for research and will not be sold. If any new scientific or medical testing products are developed using the results of the research done on your samples, you will not be paid, even if these medical products are sold.

COMPENSATION FOR RESEARCH-RELATED INJURIES

If you are in the screened group: In the unlikely event of physical injury from drawing a blood sample, you will be provided with immediate medical treatment. No payment of medical treatment is available from the National Cancer Institute for any such injury.

INFORMATION ON NEW FINDINGS

As a participant in the PLCO Trial, you will be contributing to an important scientific investigation and will be notified about important scientific findings of these studies. This notification will be of results for all participants together. No individual results will be provided from these additional studies.

CONFIDENTIALITY

The information concerning your participation in the study will be kept confidential and used only for scientific purposes, in accordance with applicable state and federal laws.

As genetic research may discover genetic information about you, this information will be kept confidential to the full extent permitted by law. Your blood, and urine samples will not be labeled with your name and no one who works with these samples will have access to that information. Your name and any identifying information will not be used in any reports.

RIGHT TO WITHDRAW

Your participation in the additional medical research is voluntary and you may refuse to participate and/or withdraw your consent and discontinue participation at any time without penalty or loss of benefits to which you are otherwise entitled.

You may participate in the screening part of the PLCO Trial and yet decline to have biologic samples (blood and urine) stored for research purposes. Further, if you initially decide to have your biologic samples stored for research purposes, but later change your mind by written notification of (principal investigator) at (the screening center), whatever remains of your biologic samples will then be destroyed. Your decision will not affect your care.

CERTIFICATION

I have read this form or it has been read to me and I understand its contents. Any questions concerning the research or the rights of the participants involved have been and will be answered by NAMES, TITLES, PHONE NUMBERS. A copy of this consent form has been given to me. My signature(s) below means that I freely agree to participate in this additional medical research. **My consent to participate in the screening part of the PLCO Trial has been provided in a separate document**

By signing this consent, I agree to have biologic samples (some of my blood, and urine, and possibly surgical tissue) stored for **future research on cancer**. I understand that it may involve genetic and biochemical studies.

PARTICIPANT'S NAME (PRINT)	DATE
PARTICIPANT'S SIGNATURE	WITNESS' SIGNATURE

By signing this consent, I agree to have biologic samples (some of my blood, and urine, and possibly surgical tissue) stored for future research **other diseases that effect my age group**. I understand that it may involve genetic and biochemical studies.

PARTICIPANT'S NAME (PRINT)	DATE
PARTICIPANT'S SIGNATURE	WITNESS' SIGNATURE

If you have questions about this document or about the research and your right's as a participant in this research, please call:

Screening Center Representative
Telephone number

PROTOTYPE DOCUMENT C:INFORMED CONSENT FOR COMPLETED SCREENING ARM PARTICIPANTS

Prostate, Lung, Colorectal and Ovarian Cancer Screening Trial

RESEARCH ON THE DEVELOPMENT AND CAUSES OF CANCER AND OTHER DISEASES
CONSENT FOR PARTICIPATION
NAME OF CLINICAL CENTER

DESCRIPTION OF ADDITIONAL STUDIES ON THE DEVELOPMENT AND CAUSES OF CANCER AND OTHER DISEASES

The purpose of the Prostate, Lung, Colorectal and Ovarian (PLCO) Cancer Screening Trial is to determine if certain screening tests can lead to early diagnosis of prostate, lung, colorectal and ovarian cancer and so reduce deaths caused by these cancers.

In addition to evaluating screening tests, the PLCO Trial also seeks to study factors which may cause these cancers to develop and progress. Additional research on cancer and other diseases which occur in your age group will be carried out among PLCO Trial participants who volunteer for these additional studies. **We are requesting your consent to participate in these additional studies of cancer and other diseases that occur in your age group**.

ADDITIONAL STUDY PROCEDURES

If you volunteer for these additional studies, blood not used for the prostate cancer screening test (PSA) or the ovarian cancer screening test (CA-125) test will be stored and used in future medical research. (If you are a women without ovaries and will not receive the CA-125 blood screening test, you may still choose to participate in this additional medical research by signing this document.). These future studies will include investigations to determine if genetic (inherited) factors and chemicals in blood, such as dietary nutrients and hormones, are related to the risk of developing cancer and other diseases that occur in your age group.

If you volunteer for these additional studies and you later have surgery for diagnosis or treatment of cancer or a related condition, the study investigators may then ask your doctors and the hospitals where you were treated to provide a small sample of the surgical material. This material may be used by the study investigators for medical research about genetic (inherited) factors and chemical changes that lead to the development of cancer and other diseases that occur in your age group.

These materials will be stored at a National Cancer Institute research storage facility for up to 25 years and used to help scientists learn what causes cancer and how to prevent its progression. It is believed that cancer may be caused by both environmental and genetic factors. Therefore, the samples which you contribute may be used in biochemical and genetic studies to identify these causes.

BENEFITS OF ADDITIONAL STUDIES

The additional studies will not provide direct benefit to you other than the satisfaction of participating in this research for the possible benefit of future generations. However, your participation in these additional studies will help answer questions related to the health and longevity of persons in your age group and will help establish a scientific understanding of the factors which influence the development and progression of cancer.

RISKS OF ADDITIONAL STUDIES

There are certain risks and discomforts that might be associated with the additional procedures.

- When blood is drawn, there may be local bruising or bleeding at the puncture site. There may also be uneasiness associated with needles.

NOTIFICATION AND COSTS

The additional biologic samples are for medical research only and the research results are not suitable for use as clinical tests for your medical care. The scientific studies require only looking at all lab results together. Therefore, the results of these additional studies will not be available to you.

The samples of blood will only be used for research and will not be sold. If any new scientific or medical testing products are developed using the results of the research done on your samples, you will not be paid, even if these medical products are sold.

COMPENSATION FOR RESEARCH-RELATED INJURIES

If you are in the screened group: In the unlikely event of physical injury from drawing a blood sample, you will be provided with immediate medical treatment. No payment of medical treatment is available from the National Cancer Institute for any such injury.

INFORMATION ON NEW FINDINGS

As a participant in the PLCO Trial, you will be contributing to an important scientific investigation and will be notified about important scientific findings of these studies. This notification will be of results for all participants together. No individual results will be provided from these additional studies.

CONFIDENTIALITY

The information concerning your participation in the study will be kept confidential and used only for scientific purposes, in accordance with applicable state and federal laws.

As genetic research may discover genetic information about you, this information will be kept confidential to the full extent permitted by law. Your blood samples will not be labeled with your name and no one who works with these samples will have access to that information. Your name and any identifying information will not be used in any reports.

RIGHT TO WITHDRAW

Your participation in the additional medical research is voluntary and you may refuse to participate and/or withdraw your consent and discontinue participation at any time without penalty or loss of benefits to which you are otherwise entitled.

You may participate in the screening part of the PLCO Trial and yet decline to have biologic samples (blood) stored for research purposes. Further, if you initially decide to have your biologic samples stored for research purposes, but later change your mind by written notification of (principal investigator) at (the screening center), whatever remains of your biologic samples will then be destroyed. Your decision will not affect your care.

CERTIFICATION

I have read this form or it has been read to me and I understand its contents. Any questions concerning the research or the rights of the participants involved have been and will be answered by NAMES, TITLES, PHONE NUMBERS. A copy of this consent form has been given to me. My signature(s) below means that I freely agree to participate in this additional medical research. **My consent to participate in the screening part of the PLCO Trial has been provided in a separate document**

By signing this consent, I agree to have biologic samples (some of my blood, and urine, and possibly surgical tissue) stored for **future research on cancer**. I understand that it may involve genetic and biochemical studies.

PARTICIPANT'S NAME (PRINT)	DATE
PARTICIPANT'S SIGNATURE	WITNESS' SIGNATURE

By signing this consent, I agree to have biologic samples (some of my blood, and urine, and possibly surgical tissue) stored for future research **other diseases that effect my age group**. I understand that it may involve genetic and biochemical studies.

PARTICIPANT'S NAME (PRINT)	DATE
PARTICIPANT'S SIGNATURE	WITNESS' SIGNATURE

If you have questions about this document or about the research and your right's as a participant in this research, please call:

Screening Center Representative
Telephone number

PROTOTYPE DOCUMENT D:INFORMED CONSENT FOR RANDOMIZED CONTROL ARM PARTICIPANTS

Prostate, Lung, Colorectal and Ovarian Cancer Screening Trial

RESEARCH ON THE DEVELOPMENT AND CAUSES OF CANCER AND OTHER DISEASES
CONSENT FOR PARTICIPATION
NAME OF CLINICAL CENTER

DESCRIPTION OF ADDITIONAL STUDIES ON THE DEVELOPMENT AND CAUSES OF CANCER AND OTHER DISEASES

The purpose of the Prostate, Lung, Colorectal and Ovarian (PLCO) Cancer Screening Trial is to determine if certain screening tests can lead to early diagnosis of prostate, lung, colorectal and ovarian cancer and so reduce deaths caused by these cancers.

In addition to evaluating screening tests, the PLCO Trial also seeks to study factors which may cause these cancers to develop and progress. Additional research on cancer and other diseases which occur in your age group will be carried out among participants in the PLCO Trial who have been chosen for the control comparison arm of the PLCO Trial and who volunteer for these additional studies. **We are requesting your consent to participate in these additional studies of cancer and other diseases that occur in your age group**.

ADDITIONAL STUDY PROCEDURES

If you volunteer for these additional studies, you will provide a sample of saliva which will be stored and used in future medical research. Such studies will include investigations to determine if genetic (inherited) factors are related to the risk of developing cancer and other diseases that occur in your age group. Briefly, the collection of saliva requires that, at home, you brush the inside of your mouth with a soft bristle brush and then spit the saliva into a small cup. You will be provided a kit and instructions for the collection and pre-paid return mailing of this sample to the researchers.

If you volunteer for these additional studies and you later have surgery for diagnosis or treatment of cancer or a related condition, the study investigators may then ask your doctors and the hospitals where you were treated to provide a small sample of the surgical material. This material may be used by the study investigators for medical research about genetic (inherited) factors and chemical changes that lead to the development of cancer and other diseases that occur in your age group.

These materials will be stored at a National Cancer Institute research storage facility for up to 25 years and used to help scientists learn what causes cancer and how to prevent its progression. It is believed that cancer may be caused by both environmental and genetic factors. Therefore, the samples which you contribute may be used in biochemical and genetic studies to identify these causes.

BENEFITS OF ADDITIONAL STUDIES

The additional studies will not provide direct benefit to you other than the satisfaction of participating in this research for the possible benefit of future generations. However, your participation in these additional studies will help answer questions related to the health and longevity of persons in your age group and will help establish a scientific understanding of the factors which influence the development and progression of cancer.

RISKS OF ADDITIONAL STUDIES

There are certain risks and discomforts that might be associated with the additional procedures.

- When saliva is collected, (by brushing sites in the mouth with a small soft brush) there may be a feeling of pressure on the inside of the cheeks. This may include a small amount of bleeding.

NOTIFICATION AND COSTS

The additional biologic samples are for medical research only and the research results are not suitable for use as clinical tests for your medical care. The scientific studies require only looking at all lab results together. Therefore, the results of these additional studies will not be available to you.

The samples of saliva will only be used for research and will not be sold. If any new scientific or medical testing products are developed using the results of the research done on your samples, you will not be paid, even if these medical products are sold.

COMPENSATION FOR RESEARCH-RELATED INJURIES

No physical injury is likely from the collection of a saliva sample. No payment of medical treatment is available from the National Cancer Institute for any such injury.

INFORMATION ON NEW FINDINGS

As a participant in the PLCO Trial, you will be contributing to an important scientific investigation and will be notified about important scientific findings of these studies. This notification will be of results for all participants together. No individual results will be provided from these additional studies.

CONFIDENTIALITY

The information concerning your participation in the study will be kept confidential and used only for scientific purposes, in accordance with applicable state and federal laws.

As genetic research may discover genetic information about you, this information will be kept confidential to the full extent permitted by law. Your saliva samples will not be labeled with your name and no one who works with these samples will have access to that information. Your name and any identifying information will not be used in any reports.

RIGHT TO WITHDRAW

Your participation in the additional medical research is voluntary and you may refuse to participate and/or withdraw your consent and discontinue participation at any time without penalty or loss of benefits to which you are otherwise entitled.

If you initially decide to have your biologic samples stored for research purposes, but later change your mind by written notification of (principal investigator) at (the screening center), whatever remains of your biologic samples will then be destroyed. Your decision will not affect your further participation as a control group member in the screening component of the PLCO Trial.

CERTIFICATION

I have read this form or it has been read to me and I understand its contents. Any questions concerning the research or the rights of the participants involved have been and will be answered by NAMES, TITLES, PHONE NUMBERS. A copy of this consent form has been given to me. My signature(s) below means that I freely agree to participate in this additional medical research. **My consent to participate in the screening part of the PLCO Trial has been provided in a separate document**

By signing this consent, I agree to have biologic samples (some of my saliva, and possibly surgical tissue) stored for **future research on cancer**. I understand that it may involve genetic and biochemical studies.

PARTICIPANT'S NAME (PRINT)	DATE
PARTICIPANT'S SIGNATURE	WITNESS' SIGNATURE

By signing this consent, I agree to have biologic samples (some of my saliva, and possibly surgical tissue) stored for future research **other diseases that effect my age group**. I understand that it may involve genetic and biochemical studies.

PARTICIPANT'S NAME (PRINT)	DATE
PARTICIPANT'S SIGNATURE	WITNESS' SIGNATURE

If you have questions about this document or about the research and your right's as a participant in this research, please call:

Screening Center Representative
Telephone number

III

SCREENING FOR SPECIFIC CANCERS

7

Breast Cancer

S. M. Moss
Institute of Cancer Research, Sutton, Surrey, England

I. INTRODUCTION

The high incidence of breast cancer worldwide, its high mortality rate despite recent advances in therapy, and the fact that, compared with some other common cancers, breast cancer occurs in women at relatively young ages have generated considerable interest in screening for breast cancer and in improving techniques and effectiveness of screening.

II. EPIDEMIOLOGY AND ETIOLOGY

A. Incidence

Breast cancer is the most common cancer worldwide among women, with an estimated 719,100 new cases in 1985 (1). It is the most common cancer in women in all developed countries other than Japan, with age-standardized incidence rates ranging from 84.8 per 100,000 in North America to 11.1 per 100,000 in western Africa. The lifetime risk of breast cancer development is 1 in 12 in the United Kingdom.

Incidence rates of the disease in a number of countries in recent years have been affected by increasing levels of screening (2–4), leading to increases in incidence rates in women in the age groups targeted for screening. However, in the United Kingdom, for example, there is evidence that incidence rates were increasing before the introduction of nationwide screening in 1988–1989 (5). In the United States, age-standardized rates of invasive disease have stabilized since 1987 (6).

Breast cancer incidence rates increase with increasing age, with generally a slower rate of increase after age 50 years. It is hypothesized that this difference may be due to influence of different hormones (7), which may be of relevance to the possible differential benefit of screening at different ages.

B. Mortality

Breast cancer is the leading cause of death from cancer in women worldwide, with an estimated 308,000 dying of the disease in 1985 (8). Again, highest rates occur in developed countries, except Japan. Average age-standardized rates in 1985 were 18.1 per 100,000 in developed countries and 10.9 per 100,000 in developing countries. Rates have generally been increasing, but there has recently been evidence from some countries that rates are leveling, or beginning to decrease (9). In the United Kingdom, it seems unlikely that screening will yet have had a considerable effect, and improvements in treatment have been suggested to be the most likely cause of the recently observed decrease (3).

In the United States, where a decrease in the age-adjusted breast cancer mortality rate of 6.8% between 1989 and 1993 has been observed, it has been proposed that both improvements in therapy and earlier detection by screening are responsible (6).

C. Risk Factors

Geographical variations in incidence and migrant studies, in which people migrating from low-to high-risk areas develop an increased risk, suggest that the risk of breast cancer is largely determined by environmental factors. Dietary factors that have been associated with an increased risk of breast cancer include obesity in postmenopausal women and high dietary fat intake (10). However, whereas case-control studies as well as international variations in incidence and migrant studies support an association of increased risk with fat intake, results of prospective studies are inconsistent and less clearly supportive (11). There is evidence that a diet high in fiber, fruits, and vegetables is associated with a decreased risk. The role of alcohol is not clear, with some but not all studies showing a weak association. There is also some evidence that increased levels of physical activity may have a protective effect.

The risk of breast cancer is increased with early age at menarche, late age at menopause, and late age at first pregnancy, all of which suggest a role of endogenous hormones in its etiology (12). It appears likely that exposure to estrogens, particularly estradiol, results in an increased risk.

A number of studies have suggested an increased risk of breast cancer with use of oral contraceptives. A recent collaborative reanalysis of data from 54 studies found that women were at a small increased risk of having breast cancer diagnosed while taking combined oral contraceptives and in the 10 years after stopping, with a relative risk in current users of 1.24 (95% CI 1.15–1.33) (13). Ten or more years after stopping oral contraceptives, there was no evidence of increased risk. In addition, breast cancers diagnosed in ever users were more likely to be localized compared with nonusers, and this difference persisted beyond 10 years after cessation of use.

The effect of use of hormone replacement therapy on breast cancer risk remains unclear. A cohort study of 23,000 Swedish women has recently shown an increasing relative risk of breast cancer incidence with time after exposure to an estrogen-progestin combined brand, with a standardized incidence ratio of 1.4 (95% CI 1.1–1.8) after 10 years of follow-up (14). One meta-analysis of 28 studies found an increased risk of less than 10% (15), and another found an increased risk only in women with more than 15 years of exposure (16).

However, the increasing number of women using hormone replacement therapy (HRT), together with the fact that such women will be older and having higher incidence of breast cancer than those using oral contraceptives, means that this may be increasingly important. In addition, it has been suggested that the use of HRT may affect the sensitivity of mammography (see Section V. B.1).

It is estimated that between 5% and 10% of all breast cancer cases may be due to inherited cancer predisposition genes. Women with a first degree relative with breast cancer have a relative risk of 1.5–2. The risk is increased with number of affected relatives, and younger age at diagnosis (17). Recently, two breast cancer genes BRCA1 and BRCA2 have been identified which are likely to account for about 85% of all familial breast cancers.

Some benign breast conditions are associated with an increased risk of subsequent development of breast cancer. One study has shown that the relative risk in women with proliferative compared with nonproliferative benign lesions is 1.9, with a relative risk of 5.7 in those with atypical hyperplasia (18), and the diagnosis of fibroadenoma has also been shown to be associated with an increased long-term risk of breast cancer (19).

D. Primary Prevention

Many of the risk factors for breast cancer are not amenable to intervention. However, trials of chemoprevention are in progress in high-risk women to determine whether use of tamoxifen results in a decreased risk (20). In the United States, a dietary intervention study is taking place, also in high-risk women, with the intervention group randomized to a very low-fat diet (21).

III. NATURAL HISTORY

As discussed previously, some benign breast conditions are markers for the subsequent development of cancer, but none has been identified as a precursor lesion. It has been recognized for some time that tumor size is related to the probability of both lymph node involvement and distant metastasis (22). It is this that has led to the concentration on detection of small invasive cancers as the target for breast cancer screening in order to maximize the resulting reduction in mortality rate. There is considerable variation in the growth rates of breast tumors. Estimates of the mean duration of the asymptomatic, preclinical phase have been made using data from screening studies, and the sojourn time appears to be shorter in younger women. For example, Tabar et al. (23) estimated mean sojourn times of 1.7, 3.3, 3.8, and 2.6 years in women aged 40–49, 50–59, 60–69, and 70–79, respectively.

The role of carcinoma in situ in the natural history of breast cancer is still the subject of some debate and merits further attention because 15%–20% of cancers detected by mammographic screening may be in situ. Recent estimates suggest that 50% of ductal carcinoma in situ may progress to invasive disease (24). The progression of lobular carcinoma in situ (LCIS), which is usually an incidental finding and not detectable on mammography, is also unclear. Follow-up studies of patients with LCIS have found relative risks of 5.4–12.0 of development of invasive cancer relative to the general population (25).

Another issue of debate in the natural history of breast cancer is the concept of ''dedifferentiation,'' by which a tumor classified as grade I or II when small may progress to grade III if allowed to grow. This was suggested by data from the Swedish Two-County screening study, which showed that, if cases detected at first screen were excluded, the malignancy grade of the remaining cancers in the study group was more favorable than in the control group (26); although an analysis of data from screening in Finland failed to confirm those findings (27). Tabar et al. (28) have recently reviewed the evidence for and against this hypothesis, and have developed a model suggesting that, in women aged 40–54, 91% of ductal tumors have the potential to dedifferentiate, whereas in women aged 55–69, 38% of ductal tumors have such a potential.

IV. TREATMENT AND PROGNOSIS

The treatment options for breast cancer are surgery, radiotherapy, hormone therapy, and chemotherapy. Most cases are treated by surgery, usually with adjuvant therapy by another modality. An overview of trials of adjuvant treatment in early breast cancer showed a statistically significant improvement in prognosis with the use of tamoxifen or chemotherapy (29); tamoxifen appears to be more effective in estrogen receptor positive patients. For treatment of early breast cancer there is a trend toward the use of breast-conserving surgery, and a recent development is the use of systemic therapy as a primary treatment to reduce the size of the tumor before surgery.

The 5-year relative survival of breast cancer patients in the United Kingdom in the 1980s was approximately 65%. Differences in survival between countries have been noted (30); however, there a number of factors will influence these findings, including differences in completeness of case ascertainment and stage at presentation.

The main predictor of survival of breast cancer patients is tumor stage at diagnosis. For example, of cases diagnosed in England and Wales during 1981–1985, five-year survival of stage I cases was 84%, and of stage IV cases was 18%. Of the components of stage, tumor size is probably the most important. Tumor grade is also an important predictor of prognosis, although its usefulness can be reduced by observer variability in reporting.

V. SCREENING TESTS FOR BREAST CANCER

X-ray mammography and physical examination (by a trained professional) are the recognized screening tests for breast cancer, together with self-screening by breast self-examination (BSE). Other imaging tests that have been considered include thermography, which has been shown to have poor sensitivity and specificity (31). The main use of ultrasound is now as a diagnostic tool in the evaluation of subjects found positive on initial screening. Magnetic resonance imaging (MRI) is a more recent development currently under evaluation for screening in high-risk women.

A. Physical Examination

Physical examination of the breast was in use as a screening test for breast cancer before the introduction of mammography, and early screening trials such as the Health Insurance Plan of New York (HIP) study included physical examinations as part of the screening test. However, there have been no randomized trials of physical examination alone, so that measurements of its effectiveness must be derived from comparisons with mammography.

In the HIP study, physical examination was estimated to have detected 67% of cancers diagnosed at screening, and 45% were detected by physical examination alone (32). From the UK Trial of Early Detection of Breast Cancer (TEDBC), the specificity of physical examination was estimated to be 95% at the ''clinical only'' screening rounds (33). Clinical examination detected 70% of cancers diagnosed at the first round and 44% of those diagnosed at later rounds (34), wherein both modalities were used; the percentages diagnosed by physical examination alone were 6% and 9%, respectively. Using the proportional incidence method, the sensitivity of physical examination alone was estimated to be 66%, although this is likely to be an overestimate because some cases will have been detected by mammography at the previous screening round.

B. Mammography

The use of mammography as a screening test has developed since the late 1930s (35). Successful mammography requires compression of the breast to prevent movement and to separate the breast tissues; this compression can cause discomfort and even pain. For modern day mammography, films have been produced that enable exposure time and hence radiation dose to be reduced. Good resolution is achieved by use of an X-ray tube with a small focal spot, fixed at a distance of about 45 cm form the breast; the use of grids help reduce scatter and improve contrast. Dedicated processors enable films to be processed under controlled conditions, and automatic roll-on display cabinets assist optimal viewing conditions.

The two standard projections that were initially employed were the mediolateral and craniocaudal views. However, in 1976, Lundgren proposed the use of the single mediolateral oblique view for screening (36). The single view was adopted in a number of screening trials and was the recommended practice at the start of the UK National Health Service Breast Screening Program (NHSBSP). However, a randomized trial comparing single-and two-view mammography at the first (prevalent) screen demonstrated not only a significantly higher cancer detection rate with two views but also a significantly lower rate of recall for assessment (37). Analysis of data from the NHSBSP has subsequently suggested that this may be largely due to an increase in the detection of small invasive cancers (38). The UK NHSBSP has recently changed to a policy of two-view mammography at prevalent screens.

The use of relatively high film densities has been shown to be correlated with an increased detection of small invasive cancers.

Another difficult factor to evaluate is the double reading of films, either by two radiologists or by a radiologist and, for example, a radiographer. Analysis of the effect of double reading on cancer detection rates and recall rates is hampered by the variety of protocols employed in deciding which women to recall; for example, recalling on suspicion by either reader is likely to result in a lower specificity but higher sensitivity than the use of a consensus verdict.

1. *Sensitivity and Specificity of Mammography*

The sensitivity and specificity of mammography have been estimated from a number of screening trials. In general, there is a tradeoff between the two, with increased sensitivity resulting in poor specificity.

Estimates of sensitivity depend on the method of calculation used. In the HIP study, sensitivity was estimated at 73.3% (132 cases detected on screening, divided by 132 + 47 interval cases) (32). Of the 132 screen-detected cancers, 44 were reported to be detected by mammography alone and 59 by clinical examination alone.

From the Swedish Two-County study, using the rate of interval cancers in a given time period as a proportion of the expected incidence (39), the sensitivity of mammography was estimated as 85% for women aged 50–69 years at entry for the first 12-month period, and as 70% at 13–24 months and 55% at 25–33 months (40). For women aged 40–49 years at entry, the sensitivity was estimated as 62% for the first 12-month period and 32% for the 13 to 24 month period.

The sensitivity of mammography in the Canadian National Breast Screening Study (NBSS) has been estimated as 75%, based on the percentage of ''potentially detected cases'' detected at the first screen (41). The specificity was 94%. In the UK TEDBC, mammography detected 94% of cancers diagnosed at first screen, and 91% of those diagnosed at later screens in which both modalities were used.

One study, based on small numbers, has suggested that the sensitivity and specificity of mammography are reduced in women using hormone replacement therapy (42).

C. Magnetic Resonance Imaging

Recently, magnetic resonance imaging (MRI) has been proposed as an alternative to mammography in screening for breast cancer, particularly in younger women when the sensitivity of mammography may be lower. Magnetic resonance imaging is a rapidly developing technique; the use of a contrast agent taken up by abnormal tissue has been found to enhance its sensitivity. However, the greater accuracy of MRI compared with mammography has not yet been clearly demonstrated; in particular, the sensitivity of MRI in detecting in situ disease needs further assessment. In addition, at present at least, the technique is too expensive to be considered as a screening test for the general population. A trial of MRI screening in young high-risk women, comparing the specificity and sensitivity of MRI with that of mammography over 5 years, has just started in the United Kingdom.

VI. EFFECTIVENESS OF SCREENING BY MAMMOGRAPHY

A. Evidence from Randomized Trials

There have been a number of randomized trials of breast screening by mammography, with or without physical examination, carried out in Europe, the United States, and Canada. The design of the various studies is summarized in Table 1, and Table 2 gives the main results.

1. *The Health Insurance Plan of New York Study*

The first randomized trial of screening for breast cancer was the Health Insurance Plan of New York (HIP) study carried out in New York in the 1960s.

Table 1 Design of Randomized Trials

Trial	Population size		Age range (years)	Period of field work	Screening intervals (months)	No. of round	Physical examination	Mammography (no. of views)
	Study group	Control group						
HIP (32)	31,000	31,000	40–69	1966–70	12	4	√	2
Swedish Two-Counties (44)	77,000	56,000	40–74	1977–85	24 (40–49) 33 (50–74)	4 2–3		1
Malmo (46)	21,000	21,000	45–70	1976–86	18–24	5		2 (Rounds 1 & 2) then 1 or 2
Stockholm (47)	40,000	19,000	40–64	1981–84	30	2		1
Gothenburg (48)	22,000	30,000	40–59	1983–90	18	3–5		2
Edinburgh (56)	28,000	26,000	45–64	1979–87	24 (mammography) 12 (Physical examination)	7	√	2/1
Canadian NBSS (49) (1)	25,000	25,000	40–49	1980–88	12	4–5	√	2
Canadian NBSS (2)	20,000	20,000	50–59	1980–88	12		(study + control group)	2

Table 2 Results of Randomized Trials

Trial	Follow-up (years)	RR (95% CI)
HIP (32)	18	0.77 (0.61–0.97)
Swedish Two-Counties (44)	11	0.70 (0.58–0.85)
Malmo (46)	8.8	0.96 (0.68–1.35)
Stockholm (47)	7.4	0.71 (0.4–1.2)
Gothenburg (48)	10	0.86 (0.54–1.37)
Edinburgh (56)	10	0.82 (0.61–1.11)
Canadian NBSS (50) (1) (40–49)	8.5	1.36 (0.84–2.21)
	10[a]	1.10 (0.78–1.54)
Canadian NBSS (52) (2) (50–59)	8.3	0.97 (0.62–1.52)

[a] Provisional results (51)

Approximately 62,000 women aged 40–69 years who were participants in the Health Insurance Plan were randomized in equal numbers to a study group and control group. Those in the study group were offered four annual screens by two-view mammography and clinical examination at annual intervals between 1966 and 1970 (32). Sixty-seven percent of women attended for the first examination, and 40% attended all four. Screening detected 2.73 cancers per 1000 women at first screen and 1.40 per 1000 at subsequent screens. After 7 years of follow-up, there was a 30% reduction in breast cancer mortality rate in the study group relative to the control group (43), the difference beginning to emerge after 3–4 years of follow-up. Following an initial excess of breast cancers in the study group resulting from earlier diagnosis by screening, numbers of cancers in the two groups became equal at around year 5. A reduction in breast cancer mortality rate of 23% was still observed after 18 years of follow-up, although this effect is diluted if cases diagnosed more than 7 years after entry are included.

First analysis of the results for different age groups at entry suggested that the benefit of screening was primarily observed in those aged 50 years and older at start of screening. However, in the longer term follow-up, a benefit in the study group aged 40–49 years at entry emerged, but only after 5–8 years of follow-up (32). There was evidence that women at higher risk were most likely to attend for screening, with a lower breast cancer incidence rate observed in the nonresponders than in the control group.

2. *Swedish Studies*

More recently, a number of randomized controlled trials of screening by mammography alone have been conducted. The largest of these was the Two-County

Study, carried out in Kopparberg and Ostergötland, starting in 1977. Women aged 40–74 years were allocated to a study group of 77,092 and a control group of 56,000; the randomization was of small geographical units (of approximately 3000 women) rather than on an individual basis, and it took place between 1977 and 1980. Women were offered screening by single-view mammography; the average interval between screens was 24 months in those aged 40–49 years and 33 months in those aged 50–74 years. The acceptance rate of the first screen was 89% and 83% of attenders also attended the second screen. The cancer detection rate at first screen was 6 per 1000, increasing from 2.1 per 1000 in women aged 40–49 years to 12.1 in women aged 70–74 years (44). At rescreening, the cancer detection rate was 3 per 1000. After 7 years, a reduction of 31% in breast cancer mortality rate was observed in the study group relative to the control group (45). After the publication of these results in 1985, screening was offered to the control group. By this time, women aged 40–49 years had had four rounds of screening, those aged 50–59 years had three, and those aged 70–74 years had two screens. After an average follow-up of 11 years, the mortality rate reduction of 30% was maintained (RR 0.70, 95% CI 0.58–0.85). There was little evidence of a benefit in women aged 40–49 years at entry.

Randomized trials have also been conducted in Malmo, Gothenburg, and Stockholm. The Malmo study randomized 42,283 women aged 45–70 years by year of birth cohort (46). Those in the study group were offered screening at 18–24 months, with two-view mammography for the first two rounds, and the number of views at subsequent rounds depending on the mammographic density of the breast. Seventy-four percent of women attended the first screen, and 70% attended subsequent screens. After an average of 8.8 years of follow-up, there was no significant reduction in breast cancer mortality rate in the study population (RR 0.96, 95% CI 0.68–1.35). In fact, in the first six years, there was a nonsignificant increase in breast cancer deaths in the study group, observed primarily in women younger than 55 years at entry. In women aged 55 and older, there was a 20% reduction at 8.8 years (RR 0.79, 95% CI 0.51–1.24).

As well as the fairly small sample size, the lack of significant effect in this study has also been attributed to a high level of screening in the control group, with 24% of a random sample of 500 having had mammography during the course of the trial, although most only once.

The Stockholm study, which started in 1981, randomized 60,261 aged 40–64 years by birth date, with 40,318 invited for screening by single-view mammography, with allocation to study and control groups by birth date. The study included only two rounds of screening with a 2½-year interval, after which the control group was screened (47). Attendance rates were more than 80%. The cancer detection rate was 4.0 per 1000 at the first round, decreasing to 3.1 per 1000 at the second round. After an average follow-up of 7.4 years, there was a nonsignificant reduction in breast cancer mortality rate (RR 0.71,

95% CI 0.4–1.2), with a 43% reduction in those aged 50 years and older (RR 0.57, 95% CI 0.3–1.1) and no evidence of a reduction in those younger than 50 years.

In Gothenburg, the study included approximately 52,000 women aged 40–59 years, of whom 22,000 were invited for screening by two-view mammography every 18 months. After an average 10-year follow-up, there was a nonsignificant 14% reduction in mortality rate from breast cancer in the study group (RR 0.86, 95% CI 0.54–1.37) (48).

In 1993, an overview of the Swedish trials was carried out, which showed a 23% reduction in breast cancer mortality rate overall in women offered screening (RR 0.77, 95% CI 0.67–0.88). Analysis of the results according to age at entry showed a nonsignificant reduction of 13% in women aged 40–49 years at entry (RR 0.87, 95% CI 0.63–1.20) (48).

3. *Canadian Trials*

In 1980, a trial was started in Canada with different protocols in two different age groups and involving 15 centers throughout Canada (49). In both age groups, women were volunteers, recruited through screening centers by a number of mechanisms. A total of 50,430 women aged 40–49 years all received a baseline physical examination and were then randomized to a study group offered annual screening by mammography for four or five screens and a control group receiving usual care only. All women were educated in breast self-examination. A total of 39,405 women aged 50–59 years were randomized to annual screening either by physical examination alone or by physical examination plus mammography. After an average 8.5 years of follow-up, there was a nonsignificant excess (36%) of breast cancer deaths in the study group in the 40-to 49-years arm (RR 1.36, 95% CI 0.84–2.21) (50). An interim but still provisional update of this analysis found a relative risk of 1.10 (95% CI 0.78–1.54) at 10 years of follow-up (51). In the 50-to 59-year arm, the cancer detection rate at first screen was 7.20 per 1000 in the mammography group, compared with 3.45 per 1000 in the physical examination group, but no difference in breast cancer mortality rate was observed after an average of 8.3 years of follow-up (RR 0.97, 95% CI 0.62–1.52) (52).

A number of criticisms have been made of the Canadian study. It has been claimed that the quality of mammography was poor because of lack of training and outdated equipment (53). The fact that, in the 40-to 49-year-old study, a significantly greater number of cases of advanced cancer were found initially in the screening arm than in the control arm, leading to queries concerning the randomization process, which took place after the initial clinical breast examination. Both these criticisms have been refuted by the investigators (54). An investigation into the possible alteration of names in the allocation registers has found that, even though some alterations were made, there was no evidence that these had been deliberately concealed, and that they could in any case have had only a trivial effect on the reported findings of the study (55). However, it is certainly the case that the sample size was too small to demonstrate an effect in the planned time scale. In addition, the fact that the study used a mainly volunteer population, together with the baseline screening in the control group, has been used as a reason for excluding it from a number of meta-analyses, although the validity of this is arguable, and several authors have performed analyses both including and excluding the Canadian trial.

4. *Edinburgh*

In 1979, a randomized trial began in Edinburgh. Women aged 45–64 years were included, and randomization was by general practice. The study group of 25,000 women also formed part of the study population of the UK Trial of Early Detection of Breast Cancer. Women in the study group were offered annual screening, with mammography and physical examination every 2 years and physical examination only in the intervening years. After 7 years, there was an 18% reduction in breast cancer mortality rate in the study group (56) and, after 10 years, an 18% reduction (RR 0.82, 95% CI 0.61–1.11) for breast cancer as underlying cause of death (57). This study has suffered from biased cluster randomization, with women in the study group having a lower all-cause mortality rate than the control group, and thus probably a higher underlying risk of death from breast cancer (58), although adjustment for this has been made in the analysis.

B. Evidence from Nonrandomized Studies

1. *UK Trial of Early Detection of Breast Cancer*

The UK TEDBC, which began in 1979, included eight geographically separate districts in England and Scotland and identified a total of 190,753 women aged 45–64 years (59). In two districts, Guildford and Edinburgh (the study arm of the randomized trial), women were invited for screening as described in the Edinburgh study above. In another two districts, the women were invited for an education session of breast self-examination, with open access clinics available to those detecting any abnormality. The remaining four districts acted as comparison centers. After 7 years, there was an 18% reduction in breast cancer mortality rate in the two study groups relative to the comparison centers, increasing to 20% when an adjustment for pretrial mortality rate was made (60). This mortality rate reduction was maintained after 10 years of follow-up; the observed reduction was similar in the two study arms (61). (Results from the BSE centers are discussed in Section VIII). There was no evidence of a differential effect of age at entry in an analysis by 5-year age bands, although the lowest observed effect was in those aged 50–54 years at entry.

2. *Breast Cancer Detection Demonstration Projects*

The Breast Cancer Detection Demonstration Project (BCDDP) enrolled a total of 283,222 women between 1973 and 1977 at 29 centers in the United States (62). The program comprised five annual screening examinations by clinical examination, two-view mammography, and thermography, although the latter was discontinued because of poor sensitivity, and the use of mammography in women younger than 50 years was reduced in the later screening rounds. A subset of 55,053 white women aged 35–74 years was followed up for 9 years after entry. In this subset, the cumulative incidence of breast cancer was 1.34 times the expected number based on U.S. Surveillance, Epidemiology and End Results program (SEER) data, and the cumulative mortality rate from breast cancer was 79.6% of that expected, again using SEER data to calculate expected mortality rate in women without diagnosed breast cancer at entry (62).

3. *Evidence from Case-Control Studies*

Case-control studies have been used to estimate the effectiveness of screening for breast cancer in a nonrandomized setting. In two studies from the Netherlands, in Nijmegen (63) and Utrecht (64), cases were deaths from breast cancer, and age-matched controls were drawn from the general population. In Nijmegen, an odds ratio of 0.48 (95% CI 0.23–1.00) was reported in ever-screened women aged 35 years and older compared with never-screened women, whereas the study in Utrecht found an odds ratio of 0.30 (95% CI 0.13–0.70) in ever-screened women aged 50–64 years at the start of screening. Another study from Florence, Italy (65), found an odds ratio of 0.53 (95% CI 0.29–0.95) in ever-screened women aged 40–70 years. Differences in the design of these studies make the results difficult to compare, and despite attempts in the Florence study to adjust for potential confounders, the results are still liable to be subject to selection bias. A case-control study carried out within the UK TEDBC (66) found an odds ratio (estimate of relative risk) of 0.51 in ever-screened versus never-screened women, compared with an observed relative risk of 0.79 in the prospective cohort analysis.

C. The Effectiveness of Screening Women Younger than 50 Years

The effectiveness of screening by mammography in women younger than age 50 years remains one of the most controversial areas of the subject. Reasons proposed for a lesser effect include lower sensitivity resulting from a tendency for younger women to have more dense breasts and a faster average growth rate in young women, meaning screening would need to be carried out more frequently for it to be effective. The first results from the HIP and Swedish Two-County study showed little evidence of any effect on mortality rate in women aged 50 years at entry to the study. However, neither the HIP nor any of the Swedish studies was designed to investigate the effect of screening in different age groups. The aforementioned Canadian study was designed to address this question, but its sample size was too small, and it also included women aged up to 49 years at entry.

In recent years, a number of meta-analyses of the randomized trials have been performed, primarily with the aim of investigating age-specific effectiveness. The results of these meta-analyses vary, mainly depending on which studies are included (particularly whether or not the Canadian study is included) and the length of follow-up of the various studies used. A meta-analysis performed in 1993 found a reduction in mortality rate from breast cancer of 15% in women aged 40–49 years invited to screening, excluding the Canadian trial (RR 0.85, 95% CI 0.68–1.08), compared with a reduction of 24% (RR 0.76, 95% CI 0.67–0.87) in women aged 50–74 years (67). If the Canadian trial was included, the relative risk in the 40–49 year age group was 0.93 (95% CI 0.76–1.15). A more recent meta-analysis by Smart et al.(68) found a significant 24% reduction in mortality rate in women aged 40–49 years (RR 0.76, 95% CI 0.62–0.95) if the Canadian trial was excluded, with a 16% reduction (RR 0.84% CI 0.69–1.02) if it was included. The report from a meeting in Falun, Sweden, in 1996 (51) includes an updated estimate of the Swedish overview results, with a relative risk of 0.77 (95% CI 0.59–1.01) in women aged 40–49 years at entry.

A number of investigators interpret these results as increasing evidence of a benefit of screening in women younger than age 50 years. However, a number of points need to be considered in interpreting these results. First, two of the randomized studies (in Edinburgh and Malmo) only contribute women aged 45–49 years at entry. Second, because the analyses include women up to age 49 years at entry, it is difficult to ascertain how much of the benefit has been due to screening in these women after age 50 years. Analyses by age at diagnosis, rather than age at entry, are hampered by the effect of lead time. De Koning et al. (69) attempted to use the MISCAN simulation model to address this problem. They simulated the results of the Swedish overview under two hypotheses; first that the prognosis of screen-detected cancers in women younger than and older than age 50 years was similar and second that the prognosis of screen-detected cancers in women younger than age 50 years was not improved (i.e., that all the observed benefit derived from screening in women older than age 50 years). However, even though the former model gave predicted results closer to the observed relative risks, the result from the latter scenario was also within the observed 95% confidence interval.

A meta-analysis by Kerlikowske et al. (70) has analyzed the results in women aged 40–49 years and 50 and older according to different attributes of the trials. The results suggested greater benefit in women aged 40–49 years with larger duration of follow-up and screening, differences that were not apparent for women aged 50–74 years.

It is hoped that an ongoing trial in the UK, which started in 1991, will provide an answer to this question. The trial aims to randomize 195,000 women aged 40–41 years at entry to a study group of 65,000 women offered annual mammography and a control group of 130,000 women who are not offered screening until they reach age 50 years and become eligible for the national breast screening program. A multicenter European trial with a very similar protocol has been proposed.

D. Screening in Older Women

Evidence from the randomized trials on the effectiveness of screening in women older than age 70 years is limited. However, there appears to be no evidence of a lesser effect, in terms of mortality rate reduction, in women aged 65–69 years at the start of screening. In a recent analysis of the Swedish Two-County study, the relative risk of breast cancer mortality rate in the study group was 0.58 (95% CI 0.39–0.86) in women aged 65–69 years at entry and 0.78 (95% CI 0.53–1.20) in those aged 70–74 years at entry (22). The difference between the relative risks in the two age groups was not statistically significant.

Arguments against screening older women include possible lower uptake and less cost effectiveness because fewer life-years are saved.

Most guidelines in the United States include women up to at least age 75 years. However, there appear to be a number of factors associated with lower uptake in older women, including both psychosocial and practical reasons for lower utilization (71).

E. Screening in High-Risk Women

It is currently believed that between 5% and 10% of breast cancers occur in women with a genetic predisposition and such women may be offered screening at a young age. The cloning of the BRCA1 and BRCA2 breast cancer genes has opened up the possibility of genetic testing for such predisposition. However, this possibility raises further issues, in particular the woman's need for counseling and what options should be offered to women found positive on such testing (72). The benefit of testing and the effectiveness of screening such women in terms of reducing mortality rate has not been proven.

VII. EFFECTIVENESS OF PHYSICAL EXAMINATION

There have been no randomized trials looking at the effectiveness of screening by physical examination alone. The Canadian study in women aged 50 years and older was designed to compare the effectiveness of annual mammography plus physical examination with that of physical examination alone, with women in both groups receiving education in BSE. As discussed, there was no observed difference in breast cancer mortality rate between the two groups, but the extent to which this is due to an effect of physical examination is not clear. In the UK TEDBC, in which screening was carried out by physical examination alone every other year, the survival of patients with cancer detected at these rounds was similar to that of cases in the comparison centers, after adjustment for size and nodal status (73). The finding that physical examination adds little to the sensitivity of modern mammography has led to its omission from most population screening programs, although it is argued by some that cancers detected by mammography alone may have less malignant potential (74).

The effectiveness of clinical examination depends on the training of the practitioner, the technique employed, and the thoroughness of and time taken for the examination.

VIII. EFFECTIVENESS OF BREAST SELF-EXAMINATION

The effectiveness of breast self-examination is more difficult to evaluate than that of screening by mammography, primarily because the practice of BSE and whether a cancer is in fact detected by this means is difficult to determine.

A. Evidence from Randomized Controlled Trials

Two randomized controlled trials of BSE have been undertaken. The first of these was initiated by the World Health Organization in Leningrad and Moscow in 1985, with the aim of determining the effect of a BSE program on mortality rate from breast cancer. More than 200,000 women aged 40–64 years have been enrolled through health clinics in Leningrad and factories or other institutions in Moscow and randomized to study and control groups (75). The BSE intervention consists of education sessions and some individual instruction. The study is planned to last for 15 years. No mortality results are yet available.

Preliminary results recently published from a trial conducted in Shanghai, China, identified 267,000 women aged 30–65 years who were current or retired employees of 520 textile factories (76). Randomization to a BSE instruction group or a control group, with nearly equal numbers, was by factory, after stratification for total number of employees and hospital of affiliation. Between October 1989 and October 1991, a baseline questionnaire on risk factors, prior breast cancer, and prior screening was administered to all women. The instruction in the BSE instruction group included an initial BSE instruction class, observation of regular BSE practice by medical workers, and reinforcement activities. There was a high level of participation in the instruction group. After 5 years of follow-up, mortality rate from breast cancer in the two groups was similar (30.9 and 32.7 per 100,000 in study and control groups, respectively). The number of breast cancers diagnosed in each group was similar, but a much greater number of benign breast lesions were detected in the instruction group.

B. Evidence from Prospective Studies

1. *UK TEDBC*

In the UK Trial of Early Detection of Breast Cancer, described earlier, two districts invited a total of 63,571 women aged 45–64 years to sessions providing education about BSE, including how to practice regular BSE, signs to look for, and what to do if any abnormality was suspected. Open-access diagnostic clinics were set up to which women were able to self-refer. Acceptance of invitations to the education sessions was 55% in one district and 31% in the other. A study of a random sample of 800 women in one district found the proportion performing BSE at least once a month, and reporting an appropriate technique increased overall from 24% before the start of the study to 41% a year later, with an increase from 28% to 47% in those attending the class (77).

The analysis of mortality rate at 10 years of follow-up showed no difference between the two BSE districts combined and the comparison centers (RR 1.07, 95% CI 0.93–1.22) (61). However there was a significant difference between the two districts, surprisingly with a lower breast cancer mortality rate in the district with the lower uptake of education sessions. Because of the nonrandomized design of the trial, it is difficult to draw firm conclusions from this on the effectiveness of BSE, although differences in treatment policies between the two districts did not appear to explain the mortality difference (78).

2. *Finnish MAMA program*

Between 1973 and 1975, a study population of 56,177 women aged 20 years and older was identified from those belonging to selected women's clubs (79). Women were given group education sessions and provided with calendars to record date and findings of BSE. A total of 29,018 women returned calendars after 2 years, and these were regarded as BSE compliers. Women with breast cancer diagnosed before or during the first 2 years of the study were excluded from the analysis. The observed number of breast cancers diagnosed up to the end of 1986 was similar to the expected number calculated from Finnish national rates and the stage distribution was similar to that from the Finnish Cancer Registry for 1980. The number of observed breast cancer deaths was significantly lower than the expected calculated from incidence and case-fatality rates (O/E = 0.71, 95% CI 0.57–0.87); however, a similar reduction was observed in all-cause mortality rate, and the education level of the study cohort was significantly higher than that of the general population. These latter two facts suggest a role of selection bias in the apparent effect on breast cancer mortality rate, which is difficult to quantify.

C. Evidence from Case-Control Studies

A case-control study, in which cases were women diagnosed with advanced breast cancer, compared self-reported practice of BSE in cases with that in age-matched controls, up to a similar reference date (80). To attempt to take account of selection bias, the analysis was performed adjusting for a number of potential confounding variables. There was no evidence of a decreased risk of advanced breast cancer in those practicing BSE (odds ratio 1.15, 95% CI 0.73–1.81) and no evidence of an effect on frequency of BSE practice. There was a suggestion of decreased risk in a small group of women reporting high proficiency scores, but the overall conclusion was of little benefit of BSE as it was generally practiced.

IX. DISADVANTAGES OF SCREENING

In addition to the financial costs of screening, which are discussed later, there are other detrimental effects that need to be considered.

A. Anxiety

The particular responsibility of initiating screening in apparently healthy people has long been recognized. The invitation to screening is itself likely to raise awareness of, and possibly anxiety about, breast cancer. However, two studies in the United Kingdom have found no evidence of increased anxiety in women attending for routine screening, compared with the levels in the same women before receiving the invitation to attend (81,82). However, it is those women who are recalled for further assessment following an abnormal or suspicious mammogram who are most likely to suffer from increased anxiety, and it is this group that has been most widely studied. In one study in the United Kingdom, Ellman et al. found increased levels of anxiety in a group of women attending for assessment after referral because of a positive mammogram, compared with those attending for routine screening (83). However, 3 months later, results of a postal questionnaire showed no persistent anxiety in these women. At clinic attendance, a third group of symptomatic women had higher anxiety levels that persisted after 3 months. A study carried out in New Jersey of 408 women questioned 3 months after mammography found that 47% of those with ''high suspicion'' initial mammogram reported substantial mammography-related anxiety, compared with 40% of those with ''low suspicion'' findings and 28% of those with normal findings (84). However, there was no baseline measurement and no information on the time taken to couple diagnostic follow-up in those with suspicious findings, so that the results may reflect fairly short-term anxiety levels.

B. Discomfort and Pain

Mammography causes a certain level of discomfort; in one study of 597 screened women in the United Kingdom, 35% reported some discomfort and 6% reported pain, but, in most cases, these were short-lived, and more than two-thirds of women ranked having a cervical smear as more uncomfortable (85). A study in the United States of more than 1800 women found similar levels of reported discomfort (86).

C. Radiation Risk

The potential harmful effect of mammography in causing radiation-induced cancers has received considerable attention, and in the early years of mammography, there were claims that screening might induce as many cancers as were detected (87). Since then, reductions in the average dose from mammography have largely allayed these concerns, but the fact that the risk posed by radiation is greater at younger ages of exposure has meant that this has been a particular issue in the debate on mammographic screening in women younger than age 50 years. However, most recent estimates of the benefit-to-risk ratio of screening conclude that, even in younger women, the benefit is likely to considerably outweigh the risks (88,89), although Law (89) suggests that there will be a small subgroup of women with large, thick breasts in whom the ratio may be reversed.

Another concern is for the subgroup of women at increased risk of breast cancer because of genetic predisposition. A number of studies have shown heterozygous carriers of the ataxia telangiectasia gene (ATM) to be at increased risk of breast cancer (90), although there is conflicting evidence (91), and these women may also have increased sensitivity to radiation (92). Experimental work in mice has recently shown that BRCA2 gene products interact with those of RAD51, which is involved in DNA repair (93), and this may have implications for radiation sensitivity.

X. COST EFFECTIVENESS OF SCREENING

The cost effectiveness of screening for breast cancer is often questioned by those who consider the resources used could be better employed elsewhere (94).

A study of the cost of screening in Australia has estimated the cost per woman screened as Australian $11,770 and that per cancer detected as Australian $11,550 (95), whilst an update of a study in Edinburgh, UK, the 1992–1993 figures estimated the costs as £22.60 per woman screened (including assessment) and £4,300 per cancer detected.

Cost effectiveness of screening is most commonly measured in terms of quality-adjusted life-years (QALYs). A number of models have been used to estimate the cost effectiveness of different screening policies. Using the MISCAN screening model, the cost effectiveness of screening women aged 50 years and older with a 2 or 3 yearly screening interval, in the Netherlands has been estimated as ranging between $4000 and $5000 per life-year gained (96). Using the same model, the variation of cost effectiveness within EC countries has also been estimated, with a predicted two-fold greater cost per life-year gained in Spain than in the Netherlands or the United Kingdom (97).

A Markov model developed in the United States (98), which was used to investigate different screening strategies, concluded that a policy of biennial mammography in women aged 50–79 years was the most cost effective, at $16,000 per life-year saved; the addition of annual mammography for women aged 40–49 years increased this amount to $20,200, but was more cost effective than annual mammography in women aged 50–79 years.

XI. CURRENT BREAST SCREENING PRACTICES AND RECOMMENDATIONS

Following the publication of results from the randomized trials, a number of countries have implemented or are considering nationwide screening programs. In the United Kingdom, a national program was started in 1988 in which women aged 50–64 years are invited for screening by mammography every 3 years. Women aged 65 years and older are not specifically invited, but they can self-refer for screening. The NHSBSP involves 97 programs throughout the United Kingdom. All programs have completed their first round of screening. The initial policy of screening by single view mammography was changed in 1995 to a policy of two views at a woman's first screen. Eighty-eight percent of the programs perform double reading to some extent, but the protocols vary (99).

In the Netherlands, a national program has also been established with 2-yearly screening for women aged 50–69 years. The whole population was expected to be covered by the end of 1997.

In Finland, a nationwide screening program was introduced in 1987, covering women aged 50–59 years, with screening every 2 years and continuing to age 64 years. The program in Finland was designed so that women in different birth cohorts were first invited for screening in different years. A recent comparison of breast cancer mortality rates in different birth cohorts found a standardized mortality ratio of 0.76 (95% CI 0.53–1.09) in women invited for screening compared to the controls (100).

In Sweden, policy for breast screening is decided by individual county councils. By 1994, 21 of 24 countries had started population-based screening programs (101). Although the national guidelines cover women aged 40–79 years, until recently, only about half the counties included women aged 40–49 years.

Elsewhere in Europe, pilot projects have been set up in a number of countries with funding from the European Commission (102).

In Australia, a national program is planned with a 2-yearly screening interval for women aged 50–69 years; screening is being introduced on a pilot basis. Likewise in New Zealand, pilot projects have been established.

In the United States, guidelines for breast cancer screening are produced by a number of different organizations. Even though there is a consensus on routine mammography in women older than 50 years, there is some variation in the recommended frequency, with a number of organizations, including the American Cancer Society and the American College of Radiology, recommending annual screening, and others, including the National Cancer Institute, recommending screening at 1-to 2-year intervals (103). Physical examination is also recommended at varying intervals. Joint guidelines drawn up in 1988 recommended mammography at 1-to 2-year intervals in women aged 40–49 years (104). In 1993, the National Cancer Institute decided it could no longer support this recommendation; however, it has recently reversed this decision. Findings from the National Health Interview Survey carried out in 1993 showed that 60% of women aged 40–49 years and 65% of those aged 50–64 years had had a mammogram in the previous 2 years.

With the growing implementation of population screening for breast cancer, quality assurance issues become of increasing importance. Quality assurance needs to cover a whole range of aspects, including technical aspects to monitor image quality, radiology, pathology, and data collection and epidemiology. Comprehensive guidelines for quality assurance in mammography screening have recently been published by the European Commission (105). The guidelines for technical parameters include ranges of acceptable values and frequency of measurement for a wide range of parameters. Guidelines for pathology cover both specimen sampling and reporting. In the United States, the Food and Drug Administration monitors screening mammography standards and performance under the authority of the Mammography Quality Standards Act (MQSA).

XII. CONCLUSION

The effectiveness of screening by mammography in women older than 50 years has been clearly demonstrated by a number of randomized trials, and population screening has become established policy in many countries. The full impact of population screening on breast cancer mortality rate, however, will not become evident for a number of years after the introduction of screening.

The effectiveness of mammographic screening in women younger than 50 years remains an area of controversy. The effectiveness of physical examination is not known, but its benefit in addition to mammography is likely to be small. Routine breast self-examination has not so far been demonstrated to be of benefit. The value of thermography as a screening test has been disproved; the effectiveness of MRI is currently under investigation.

Further improvements in mammography, attention to quality assurance, and careful monitoring of screening results may contribute to improved effectiveness of screening.

REFERENCES

1. Parkin DM, Pisani P, Ferlay J. Estimates of the worldwide incidence from eighteen major cancers in 1985. Implications for prevention and projections of future burden. Int J Cancer 1993; 55:891–903.
2. Feuer EJ., Wun LM. How much of the recent rise in breast cancer incidence can be explained by increases in mammography utilization? A dynamic population model. Am J Epidemiol 1992; 136(12):1423–1436.
3. Quinn M, Allen E, on behalf to the United Kingdom Association of Cancer Registries. Changes in incidence of and mortality from breast cancer in England and Wales since introduction of screening. BMJ 1995; 311:1391–1395.
4. McCredie M, Coates M, Bilous M, Kricker A, Hoyer A. Rising incidence of breast cancer in New South Wales, Australia: real increase or earlier detection? J Epidemiol Biostats 1996; 1(1):25–29.
5. Prior P, Woodman, CBJ, Wilson S, Threlfall AG. Reliability of underlying incidence rates for estimating the effect and efficiency of screening for breast cancer. J Med Screening 1996; 3:119–122.
6. Chu KC, Tarone RE, Kessler LG, Ries LAG, Hankey BF, Miller BA, Edwards BK. Recent trends in U.S. breast cancer incidence, survival, and mortality rates. J Natl Cancer Inst 1996; 88:1571–1579.
7. Pike MC, Spicer DV, Dahmoush L, Press MF. Estrogens, progestrogens, normal breast cell proliferation and breast cancer risk. Epidemiol Rev 1993; 15:17–25.
8. Pisani P, Parkin DM, Ferlay J. Estimates of the worldwide mortality of eighteen major cancers in 1985. Int J Cancer 1993; 54:594–606.
9. Hermon C, Beral V. Breast cancer mortality rates are levelling off or beginning to decline in many Western countries: analysis of time trends, age-cohort and age-period models of breast cancer mortality in 20 countries. Br J Cancer 1996; 73: 955–960.
10. Berrino F, Panico S, Muti P. Dietary fat, nutritional status and endocrine-associated cancers. In: Miller AB, ed. Diet and the aetiology of cancer. Berlin, Springer-Verlag, 1989:3–12.
11. Willett WC, Hunter DJ. Prospective studies of diet and breast cancer. Cancer 1994; 74:1085–1089.
12. Bernstein L, Ross RK. Endogenous hormones and breast cancer risk. Epidermiol Rev 1993; 15:48–65.
13. Collaborative Group on Hormonal Factors in Breast Cancer. Breast cancer and

hormonal contraceptives: collaborative reanalysis of individual data on 53297 women with breast cancer and 100239 women without breast cancer from 54 epidemiological studies. Lancet 1996; 347: 1713–1727.
14. Persson I, Yuen J, Bergkvist L, Schairer C. Cancer incidence and mortality in women receiving estrogen and estrogen-progestin replacement therapy—long-term follow-up of a Swedish cohort. Int J Cancer 1996; 67:327–332.
15. Dupont WD, Page DL. Menopausal estrogen replacement therapy and breast cancer. Arch Intern Med 1991; 151:67–72.
16. Steinberg KK, Thacker SM, Smith SJ, Stroup DF, Sack MM, Flanders WD, Berkelman RL. A meta-analysis of the effect of estrogen replacement therapy on the risk of breast cancer. JAMA 1991; 265(15):1985–1990.
17. Houlston RS, McCarter E, Parbhoo S, Scurr JH, Slack J. Family history and risk of breast cancer. J Med Genet 1992; 29:154–157.
18. Dupont WD, Page DL. Risk factors for breast cancer in women with proliferative breast disease. N Engl J Med 1985; 312:146–151.
19. Dupont WD, Page DL, Parl FF, et al. Long-term risk of breast cancer in women with fibroadenoma. N Engl J Med 1994; 331:10–15.
20. Bush TL, Helzlsouer KJ. Tamoxifen for the primary prevention of breast cancer: a review and critique of the concept and trial. Epidemiol Rev 1993; 15(1):233–243.
21. Greenwald P, Kramer B, Weed D. Expanding horizons in breast and prostate cancer prevention and early detection. J Cancer Educ 1993; 8:91–107.
22. Tubiana M, Koscielny S. Natural history of human breast cancer: recent data and clinical implications. Breast Cancer Res Treat 1991; 18:125–140.
23. Tabar L, Fagerberg G, Chen HH, Duffy SW, Smart CR, Gad A, et al. Efficacy of breast cancer screening by age. New results from the Swedish Two-County Trial. Cancer 1995; 75:2507–2517.
24. Ernster VL, Barclay J, Kerlikowske K, Grady D, Henderson C. Incidence of and treatment for ductal carcinoma in situ of the breast. JAMA 1996; 275(12):913–918.
25. Bodian CA, Perzin KH, Lattes R. Lobular neoplasia. Long term risk of breast cancer and relation to other factors. Cancer 1996; 78(5):1024–1034.
26. Duffy SW, Tabar L, Fagerberg G, Gad A, Grontoft O, South MC, Day NE. Breast screening, prognostic factors and survival—results from the Swedish Two County study. Br J Cancer 1991; 64(6):1133–1138.
27. Hakama M, Holli K, Isola J, Kallioniem O-P, Karkkainein A, Visakorpi T, et al. Aggressiveness of screen-detected breast cancers. Lancet 1995; 345:221–224.
28. Tabar L, Fagerberg C, Chen HH, Duffy SW, Gad A. Tumour development, histology and grade of breast cancers: prognosis and progression. Int J Cancer 1996; 66(4):413–419.
29. Early Breast Cancer Trialists' Collaborative Group. Systemic treatment of early breast cancer by hormonal, cytotoxic, or immune therapy. Lancet 1992; 339:1–15.
30. Berrino F, Sant M, Verdecchia et al. Survival of cancer patients in Europe. IARC Scientific Publications, 1995; No. 32, Lyon.
31. Lloyd Williams K, Phillips PH, Jones PA, Beaman SA, Fleming PJ. Thermography for screening for breast cancer. J Epidemiol Community Health 1990; 44(2):112–113.

32. Shapiro S, Venet W, Strax P, Venet L. Periodic screening for breast cancer. In: The Health Insurance Plan Project and its sequelae, 1963–1986. Baltimore: Johns Hopkins University Press, 1988.
33. UK Trial of Early Detection of Breast Cancer Group. Specificity of screening in United Kingdom trial of early detection of breast cancer. BMJ 1992; 304:346–349.
34. Moss SM, Coleman DA, Ellman R, Chamberlain J, Forrest APM, Kirkpatrick AE, et al. Interval cancers and sensitivity in the screening centres of the UK Trial of Early Detection of Breast Cancer. Eur J Cancer 1993; 29A:225–258.
35. Gershon-Cohen J, Strickler A. Roentgenological examination of the normal breast: its evaluation in demonstrating early neoplastic change. Am J Roentgenol 1938; 40:189–120.
36. Lundgren B, Jakobsson S. Single view mammography; a simple and efficient approach to breast cancer screening. Cancer 1976; 38:1124–1129.
37. Wald NJ, Murphy P, Major P, Parkes C, Townsend J, Frost C. UKCCCR multicentre randomised controlled trial of one and two view mammography in breast cancer screening. BMJ 1995; 311:1189–1193.
38. Blanks RG, Day NE, Moss SM. A comparison of two view and one view mammography in the detection of small invasive cancers: results from the National Health Service breast screening programme. J Med Screening 1996; 3:200–203.
39. Day NE. Estimating the sensitivity of a screening test. J Epidemiol Community Health 1985; 39:364–366.
40. Tabar L, Fagerberg G, Day NE, Holmberg L. What is the optimum interval between mammographic screening examinations? An analysis based on the latest results of the Swedish Two-County breast cancer screening trial. Br J Cancer 1987; 55:547–551.
41. Baines CJ, McFarlane DV, Miller AB. Sensitivity and specificity of first screen mammography in 15 NBSS centres. J Can Assoc Radiol 1988; 39:273–276.
42. Laya MB, Larson EB, Taplin SH, White E. Effect of estrogen replacement therapy in the specificity and sensitivity of screening mammography. J Natl Cancer Inst 1996; 88:643–649.
43. Shapiro S. Evidence on screening for breast cancer from a randomized trial. Cancer 1977; 39 (suppl 6):2772–2782.
44. Tabar L, Fagerberg G, Duffy SW, Day NE, Gad A, Grontoft O. Update of the Swedish Two-County Program of Mammographic Screening for Breast Cancer. Radiol Clin North Am 1992; 30:187–210.
45. Tabar L, Gad A, Holmberg LH, Ljunquist U, Fagerberg CJG, Baldetorp L, et al. Reduction in mortality from breast cancer after mass screening with mammography. Lancet 1985; 1:829–832.
46. Andersson I, Aspegren K, Janzon L, Landberg T, Lindholm K, Linell F, et al. Mammographic screening and mortality from breast cancer: the Malmo mammographic screening trial. BMJ 1988; 297:943–948.
47. Frisell J, Eklund G, Hellstrom L, Lidbrink E, Rutqvist L-E, Somell A. Randomised study of mammography—preliminary report on mortality in the Stockholm trial. Breast Cancer Res Treat 1991; 18:49–56.

48. Nystrom L, Rutqvist LE, Wall S, Lindgren A, Lindqvist M, Ryden S, et al. Breast cancer screening with mammography: overview of Swedish randomised trials. Lancet 1993; 341:973–978.
49. Miller AB, Howe GR, Wall C. The national study of breast cancer screening: protocol for a Canadian randomised controlled trial of screening for breast cancer in women. Clin Invest Med 1981; 4:227–258.
50. Miller AB, Baines CJ, To T, Wall C. Canadian National Breast Screening Study: 1. Breast cancer detection and death rates among women aged 40 to 49 years. Can Med Assoc J 1992; 147:1459–1476.
51. Tabar L, Larsson LG, Andersson I, Duffy SW, et al. Report of the Organizing Committee and Collaborators, Falun Meeting, Falun Sweden (March 1996). Breast-cancer screening with mammography in women aged 40–49 years. Int J Cancer 1996; 68:693–699.
52. Miller AB, Baines CJ, To T, Wall C. Canadian National Breast Screening Study: 2. Breast cancer detection and death rates among women aged 50 to 59 years. Can Med Assoc J 1992; 147:1477–1488.
53. Kopans DB, Feig SA. The Canadian National Breast Screening Study: a critical review. Am J Radiol 1993; 161:755–760.
54. Baines CJ. The Canadian National Breast Screening Study: a perspective on criticisms. Ann Intern Med 1994; 120:326–334.
55. Bailar JC III, MacMahon B. Randomization in the Canadian National Breast Screening Study: a review for evidence of subversion. Can Med Assoc J 1997; 156(2):193–199.
56. Roberts MM, Alexander FE, Anderson TJ, Chetty U, Donnan PT, Forrest P, et al. Edinburgh trial of screening for breast cancer: mortality at seven years. Lancet 1990; 335:241–246.
57. Alexander FE, Anderson TJ, Brown HK, Forrest AP, Hepburn W, Kirkpatrick AE, et al. The Edinburgh randomised trial of breast cancer screening: results after 10 years of follow-up. Br J Cancer 1994; 70:542–548.
58. Alexander FE, Anderson TJ, Brown HK, et al. The Edinburgh randomised trial of breast cancer screening: results after 10 years of follow-up. Br J Cancer 1994; 70: 542–548.
59. UK Trial of Early Detection of Breast Cancer Group. Trial of early detection of breast cancer: description of method. Br J Cancer 1981; 44:618–627.
60. UK Trial of Early Detection of Breast Cancer Group. First results on mortality reduction in the UK Trial of Early Detection of Breast Cancer. Lancet 1988; ii: 411–417.
61. UK Trial of Early Detection of Breast Cancer Group. Breast cancer mortality after 10 years in the UK trial of early detection of breast cancer. The Breast 1993;2:13–20.
62. Morrison AS, Brisson J, Khalid N. Breast cancer incidence and mortality in the Breast Cancer Detection Demonstration Project. J Natl Cancer Inst 1988; 80:1540–1547.
63. Verbeek ALM, Hendricks JHCL, Holland R, Mravunac M, Sturmans F, Day NE. Reduction in breast cancer mortality through mass screening with modern

mammography (First results of the Nijmegan Project 1975–81). Lancet 1984; i: 1222–4.
64. Collette HJA, Day NE, Rombach JJ, de Waard F. Evaluation of screening for breast cancer in a non-randomised study (the DOM Project) by means of a case-control study. Lancet 1984; i:1224–1225.
65. Palli D, Rosselli del Turco M, Buiatti E, Carli S, Ciatto S, Toscani L, Maltoni G. A case-control study for the efficacy of a non-randomized breast cancer screening program in Florence (Italy). Int J Cancer 1986; 38:501–504.
66. Moss SM, Summerley ME, Thomas BT, Ellman R, Chamberlain JOP. A case-control evaluation of the effect of breast cancer screening in the United Kingdom trial of early detection of breast cancer. J Epidemiol Community Health 1992; 46: 362–364.
67. Wald NJ, Chamberlain J, Hackshaw A. European Society of Mastology Consensus Conference on Breast Cancer Screening, Paris, 4–5 February 1993: Report of the Evaluation Committee. Clin Oncol 1994; 4:261–268.
68. Smart CR, Hendrick RE, Rutledge JHI, Smith RA. Benefit of mammography screening in women ages 40 to 49 years. Cancer 1995; 75:1619–1626.
69. de Koning HJ, Boer R, Warmerdam PG, Beemsterboer PMM, van der Maas PJ. Quantitative interpretation of age-specific mortality reductions from the Swedish breast cancer-screening trials. J Natl Cancer Inst 1995; 87:1217–1223.
70. Kerlikowske K, Grady D, Rubin SM, Sandrock C, Ernster VL. Efficacy of screening mammography. A meta-analysis. JAMA 1995; 273:149–154.
71. Constanza ME. The extent of breast cancer screening in older women. Cancer 1994; 74:2046–2050.
72. Eeles RA, Stratton MK, Goldgar DE, Easton DF. The genetics of familial breast cancer and their practical implications. Eur J Cancer 1994; 30A:1383–1390.
73. Moss SM, Ellman R, Coleman D, Chamberlain J, for the United Kingdom Trial of Early Detection of Breast Cancer Group. Survival of patients with breast cancer diagnosed in the United Kingdom trial of early detection of breast cancer. J Med Screening 1994; 1:193–198.
74. Mittra I. Breast screening: the case for physical examination without mammography. Lancet 1994; 343;342–344.
75. Semiglazov VF, Sagaidak VN, Moiseyenko VM, Mikhailov EA. Study of the role of breast self-examination in the reduction of mortality from breast cancer. The Russian Federation/World Health Organization Study. Eur J Cancer 1993; 29A(14): 2039–2046.
76. Thomas DB, Gao DL, Self SG, Allison CJ, Tao Y, Mahloch J, et al. Randomized trial of breast self-examination in Shanghai: methodology and preliminary results. J Natl Cancer Inst 1997; 89:355–365.
77. Calnan MW, Chamberlain J, Moss S. Compliance with a class teaching breast self-examination. J Epidemiol Community Health 1983; 37:264–270.
78. Ellman R, Moss SM, Coleman D, Chamberlain J. Breast self-examination programmes in the trial of early detection of breast cancer: 10 year findings. Br J Cancer 1993; 68:208–212.

79. Gastrin G, Miller AB, To T, Aronson KJ, Wall C, Hakama M, et al. Incidence and mortality from breast cancer in the Mama Program for breast screening in Finland, 1973–1986. Cancer 1994; 73:2168–174.
80. Newcomb PA, Weiss NS, Storer BE, Scholes D, Young BE, Voigt LF. Breast self-examination in relation to the occurrence of advanced breast cancer. J Natl Cancer Inst 1991; 83:260–265.
81. Walker LG, Cordinier CM, Gilbert FJ, Needham G, et al. How distressing is attendance for routine breast screening? Psycho-Oncology 1994; 3:299–304.
82. Sutton S, Saidi G, Bickler G, Hunter J. Does routine screening for breast cancer raise anxiety? Results from a three wave prospective study in England. J Epidemiol Community Health 1995; 49:413–418.
83. Ellman R, Angeli N, Christians A, Moss S, Chamberlain J, Maguire P. Psychiatric morbidity associated with screening for breast cancer. Br J Cancer 1989; 60:781–784.
84. Lerman C, Trock B, Rimer BK, Boyce A, Jepson C, Engstrom PF. Psychological and behavioral implications of abnormal mammograms. Ann Intern Med 1991; 114: 657–661.
85. Rutter DR, Calnan M, Vaile MSB, Field S, Wade KA. Discomfort and pain during mammography: description, prediction and prevention. BMJ 1992; 305:443–445.
86. Stomper PC, Kopans DB, Sadowsky NL, Sonnenfeld MR, Swann CA, Gelman RS, et al. Is mammography painful? A multicentre patient survey. Arch Intern Med 1988; 148:521–524.
87. Bailar JC. Mammography: a contrary view. Ann Intern Med 1976; 84:77–84.
88. Feig SA. Mammographic screening of women aged 40–49 years: benefit, risk and cost considerations. Cancer 1995; 76:2097–2106.
89. Law J. Variations in individual radiation dose in a breast screening programme and consequences for the balance between associated risk and benefit. Br J Radiol 1993; 66:691–698.
90. Lavin MF, Bennett I, Ramsay J, Gardiner RA, et al. Identification of a potentially radiosensitive subgroup among patients with breast cancer. J Natl Cancer Inst 1994; 86:1627–1634.
91. Athma P, Rappaport R, Swift M. Molecular genotyping shows that ataxia-telangiectasia heterozygotes are predisposed to breast cancer. Cancer Genet Cytogenet 1996; 52:130–134.
92. FitzGerald MG, Bean JM, Hegde SR, Unsal H, MacDonald DJ, Harkin DP, et al. Heterozygous ATM mutations do not contribute to early onset of breast cancer. Nature Genet 1997; 15:307–310.
93. Sharan SK, Morimatsu M, Albrecht U, Lim D-S, et al. Embryonic lethality and radiation hypersensitivity mediated by RAD51 in mice lacking BRCA2. Nature 1997; 386:804–810.
94. Wright CJ, Mueller CB. Screening mammography and public health policy: the need for perspective. Lancet 1995; 346:29–32.
95. Hurley SF, Livingstone PM, Thane N, Quang L. Mammographic screening: measurement of the cost in a population based programme in Victoria, Australia. J Epidemiol Community Health 1994; 48(4):391–399.

96. de Koning HJ, van Ineveld BM, van Oortmarssen GJ, de Haes JCJM, Collette HJA, Hendricks JHCL, et al. Breast cancer screening and cost-effectiveness; policy alternatives, quality of life considerations and the possible impact of uncertain factors. Int J Cancer 1991; 49:531–537.
97. van Ineveld BM, van Oortmarssen GJ, de Koning HJ, Boer R, van der Maas PJ. How cost-effective is breast cancer screening in different EC countries? Eur J Cancer 1993; 12:1663–1668.
98. Lindfors KK, Rosenquist CJ. The cost-effectiveness of mammographic screening strategies. JAMA 1995; 274(11):881–884.
99. Gerard K, Brown J, Johnston K. UK breast screening programme: how does it reflect the Forrest recommendations? J Med Screening 1997;4:10–15.
100. Hakama M, Pukkala E, Heikkila, Kallio M. Effectiveness of the public health policy for breast cancer screening in Finland: population based cohort study. BMJ 1997; 314:864–867.
101. Tornberg S, Carstensen J, Hakulinen T, Lenner P, Hatschek T, Lundgren B. Evaluation of the effect on breast cancer mortality of population based mammography screening programmes. J Med Screening 1994; 1:184–187.
102. de Waard F, Kirkpatrick A, Perry NM, Tornberg S, Tubiana M, de Wolf C. Breast cancer screening in the framework of of the Europe against Cancer programme. Eur J Cancer Prev 1994; 3(supp 1):3–5.
103. Leitch AM. Controversies in breast cancer screening. Cancer 1995; 76:2064–2069.
104. Joint statement on mammography guidelines. ARC Bull 1989; 45:4.
105. de Wolf CJM, Perry NM, eds. European Guidelines for Quality Assurance in Mammography Screening. Europe Against Cancer. 2d ed. Luxembourg: Office for Official Publications of the European Communities, 1996.

8

Bladder Cancer

Paul F. Schellhammer and Richard Moriarty
Eastern Virginia Medical School, Norfolk, Virginia

Several facts about transitional cell carcinoma deserve initial comment. Cigarette smoking is the most critical risk factor in the Western world. Although heralded by the unmistakable sign and disturbing symptom of urinary bleeding, the diagnosis of transitional cell carcinoma is often delayed because of incomplete evaluation of hematuria. Transitional cell carcinoma is characterized by two phenotypes: one is noninvasive, usually papillary, and frequently recurrent, presenting a nuisance but rarely a mortality risk; the other is muscle invasive, usually sessile, and often rapidly lethal. Over the past decade, while the incidence of transitional cell carcinoma has increased, the mortality rate has declined. This trend may be attributable to the unique success of intravesical immunotherapy with Bacillus Calmette Guérin (BCG).

I. EPIDEMIOLOGY

Transitional cell carcinoma has been the subject of intense epidemiological study. As a result, many of the causal agents are known and lifestyle, occupational, iatrogenic, infectious, and genetic risk factors have been identified. It is likely that most bladder cancers result from a complex interaction between multiple environmental factors and genetic susceptibility.

A. Lifestyle

In the United States, tobacco use is the dominant risk factor. Nearly half of all new cases of bladder cancer occur in current or former smokers (1,2). Smokers are at a three-fold greater risk of development of transitional cell carcinoma than nonsmokers, and heavy smokers (greater than 40 pack-years) are at five-fold risk (3–5). Smoking at an early age increases risk (6). The major carcinogen in tobacco is 4-aminobiphenyl, but numerous other carcinogens are found in tobacco, including the polycyclic aromatic hydrocarbons, aromatic amines, acrolein, and oxygen-free radicals, demonstrating the powerful cocarcinogenic risk of tobacco use (6–9). Cessation of smoking leads to a reduction in bladder cancer risk (2,8,10).

Lifestyle risks include dietary factors. Even though they have been implicated in the past, saccharine cyclamate, aspartame, and coffee have not been identified as risk factors (11–14).

B. Occupational Exposure

Approximately 25% of bladder cancers occurring in American white males are attributable to occupational exposure (15,16). In 1895, aniline chemical dyes were replacing vegetable dye for the coloring of cloth. Rehn (17) observed that bladder cancer occurred frequently among industrial workers in the aniline dye factories of Germany. Multiple studies since have confirmed the carcinogenicity of the benzine ring compounds. In 1938, Hueper et al. (18) established the carcinogenicity of 2-arylamine naphthylamine by showing that it induced bladder cancers in dogs identical to those found in humans (18). A large study of the British chemical industry by Case et al. (19) reported a 30-fold increase of bladder cancer associated with arylamine exposure. Similarly, the Drake Health Registry Study, which follows up workers employed in the manufacture of beta-naphthylamine, has calculated a 20-to 30-fold excess mortality rate from bladder cancer in this population (20). Wakayama City, Japan, was the center for benzidine and beta-naphthylamine production between 1921 and 1972. Recognition of the exposure risk of the worker population prompted a screening registry that included 1085 male subjects, of which, urothelial cancer had developed in 112 (10.3%) (21). The risk appeared greater among those patients exposed to benzidine (12%) versus those exposed to beta-naphthylamine (3.3%). The mean interval from initial exposure to diagnosis was 24 ± 9.4 years. A recently identified industrial exposure risk has been reported in the aluminum refining industry in Canada and the rubber chemicals manufacturing industry in the United States (22,23).

In Taiwan, transitional cell carcinoma is associated with a peripheral vascular disease termed “blackfoot disease” (24). Both are deemed secondary to arsenic contained in artesian well water.

C. Iatrogenic

Cyclophosphamide, an alkylating agent, causes urothelial cancer through the action of a toxic metabolite, acrolein (25). The risk for bladder cancer is related to cumulative dose exposure and has been reported to be as high as 11% after a latency period of 6 to 13 years (26–28). Because of the long latency period for cancer induced by alkylating agents, the patients most at risk for the sequelae of drug-induced cancer are those who have been cured from their primary disease by these very agents. Ifosfamide is an alkylating agent which also causes acute bladder toxicity and, therefore, could possibly be a bladder carcinogen (29).

Phenacetin is an established cause of transitional cell carcinoma, and risk is related to increasing consumption (30, 31).

Patients who have received radiation therapy for treatment of carcinoma of the cervix have an increased risk of transitional cell carcinoma (32, 33).

D. Infectious

Squamous cell carcinoma caused by the chronic inflammation secondary to *Schistosoma haematobium* infection is the most common malignancy in Egypt (34). Chronic cystitis secondary to calculi and indwelling catheters increase the risk for squamous cell carcinoma among spinal cord injury patients (35). Ochratoxin is a mycotoxin that has been implicated as a causal agent of Balkan endemic nephropathy, which is also associated with an increased risk for transitional cell carcinoma of both the upper and lower urinary tract (36, 37).

E. Genetic Factors

The human liver enzyme N-acetyltransferase (NAT) may be involved in the detoxification of aromatic amines. The activity of this enzyme is genetically regulated, and individuals who are homozygous for this autosomal recessive gene acetylate slowly. Patients with bladder cancer are more likely to be slow acetylators. Because they are unable to detoxify aromatic amines efficiently, they are at greater risk from exposure (38, 39).

Of interest is the potential risk secondary to endogenous carcinogens. Tryptophan and its metabolites have been implicated. The normal degradation products of the amino acid tryptophan have structures similar to aromatic amines and have been shown to cause bladder papilloma and carcinoma in mice (40–42).

II. THE BURDEN OF SUFFERING

More than 52,900 new cases and more than 11,700 deaths were predicted to occur in 1996 in the United States (43). From 1973 to 1990, the incidence of bladder cancer in the United States increased at a rate of 0.7% a year; however, the mortality rate decreased at an annual rate of 1.8% (44). The incidence of bladder cancer is three to four times greater in men than in women and approximately two times greater in white men compared to black men. The risk of cancer increases steeply with age. Among white men, the annual incidence of bladder cancer after the age of 65 years is approximately 2 per 1000 versus 0.1 per 1000 in men younger than 65 years (45). In the United States, the lifetime probability of bladder cancer development is 3% and of dying of bladder cancer is less than 1% (46). Worldwide, bladder cancer accounts for 3.2% of all newly diagnosed cases and 2.4% of cancer deaths. In 1985, it was estimated that 243,000 new cases and 112,700 deaths were secondary to bladder cancer (47, 48).

As discussed, although the incidence of bladder cancer is increasing, the mortality rate is declining (44). This has, in part, been attributed to the effective use of intravesical BCG in patients with stage Ta, Tis, and T1 (Table 1) transitional cell carcinoma. Several studies have demonstrated statistical decrease in tumor progression after induction course of BCG (49–51). Maintenance therapy with BCG may further decrease progression to muscle invasive disease (52), which is the morbid and lethal event associated with bladder cancer.

III. NATURAL HISTORY OF BLADDER CANCER

Transitional cell carcinoma arising in the urothelial lining of the bladder comprises 95% of bladder cancers in the United States. Squamous cell and adeno-

Table 1 Pathological Staging of Bladder Carcinoma

Depth of invasion	AICC/UICC
Noninvasive, papillary	Ta
Noninvasive, flat	TIS
Lamina propria	T1
Superficial muscularis propria	T2
Deep muscularis propria	T3a
Perivesicle fat	T3b
Adjacent structures	T4
Lymph node metastases	N1–N3[a]
Distant metastases	M1

[a] N1, regional lymph node, <2 cm; N2, regional lymph nodes, 2–5 cm; N3, regional lymph nodes, >5 cm or other lymph nodes.

carcinoma comprise the remaining 5%. The TNM staging system is provided in Table 1.

Precise staging of transitional cell carcinoma from tissue obtained either by cold cup biopsy or loop resection obtained at cystoscopy is often difficult because of sampling error, difficulty in proper specimen orientation, or the absence of adequate muscle to document invasion. In the past, definitions of grading have also been inconsistent. This has been rectified by the establishment of grading guidelines by the World Health Organization (WHO) (53). An accurate and consistent study of the natural history of well-defined subgroups has been compromised by these problems. In addition, treatment interventions by resection, fulguration, intravesical immunotherapy with BCG, and/or intravesical chemotherapy alter the natural progression of superficial lesions and occasionally cause difficulty with posttreatment histological and cytological interpretation.

A. Stage Ta

Papillary carcinoma is limited to the mucosa, usually is of low-grade, and has a low incidence of progression. Some pathologists refer to grade 1 Ta lesions as papilloma and do not classify them as malignant. Ta lesions vary in size and are often multiple. Three grades based on the degree of cellular anaplasia are recognized by the WHO. In a multivariate analysis, tumor grade has been shown to be the dominant prognostic factor for likelihood of recurrence, invasion, and cause-specific death of superficial bladder cancer (54).

B. Stage Tis

In 1964, Melamed et al. (55) described the prognostic significance of flat carcinoma in situ and separated it from the other superficial phenotype, noninvasive papillary carcinoma. Carcinoma in situ, which frequently coexists with papillary carcinoma, is a high-grade lesion with a propensity for recurrence, invasion, and cause-specific death. Because of the lack of cellular cohesion, cytology is of considerable diagnostic value in detecting an abnormality, which may be cystoscopically invisible.

C. Stage T1

Carcinoma penetrating the basement membrane and involving the subepithelial tissue (i.e., lamina propria involvement) is classified as T1 and has a likelihood of progression and cancer-specific death intermediate between Ta/Tis and muscle invasive tumors (54).

The incidence of recurrence progression and cancer-specific death by stage and grade are illustrated in Figures 1 and 2. In the next decade, molecular markers promise to identify subgroups of patients within stage and grade at

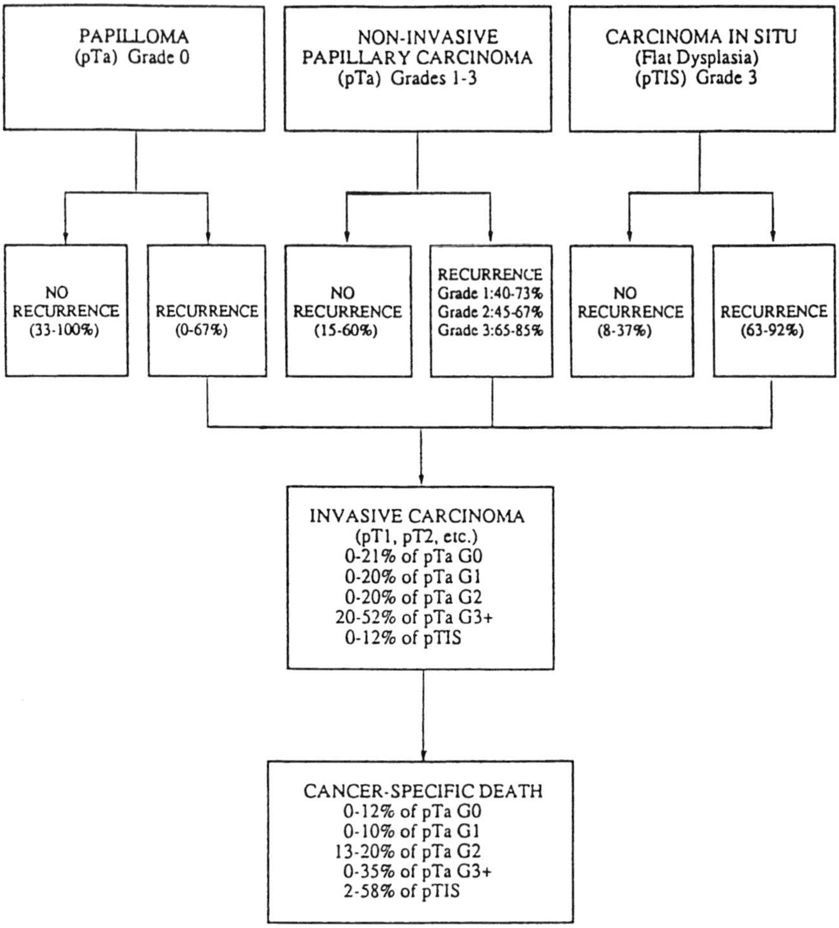

Figure 1 Recurrence and progression in predominantly surgically treated noninvasive TCC (data from Table 1).

particularly high risk for progression and metastasis. Molecular markers associated with poor prognosis include p53 and Rb1 mutation and epidermal growth factor overexpression (56–58).

If transitional cell carcinoma is diagnosed when still confined to the urothelium or subepithelial connective tissue (Ta/Tis/T1), it can often be managed successfully with resection, fulguration, and, in some cases, intravesical instillation of chemotherapeutic or immunotherapeutic agents. Approximately 80% of bladder cancers present in superficial form, and the 5-year progression-free

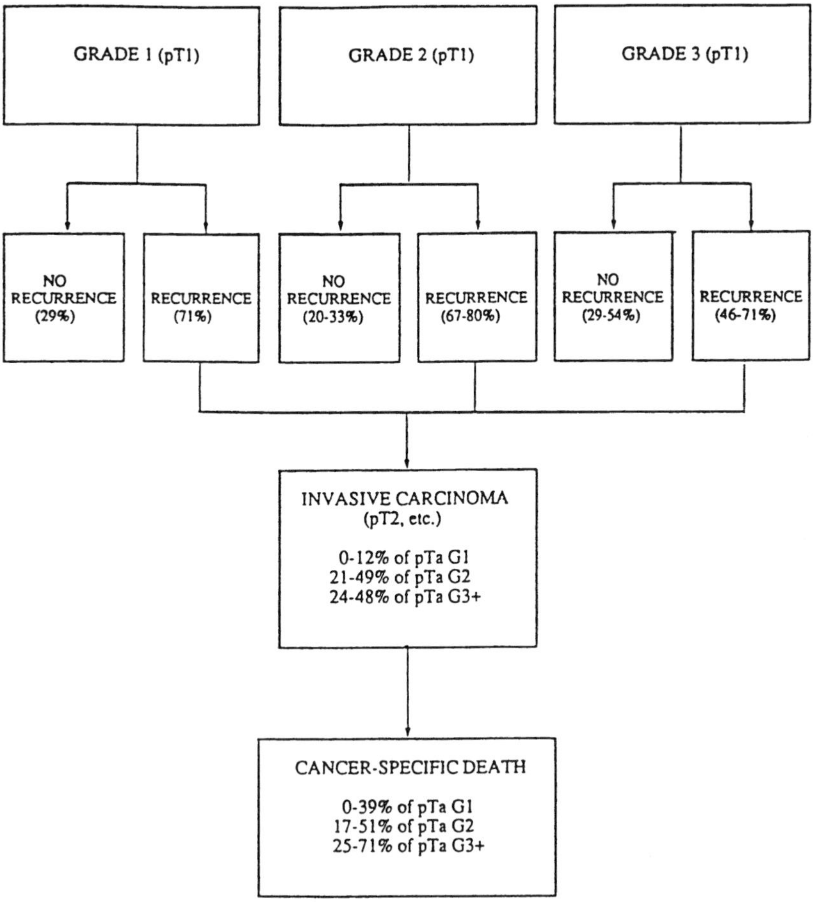

Figure 2 Recurrence and progression in predominantly surgically treated TCC with lamina propria invasion (data from Table 2).

rate for patients with superficial bladder cancer, based on presenting stage and grade, ranges between 50% and 95%. Approximately 20% of bladder cancers present with initial evidence of muscle invasion without having first evidenced a superficial phenotype (59,60). Presumably, this superficial phenotype existed in an undetected state, and belated symptoms or investigations resulted in an advanced stage being found at first diagnosis. The rationale for screening and early detection largely resides in the earlier detection of these bladder cancers destined to invade muscle while still superficial and still appropriate for resection, fulguration, and intravesical therapy. This strategy of early detection assumes that tumors with muscle invasion at the time of presentation were superficial for some period of time during which the opportunity existed for more successful and less invasive treatment. This assumption is logical, based on the recognized urothelial (surface) origin of transitional cell carcinoma of the bladder, but it is, as yet, not tested in a randomized controlled clinical trial setting.

IV. REQUIREMENTS FOR SCREENING SUCCESS

A. General Screening

In 1968, the World Health Organization stated three criteria necessary for consideration of screening for a disease process (61). They are summarized as follows: The disease must be a serious health problem that causes significant morbidity, mortality, or both in the general or a well-defined population; an effective treatment that decreases morbidity and mortality must be available for early stage disease; the screening test used must be simple, safe, rapid, inexpensive, and reproducible and must have acceptable performance characteristics. The outcomes of an effective screening would include the following: The detected tumors are smaller, of lower clinical and pathological stage, with fewer having metastasized; survival time following diagnosis is prolonged, and the disease-related mortality rate of the screened population is lower than that of a comparable unscreened population. These outcomes can be rigorously tested only through prospective randomized trials. Prospective randomized trials minimize lead time bias (earlier diagnosis and prolonged survival time with unchanged time to death), length time bias (a predilection by screening to detect slowly growing rather than rapidly growing tumors), overdetection bias (the detection of tumors, which, although present, will never manifest disease-related morbidity or mortality), and selection bias (the possibility that a group of patients seeking screening through personal initiative are likely to differ from the general population).

No prospective randomized trials have tested screening for bladder cancer. However, present knowledge of the natural history of the disease, documented success of treatment modalities, and preliminary observational studies make it possible to speculate that bladder cancer might fulfill screening criteria and be a candidate for the labor and cost-intensive effort of a large phase 3 trial.

B. Bladder-Specific Screening

The three criteria for productive screening are met by bladder cancer demographics as follows:

1. Bladder cancer, the fourth leading cause of cancer deaths in the United States and responsible for 11,700 deaths in 1996, is a serious health problem.

2. Because virtually all bladder cancers are transitional in origin and arise on the luminal surface, they are immediately accessible to diagnosis by positive Hemastix, cytology, and cystoscopy and to therapy by resection, fulguration, and adjuvant intravesical therapy. These are effective treatments associated with acceptable morbidity and virtually no mortality. In contrast, muscle invasive disease is associated with a high mortality rate, and the intervention of cystectomy and urinary diversion is morbid and often unsuccessful.

3. Although hematuria is related to neither bladder tumor size nor number and may be intermittent and minimal, the most tumors do produce hematuria. Microhematuria testing by Hemastix represents a screening test that is sensitive for blood and is safe, rapid, inexpensive, noninvasive, and simple. Furthermore, it can be performed repetitively and accurately by the subjects themselves.

The observational hematuria screening studies of Messing (62–65) and Britton (66, 67) have demonstrated the shift of clinical and pathological stage of disease to less advanced stage. They have suggested a longer survival time and, therefore, the potential for decrease in disease-related mortality rate.

The delay in detection and evaluation of hematuria may be the explanation for muscle invasion at the time of first detection of some tumors. The opportunity to diagnose before progression to muscle invasion would represent a major step toward applying less morbid therapy with greater success. Therefore, the groundwork studies have been done to identify bladder cancer as a disease suited to a screening effort. Until a randomized trial is performed to test the hypothesis, recommendations for bladder screening must be held in abeyance. However, for high-risk populations, screening may represent a reasonable strategy for early detection and has been implemented (Drake Registry, Wakayama City). Currently, high-risk populations include men older than age 60 years who are active smokers or who have a long smoking history and subjects with exposure to specific high-risk environmental carcinogens in the work place setting.

Urinalysis with Hemastix testing is often routinely done at the time of annual health check evaluations. ''Opportunistic'' screening for hematuria is, therefore, already being performed to some extent. However, the follow-up urological evaluation to detect a source is often omitted or delayed.

V. SCREENING TESTS AND OPERATING CHARACTERISTICS

The gold standard for detection/diagnosis of bladder cancer is cystoscopy, which provides visualization of any abnormalities and an avenue for direct biopsy of these abnormalities for histological diagnosis. However, cystoscopy is expensive, requires technical expertise, is invasive and, for these reasons, cannot be used as a general screening tool of bladder cancer. However, it is the current standard against which other early detection strategies must be measured.

Therefore, when alternative tests for screening and detection have been investigated, their sensitivity and specificity have been measured against cystoscopic evaluation.

The two most thoroughly investigated tests for early detection of transitional cell carcinoma in asymptomatic patients are examination of the urine for microhematuria or for abnormal cytology. Recently, other immunologically based testing has been proposed and is briefly discussed at the end of this section.

A. Microhematuria

The definition of microhematuria is generally accepted as more than three red blood cells (RBCs) per high powered field (68). Some advocate the more stringent criteria of no more than one red cell per high powered field (69,70). Hemastick evaluation has supplanted microscopic analysis of urine because of its excellent sensitivity (>90%) for hemoglobin (71–74), its convenience, reproducibility, and economy, and its ease of application in both laboratory and patient home testing settings. Hemastix depends on the reaction of peroxidase activity with hemoglobin to record a colorimetric reaction (70). Urine specific gravity of 1.010 or less may cause lysis of RBCs. For example, in urine of specific gravity equal to or less than 1.008, quite common in well-hydrated person, more than 96% of RBCs will lyse (75). However, because it measures hemoglobin, the hemastick remains positive. The complexity of reagent strip testing must not be overlooked. Although they are easily used, reagent strips represent multiple state-of-the-art chemical reactions, and directions must be followed appropriately (76). False-negative reactions may result from high concentrations of ascorbic *acid* or exposure of the strip to air (76). The test lacks specificity for a number of reasons; false-positive results may occur secondary to chemical interference by oxidizing agents, myoglobin, contamination of a urine specimen with iodine (77), substances that discolor urine (e.g., azo dyes, nitrofurantoin, riboflavin); equally important is that false-positive results occur secondary to multiple other processes in the urinary tract that cause hematuria (e.g., infection, benign prostatic hyperplasia, prostatitis, calculus disease, renal disease, nonsteroidal anti-inflammatory drugs, prolonged recumbency, vigorous physical activity) (78,79). Even though microscopic urinalysis clearly provides substantial information beyond whether an abnormal amount of hemoglobin is present in urine, for the purpose of looking for urothelial malignancies, these additional pieces of information are not essential.

Despite the false-positive confound, a number of population-based studies have concluded that asymptomatic microhematuria requires evaluation (80, 81). Evaluation includes upper tract imaging by intravenous pyelography (IVP) and cystoscopic evaluation of the bladder. The small risk of contrast reaction with

IVP can be eliminated by substituting renal ultrasound, and the discomfort of cystoscopy has been markedly reduced by use of the flexible fiberoptic instrument. In some studies, no source of hematuria is ever identified in 50% of patients testing positive (67, 82). However, in a recent study of patients with microhematuria, a source was identified in 85% (64). As diagnostic procedures improve, fewer unresolved sources of hematuria are likely.

B. Cytology

Cytologic examination of urine was introduced by Papanicolaou and Marshall in 1945, and it has been successfully used to screen for carcinoma of the bladder (83). Voided urine cytology is a noninvasive test. The first voided morning specimen, although most cellular, is not a hospitable medium for desquamated cells because it is hyperosmotic and has a low acid pH. The cells are often poorly preserved and sometimes difficult to assess because of nuclear and cytoplasmic degenerative changes. It is a tenet of cytology that a malignant diagnosis should never be rendered on cells with degenerative changes in which the chromatin is not well preserved and clearly visible. To minimize these problems, a specimen other than the first void should be submitted. Three consecutive daily specimens are recommended. Of the neoplasms that can be diagnosed with voided urine cytology, about 80% are with the first, 15% with the second, and the remainder with the third (84). There is no difference in yield of benign or malignant cells from the initial, midstream, or terminal part of the void (85).

A voided urine is easy to obtain, it samples the entire urinary tract, and it has good sensitivity for high-grade tumors. The disadvantages include cellular degeneration, a low number of cells, particularly in low-grade tumors, and possible contaminations from the female genital tract.

A catheterized urine specimen yields better preserved, more cellular, and less contaminated specimens than voided urines. However, the catheter tip can dislodge clusters of urothelial cells that can have a pseudopapillary appearance that can be confused with papillary carcinoma. In addition, this technique yields more surface umbrella cells, which, because of their larger nucleus and abundant cytoplasm, may be misinterpreted as atypical. These umbrella cells can also become multinucleated. The presence of the large nuclei reflects the tendency of the urothelium to form tetraploid or even octaploid nuclei (86). This tendency to polyploidy appears to be the pattern of normal urothelial differentiation. The nuclear enlargement that results is more evident in specimens obtained by means of catheterization than in voided specimens (87).

Saline bladder wash specimens may provide further diagnostic superiority because they are more cellular. Even though catheterized or bladder wash urine is superior to voided urine in diagnosis, it is inconvenient, uncomfortable, more expensive, and carries a risk of infection.

Specimens may be submitted fixed or unfixed. Unfixed fresh specimens are preferred in some laboratories where there will not be a delay in delivery to the cytology laboratory. As an alternative, all specimens are preserved immediately in a fixative, usually some modification of 50%–70% ethanol.

Koss (88) succinctly discusses and lists the principal indications for use of the cytology in disorders of the lower urinary tract. Cytology has important limitations in reference to the diagnosis of well-differentiated (low-grade) papillary tumors, which, as a general rule, cannot be reliably identified in cytologic material. However, cytology is accurate in the identification of high-grade tumors and flat carcinoma in situ.

Urinary cytology is fraught with diagnostic traps and pitfalls. There are many situations that can lead to false-positive or false-negative interpretations. Transitional cell epithelium has a marked tendency to exfoliate with the slightest trauma to the mucosa. Catheterization, bladder washing, other instrumentation, or even vigorous exercise can result in detachment of pseudopapillary tissue fragments. Moreover, stones, postsurgical trauma, ulceration, or other inflammatory processes can lead to exfoliation of pseudopapillary aggregates with cytologic atypia (89). Because of these disadvantages and the expense of single and repetitive testing, cytology is not a practical first screen strategy.

VI. MICROHEMATURIA SCREENING STUDIES

Observational studies have investigated microhematuria as an early detection test by using one-time hematuria screen versus multiple hematuria screens. Because hematuria is intermittent, one-time screening has been least sensitive for detection of bladder cancer. In a retrospective review of 20,000 men and women, only one bladder cancer was found (90).

Two centers have reported initial pilot studies followed by larger confirmatory studies of hemastick screening. Britton et al. (66,67) conducted two studies in Leeds, England. Men between the ages of 60 and 85 years were invited to attend a health center for urine screening as part of a health check. The pilot study tested 578 men with a single reagent strip evaluation followed by 10 weekly tests; 22% of subjects tested positive at least once. Of these, 87 were investigated by upper tract evaluation and cystoscopy. Four were found to have bladder cancer (3 G1pTa, 1 G2pT1) for a prevalence of (4/578) 0.7%. No correlation with degree or frequency of hematuria was noted. A larger follow-up study of 2356 male subjects, similarly solicited from the Leeds population, was reported 3 years later and confirmed the incidence of hematuria (20%) and prevalence of bladder cancer (0.72%). Of the 17 tumors, nine were G1pTa1, seven were G2pT1, and one carcinoma in situ. Cytology was positive in all tumors with grade greater than 1. In neither study were all men with hematuria tested. The results, therefore, represent minimum prevalence but may also overestimate sensitivity and specificity. A 3-year follow-up (91) reported progression of only one of the T1G2 tumors, which is a rate lower than that predicted (see Figure 2).

In 1987, Messing et al. reported a pilot study of 235 men 50 years and older who were recruited from the University of Wisconsin Health Maintenance System. Patients with a history of hematuria, urinary tract infection, or instrumentation were excluded. Their urine was tested by reagent strip for 5 consecutive days and then once weekly for 1 year (62). Any one episode of hematuria was investigated by upper urinary tract study, cytology, and cystoscopy; five of the 235 men were found to have a urothelial malignancy. In 1992, this study was extended to a larger population of men recruited from four clinics representing a range of private health care in south central Wisconsin, again excluding subjects whose records indicated a prior history of or predisposition to hematuria (64). In this study, subjects were asked to test their urine daily for 14 consecutive days. Again, those with any positive test were evaluated by IVP (renal ultrasound and retrograde pyelography if allergy or elevated creatinine) and bladder lavage for cytology and cystoscopy. Of the 2982 men solicited, 1640 (55%) agreed to participate; 77% (1340) of the participants completed the testing; 21.1% (270) of these had at least one episode of hematuria, and 73% of these had a urologic evaluation as described. Sixteen men were diagnosed with urologic cancers: nine transitional cell carcinoma of the bladder, six prostatic carcinoma, and one renal carcinoma. The prevalence of bladder cancer was 0.67% (9/1340). Of the nine bladder cancers detected, three were G1pTa, two G2pTa, and four G3 (2pTa, 1 Tis, and 1pT1). A cause for hematuria was determined in 85% of the subjects evaluated.

Messing et al. (64a) investigated the interval that might be appropriate for hematuria screening. Eight hundred fifty-six men who had had 14 negative tests for hematuria in the initial study repeated the 14-day testing protocol 9 months later. Of the 50 (5.8%) men who had a positive test, 38 were evaluated and seven had transitional cell carcinoma. The incidence rate was, therefore, 0.82% (7/856). The authors suggest that this very similar yield of transitional cell carcinoma among patients who had a negative test 9 months previous to that of the initial testing screen indicates a brief preclinical duration of disease and supports annual screening. Alternative explanations as to why the bladder cancers detected in the second screen were not detected in the first include noncompliance, lack of sensitivity of the agent strip, inappropriate testing, and intermittent bleeding. A clear advantage to the second screen was the reduction of the confounding influence of benign causes of hematuria from this population, which had previously tested negative. Only 5.8% were hematuria-positive compared to 21.1% in the initial screen. This increase in specificity reduced invasive and unnecessary testing and cost of cancer detection. The expense of a complete urologic evaluation for the first screen was $522.00. The average expense of screening and evaluation of all first screen subjects was $90.00. The average cost per cancer detected was $7,618.00. In contrast, the average expense for screening all patients in the repeat screen was reduced from $90.00 to $22.76, and the average cost per bladder cancer detected was reduced from $7,618.00 to $2,783.57. The authors also point out that the economies of screening could be further increased by identifying risk groups, namely patients older than 60 years of age with a history of cigarette smoking. Even though this might miss some cancers, it would increase the yield of screening and, therefore, decrease cost. High levels of compliance indicate that reagent strip testing is applicable to the general population; 97.7% of participants completed at least 10 of the 14 tests, and 88.7% used correct testing techniques.

The detection of 16 (9 + 7) cancers among 320 (270 + 50) subjects with hematuria (5.3%) can be compared to the detection rate reported for a population of men referred for evaluation of hematuria found on a routine urinalysis or reagent strip test. Carson et al. (80) found 22 cancers of the bladder among 200 men (11%), and Golin et al. (81) found 16 bladder cancers among 246 men (6.5%) referred because hematuria was found on routine urinalysis. These similar rates are used to argue against the concern of overdiagnosis driven by screening. There were 1940 subjects who were solicited but opted not to participate in the screening studies. They were tracked for the diagnosis of bladder cancer over the same time period; 23 of 1940 (1.2%) were identified in whom bladder cancer developed. Transitional cell carcinoma was detected in 16 of 1340 (1.2%) participants in the two testing periods 9 months apart. This identical incidence is also used to argue against overdiagnosis. Evidence against overdiagnoses is the fact that bladder cancer is an unusual incidental finding at autopsy (92, 93), even among studies in which concerted efforts have been made to inspect the entire bladder urothelium. A study of 313 autopsy cases identified one case of carcinoma in situ. These observations indicate that there is not a large pool of subclinical disease and that the biological behavior of bladder tumor requires relatively frequent repeat testing. However, screening for hematuria in asymptomatic patients does not detect more cancers than in a nonscreened population and can only be tested in a randomized trial with defined diagnostic interventions at specific intervals.

As previously stated, a requirement for a successful screening trial is the detection of disease at an earlier stage with the potential for more successful therapy to reduce cancer morbidity and mortality rates. To gain insight into these questions, Messing et al. extended the comparison of the hematuria-screened population with nonscreened populations by investigating bladder tumor grade, stage, and mortality. The unscreened population data was derived from two sources: the group of patients solicited for screening who did not participate in the protocol and a group of patients recorded in the Wisconsin Tumor Registry.

The 16 cancers detected over a 9-month interval from the large study were combined with the five cancers detected in the original pilot study; 527 cancers were reported from the Wisconsin Cancer Registry over this time period. All *21* of the histology slides from the screening studies and 89.3% of the slides from the registry were available for review by the same pathologist without knowledge of the original report.

The incidence of high-grade cancer in the screened population (47.6%) was virtually identical to the unscreened population (43.2%) (65). However, muscle invasive tumor was present in 23.9% of the cancer tumor registry population versus 4.8% of the screened cases. Unscreened high-grade tumors were muscle invasive much more frequently than screened high-grade tumors. At 24 months' follow-up after the diagnosis of bladder cancer, 16.4% of the unscreened Wisconsin Tumor Registry patients, but none of the screened patients, had died of bladder malignancy. Of the patients who were solicited but did not participate in screening, 8.7% died of malignancy within 24 months of diagnosis. Bladder cancer mortality rate in the Wisconsin Tumor Registry Group (16.4%) was similar to the bladder cancer mortality rate (17.7%) reported in the United States for men within the 2 years following diagnosis.

There is evidence from other studies that screening might reduce mortality rate. In the Wakayama City study II (21), patients whose primary bladder carcinoma was diagnosed by symptoms were compared to those whose primary bladder tumor was diagnosed by screening. There was a statistically higher proportion of grade 3 tumors observed in the symptomatic group. A statistically greater proportion of patients in the screening group were suitable for treatment by cystectomy. The 5-and 10-year survival rates of 91% and 75%, respectively, for the screened patients were statistically superior to the 69% and 55% survival rates for the symptomatic patients. However, in the absence of a randomized trial, these apparent benefits may be secondary to lead time and length time bias.

Finally, Messing compared his study participant demographics with those from a population-based survey for the Wisconsin male population, aged 50 years and older. Men screened had a higher level of education and were less likely to be currently smoking cigarettes. These and other unrecognized characteristics that might exist between the two groups emphasizes the importance of a prospective randomized trial to eliminate population selection biases (e.g., access to optimal health care, differences in occupational or carcinogen exposure, lead and length time between screened and unscreened populations). Only in this way can the critical question be answered as to whether bladder cancer mortality rate is decreased by hematuria screening. However, the low prevalence of bladder cancer, the low positive predictive value for detecting urothelial cancer by hematuria screening, and the cost and inconvenience of subsequent evaluation are obstacles to the conduct of such a trial.

VII. POTENTIAL OF NEW METHODOLOGY FOR DETECTION OF BLADDER CANCER

The bladder tumor antigen (BTA) test detects the presence of basement membrane complexes in voided urine samples. It is rapidly performed within 3 to 5 minutes. The mechanism of the test is based on tumor secretion of proteolytic enzymes that disrupt the basement membrane into fragments. These are discharged into the bladder lumen, aggregate into high molecular weight complexes, which are detected by latex agglutination. Several centers have tested the sensitivity of BTA in comparison to either bladder wash or voided urine cytology. A relatively small study involving 65 bladder cancer patients from a single center (94) and a large 499-patient multicenter trial found a statistically significant increase in sensitivity for BTA versus voided urine cytology (95). For healthy volunteers, the specificity was 95.9%. Specificities for patients with urologic conditions other than bladder cancer was 95% for benign prostatic hypertrophy (BPH) and cystitis, 85% for benign renal disease, and 86% for prostate and renal cancer. For renal stones, the specificity decreased to 81.8%. However, the specificity of this test in the population for which it is used, namely those with some evidence of other genitourinary disease rather than a completely healthy population, is only in the 80%–85% range. A high incidence of false-positive studies has limited the application of flow cytometry to screening (96, 97). Furthermore, the cost and labor-intensive characteristics of flow cytometry limit their application and certainly eliminate them from point of contact office performance.

An immunoassay that detects a nuclear matrix protein known as NMP22 has been tested on voided urine of patients approximately 10 days after transurethral resection of bladder cancer (98). Postsurgical patients with an elevated NMP22 test have a 70% chance of recurrence at subsequent cystoscopy. Because the negative predictive value of the test is 86%, the frequency of invasive follow-up may be modified by NMP22 test results. However, the operating characteristics of this test as a primary screening tool have not been determined. An immunoassay to detect hyaluronidase and hyaluronic acid in urine has been reported to have a sensitivity and specificity that exceeds 90% (99).

Based on the fact that multiple genetic changes occur during the development of primary bladder cancer, a testing strategy to detect these genetic changes before the clinical manifestation of tumor is logical. Using bladder cancer tissue, Mao et al. (100) identified microsatellite marker alterations that were commonly found in these cancers. In addition to these alterations, primary tumors often contained chromosomal deletions (loss of heterozygosity). Using a test for microsatellite alterations and loss of heterozygosity, 19 of 20 patients with documented transitional cell carcinoma were identified. Cytology was able to detect only 50% of the cancers. Five patients without neoplasia were tested, and no abnormalities were noted. This limited experience needs to be tested further. However, the excellent sensitivity and specificity and the potential for testing at a third of the cost of cytology make such molecular testing promising for future research. Parenthetically, the power of molecular biology as an avenue to the early detection of bladder cancer was dramatically demonstrated in a report by Hruban et al. (101). A chronology of the bladder cancer history of the late Senator and Vice President Hubert Humphrey was outlined as follows: Using Polymerase chain reaction technology, a mutation of the p53 tumor suppressor gene was found in the paraffin-embedded tissue block of the infiltrating bladder cancer that was resected in 1976 from Mr. Humphrey. This same mutation was identified in a preserved urine cytology specimen collected in 1967 when he had been first evaluated for hematuria. Therefore, the same cells associated with an eventually lethal bladder cancer in 1978 were present 11 years earlier in a voided urine specimen. The authors make several important observations: p53 mutations occur at various sites, and many bladder cancers do not harbor the p53 mutation. Therefore, p53 mutations may not be the best molecular marker for early detection. Nevertheless, the power of PCR-based technology is illustrated in this historical case. Other molecular-based markers will undoubtedly be found with greater screening potential (99,102).

VIII. RECOMMENDATIONS

Currently no major organization recommends screening for bladder cancer in asymptomatic adults. The Canadian Task Force on the Periodic Health Examination recommends against routine screening in asymptomatic individuals and concludes that there is insufficient evidence for or against screening in specific high-risk groups (103). The American Cancer Society and the National Cancer Institute (NCI) have not issued any specific guidelines on screening for bladder cancer.

Routine screening for bladder cancer with urine Hemastix, microscopic urinalysis, or urine cytology is not currently recommended in asymptomatic persons. The level of evidence for screening is based on cohort studies, which, although suggestive, are not conclusive.

All patients who smoke tobacco should be routinely counseled to cease smoking. This constitutes effective preventive action. In addition, persons working in high-risk occupational exposure may be considered target populations for screening. For the general population, however, the low positive predictive value of a positive hematuria screen for bladder cancer would subject many asymptomatic patients to cost, risks, and discomforts of urologic evaluation without clear evidence of benefit.

Although Messing's data suggest that bladder cancer patients detected by screening have lower stage, longer survival, and less cancer mortality, lead and length time bias cannot be eliminated, and only a phase 3 trial can resolve these issues. Presently there are none in planning.

REFERENCES

1. National Cancer Institute. Cancer of the Bladder [NIH Publication No. 90–722]. Bethesda: National Cancer Institute, 1990.
2. Hartge P, Silverman D, Hoover R, et al. Changing cigarette habits and bladder cancers risk: a case-control study. J Natl Cancer Inst 1987; 78:1119–1125.
3. Wynder EL, Goldsmith R. Epidemiology of bladder cancer: a second look. Cancer 1977; 40:1246–1253.
4. Hartge P, Silverman D, Hoover R, et al. Changing cigarette habits and bladder cancer risk: a case-control study. J Natl Cancer Inst 1987; 78:1119–1125.
5. Burch JD, Rohan TE, Howe GR, et al. Risk of bladder cancer by source and type of tobacco exposure: a case-control study. Int J Cancer 1989; 44:622–628.
6. Vineis P: Black (air-cured) and blond (flue-cured) tobacco and cancer risk. I: bladder cancer (review). Eur J Cancer 1991; 27:1491.
7. Talaska G, Al-Juburi AZSS, Kadlubar FF. Smoking-related carcinogen-DNA adducts in biopsy samples of human urinary bladder: identification of N-(deoxyguanosin-8-yl)-4-aminobiphenyl as a major adduct. Proc Natl Acad Sci U S A 1991; 88:5350.
8. Augenstein A, Herbert JR, Kabat GC, et al. Bladder cancer in relation to cigarette smoking. Cancer Res 1988; 48:4405.
9. International Agency for Research on Cancer. Overall evaluations of carcinogenicity: and updating of IARC monographs. In: IARC Monographs on the Evaluation of Carcinogenic Risks to Humans. Vols. 1–42. 1987; 7(suppl):1–440.
10. Burch JD, Rohan TE, Howe GR, et al. Risk of bladder cancer by source and type of tobacco exposure: a case-control study. Int J Cancer 1989; 44:622.
11. Armstrong BK. Saccharin/cyclamates: epidemiological evidence. IARC Scientific Publications 1985; 65:129–143.
12. Chappel CI. A review and biological risk assessment of sodium saccharin. Regul Toxicol Pharmacol 1992; 15:253–270.
13. Council on Scientific Affairs: Aspartame: review of safety issues. JAMA 1985; 254:400–402.
14. Viscoli CM, Lachs MS, Horwitz RI. Bladder cancer and coffee drinking: a summary of case control research. Lancet 1993; 341:1432–1437.
15. Silverman D, Levin L, Hoover R, Hartge P. Occupational risks of bladder cancer in the United States. I. White men. J Natl Cancer Inst 1989; 81:1472–1480.
16. Silverman D, Levin L, Hoover R. Occupational risks of bladder cancer in the United States. II. Nonwhite men. J Natl Cancer Inst 1989; 81:1480–1483.
17. Rehn L: Blasengeschwulste bei Fuchsin-Arbeitern. Arch Klin Chir 1895; 50:588–600.
18. Hueper WC, Wiley FH, Wolfe HD. Experimental production of bladder tumors in dogs by administration of beta-naphthylamine. J Indust Hyg Toxicol 1938; 20:46–84.

19. Case RAM, Hosker ME, McDonald DB, Pearson JT. Tumors of the urinary bladder in workmen engaged in the British chemical industry. Role of aniline, benzidine, alpha-naphthylamine, and beta-naphthylamine. Br J Indust Med 1954; 11:75–104.
20. Marsh GM, Leviton LC, Talbott EO, Callahan C, Pavlock D, et al. Drake Chemical Workers' Health Registry Study: I. Notification and medical surveillance of a group of workers at high risk of developing bladder cancer. Am J Indust Med 1991; 19: 291–301.
21. Shinka T, Sawada Y, Morimoto S, Fujinaga T, Nakamura J, Ohkawa T. Clinical study of urothelial tumors of dye workers in Wakayama City. J Urol 1991; 146: 1504–1507.
22. Theriault G, Tremblay C, Cordier S, et al. Bladder cancer in the aluminum industry. Lancet 1984; 1:947–950.
23. Ward EM, Sabbioni G, DeBord DG, Teass AW, Brown KK, Talaska GG, Roberts D R, Ruder A M, Streicher R P. Monitoring of aromatic amine exposures in workers at a chemical plant with a known bladder cancer excess. J Natl Cancer Inst 1996; 88:1046–1052.
24. Chen C-J, Chuang Y-C, You S-L, et al. A retrospective study on malignant neoplasms of bladder, lung and liver in blackfoot disease endemic area in Taiwan. Br J Cancer 1986; 53:399–405.
25. Levine LA, Richie JP. Urological complications of cyclophosphamide. J Urol 1989; 141:1063–1069.
26. Pederson-Bjergaard J, Ersboll J, Hansen V, et al. Carcinoma of the urinary bladder after treatment with cyclophosphamide for non-Hodgkin's lymphoma. N Engl J Med 1988; 318:1028–1032.
27. Catalona W. Urothelial tumors of the urinary tract. In: Walsh PC, Retik AB, Stamey TA, Vaughan ED, eds. Campbell's Urology. 6th ed. Philadelphia: WB Saunders, 1992.
28. Stillwell TJ, Bensaon RC Jr. Cyclophosphamide induced hemorrhage cystitis. Cancer 1988; 61:451.
29. Araujo CE, Tessler J. Treatment of ifosfamide-induced urothelial toxicity by oral administration of sodium 2-mercaptoethane sulphonate (MESNA) to patients with inoperable lung cancer. Eur J Cancer Clin Oncol 1983; 19(2):195–201.
30. Maclure M, MacMahon B. Phenacetin and cancer of the urinary tract. N Engl J Med 1985; 313:1479–1481.
31. Piper JM, Tonascia J, Matanowski GM. Heavy phenacetin use in bladder cancer in women aged 20 to 49 years. N Engl J Med 1985; 313:292–295.
32. Boice JD, Engholm G, Kleinerman RA, et al. Radiation dose and second cancer risk in patients treated for cancer of the cervix. Radiat Res 1988; 116:3–55.
33. Sella A, Dexeus FH, Chong C, Ro JY, Logothetis CJ. Radiation therapy-associated invasive bladder tumors. Urology 1989; 33(3) :185–189.
34. El-Bolkainy MN, Mokhtar NM, Ghoneim MA, Hussein MH. The impact of schistosomiasis on the pathology of bladder carcinoma. Cancer 1981; 48:2643–2648.
35. Broecker BH, Klein FA, Hackler RH. Cancer of the bladder in spinal cord injury patients. J Urol 1981; 125:196–197.

36. Krogh P. Role of ochratoxin A in disease causation. Food Chem Toxicol 1982; 30: 213–224.
37. Bozic Z, Duanicic V, Belicza M, et al. Balkan endemic nephropathy: still a mysterious disease. Eur J Epidemiol 1995; 11:235–238.
38. Wolf H, Loer GM Jr, Bryan GT. Role of N-acetyl transferase phenotype in human susceptibility to bladder carcinogenic arylamines. Scand J Urol Nephrol 1980; 14: 161.
39. Cartwright RA, Rogers HJ, Barham-Hall D, et al. Role of N-acetyltransferase phenotypes in bladder carcinogenesis: a pharmacological epidemiological approach to bladder cancer. Lancet 1982; 2:842.
40. Brown RR, Price JM, Friedell GH. Tryptophan metabolism in patients with bladder cancer: geographical differences. J Nat Cancer Inst 1969; 43:295.
41. Boyland E, Williams DC. Metabolism of tryptophan in patients suffering from cancer of the bladder. Biochem J 1956; 64:578.
42. Allen MJ, Boyland E, Dukes C, et al. Cancer of the urinary bladder induced in mice with metabolites of aeromatic amines and tryptophan. Br J Cancer 1957; 11: 212.
43. Wingo PA, Tong T, Bolden S. Cancer statistics, 1995. CA Cancer J Clin 1995; 45: 8–30.
44. Hankey BF, Silverman DT, Kaplan R. Urinary bladder. In: Miller BA, Ries LAG, Hankey BF, et al, eds. SEER cancer statistics review 1973–1990 [NIH Publication No. 93–2789]. Bethesda, MD: National Cancer Institute, 1993:26.1.
45. Ries LAG, Miller BA, Hankey BF, et al, eds. SEER cancer statistics review 1973–1991: tables and graphs [NIH Publication No. 94-2789]. Bethesda, MD: National Cancer Institute, 1994.
46. U.S. Preventive Services Task Force. Guide to clinical preventive services. 2d ed. Baltimore: Williams & Wilkins, 1996: 181–186.
47. Pisani P, Parkin DM, Ferlay J. Estimates of the world mortality from eighteen major cancers in 1985; implication and projections of future burden. Int J Cancer 1993; 55:891.
48. Parkin DM, Pisani P, Ferlay J. Estimate of the world incidence of eighteen major cancers in 1985. Int J Cancer 1993; 54:594.
49. Herr HW, Laudone VP, Badalament RA, et al. BCG therapy alters progression of bladder cancer. J Clin Oncol 1988; 6:1451.
50. Pagano F, Bassi P, Milani C, Meneghini A, Maruzzi D, Garbeglio A. A low dose Bacillus Calmette-Guerin Regimen in superficial bladder cancer therapy: is it effective? J Urol 1991; 146:32–35.
51. Herr HW, Schwab DM, Zhang ZF, et al. Intravesical BCG therapy prevents tumor progression and death from superficial bladder cancer: 10-year follow-up of a prospective randomized trial. J Clin Oncol 1995; 13:1404.
52. Lamm DL, Crawford, ED, Blumenstein, BA, et al. Maintenance BCG immunotherapy of superficial bladder cancer: a randomized prospective study of the Southwest Oncology Group (abstr). J Urol 1992; 147:274A.
53. Sobin LH. The WHO histological classification of urinary bladder tumors. Urol Res 1978; 6:193–195.

54. Bostwick DG. Natural history of early bladder cancer. J Cell Biochem 1992; 161 (suppl) :31–38.
55. Malamed MR, Voursa NG, Grabstald H. Natural history and clinical behavior of in situ carcinoma of the human urinary bladder. Cancer 1964; 17:1533–1545.
56. Sidransky D, Von Eschenbach A, Tsai YC, et al. Identification of p53 gene mutations in bladder cancers and urine samples. Science 1991; 252:706.
57. Cordon-Cardo C, Wartinger D, Petrylak D, et al. Altered expression of the retinoblastoma gene product is a prognostic indicator in bladder cancer. J Natl Cancer Inst 1992; 84:1251.
58. Messing EM. Clinical implications of the expression of epidermal growth factor receptors in human transitional cell carcinoma. Cancer Res 1990; 50:2530–2537.
59. Kaye KW, Lange PH. Mode of presentation of invasive bladder cancer: reassessment of the problem. J Urol 1982; 128:31–33.
60. Hopkins SC, Ford KS, Soloway MS. Invasive bladder cancer: support for screening. J Urol 1983; 130:61–64.
61. Wilson JMG, Junger G. Principles and practice of screening for disease. Public Health Papers 1968; 34:7–151.
62. Messing EM, Young TB, Hunt VB, Emoto SE, Wehbie JM. The significance of asymptomatic microhematuria in men 50 or more years old: findings of a home screening study using urinary dipsticks. J Urol 1987; 137:919–922.
63. Messing EM, Young TB, Hunt VB, Wehbie JM, Rust P. Urinary tract cancers found by home screening with hematuria dipstick in healthy men over 50 years of age. Cancer 1989; 64:2361–2367.
64. Messing EM, Young TB, Roecker EB, et al. Home screening for hematuria: results of a multiclinic study. J Urol 1992; 148:289–292.
64a. Messing EM, Young TB, Hunt VB, et al. Hematuria home screening: repeat testing results. J Urol 1995; 154:57–61.
65. Messing EM, Young TB, Hunt VB, et al. Comparison of bladder cancer outcome in men undergoing humaturia home screening versus those with standard clinical presentations. Urol 1995; 45(3) :387.
66. Britton JP, Dowell AC, Whelan P. Dipstick hematuria and bladder cancer in men over 60: results of a community study. BMJ 1989; 299:1010–1012.
67. Britton JP, Dowell AC, Whelan P, Harris CM. A community study of bladder cancer screening by the detection of occult urinary bleeding. J Urol 1992; 148:788–790.
68. Woolhandler S, Pels RJ, Bor DH, et al. Dipstick urinalysis screening of asymptomatic adults for urinary tract disorders: I. Hematuria and proteinuria. JAMA 1989; 262:1214–1219.
69. Kennedy TJ, McConnell JD, Thal ER. Urine dipstick vs. microscopic urinalysis in the evaluation of abdominal trauma. J Trauma 1988; 28:615–617.
70. Grinstead GF, et al. The Ames Clinitek 200/multistix 9 urinalysis methods compared with manual and microscopic methods. Clin Chem 1987; 33:1660–1662.
71. Loo SY, Scottolini S, et al. Urine screening strategy employing dipstick analysis and selective culture: an evaluation. Am J Clin Pathol 1981; 81:634–642.

72. Sewell DL, Burt SP, Gabbert NJ, et al. Evaluation of the Chemstrip 9 as a screening test for urinalysis and urine culture in men. Am J Clin Pathol 1985; 83:740–743.
73. Shaw ST, Poon SY, Wong ET. Routine urinalysis: is the dipstick enough? JAMA 1985; 253:1596–1600.
74. Smalley DI, Bryan JA. Comparative evaluation of biochemical and microscopic urinalysis. Am J Med Technol 1983; 49: 237–239.
75. Vaughn ED, Wyker AW. Effect of osmolality on the evaluation of microscopic hematuria. J Urol 1971; 105:709–711.
76. Bayer Manual. Modern urine chemistry manual. Tarrytown, NY: Bayer Corporation, Diagnostic Division, 1996.
77. Said R. Contamination of urine with povidone-iodine. Cause of false positive test for occult blood in urine. JAMA 1979; 242:748–749.
78. Kincaid-Smith P. Hematuria and exercise related haematuria. BMJ 1982; 285:1595–1596.
79. Kraus SE, et al. Hematuria and the use of nonsteroidal anti-inflammatory drugs. J Urol 1984; 132:288–290.
80. Carson CC III, Seagura JW, Greene LF. Clinical importance of microhematuria. JAMA 1979; 241:149–150.
81. Golin AL, Howard RS. Asymptomatic microscopic hematuria. J Urol 1980; 124: 389–391.
82. Muramaki S, Igarashi T, Hara S, Shimazaki J. Strategies for asymptomatic microsocopic hematuria: a prospective study of 1,034 patients. J Urol 1990; 144:99–101.
83. Papanicolaou GN, Marshall VF. Urine sediment smears as a diagnostic procedure in cancers of the urinary tract. Science 1945; 101:519–520.
84. Koss LG, Deitch D, Ramanathan R, et al. Diagnostic value of cytology of voided urine. Acta Cytol 1985; 29:810–816.
85. Hastie KJ, Ahman R, Moisey CU. Fractionated urinary cytology in the follow-up of bladder cancer. Br J Urol 1990; 66:40–41.
86. Levi PE, et al. Analysis of DNA content, nuclear size and cell proliferation of transitional cell carcinoma in man. Cancer 1969; 23:1074–1085.
87. Badalament RA, Hermansen KK, Kimmel M, et al. The sensitivity of bladder wash flow cytometry, bladder wash cytology and voided cytology in the detection of bladder carcinoma. Cancer 1987; 60:1423–1427.
88. Koss LG. Diagnostic Cytology of the Urinary Tract with Histopathologic and Clinical Correlations. New York: Lippincott-Raven, 1996.
89. Farrow GM. Urine cytology in the detection of bladder cancer: a critical approach. J Occupat Med 1990; 32:817–821.
90. Hiatt RA, Ordonez JD. Dipstick urinalysis screening, asymptomatic microhematuria, and subsequent urological cancers in a population-based sample. Cancer Epidemiol Biomarkers Prev 1994; 3:1–5.
91. Whelan P, Britton JP, Dowell AC. Three-year follow-up of bladder tumours found on screening. Br J Urol 1993; 72:893–896.
92. Marshall VF. Current clinical problems regarding bladder tumors. Cancer 1956; 3: 543–550.

93. Shirai T, Fukushima S, Hirose M, Ohshima M, Ito N. Epithelial lesions of the urinary bladder in three hundred and thirteen autopsy cases. Jpn J Cancer Res 1987; 78(10):1073–1080.
94. D'Hallewin M-A, Baert L. Initial evaluation of the bladder tumor antigen test in superficial bladder cancer. J Urol 1996; 155:475–476.
95. Sarosdy MF, White, RWD, Soloway MS, Sheinfeld J, Hudson MA, Schellhammer P F, Jarowenko M V, Adams G, Blumenstein B A. Results of a multicenter trial using the BTA test to monitor for and diagnose recurrent bladder cancer. J Urol 1995; 154:379–384.
96. Konchuba AM, Schellhammer PF, Alexander JP, et al. Flow cytometric study comparing paired bladder washing and voided urine for bladder cancer detection. Urology 1989; 33:89–96.
97. Hermanse DK, Badalament RA, Bretton PR, et al. Voided urine flow cytometry in screening high-risk patients for the presence of bladder cancer. J Occup Med 1990; 32:894–897.
98. Soloway MS, Briggman JV, Carpinito GA, Chodak GW, Church PA, Lamm DL, Lange PH, Messing EM, Pasciak RM, Reservitz GB, Rukstalis DB, Sarosdy MF, Stadler WM, Thiel RP, Hayden CL. Use of a new marker, urinary NMP22, in the detection of occult or rapidly recurring transitional cell carcinoma of the urinary tract following surgical treatment. J Urol 1996; 156:363.
99. Lokeshwar VB, Öbek C, Soloway S, Block N. Tumor-associated hyaluronic acid: a new sensitive and specific urine marker for bladder cancer. Cancer 1997; 57:773–777.
100. Mao L, Schoenberg MP, Scicchitano M, Erozan YS, Merlo A, Schwab D, Sidransky D. Molecular detection of primary bladder cancer by microsatellite analysis. Science 1996; 271:659–662.
101. Hruban RH, van der Riet P, Erozan YS, Sidransky D. Brief report: molecular biology and the early detection of carcinoma of the bladder—the case of Hubert H. Humphrey. N Engl J Med 1994; 330:1276–1278.
102. Yoshida K, Sugino T, Tahara H, Woodman A, Bolodeoku J, Nargund V, Fellows G, Goodison S, Tahara E, Tarin D. Telomerase Activity in Bladder Carcinoma and Its Implications for Noninvasive Diagnosis by Detection of Exfoliated Cancer Cells in Urine. Atlanta: American Cancer Society, 1997.
103. Canadian Task Force on the Periodic Health Examination. Canadian Guide to Clinical Preventive Health Care. Ottawa: Canada Communication Group, 1994:826, 836.

9

Cervix Cancer

Anthony B. Miller
University of Toronto, Toronto, Ontario, Canada

I. EPIDEMIOLOGY AND BURDEN OF THE DISEASE

Cervical cancer is the second most important cancer in women worldwide, with about 437,000 new cases diagnosed worldwide each year, and it is the most important cancer in developing countries (1). In most countries in North America and western Europe, the incidence of cervical cancer has been decreasing, recently at a much slower rate (2). In many developing countries, however, incidence of cancer of the cervix has changed little, except in those countries that have achieved the demographic transition with increasing affluence from industrialization. In such countries, there has been a decrease in incidence of cancer of the cervix, and an increase in incidence in cancer of the breast, changes that occurred in North America and western Europe in the early part of the 20th century.

Recently, Parkin et al. (1). estimated that the number of 5-year prevalent cervical cancer cases is around 1,558,000 in the world compared to 2,810,000 for breast cancer, but of those prevalent cases, more than 1 million were in developing countries, slightly more than for cancer of the breast, and only 320,000 were in developed countries.

Data are available internationally on trends and incidence of cancer of the cervix and, with some notable exceptions, tend to show declines (3). This is true for nearly all registries in the Americas, Asia, Australasia, Hawaii, and Europe. The reductions have been striking in Hawaii, Denmark, Finland, Sweden, Myagi in Japan, and more recently in the Maoris of New Zealand, but also in Cali in Columbia and Puerto Rico. Reductions have been small in many countries, with low incidence seen in the early 1960s, in Canada, many parts of the United

States, and the Whites of New Zealand, and even an increase seen in Birmingham, United Kingdom. In Canada, both incidence and mortality rate stopped declining about 10 years ago, although in Ontario, Canada, there is some evidence of continued decline at older, but not younger ages.

In the United States, the incidence of the disease has been decreasing in both blacks and whites, but it is still approximately twice as frequent in blacks than whites (4). Rates in whites are hardly declining, although blacks show continuing declines toward the levels of whites. A suggestion in the United States that incidence was increasing in younger women (5) may be an overinterpretation of the data potentially confounded by substantial increases in the diagnosis of carcinoma in situ, which, in cancer registries, is often difficult to distinguish from invasive cancer. In whites, the lifetime incidence is less than 1%.

Most data come from registries from areas where there are good services, which is why there are good cancer registries and, in general, screening programs. The fact that increase in early detection activity and improvement in treatment leads to reduction in mortality rate from the disease has been recognized for some time and was recently reinforced by a study in Sweden (6).

High-risk groups in all races include those with an early age at first intercourse, multiple sexual partners, monogamous women with husbands with multiple sexual partners, multiparous women, and those with a history of sexually transmitted diseases. Cervical cancer is associated with poverty and occurs more often in smokers than nonsmokers (7). Most of these associations are explained by the disease being a result of a sexually transmitted agent; it is generally believed that the oncogenic types of the human papilloma virus (HPV) are the relevant agents (8). This knowledge will eventually be incorporated into vaccines for the oncogenic HPV viruses; phase I safety studies of some vaccines have already commenced. However, there are many issues to be resolved before a primary prevention program based on HPV vaccination could substitute for cervical screening, especially whether vaccination would be of any value in women already infected with HPV. Therefore, screening will remain the number one cancer control measure for cervical cancer for many years, if not decades.

The contrast between Norway and Finland and other Nordic countries highlighted over a decade ago the success of the program of screening in Finland and the lack of success of opportunistic screening in Norway (9). So, although it is clear there are many gains to be made, particularly in countries that have not yet introduced screening, in those countries that have introduced screening, to continue to make an important impact requires a major shift in approaches from opportunistic to organized programs. These must be planned recognizing that even in very high incidence countries such as India and some of the countries of South America, the lifetime incidence of cancer of the cervix hardly exceeds 5%; thus, in most women, the disease is not destined to develop and therefore if screened, many women could be regarded as being screened unnecessarily. This leads to the importance of understanding the natural history of the cancer so that appropriate investments can be made.

II. NATURAL HISTORY OF THE DISEASE AND RELEVANCE TO SCREENING

To understand the natural history of cancer of the cervix, it is necessary to study the occurrence of precursors of invasive cancer in women who have been screened, preferably longitudinally rather than in a cross-sectional way. Part of the difficulty is that the terminology of the various presumed precursors has changed, with an increasing tendency to combine categories, largely on the grounds of substantial observer variation. However, in the major laboratories in Canada, there has been consistent use of the same terminology for many decades, with strict attention to quality control, hence in this section, the terminology used in these laboratories is used. Table 1 indicates the equivalence of the various terms used in most laboratories until recently.

The cervical intraepithelial neoplasia (CIN) terminology must be distinguished from the original Papanicolaou smear classes used to indicate normal smears (class I and much of class II) up to cells indicative of malignancy (class V). Although still used by some laboratories, Papanicolaou classes have been shown not to have strict histological counterparts and are no longer recommended for routine cytological reporting.

Dysplasia signifies morphological changes in the cells of the squamous epithelium of the cervix, giving them the characteristics of malignancy but without involvement of the full thickness of the epithelium by basal type neoplastic cells. When the full thickness is involved, the term carcinoma in situ (CIS) is used. Once a small amount of invasion has occurred through the basement membrane of the epithelium, the term microinvasive carcinoma is used. Occult invasive carcinoma is a term used to describe frank histological invasion in an asymptomatic woman. However, the term cannot be uniformly applied because it re-

Table 1 Terminology Equivalents

Histological descriptor	Cervical intraepithelial neoplasia system
Dysplasia, mild	CIN I
Dysplasia, moderate	CIN II
Dysplasia, severe	CIN III
Carcinoma in situ	CIN III
Microinvasive carcinoma	No CIN equivalent
Occult invasive carcinoma	No CIN equivalent

quires the availability of both clinical and histological data. Furthermore, a history of presence or absence of symptoms may not be helpful because some women with precancerous lesions may have similar symptoms to those of women with invasive cancer. Judgment is required by the pathologist regarding how extensive the changes are, and because there is a continuous sequence of changes, the boundaries between the various degrees of abnormality are not precise.

Recently, a further blurring of the categories has been proposed in which mild dysplasia (CIN I) is combined with cytologic atypia consistent with HPV infection in one category called ''low grade squamous intraepithelial lesions'' (LSIL) and moderate and severe dysplasia (CIN II + III) are combined in another category as ''high grade squamous intraepithelial lesions'' (HSIL) in order to simplify recommendations on management (10). This system is referred to as the ''Bethesda'' system. However, because it does not facilitate consideration of natural history, it is not used in this chapter.

Although there have been a few studies in which individuals were followed up to determine directly the probability of progression or regression of biopsied abnormalities (11), this has generally been considered unethical and, in any case, the process of diagnosis may have affected the natural history. Therefore, understanding of the natural history of the disease is dependent on inferences derived from the follow-up of women with cytological abnormalities and the study of the incidence and prevalence of well-characterized lesions in a defined population, preferably followed up over a long time period.

A number of studies have shown that the time of progression to invasive cancer of preclinical disease was of the order of several years if not decades, such as was found in a study carried out on records of screened women in British Columbia (12). The British Columbia laboratory serves the whole province. The program was initiated in 1949. This study of two identified cohorts of women, 15 years apart, followed up from 1949 to 1969 in what was essentially a population-based program, confirmed that a large proportion of cases of CIS regress, with regression rates being higher in younger than older women. The maximum incidence of carcinoma in situ occurs at about ages 30–34 years, whereas the maximum incidence of invasive cancer does not occur until about age 55–59 years. Thus, in addition to the fact that a majority of screen-detected precursors are not destined to develop invasive cancer, even for those cases that are, there is an interval of many years before such progression occurs.

The British Columbia data, together with data from other studies, largely conducted in Europe, were combined by a working group of the International Agency for Research on Cancer (IARC) (13). The analysis documented the expected risk of development of invasive cancer of the cervix following a negative screening test. The risk was lowest in the first 3 years after a negative test and remained low for at least 5 years. On the basis of this analysis, it was possible to compute the expected degree of benefit from different schedules of rescreening. It was clear that almost as much benefit is obtained from screening cytology tests conducted at 3-year intervals as from annual screening, at a considerable saving in terms of numbers of tests required in a lifetime and thus in use of resources. The analysis assumes that the test is of high sensitivity and that a high proportion of at-risk women attend for screening. Adequate sensitivity of the test can only be achieved by high-quality laboratory services and high compliance by an organized program, both are a prerequisite for effective screening programs (14). However, defects in either of these parameters cannot be overcome by increasing the frequency of rescreening; rather, actions have to be in place to ensure an effective and cost-effective program (15). The difficulty with annual screening, even in a wealthy country, is that it places emphasis on rescreening of women already in programs, whereas the emphasis should be on bringing women who are poorly screened, or not screened at all, into programs if failures of screening policies are to be avoided (16,17).

For some time it has been recognized that screening at older ages was critical to ensure maximum impact. Recently, however, it has been suggested that screening at older ages is far less efficient than screening at younger ages. Gustaffson et al. (18) have calculated from Swedish population data the probability of detecting carcinoma in situ and point out that, if age 30–34 years is regarded as the referent age, then at very young ages the probability of detecting CIS is low and that the probability decreases to very low levels (again) by about the age of 60 years or older (Table 2). They postulate that screening at these older ages may not be successful. This is a misinterpretation of the natural history. It has been known for some time both from the initial British Columbia

Table 2 Probability of Detecting Carcinoma In Situ

Age at test (years)	OR	95% CI
<25	0.47	0.38–0.58
25–29	0.91	0.76–1.08
30–34	1.00	Referent
35–39	0.96	0,80–1.17
40–44	0.56	0.44–0.71
45–49	0.47	0.38–0.61
50–54	0.21	0.13–0.52
55–59	0.22	0.12–0.38
60+	0.16	0.09–0.26

Source: Ref. 18.

Table 3 The Toronto Dysplasia Study

	Probability	RR
Progression within 10 years to carcinoma in situ or worse		
Controls	0.006	
Mild dysplasia	0.03	4.7
Moderate dysplasia	0.10	17.0
Severe dysplasia	0.21	35.1
Progression within 10 years to invasive cancer of the cervix		
Controls	0.0006	
Mild dysplasia	0.003	5.7
Moderate dysplasia	0.009	16.3
Severe dysplasia	0.029	52.8

Source: Ref. 21.

cohort study (12) and from its extension (19,20) that, as the incidence from invasive cancer increases with increasing age, so first the incidence of CIS decreases and then the prevalence. However, although the incidence of CIS decreases below the increasing rate of invasive cancer, prevalence does not. There are still, therefore, cases of carcinoma in situ in the population prevalent in women as yet unscreened from which new cases of invasive cancer can arise. It was these data that convinced the Canadian workshop group to recommend that screening in Canada should probably be extended through to age 69 years and that women of any age, even older than age 70 years, should still be subjected to at least two smears if they had not been screened before to detect prevalent CIS (14). Thus, Gustaffson et al. (18) have succeeded only in pointing again to the importance of screening at older ages. It may be true that the efficiency of detecting CIS at older ages is low, but these cases of CIS are more likely to progress to invasive cancer than those detected at young ages when the majority regress (12,19).

Further data on the importance of being cautious regarding overtreating precancerous disease, particularly in the young, has come from the Toronto Dysplasia Study, a study in which the records of nearly 50,000 women screened during a period when cases of dysplasia were observed cytologically unless they showed signs of progression have been evaluated in terms of long-term endpoints by linking to the Ontario Cancer Registry (21). These data have confirmed that the probability of progression of mild dysplasia to severe dysplasia or worse is very low, whereas the probability of regression of mild and moderate dysplasia is at least 50%. From these data, it has been possible to calculate the probability of progression within 10 years to carcinoma in situ or worse, showing the expected gradation of effect from mild to severe dysplasia, with an extremely low probability even for severe dysplasia of progressing to invasive cancer of the cervix (Table 3). This study reinforces the Canadian workshop recommendations (14) that it is appropriate to continue to classify the different degrees of dysplasia cytologically, that it is appropriate to observe cytologically at least cases of mild dysplasia and probably moderate as well, and that there is a substantial danger of excessive screening in young ages in terms of over-treating dysplastic abnormalities and thus increasing costs.

As emphasized previously, an important study that increased understanding of the appropriate approach to rescreening was that coordinated by the IARC providing the basis for three yearly or even less frequent repeat screening and underlining the importance of concentrating screening in women older than ages 35–64 years (13). However, there has been an underlying difficulty in that probably the most successful screening program in the world is the organized program in Finland, which used only 5-year screening, apparently achieving the levels of success that the IARC model suggested would require three-yearly screening. A reevaluation of the British Columbia data (22) has shown that, although the median duration for progression from carcinoma in situ to invasive cancer from the IARC study is of the order of 5–8 years, this includes within its computation screen-detected cancers whose prognosis is extremely good and which are affected by the lead time gained by screening. Correcting for such screen-detected cancers results in a computation of median duration from their model from carcinoma in situ to invasive cancer of 15 years. The implication of this is that screening every 5, not 3, years will give a 90% reduction in invasive cancer incidence and mortality rate.

Using these data it is possible to recompute some of the inferences from the IARC (13) model. In a low-incidence country, nearly maximal benefit is obtained from 5-yearly screening with little additional gain starting at the age of 20 years rather than at 25 or 26 years (Table 4). Extending this to high-incidence countries, even 10-yearly screening from the age of 35–64 years will

Table 4 IARC Study—Modified, Low-Incidence Country

Age and schedule of screening	Cumulative rate per 100,000	Reduction in incidence	Number of tests in a lifetime
None	1575		
20–64 years			
3 yearly	47	97%	15
5 yearly	158	90%	9
26–64 years			
3 yearly	63	96%	13
5 yearly	189	88%	8

result in a nearly 70% reduction in invasive cancer, requiring only 3 tests in a lifetime (Table 5).

These effects are achieved only with efficient laboratories and full compliance. However, 10-yearly screening with 80% compliance is more than twice as efficient and considerably more effective than 3-yearly screening with only 40% compliance (Table 6). Even 100% compliance of women at an early age, as in maternal and child health programs, is far less efficient and not as effective as 10-yearly screening with 80% compliance. Once again, this reemphasizes the need to allocate new resources to screening older women.

In terms of management of dysplasia, concern has been expressed over substantial observer disagreement in classifying less severe forms of dysplasia as well as their importance in indicating increased risk of invasive cancer. Furthermore, the role of HPV-associated lesions in management is also controversial. In the United States, the question of cytological surveillance of cases with CIN I versus immediate referral to colposcopy or triage into those carrying HPV

Table 5 IARC Study—Modified, High-Incidence Country

Age and schedule of screening	Cumulative rate per 100,000	Reduction in incidence	Number of tests in a lifetime
None	5080		
25–64 years			
3 yearly	210	96%	13
5 yearly	616	88%	8
35–64 years			
5 yearly	1037	82%	6
10 yearly	1728	68%	3

Table 6 IARC Study—Modified, High-Incidence Country, Effect of Different Compliance Levels

Age and schedule of screening	Compliance	Reduction in incidence	Number of tests in a lifetime
35–64 years			
3 yearly	40%	40%	6[a]
5 yearly	50%	45%	3[a]
10 yearly	80%	55%	2.4[a]
20–39 years			
2 yearly	100%	52%	10

[a] Averaged over the total population

oncogenic types who would have immediate colposcopy and of those without who would have cytological surveillance is being evaluated in a National Cancer Institute trial.

III. AVAILABLE EARLY DETECTION TESTS—DESCRIPTION OF TEST PERFORMANCE

A. Visual Inspection

In most developing countries, a high proportion of cases of cancer of the cervix are diagnosed at an advanced stage. There are many reasons for this, including a lack of appreciation by women of the relevance of symptoms of the disease, a fatalism toward cancer and the possibility for cure, and a lack of availability of health care in rural areas, combined with a low sense of priority for women's health issues. Although some benefit might be obtained by public education directed at informing women of the symptoms of cervical cancer, such an approach would be unlikely to make a major impact on mortality rate from the disease unless accompanied by a more active attempt to detect early disease. One approach proposed for developing countries without the laboratory facilities or resources to envisage cytological screening of all adult women for many years to come is visual inspection of the cervix in an attempt at "downstaging" (23).

Downstaging has been defined as "the detection of the disease in an earlier stage when still curable, by nurses and other nonmedical health workers using a simple speculum for visual inspection of the cervix" (24). The approach must still be regarded as experimental, although it is not clear that the term "downstaging" as such is particularly helpful. This is because certain approaches to visual inspection, described in more detail in this section, seem capable of identifying high-grade dysplastic lesions including CIN III. Therefore, it is possible that visual inspection could eventually be shown to achieve more than just the earlier detection of invasive cancer, thus explaining the intense interest it is attracting in studies in several developing countries.

Approaches to visual inspection currently being evaluated include unaided visual inspection of the cervix (aimed at detecting early cancer) and aided visual inspection of the cervix (aimed at detecting CIN III as well as early cancer).

Unaided visual inspection of the cervix involves looking at the cervix during a speculum examination to detect early stage cancer. Adequate visualization requires good light, an examination table preferably with stirrups, swabs, and a vaginal speculum.

Several studies have been conducted in India to determine the efficacy of visual inspection of the cervix in identifying cancers. Seghal et al. (25) claimed to identify 50% of early cancer through visual examination. Signs of cancer were identified as bleeding erosions, unhealthy cervix, or presence of a growth. The same group visually screened all women attending a maternal and child health clinic, referring for evaluation by cytology and colposcopy only those women presenting with high-risk signs (26). They claimed that 70% of cancers could be detected at an early stage. In a population-based study in rural India, the performance of paramedical workers was compared to that of gynecologists, with all subjects having a visual inspection and a cervical smear (27). The performance of the paramedical workers was similar to that of the gynecologists; the sensitivity for cervical cancer using a low threshold to define a positive test was 90%, but the specificity was only 43%; with a high threshold to define a positive test, these parameters were 60% and 94%, respectively. In a study in Kerala, India, visual inspection by paramedical personnel was again compared to cervical cytology (28). Once again, with a low threshold for a positive test, the specificity was low (45% of 2843 women were positive); the sensitivity for invasive cancer, however, was 92%. With a high threshold, the specificity increased to 94%, but the sensitivity decreased to 54%. These studies showed, therefore, that there is a need for mechanisms that would help to increase the specificity of visual inspection, without unduly reducing its sensitivity.

The simplest approach is to perform visual inspection of the cervix with a speculum with a good light after the cervix has been swabbed with dilute acetic acid. Areas of the cervix that are affected by dysplastic change exhibit white staining, and these "aceto-white" areas can then be biopsied. This approach was initially called "aided visual inspection," but more recently has simply been called visual inspection, or direct visual inspection (DVI) to signify the absence of any magnifying device. The term "cervicoscopy" has also been used. A series of studies are in progress reporting evaluation of DVI (which seems to be a preferable term to use), and preliminary data from studies in India, Indonesia, and South Africa suggest that the sensitivity for invasive cancer and CIN III is high, although specificity is still less than cytological study, but acceptably so.

The other mechanism uses a small, lightweight, low-powered monocular telescope to view acetic acid–treated tissue. This technique, probably best called aided visual inspection of the cervix (AVI) is being evaluated in several developing countries, but its sensitivity, specificity, and any advantage that it confers in comparison with DVI is yet to be determined.

Many rural women in developing countries are resistant to recommendations that they attend a hospital for diagnosis and treatment of suspected abnormalities, probably because they are asymptomatic and they lack background knowledge on the importance of early detection of the disease. It is therefore important that adequate programs of public and professional education accompany, and preferably precede, the introduction of screening programs for asymptomatic women, whatever screening test is used. Such programs should also be evaluated in their own right, by comparison with historical data on stage distribution of cases from the relevant areas, while the effectiveness of the screening tests should be evaluated over and above any benefit that the education programs can provide.

Earlier detection of cancer of the cervix was achieved in a number of developed countries before the introduction of cervical cytology screening. The evidence for this is indirect, but relates to data on downstaging of cancer of the cervix and reduction in mortality rate from the disease from historical records in Sweden and Canada. This downstaging did not arise from a deliberate policy of visual examination of asymptomatic women. Rather, it arose from programs of public and professional education that made women and their health care providers aware of the symptoms and signs of the disease and its potential curability if detected early. These programs therefore appeared to be successful of themselves, while also preparing the population for the successful introduction and acceptance of cervical cytology screening (6).

It is important to recognize that cervical cytology screening also involves visual inspection and that it has not been possible to determine the contribution of inspection to the effectiveness of cervical cytology screening programs. If cervical smears are to be taken in a selective fashion, discriminant cytological screening based on the findings at cervical inspection might be the answer to the high cost of mass cytology screening of all women in many developing countries.

B. Cervical Smear

Cytological screening using the Papanicolaou smear is the established method of screening. It is essential that those professionals obtaining smears should have had adequate instruction in the technique and should use an appropriate method for obtaining the smear, placing it on the slide, and fixing it. Inappropriate sampling methods are responsible for approximately two-thirds of false-negative smears (14). It is important that the woman is comfortable during the examination. It should be performed in a warm room, with a warm speculum; the woman's head should be raised so that she can see the examiner and see the equipment; she should be mostly covered and should be able to talk and ask questions during the procedure. The smear should be taken with a spatula after visual examination using a speculum; any clinical abnormality detected dictates referral for diagnostic evaluation, whatever the result of the smear. When obtaining a smear directly from the cervix with a spatula, a 360-degree rotation with firm contact is required, ending in the horizontal position (3 and 9 o'clock), and retaining the sample on the upper surface of the spatula during transfer. The material on the spatula should be carefully spread on a slide and fixed in 95% ethanol or with a commercial spray fixative immediately. Smears should not be allowed to air dry. Smears should only be read in laboratories with expertise in cytopathology. In such laboratories, smears should be screened by qualified cytotechnologists and interpreted with the input of cytopathologists. Quality assurance programs in cytopathology are an essential component of cytological screening programs aimed at reducing mortality rate from carcinoma of the cervix. The laboratory director is responsible for internal quality control within cytopathology laboratories; external quality control is the responsibility of the appropriate professional medical association in coordination with accrediting and licensing bodies.

An alternative approach to the preparation of smears currently being evaluated is the liquid sample, which is collected with a special instrument (a Papette broom) from the cervix in the normal way and then deposited in a special liquid preservative for the subsequent preparation of a cellular monolayer for examination. Although this method results in a more even distribution of cells for microscopic evaluation than a standard smear, it is more expensive and it is not yet clear that the sensitivity and specificity of the smear are greatly improved over conventional smears.

The sensitivity of the standard smear has been estimated from the large central laboratory in British Columbia as a result of a standard policy of rereading smears previously reported as negative in a woman subsequently found to have an abnormality (12). This enabled the ''reader error'' to be quantified. The other component of false-negative results, the ''taker error,'' caused by a failure of the health care professional obtaining the smear to adequately sample the cervix, can only be estimated by statistical means. The theory behind this is that over a relatively short period, the estimated incidence decreases and then plateaus before increasing slowly with increasing age. The component contributing to the initial excess incidence above baseline after the first (prevalence) smear represents the false-negative results at the first smear. Boyes et al. (12) estimated that the two components of false-negative results were approximately equal in size and that the sensitivity of the smear approximated to 75%. Although this estimate may seem to indicate low sensitivity, it is almost certainly close to the maximum currently achievable in a large laboratory with good quality control procedures. Almost all other estimates of sensitivity reported ignore the ''taker'' component, thus overestimating the true sensitivity of the technique. A number of approaches are being evaluated to increase sensitivity, including automated or computer-assisted reading of smears or liquid cytology preparations. None are yet sufficiently developed to substitute for cytotechnologists in routine practice, although two have been licensed by the Food and Drug Administration (FDA) to be used in identifying false-negative results as part of quality control procedures.

Estimates of the specificity of the test have not been published, in part because of disagreements as to what constitutes a true abnormality sufficient to dictate further evaluation (in the past it was by biopsy; currently it is almost invariably colposcopy). Confusion has increased somewhat after the Bethesda system (10) characterized a category encompassing this uncertainty as ASCUS (atypical squamous cells of undetermined significance). A trial is underway that should help to clarify this uncertainty over both significance and appropriate management. Operationally, it seems as if the specificity of the test approximates to 90%, if the majority of those specimens categorized as ASCUS are false-positive results.

C. Cervicography

Cervicography involves the inspection by trained personnel of photographs of the cervix obtained at the time of speculum examination using a special camera device. National Testing Laboratories Worldwide holds an exclusive license on the patents relating to Cervicography technology. Its sensitivity and specificity in comparison with cervical cytology has not been established, although it is believed to be of lower specificity (29,30) and special readers must be available to assess the photographs, which must be sent to National Testing Laboratories in the United States for reading.

IV. SCREENING STUDIES OF THE VARIOUS TESTS

A. Randomized

Screening for cervical cancer was introduced before there was recognition of the need to evaluate programs by randomized trials, and no such trials have ever been conducted.

B. Nonrandomized Observational and Population Experiences

Other mechanisms have therefore been used to evaluate the effectiveness of screening for this disease, including geographical comparisons of screened and unscreened populations combined with studies of the trends of incidence and mortality rates that followed introduction of screening (9, 31–44), and studies that correlated the intensity of screening with reduction in mortality rate from the disease (45–47).

Because a precursor is detected by screening (carcinoma in situ or severe dysplasia), reduction in incidence of the disease as well as reduction in mortality rate are both valid endpoints. However, very few countries have noted a dramatic reduction of incidence or mortality rate from the disease that could be directly related to the introduction of screening. The exceptions to this are largely the Nordic countries, where organized programs were introduced, and where an impressive reduction in incidence and mortality rate followed within 6 to 10 years (9,39,44). Only Norway failed to introduce an organized screening program for cancer of the cervix. All the other Nordic countries introduced organized programs, and all except Norway showed a substantial decrease in incidence, following 5–10 years after the introduction of the programs. Previously, Denmark had shown the highest incidence of cancer of the cervix. However, by the end of the 1970s, the incidence in Denmark had decreased to that of Norway, and in Finland, Iceland, and Sweden, incidence had decreased to much lower levels. Within Denmark, incidence decreased most in the areas with organized programs (39). These reductions occurred even though the frequency of screening was much less than advocated in North America. In Finland, for example, where screening occurred in response to invitations sent every 4 years, the reduction in incidence was greater than any of the Nordic countries (9), presumably because of the high compliance of those to whom the invitations were sent.

In North America and much of Europe, where organized screening was not introduced, and where screening was largely opportunistic (using approaches that depend on the physician or the woman herself), although there have been some reductions, they have not been dramatic and it has often been difficult to determine the extent that screening has been responsible for the reduction observed (33–36). More convincing have been the correlation studies (45–47), in which it was possible to show a strong association between reduction in mortality rate and screening that was not explained by census-derived risk factors (46) nor by changes in hysterectomy rates (47).

C. Case-Control Studies

A number of case-control studies have been performed evaluating the effect of screening for cancer of the cervix in reducing the incidence of the disease (48–53). The first was performed by Clarke and Anderson in Toronto (48). They interviewed 212 of 323 eligible patients aged 20–69 years with newly diagnosed cancer of the cervix and five neighborhood controls per case. Details of the screening history for the previous 5 years was obtained, confirmed as far as possible from physician's records. They found that 68% of the women had had no screening smears in the 5 years before diagnosis, compared to 44% of the control subjects for a comparable period, for a relative risk of 2.7. Control for various potential confounders did not appreciably change the relative risk.

Similar studies performed in several countries (49–53) also showed a protective effect of screening in reducing the risk of invasive cancer. Although there are methodological issues that have to be taken into consideration when case-control studies are used to evaluate screening (54–56), the consistency of the findings, and the care with which these studies were conducted, make it extremely unlikely that bias could explain the findings. Hence, it may be concluded that the effectiveness of screening in reducing both the incidence and the mortality rate from cancer of the cervix is established, and that consideration needs to be given to organizational issues to ensure the greatest effect of existing and new programs (15).

V. RISKS ASSOCIATED WITH SCREENING

Although there is little or no risk associated with administering the screening test, the major risks associated with cervical cytology screening are those related to overtreatment, especially overtreatment of cytological abnormalities that are not progressive but that would have regressed if left untreated. The difficulty, as emphasized previously, relates particularly to too frequent rescreening, as well as initiating screening at too young an age. It is impossible to escape from the incorrect view that persists in the clinical community, that every abnormality must be treated as if, in the absence of treatment, an invasive cancer will develop and the woman will die. Modern management has reduced some of the adverse consequences of treatment. With colposcopically directed biopsies and Loop Electrosurgical Excision Procedures becoming the norm, there is far less risk of stenosis of the cervix (as from cone biopsies) and thus of pregnancy and delivery complications.

VI. COST AND COST-EFFECTIVENESS ANALYSIS

The cost of screening for cervical cancer has been estimated in Canada to be about $10 (Canadian) per smear (although the cost would increase to $45 if the fee for the office visit to obtain the smear was included) (14). Additional costs must be considered when setting up an organized program. These are associated with the need for quality control, evaluation and monitoring of the program, centralization of cytology services, establishing an appropriate information system, and financing public and professional education programs.

Eddy (57) conducted a cost-effectiveness analysis that demonstrated that 3-yearly screening should be the norm. More recently, the same principle was found to be true for older women (58). It is important to recognize that, as a woman moves from less frequent screening to annual screening, the marginal costs per year of life gained escalate. As an example, it has been estimated by the U.S. Office of Technology Assessment that one-time screening at age 65 years would cost $1666 per life-year gained (58). Screening every 5 years from age 65 years would cost an added $1453 per life-year gained, but screening every 3 years would cost an additional $5956 per life-year gained over 5-yearly screening and annual screening $39,693 per extra life-year gained over 3-yearly screening. Changing to an organized system of cervical cancer screening and reducing the frequency of unnecessary rescreening is therefore likely to be cost effective.

VII. PUBLIC AND PROFESSIONAL EDUCATION: WHAT TO TELL THE PATIENTS

Major programs of public education have been run for many years advocating cervical screening for women who are sexually active. These programs have been successful in inculcating screening among women who perceive themselves to be at risk, but, in general, they have resulted in health conscious but not high-risk women being screened. Women have been recruited into screening programs by one of two mechanisms, an action initiated by a physician or a request by an informed woman that a smear be performed. The physicians that have primarily been responsible for cervical cytology screening have been family physicians and gynecologists, and they have tended to rely on the opportunities offered by premarital examinations, antenatal and postnatal care, and prescription of oral contraceptives to initiate screening. In addition, college and university medical services and family planning clinics have invariably offered screening to those women who avail themselves of their services, while the concept of the annual medical examination by physicians has encouraged continuation of regular screening for some women (14). This laissez faire approach has been moderately successful in a province such as British Columbia, Canada, where there is a central authority that encourages screening by a program of professional and public education, but has been far less successful in other parts of North America where there is no central authority prepared to take the responsibility to undertake such activities. Furthermore it is clear that such an approach tends to concentrate on normal risk women; most high-risk women, especially those past the age of childbearing but entering the age at which cancer of the cervix becomes most common, recent migrants and native women, are excluded from such an approach. The need is to examine ethnocultural population needs and to examine other factors of the community such as age distribution, socioeconomic status, migration/immigration history, geographic distribution, general demographics of the community, education levels in the community, and other similar factors relevant to compliance with screening. These are all issues that will influence the publicity that the screening service will need to reach the population at risk. Furthermore, information on cervical cancer and the role of the male in prevention as well as facilitating the participation of his partner in screening should be imparted sufficiently early in life to have an impact, and be available to males as well as females. This involves directing information on cancer of the cervix to all adolescents.

Because not all physicians are aware of the need for regular screening, an informed woman is one mechanism to ensure that screening occurs, and at the right frequency. Unfortunately, most organizations have been reluctant to enter what they regard as controversial areas relating to frequency of rescreening. Therefore, even informed women may not be aware of the desirability of avoiding unnecessarily frequent screening, although clearly they should be informed. Furthermore, women who have been screened with negative results and who are no longer in their reproductive period may be unaware that they are still at risk for the disease. Public (and professional) education programs promoting screening of women at least through to age 65 years are therefore highly desirable.

It is by no means clear that all women at risk for the disease will respond to simple invitations to attend for screening, especially if their previous experience with the medical care system has not been positive. Furthermore, many women at high risk for the disease have other health, social, or economic problems that may tend to reduce to an insignificant priority their perception of the need for a cervical cytology smear. Hence, there is a need for a coordinated program of health promotion to ensure that eligible women are screened. The approaches to health promotion should be client centered and custom built with due regard given to cultural sensitivities (e.g., of native groups and recent immigrants). In designing these approaches, key parties for participation are women, health care providers, community leaders, and those knowledgeable in other relevant sectors such as housing and education (14).

With an organized program introduced to apply at all ''at-risk'' ages, it can be anticipated that many women will be recruited into screening at relatively advanced ages with no previous screening history. To recruit such women should be regarded as a high priority if the program is to have a rapid impact on the disease, because most invasive cancer occurs at older ages and in women who have not been previously screened. If a woman is recruited into screening for the first time in her 60s, the absence of a detectable abnormality is reassurance that she is not at high risk. Nevertheless, because a second smear 1 year later may not eliminate all false-negative results at older ages, it is important to ensure that a woman who enters screening in her late 60s receives at least two smears before she is dropped from the program.

VIII. RECOMMENDATIONS FROM PROFESSIONAL ORGANIZATIONS

There have been a series of recommendations from professional organizations directed to encourage women to be screened and what those organizations regard as at an appropriate frequency. In Canada, two national task forces (16,59) and a workshop group (14) have tried to increase interest in introducing organized programs.

The essential features of organized programs have been described as follows:

The individual women in the target population are identifiable.
Measures are available to guarantee high coverage and attendance, such as a personal letter of invitation.
Adequate field facilities for obtaining the smears and adequate laboratory facilities to examine them are available.
An organized quality control program on obtaining the smears and on interpreting them is in place.
Adequate facilities exist for diagnosis and for appropriate treatment of confirmed neoplastic lesions.
A carefully designed and agreed upon referral system is in place for management of any abnormalities found and for providing information about normal screening tests.
Evaluation and monitoring of the total program is organized (15).

One of the most controversial issues on cervical cancer screening relates to the frequency of rescreening. Several years ago, the American Cancer Society (ACS) (60), largely echoing an earlier report by a Canadian Task Force (16), recommended 3-yearly screening as a norm. This proved intensely controversial, especially to a number of clinical interests, and later the recommendation was changed back to one that could be interpreted as approving annual smears, although the physician has the discretion to perform tests less frequently (61,62). Unfortunately, these changes failed to reflect the natural history of the disease.

More recently, the U.S. Preventive Services Task Force (63) has updated its previous recommendation. This now reads:

> Routine screening for cervical cancer with Papanicolaou (Pap) testing is recommended for all women who are or have been sexually active and who have a cervix. Pap smears should begin with the onset of sexual activity and should be repeated at least every 3 years. There is insufficient evidence to recommend for or against an upper age limit for Pap testing, but recommendations can be made on other grounds to discontinue regular testing after age 65 in women who have had regular previous screenings in which the smears have been consistently normal. There is insufficient evidence to recommend for or against routine screening with cervicography or colposcopy, or for screening for human papillomavirus infection, although recommendations against such screening can be made on other grounds.

This recommendation is a substantial improvement over that of the ACS, although it still fails to recognize the need for organized programs, and is still unnecessarily cautious over the recommended frequency of rescreening.

Two other features of cervical cancer screening have proven controversial, the age range over which screening should be performed and the management of the preclinical lesions found as a result of screening. Initiating screening at age 25 years is almost as effective as initiating screening at age 20 years, but postponing initiation of screening to age 35 years is less effective, although not markedly so (15). Most jurisdictions in North America, however, recommend initiation of screening at around ages 18 to 20 years for sexually active women, because of a high frequency of preclinical lesions in young women (14,63). One advantage of such a policy is that it can capitalize on the attendance of young women for contraceptive advice or antenatal care. However, many successful screening programs in Europe do not initiate screening until age 25 years (15,39). Furthermore, concentrating all screening on special health services for women is not wise because in most countries those most at risk for the disease are the older women in their 50s and 60s who have not been screened or who have lapsed from screening. It is one of the main purposes of organized programs of screening to ensure that such women are screened.

In terms of age to stop screening, although cessation of screening used to be recommended at age 60 years in Canada, recently the upper age was extended to 69 years, although with identification that this was a research issue (14). However, for women older than 65 years who have not had a number of negative smears, screening should continue until they have achieved such a record (at least two negative smears) (14,15,63).

IX. CONCLUSIONS

Cervical cancer screening currently only complies with some of the requirements for effective screening programs. These are a valid and acceptable screening test, earlier and efficient diagnosis of the disease, minimal diagnosis of nonprogressive disease such as dysplasia, effective therapy of the detected disease, and compliance of at-risk subjects with screening. With the current emphasis in North America on physician-directed screening, there are major unresolved problems over excessive diagnosis of nonprogressive disease and compliance of all at-risk groups with screening. Without major changes, incidence and mortality rate from cervical cancer, especially in the United States, may be very close to the irreducible minimum (64). The solution, as indicated by the greater success in some of the Nordic countries with far less expenditure of resources, appears to be organized programs. The components of organized programs are well recognized (15), but it is by no means clear that the will exists to bring such programs into effect.

Improvements in current understanding of the role of oncogenic HPV may lead to new tests for those truly at risk and to simpler treatments that could make much of the present structure for screening by the Papanicolaou smear obsolete. However, any new tests need to be carefully evaluated, preferably by randomized trials, and it is worthwhile to put present knowledge into effect so that further decades do not pass while truly cost-effective programs are developed.

REFERENCES

1. Parkin DM, Pisani P, Ferlay J. Estimates of the worldwide incidence of eighteen major cancers in 1985. Int J Cancer 1993; 54:594–606.
2. Beral V, Hermon C, Munoz N, Devesa SS. Cervical cancer. Cancer Surv 1994; 19–20: 265–285.
3. Coleman M, Esteve J, Damiecki P, Arsian A, Renard H. Time Trends in Cancer Incidence and Mortality [IARC Scientific Publication No. 121]. Lyon, International Agency for Research on Cancer, 1995.
4. Kosary CL, Reis LAG, Miller BA, Hankey BF, Harras A, Edwards B K, eds. SEER Cancer Statistics Review, 1973–1992: Tables and Graphs.[NIH Publication No. 96–2789]. Bethesda, MD: National Cancer Institute, 1995.
5. Larsen NS. Invasive cervical cancer rising in young white females [News item]. J Natl Cancer Inst 1994; 86:6–7.
6. Ponten J, Adami HO, Bergstrom R, Dillner J, Friberg LG, Gustafsson L, Miller A B, Parkin DM, Sparen P, Trichopoulos D. Strategies for global control of cervical cancer. Int J Cancer 1995; 60:1–26.
7. Tomatis L, Aitio A, Day NE. Heseltine E, Kaldor J, Miller AB, Parkin DM, Riboli E, eds. *Cancer: Causes, Occurrence and Control* [IARC Scientific Publication No. 100]. Lyon: International Agency for Research on Cancer, 1990.
8. IARC monographs on the evaluation of the carcinogenic risks to humans. Vol. 64. Human Papillomaviruses. Lyon: International Agency for Research on Cancer, 1995.
9. Hakama M. Trends in the incidence of cervical cancer in the Nordic countries. In: Magnus K, ed. Trends in Cancer Incidence, Causes and Practical Implications. New York: Hemisphere Publishing Corp, 1982: 279–292.
10. National Cancer Institute Workshop. The 1988 Bethesda system for reporting cervical/vaginal cytological diagnoses. JAMA 1989; 262:931–934.
11. Kinlen LJ, Spriggs AI. Women with positive cervical smears but without surgical intervention. Lancet 1978; ii:463–465.
12. Boyes DA, Morrison B, Knox EG, Draper G, Miller AB. A cohort study of cervical cancer screening in British Columbia. Clin Invest Med 1982; 5:1–29.
13. IARC Working Group on Cervical Cancer Screening. Summary chapter. In: Hakama M, Miller AB, Day NE, eds. Screening for Cancer of the Uterine Cervix [IARC Scientific Publications No. 76]. Lyon: International Agency for Research on Cancer, 1986: 133–142.
14. Miller AB, Anderson G, Brisson J, Laidlaw J, Le Pitre N, Malcolmson P, Mirwaldt P, Stuart G, Sullivan W. Report of a National Workshop on Screening for Cancer of the Cervix Can Med Assoc J 1991; 145:1301–1325.
15. Hakama M, Chamberlain J, Day NE, Miller AB, Prorok PC. Evaluation of screening programmes for gynaecological cancer. Br J Cancer 1985; 52:669–673.
16. Task Force. Cervical Cancer Screening Programs. The Walton Report. Can Med Assoc J 1976; 114:1003–1033.
17. Chamberlain J. Reasons that some screening programmes fail to control cervical cancer. In: Hakama M, Miller AB, Day NE, eds. Screening for Cancer of the

Uterine Cervix [IARC Scientific Publications No. 76]. Lyon: International Agency for Research on Cancer, 1986: 161–168.
18. Gustafsson L, Sparen P, Gustafsson M, Petersson B, Wilander E, Bergstrom R, Adami HO. Low efficiency of cytologic screening for cancer in situ of the cervix in older women. Int J Cancer 1995; 63:804–809.
19. Miller AB, Knight J, Narod S. The natural history of cancer of the cervix, and the implications for screening policy. In: Miller AB, et al, eds. Cancer screening. Cambridge, Cambridge University Press, 1991: 144–152.
20. Morrison BJ, Coldman AJ, Boyes DA, Anderson GH. Forty years of repeated screening: the significance of carcinoma in situ. Br J Cancer 1996; 74:814–819.
21. Holowaty PH. The Natural History of Cervical Dysplasia. Ph.D. thesis, University of Toronto, Toronto, Canada, 1996.
22. van Oortmarssen GJ, Habbema, JDF. Duration of preclinical cervical cancer and reduction in incidence of invasive cancer following negative Pap smears. Int J Epidemiol 1995; 24:300–307.
23. Stjernswärd J. National training of radiotherapists in Sri Lanka and Zimbabwe: priorities and strategies for cancer control in developing countries. Int J Radiat Oncol Biol Phys 1990; 19:1275–1278.
24. Stjernswärd J, Eddy D, Luthra U, Stanley K. Plotting a new course for cervical cancer screening in developing countries. World Health Forum 1987; 8:42–45.
25. Seghal A, Singh V, Bhampani S, Luthra UK. Screening for cervical cancer by direct inspection. Lancet 1991; i:282.
26. Singh V, Seghal A, Luthra UK. Screening for cervical cancer by direct inspection. BMJ 1992; 304: 534–535.
27. Nene BM, Deshpande S, Jayant K, Buduki AM, Dale PS, Deshpande DA, Chiwate AS, Malvi SG, Deokar S, Parkin DM, Sankaranarayanan R. Early detection of cervical cancer by visual inspection: a population-based study in rural India. Int J Cancer 1996; 68: 770–773.
28. Wesley R, Sankaranarayanan R, Mathew B, Cahandralekha B, Beegum AA, Amma NS, Nair MK. Evaluation of visual inspection as a screening test for cervical cancer. Br J Cancer 1997; 75:436–440.
29. Szarewski A, Cuzick J, Edwards R, Butler B, Singer A. The use of cervicography as a primary screening service. Br J Obstet Gynaecol 1991; 98:313–317.
30. Tawa K, Forsythe A, Cove JK, et al. A comparison of the Papanicolaou smear and the cervigram: sensitivity, specificity and cost analysis. Obstet Gynecol 1988; 71: 299–235.
31. Lundin FE, Christopherson WM, Mendez WM, Parker JE. Morbidity from cervical cancer. Effects of cervical cytology and socio-economic status. J Natl Cancer Inst 1965; 35:1015–1025.
32. Hill GB, Adelstein AM. Cohort mortality from carcinoma of the cervix. Lancet 1967; ii: 605–606.
33. Burns EL, Hammond EC, Percy C, Seidman H, Gorski TW. Detection of uterine cancer. Results of a community program of 17 years. Cancer 1968; 22:1108–1119.
34. Christopherson WM, Parker JE, Mendez WM, Lundin FE. Cervix cancer death rates and mass cytologic screening. Cancer 1970; 26:808–811.

35. Higgins ITT. Recent mortality from cancer of the cervix in the United States and United Kingdom. Lancet 1971; ii: 1141–1143.
36. Dickinson L, Mussey ME, Soule EH, et al. Evaluation of the effectiveness of cytologic screening for cervical cancer I. Incidence and mortality trends in relation to screening. Mayo Clin Proc 1972; 47:534–544.
37. Kinlen LJ, Doll R. Trends in mortality from cancer of the uterus in Canada and in England and Wales. Br J Prev Soc Med 1973; 27:146–149.
38. Christopherson WM, Lundin FE, Mendez WM, Parker J. Cervical cancer control. A study of morbidity and mortality trends over a twenty-one year period. Cancer 1976; 38:1357–1366.
39. Johanneson G, Geirsson G, Day N. The effect of mass screening in Iceland, 1965–1974, on the incidence and mortality of cervical carcinoma. Int J Cancer 1978; 21: 418–425.
40. MacGregor JE, Teper S. Mortality from cancer of the cervix uteri in Britain. Lancet 1978; ii:774–776.
41. Draper GJ, Cook GA. Changing patterns of cervical cancer rates. BMJ 1983; 287: 510–512.
42. Lynge E, Madsen M, Engholm G. Effect of organized screening on incidence and mortality of cervical cancer in Denmark. Cancer Res 1989; 49:2157–2160.
43. Parkin DM, Nguyen-Dinh X, Day NE. The impact of screening on the incidence of cervical cancer in England and Wales. Br J Obstet Gynaecol 1985; 92: 150–157.
44. Laara E, Day NE, Hakama M. Trends in mortality from cervical cancer in the Nordic countries: association with organised screening programmes. Lancet 1987; i:1247–1249.
45. Cramer DW. The role of cervical cytology in the declining morbidity and mortality of cervical cancer. Cancer 1974; 34:2018–2027.
46. Miller AB, Lindsay J, Hill GB. Mortality from cancer of the uterus in Canada and its relationship to screening for cancer of the cervix. Int J Cancer 1976; 17:602–612.
47. Miller AB, Visentin T, Howe GR. The effect of hysterectomies and screening on mortality from cancer of the uterus in Canada. Int J Cancer 1981; 27:651–657.
48. Clarke EA, Anderson TW. Does screening by ''Pap'' smears help prevent cervical cancer? Lancet 1979; ii:1–4.
49. Aristizabal N, Cuello C, Correa P, Collazos T, Haenszel W. The impact of vaginal cytology on cervical cancer risks in Cali, Columbia. Int J Cancer 1984; 34:5–9.
50. Raymond L, Obradovic M, Riotton G. A case-control study to evaluate cytological screening of cervical cancer. Rev Epidemiol Santé publ 1984; 32:10–15.
51. MacGregor JE, Moss SM, Parkin DM, Day NE. A case control study of cervical screening in North East Scotland. BMJ 1985; 290:1543–1548.
52. Berrino F, Gatta G, D'Alto M, Crosignani P. Efficacy of screening in preventing invasive cervical cancer: a case-control study in Milan, Italy. In: Hakama M, Miller AB, Day NE, eds. Screening for Cancer of the Uterine Cervix [IARC Scientific Publications No. 76]. Lyon: International Agency for Research on Cancer, 1986: 111–123.

53. Van der Graaf Y, Zielhuis GA, Peer PG, Vooijs PG. The effectiveness of cervical screening: a population-based case-control study. J Clin Epidemiol 1988; 41:21–26.
54. Prorok PC, Chamberlain J, Day NE, Hakama M, Miller AB. UICC workshop on the evaluation of screening programmes for cancer. Int J Cancer 1984; 34:1–4.
55. Morrison AS. Screening in chronic disease. Oxford: Oxford University Press, 1985: 102.
56. Sasco AJ, Day NE, Walter SD. Case-control studies for the evaluation of screening. J Chron Dis 1986; 39:399–405.
57. Eddy D. Screening for Cancer: Theory, Analysis, and Design. Englewood Cliffs, NJ: Prentice-Hall, 1980.
58. Muller C, Mandelblatt J, Schechter CB, Power EJ, Duffy BM, Wagner JL. Costs and effectiveness of cervical cancer screening in elderly women—background paper. US Congress, Office of Technology Assessment [OTA-BP-H-65]. Washington, DC: US Government Printing Office, 1990.
59. Walton RJ, Allen HH, Anderson GH, Boyes DA, Blanchet M, Carmichael J, Gellman DD, Hill GB, Miller AB, Thomson DW, Kassierer E. Cervical cancer screening programmes: summary of the 1982 Canadian Task Force Report. CMA Journal 1982; 127: 581–589.
60. American Cancer Society. Guidelines for the cancer-related check-up. Recommendations and rationale. CA 1980; 30:193–240.
61. Fink DJ. Change in American Cancer Society checkup guidelines for early detection of cervical cancer. CA 1988; 38:127–128.
62. Mettlin C, Dodd GD. The American Cancer Society guidelines for the cancer-related checkup: an update. CA 1991; 41:279–282.
63. Report of the U.S. Preventive Services Task Force. Guide to Clinical Preventative Services. 2d ed. Washington, DC: U.S. Department of Health and Human Services, 1996.
64. Miller AB. Editorial: failures of cervical cancer screening. Am J Public Health 1995; 85:761–762.

10

Colorectal Cancer

Jack Cuzick
Imperial Cancer Research Fund, London, England

I. EPIDEMIOLOGY AND BURDEN OF DISEASE

A. Incidence and Mortality

Colorectal cancer is the fourth most common nonskin cancer worldwide with an estimated 678,000 cases and 394,000 deaths in 1985 (1,2). In terms of mortality, its ranking is third. In the United States, Great Britain, and most other developed countries, it is the second most frequent cause of cancer death following only lung cancer. However, it is less common in developing countries, where stomach cancer currently poses a greater burden. The geographical breakdown of incidence rates is shown in Figure 1.

The age-incidence curve for both colon and rectal cancer is typical of epithelial tumors and increases approximately as the fifth to sixth power of age. (Figure 2) (3,4). In most countries, rectal cancer is about twice as common in men than in women, but the differences are less marked in the colon, especially in the proximal colon where there is little sex difference overall. For both colon and rectal cancer, the disease becomes relatively more common in women with increasing age (Figure 3). The subsite distribution within the bowel is shown in Figure 4 for England and Wales. This distribution is typical of an unscreened Western population. It is clear that the distal bowel contains a disproportionately high percentage of the disease, with 40% of cancers in the rectum and another 28% in the sigmoid colon or rectosigmoid junction. The density of cancer then becomes much lower, but increases again in the caecum where 12% of the cancers are found. There is a shift toward more proximal cancers with increasing age, in both men and women. The proportion of colorectal cancers that are found in the proximal bowel appears to be increasing, especially in the United States

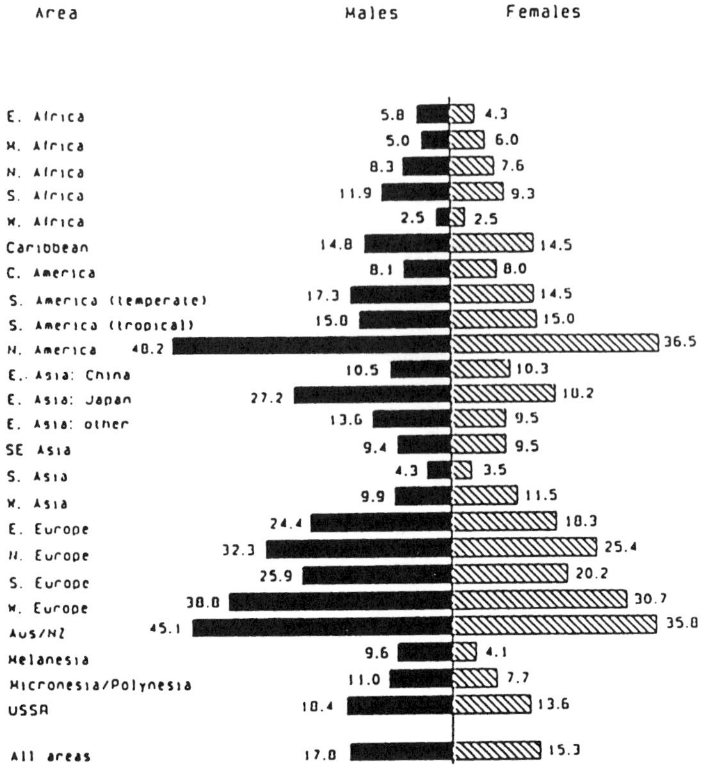

Figure 1 Estimated age-standardized rates (per 100,000) of colorectal cancer incidence by sex and area, 1985. (From Ref. 1.)

where the figure is around 50% in some recent series. It is difficult to assess whether this reflects a true change in distribution or is the result of screening, which is more intense in the distal bowel with removal of preinvasive neoplastic lesions (5–7).

Survival is highly dependent on the stage at which the cancer is detected. For Duke's stage A patients (localized disease), 5-year survival rate is about 90%, which decreases to about 60% for Duke's stage B (spread beyond bowel wall), 25% for stage C (regional nodal involvement), and less than 10% for stage D (distant metastases or irremovable tumor) (8,9). The markedly improved survival rate in early stage cancers indicates a potential place for early detection and screening. For symptomatically presenting cases, typically less than 10% of colorectal cancers present with stage A disease and about half are stages C and D.

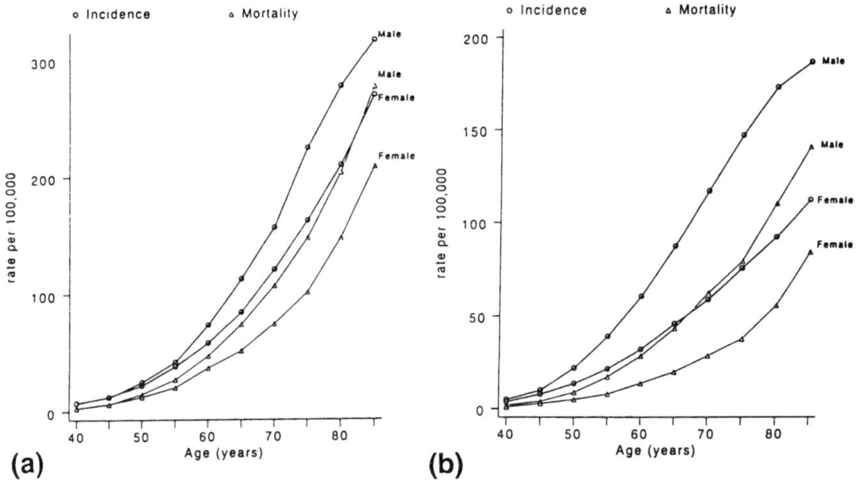

Figure 2 Incidence and mortality rates of cancer of the (a) colon and of the (b) rectum in England and Wales. Incidence rates for the period 1989–1991; mortality rates for the period 1992–1994.

B. Time Trends

Chu *et al.* (10) noted that the time trends for colorectal cancer incidence and mortality rate in the United States are different between men and women (Figure 5). For men, mortality rate was steady until 1985, but then decreased sharply from 1985–1990 at a rate of 1.72% per annum. Incidence appeared to be increasing until about 1975, was steady from 1975–1985, but has shown recent declines. For women, a declining mortality rate has been apparent since at least 1950, but has steepened since 1985 and was decreasing at a rate of 2.50% per annum from 1985–1990. Incidence rates were fairly constant until 1985, when a decrease became apparent. In the United Kingdom, incidence has been fairly constant in both men and women, as has mortality rate in men. The mortality rate has decreased by about 25% in women between 1955 and 1990. Most European countries have shown an increase in incidence from 1973–1987 both for men and women. However mortality rate decreased in this period for both sexes in western European countries (11). Colorectal cancer has been increasing in most other parts of the world where the baseline rates are generally lower. This is especially true for eastern Europe. In general, incidence is increasing faster than the mortality rate.

Studies of migrant populations support lifestyle as a major determinant of colorectal cancer risk. The most striking example is the increase of incidence in

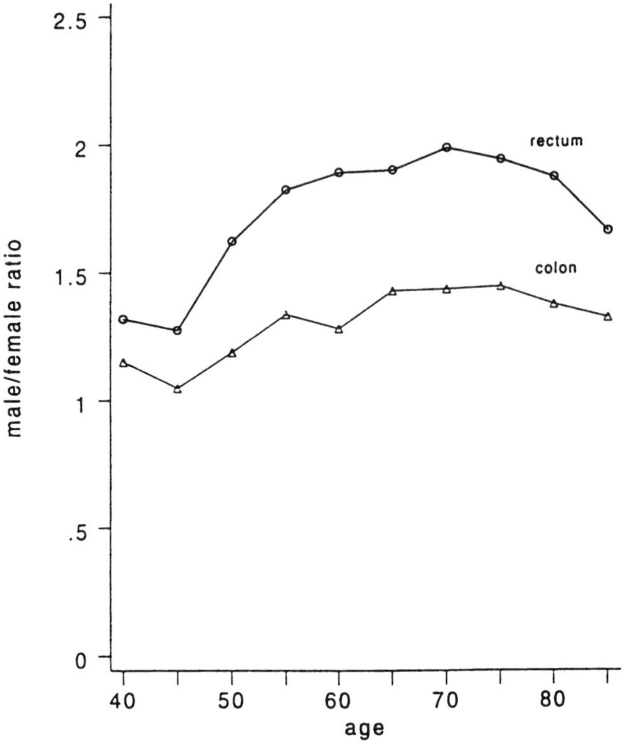

Figure 3 Sex-ratio of incidence for cancer of the colon and rectum in England and Wales, 1989–1991.

Japanese people migrating to California, where the risk approximated that of U.S. whites within about 20 years (12). Jews migrating to Israel from low-risk areas in Africa experienced an increased risk (13). Southern European immigrants to Australia have inherited the increased risk of the locally born in a duration of residence correlated way (14). Chinese men who migrate to North America have rates similar to white Americans, even though rates in mainland China are seven times lower. Rates in migrant Chinese women are intermediate between Chinese women in China and American white women. This difference may be due to duration of stay in a Western culture (15).

C. Risk Factors

1. *Diet*

Just as lung cancer epidemiology is dominated by inhaled substances (tobacco smoke), one would expect colon cancer risk factors to be very strongly related

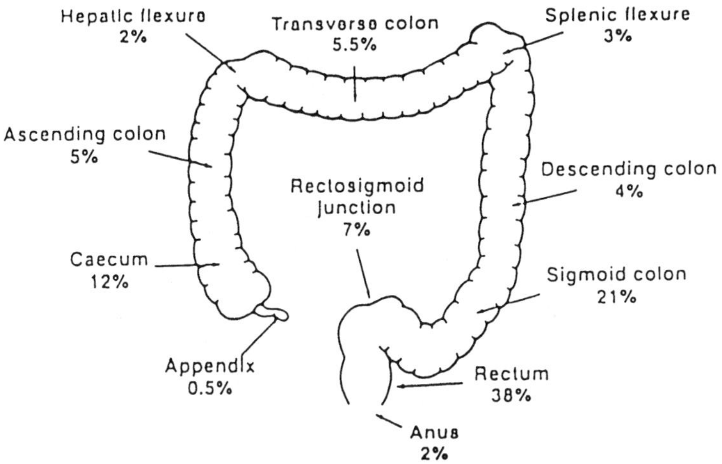

Figure 4 Percent distribution of cases by site within the large bowel (England and Wales).

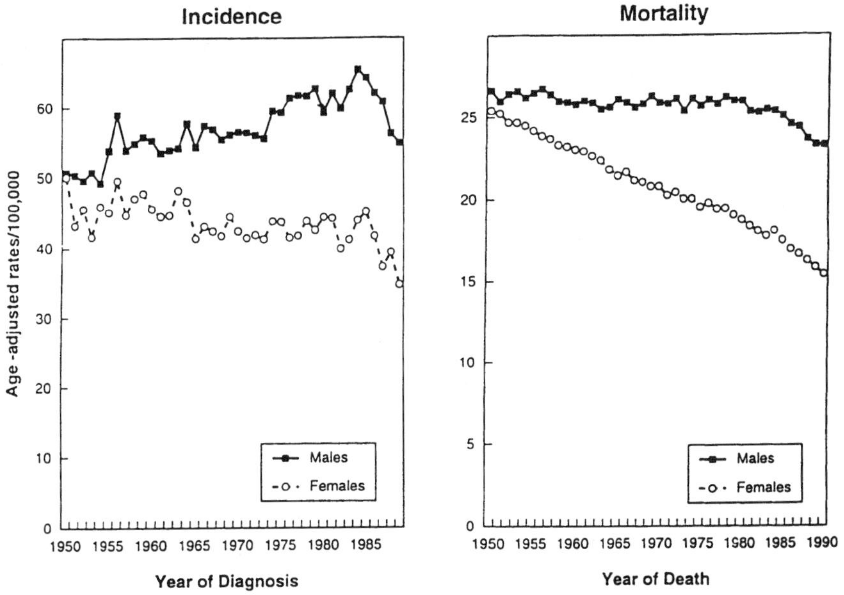

Figure 5 Time trends for colorectal cancer incidence and mortality rate in men and women in the United States.

to diet. This is true to a large extent, but the concordance is not nearly so close, and there is still not a good understanding of the specific components of diet that affect risk. A summary of dietary risk factors is given in Table 1.

The clearest results relate to the consumption of fresh fruits and vegetables. At least 20 studies have found an inverse relationship between fruits or vegetables and colon cancer (16,17). Vegetables have more often been reported as protective than fruit, but no single fruit or vegetable has emerged as particularly important, although there is some evidence that the Brassicas such as broccoli, brussels sprouts, cauliflower, and cabbage may have special properties (18). These foods contain 3-indole carbonyl, brassinin, and other isothiocyanates and phytoalexins, which have been shown to be chemoprotective in experimental systems (19). Because of the high correlation between foods and specific nutrients, it is not surprising that these results have generally been confirmed when micronutrients, either when estimated from food frequency questionnaires or when measured in serum, have been evaluated. In particular, beta-carotene and vitamin C consumption have been associated with reduced risk. Studies on the incidence of colorectal adenomas have also supported these general observa-

Table 1 Dietary Risk Factors for Colorectal Cancer

Factor	Effect	Strength of evidence[a]
Foods		
Fruits and vegetables	Protective	High
Animal fat	Harmful	Moderate
Red meat	Harmful	Moderate
Cereals	Protective/harmful	Low
Dairy products	Harmful	Low
Alcohol	Harmful	Low
Coffee	Beneficial	Low
Nutrients		
Beta-carotene	Beneficial	Low
Vitamin C	Beneficial	Moderate
Vitamin E	Beneficial	Moderate
Folate	Beneficial	Moderate
Calcium	Beneficial	Moderate
Vitamin D	Beneficial	Moderate
Fiber	Beneficial	Low
Protein	Harmful	Moderate
Carbohydrate	Harmful	Low
Fat	Harmful	Low

[a]Based on strength, quality, and consistency of published data

tions. However, the correlation could be with any number of micronutrients found in the same foods, and randomized trials have not shown any effect of beta-carotene, vitamin C, or vitamin E supplementation on the development of new adenomas in individuals with a prior adenoma (20,21). The lack of a beneficial effect of beta-carotene, and in some cases the suggestion of harm, on any cancer in these intervention studies aimed primarily at lung cancer reduction (22–24) also indicates that caution is needed in translating epidemiological observation on foods into preventative strategies based on nutrients or food supplements.

LaVecchia et al. (25) estimated that 39% of colorectal cancer in northern Italy could be attributed to low beta-carotene intake and that 14% was attributed to low vitamin C intake, but these two estimates were confounded so that together they accounted for 43% of the cancers. Again, this could easily be due to other micronutrients in the same foods, and the results are best interpreted in terms of foods rich in these nutrients and cannot be applied to vitamin supplementation.

Burkitt (26) was an early champion of the effect of fiber on colon cancer. He believed fiber was the major protective factor and worked by reducing transit times and absorbing and neutralizing potential carcinogens. Much fiber comes from fruits and vegetables, and it is difficult to separate the effect of fiber from other protective factors contained in these foods. In a meta-analysis of 13 case-control studies, Howe *et al.* (27) found a protective effect of overall fiber intake. However fiber from cereals has not been shown to be protective and increased risks have been associated with higher cereal consumption (28–30). Indeed, Wasan and Goodlad (31) have warned that fiber-supplemented foods may be harmful, because they can stimulate cell-proliferation in some circumstances.

Animal fats, protein and especially red meat, have been associated with an increased risk of colorectal cancer in many case-control studies (32). Cohort studies have been less convincing with both negative (33,34) and positive (35,36) results. Key *et al.* (37) found a nonsignificant 22% reduction in colorectal mortality rate among vegetarians in a 17-year follow-up of 11,000 individuals.

Beneficial results of the consumption of foods rich in calcium, vitamin D, folate, and vitamin E have been reported in several studies, but the results are still inconclusive and not as clear as for foods containing vitamin C and beta-carotene. Coffee may have a beneficial effect (38), possibly by reducing transit times. There is evidence for a weak positive association between alcohol consumption and colorectal cancer, especially rectal cancer (39–41), which is strongest for beer (42,43), but the data are still less than fully convincing. A report on colorectal adenomas is supportive of a risk (44).

A variety of nondietary risk factors have also been found. They are summarized in Table 2.

Table 2 Nondietary Risk Factors for Colorectal Cancer

Factor	Effect	Strength of evidence[a]
Physical activity	Beneficial	High
Height	Harmful	Moderate
Nonsteroidal anti-inflammatory drugs (e.g., aspirin)	Beneficial	Moderate
Estrogen (females)	Harmful	Low/moderate
Inflammatory bowel disease	Harmful	High
Cholecystectomy	Harmful	Low

[a] Based on strength, quality, and consistency of data

2. *Physical Activity*

Lack of physical activity has been consistently associated with an increased risk of colorectal cancer. Almost 20 studies have reported on this and the results have been summarized by Potter *et al.* (16). A more recent report is by White *et al.* (45). Typically men who have sedentary occupations have been found to have a 50%–100% higher risk of colorectal cancer than those who do not. There is less information for women, but the results are generally similar. Some studies have looked at total activity via a physical activity index, or sporting and leisure activities, and again a protective picture emerges. Long-term or lifetime patterns of exercise appear to be most predictive.

3. *Body Size*

Height seems to be more clearly related to colon cancer than weight (46). Being overweight may also be associated with risk, but the results have been conflicting (16).

4. *Endogenous and Exogenous Estrogens*

The possibility that estrogens might play a role in colon cancer has been suggested by a number of studies. The first indications were reported by Fraumeni *et al.* (47) who noted an increased risk in nuns. McMichael and Potter (48) have developed the concept, noting that the male-to-female sex ratio for colon cancer is near unity in persons younger than age 50 years, but increases to nearly 1.5 at older ages when most women are postmenopausal, suggesting that estrogen withdrawal may be protective (see Figure 3). Several case control studies have found a higher risk in nulliparous women. However, age at first childbirth does not appear to affect risk as it does for breast and ovarian cancer. Use of oral contraceptives does not appear to affect colon cancer, but in contrast to the changing sex ratio when women become menopausal, a number of studies have suggested protective effects for hormone replacement therapy (49).

5. *Medical History*

Ulcerative colitis has long been known to predispose a person to colon cancer with a risk on the order of 10-fold and a cumulative lifetime incidence of about 5%. However, it is so rare that this accounts for less than 1% of colon cancers. The risk is related to the extent, severity, and duration of disease. More details can be found in the reviews of Calkins and Mandel (50) and Glickman (51). A distinctive feature of these cancers is that they do not appear to arise from adenomas, in contrast to most other colorectal cancers, but evolve from areas of dysplasia. An increased risk is also found following Crohn's disease, another chronic inflammatory condition, but the increase is generally much less than for ulcerative colitis and appears to be concentrated in the young (52,53). However, extensive disease has a similar risk to that of ulcerative colitis.

Several reports have suggested a risk associated with cholecystectomy, but this has not been consistent (16).

6. *Nonsteroidal Anti-Inflammatory Drugs*

Several lines of evidence support the notion that aspirin and other nonsteroidal anti-inflammatory drugs (NSAIDs) may prevent colorectal cancer. Most notably, a number of case-control and cohort studies have found a 30%–50% reduction in incidence. Long-term use appears to lead to greater benefit. Intervention studies have not yet shown any benefit, and although follow-up is still short, the NSAID Sulindac has been shown to cause regression of small polyps and reduce the incidence of new polyps in high-risk groups. These drugs are potent inhibitors of cyclo-oxygenases, which catalyse the synthesis of prostaglandins from arachidonic acid, suggesting a mechanism of protection, and they have been shown to be effective chemoprevention agents in animal models. The importance of inhibiting the cyclo-oxygenases has been directly demonstrated by Oshima *et al.* (54). However, they have a number of side effects in humans, which may limit their role as a chemopreventive agent in the bowel (55). These issues have been explored in greater detail by Greenberg and Baron (56).

7. *Occupation*

A variety of occupations and exposures have been suggested as risk factors, but none have been consistently reported, apart from sedentary occupations discussed previously.

8. *Family History*

A familial basis for colon cancer is well established and the genetic basis is discussed further below. A rare but striking case is familial polyposis coli, which is a dominantly inherited condition with penetrance approaching 100%. It is associated with a defect in a gene known as APC (adenomatous polyposis coli). In affected individuals, adenomas usually develop around age 20–30 years and, if left untreated, on the order of 1000 adenomas typically develop. Cancer is inevitable and usually occurs about 10 years later. Another well-recognized entity is hereditary non-polyposis colon cancer (HNPCC), which is defined as the presence of three or more family members with non-polyposis colorectal cancer in at least two successive generations with at least one member diagnosed before age 50 years (Amsterdam criteria). Specific genetic defects are found in the majority of cases of HNPCC. They involve mutations in genes responsible for repair of mismatches in complementary DNA nucleotide bases that can occur during normal DNA replication for cell division.

In considering the general population, a more than doubling of risk is found when a first degree relative has had the bowel cancer, and this risk is modified by the age at which the cancer occurs in the proband, being about four-fold for diagnosis in persons younger than age 45 years, three-fold for age 45–55 years, and two-fold for persons older than age 55 years (57). When two or more first degree relatives are affected, the risk is doubled again (58).

II. NATURAL HISTORY

A. Adenoma-Carcinoma Sequence

There is very strong evidence that most carcinomas arise out of preexisting adenomas. The few exceptions seem to be among patients with chronic inflammatory bowel diseases, such as ulcerative colitis, in whom a high cell turnover rate may provide an alternative proliferative background. The adenoma-carcinoma sequence has important consequences for screening, because detection of preinvasive precursor lesions offers a much greater prospect for large mortality rate reductions than early detection of invasive cancers, a fact that the comparison of screening for breast and cervix cancer amply illustrates. Most adenomas appear to take between 10 and 25 years to develop into cancer, although many never do so and remain arrested at some intermediate stage (Figure 6).

The importance of this progression justifies a detailed account of the evidence supporting it. The most striking example is familial polyposis coli, a dominantly inherited rare condition caused by a mutation in the APC gene. Hundreds of adenomas develop in the colon and rectum of the affected individuals. If the condition is left untreated, cancer invariably ensues, usually with a latency on the order of 10 years (59).

More direct evidence for the adenoma-carcinoma sequence comes from the finding of contiguous lesions. Morson (60) showed that the incidence of a benign component of a large bowel cancer depended on the extent of spread of the tumor. Adenomatous tissue contiguous to the cancer was found in only 7% of cases where the tumor had spread through the bowel wall, but the rate increased to 20% when spread was limited to the bowel wall and to 60% with invasion

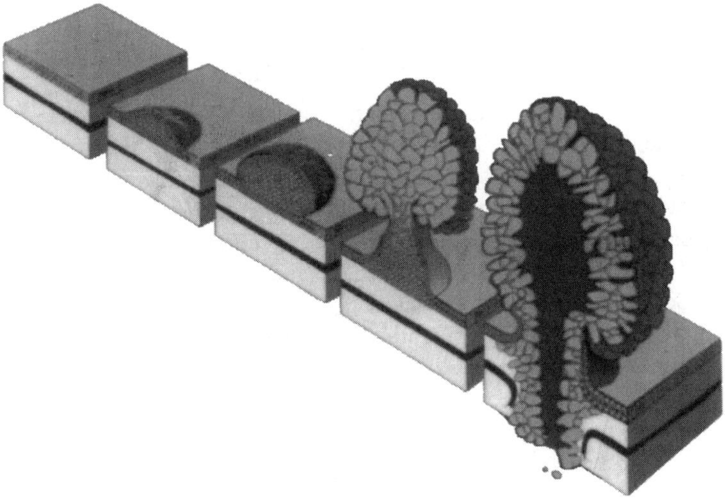

Figure 6 Most adenomas appear to take between 10 and 25 years to develop into cancer, although many will never do so and remain arrested at some intermediate stage.

of only the submucosal area, suggesting that most cancers grow out of and engulf the adenoma from which they arose. In addition, adenomas larger than 2 cm have a focus of invasive cancer in about 30% of cases. Adenomas also have a range of histological types, from tubular, to tubulovillous, to villous, and Muto *et al.* (61) noted that the villous type was more often associated with a concurrent cancer. Other factors are associated with malignant potential, the most important of which is adenoma size, but multiplicity and dysplasia grade are also important. Usually the other high-risk factors are associated with at least one large (>1 cm) adenoma so this is the simplest and most useful single criterion.

Stryker *et al.* (62) demonstrated the malignant potential of a large (≥1 cm) adenoma if left untreated. In his series, the cumulative risk of cancer at the polyp site after 5,10, and 20 years was 2.5%, 8%, and 24%, respectively. Eide (63) has indirectly estimated the conversion rate of an adenoma to cancer using data from an autopsy series of 271 cases. Assuming all colorectal cancers develop from adenomas, Eide estimated an annual conversion rate of 0.25% for any adenoma, but for adenomas ≥1 cm, the rate was suggested to be 3% per annum. Several authors have shown that the finding of an index adenoma in the distal bowel predisposes the patient to concurrent proximal adenomas, subsequent adenomas, or cancer. The risk is particularly high when large or villous adenomas are found (64,65), but there does not seem to be any increased risk of cancer when only a single tubular adenoma of diameter less than 1 cm or hyperplastic polyp is found (65). This is the most common finding in asymptomatic patients.

The yield of polyps in asymptomatic patients has been studied in autopsy series and screening studies. Rex *et al.* (66) reported that 25% of average risk subjects aged 50–75 years had an adenoma, of which, 13% were in the distal bowel. Rates are somewhat higher in men than in women. Similar numbers have been reported in autopsy series, although some geographical variation was noted (67–69). Figure 7 gives a summary of the available data separately for men and women. Screening studies of the distal bowel by sigmoidoscopy also typically report about 10% of subjects have a distal adenoma. Of these, about half, or 5% of the general population, have a large (≥1 cm) or high-risk adenoma. The prevalence of adenomas increases with age, but for distal adenomas appears to plateau around age 60 years (65), whereas the rates continue to increase for proximal adenomas, leading to a shift in distribution with age (69). The site distribution is similar to that for carcinoma, except in the rectum, where adenomas are relatively less common but hyperplastic polyps are more common. Distal adenomas become common about 10 years before bowel cancer rates increase, indicating a window of opportunity for early detection and consistent with a 10- to 25-year interval for the development of cancer from adenomas (Figure 8).

B. Genetic Events

The age distribution of colon cancer suggests that many genetic events are needed before cancer develops. Direct evidence for this is more clearly available for colorectal cancer than at any other site. A number of genes have been consistently associated with colorectal cancer (Table 3). Vogelstein *et al.* (70) have proposed a specific scheme that has some empirical support. An elaboration of their scheme has been developed by Jessup and Gallick (71) (Figure 9).

1. *Initiation*

The most clearly established fact is that loss of the APC gene, a tumor suppressor on chromosome 5q21, is an early event that is associated with hyperproliferation and adenoma development. This is the gene that predisposes a person to familial polyposis coli and is mutated or lost in about 60% of sporadic cancers and even small sporadic adenomas (72). The APC gene has therefore been referred to as a ''gatekeeper'' for colorectal cancer development. Development of adenomas in APC-deficient individuals appears to require the action of the cyclo-oxygenase 2 (Cox 2) gene, which catalyzes the conversion of ar-

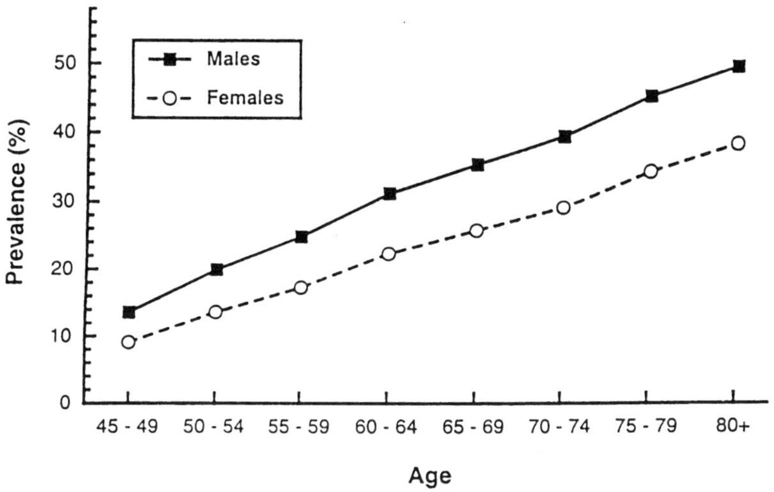

Figure 7 Prevalence of colorectal adenomas by age in males and females separately. (Data from Refs. 66 and 69 by interval censored analysis.)

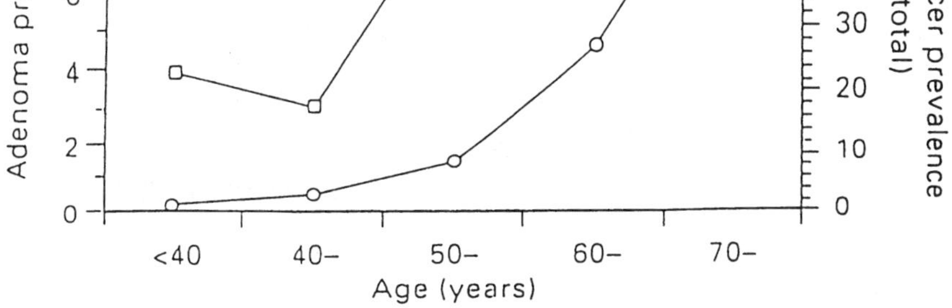

Figure 8 Age at diagnosis of distal adenomas and colorectal cancers. Cancer incidence based on UK rates for 1987 (circular dots) and adenoma prevalence. (Based on combining the results of Refs. 174, 175, and 176 (square dots).)

Table 3 Genes Associated with Colorectal Cancer

Gene	Effect	Strength of evidence[a]
APC	Harmful	Very high
Mismatch Repair (MLH1, MsH2, PMS1, PMS2)	Harmful	Very high
H-RAS	Harmful	Moderate
DCC	Harmful	Moderate
MCC	Harmful	Moderate
P53	Harmful	High
NAT1, NAT2	Harmful (rapid acetylator)	Low
GSTM1-null	Harmful	Low

[a] Based on strength, quality, and consistency of data

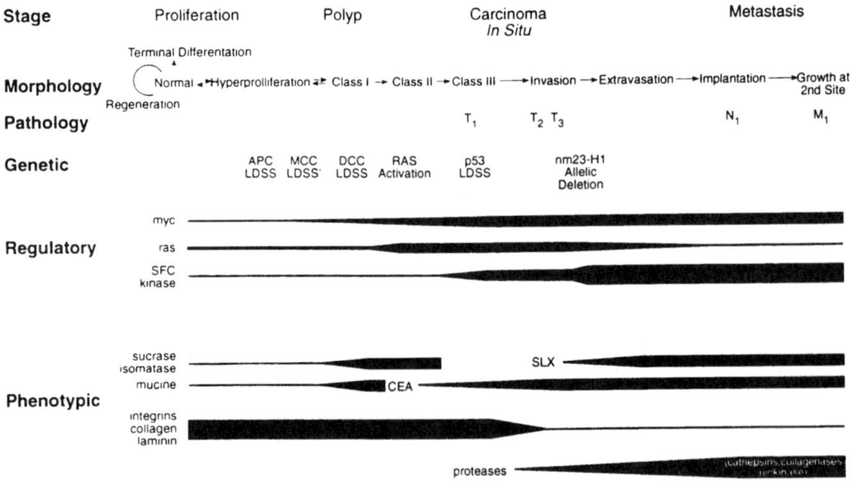

Figure 9 Clinical, genetic, and biochemical events during malignant transformation and progression to metastasis of colonic and rectal epithelial cells. Various stages of colon cancer development and, below, approximate time at which molecular changes occur. Genetic changes are depicted at the time they become frequent. Regulatory and biochemical changes are illustrated by increasing and decreasing levels or activity. (From Ref. 71.)

achidonic acid to prostaglandins H_2. Far fewer adenomas develop in Cox-2 knockout mice with an APC mutation (54).

2. *Development*

Several genetic alterations have been associated with the development of large adenomas and invasive cancer from small adenomas, but none appear to be necessary, in that their prevalence is well below 100%, even in invasive cancer. Three genes have been most clearly implicated.

a. MCC. The MCC (mutated in colorectal cancer gene) is also on chromosome 5 very near APC and has a similar structure. The prevalence of mutations increases as adenomas become larger and more villous.

b. DCC. Loss of the tumor suppresser gene DCC (deleted in colorectal cancer) on chromosome 18q21 appears to occur later in the natural history of colon cancer development. About 70% of cancers show loss of this gene. Most studies suggest that it has a role in causing small adenomas to grow (70), but others have suggested a later role in metastasis (73). The predicted protein sequence of DCC has strong homology with the neural cell adhesion molecule (NCAM), again suggesting a role in the dissemination of established cancers.

c. K-ras. Mutations in one of the K-ras genes are found in about 40% of colon cancers, mostly in codon 12 of the K-ras gene (74,75). Lower rates are found in adenomas, suggesting that K-ras is important in adenoma development and/or malignant conversion.

3. *Late Stage*

Loss of function of the P53 gene on chromosome 17p13 is a late event and is found in about 75% of cancers, but is rare in adenomas. Its role appears to be in malignant conversion or spread of tumors.

4. *Mismatch Repair Genes*

A familial pattern of disease with variable age at onset has been termed hereditary non-polyposis colon cancer (HNPCC). Onset is still typically younger than for sporadic cancers, but not usually as early as for familial polyposis coli. These cancers are associated with defects in mismatch repair genes, leading to widespread replication errors throughout the genome that can be detected by analyzing random microsatellite repeat loci. Four such genes have been identified, the two most common being MSH2 on chromosome 2p15-16 and MLH1 on chromosome 3p21. Two other mismatch repair genes, PMS1 and PMS2, are much more rare. Hereditary non-polyposis colon cancer accounts for about 5%–10% of all colorectal cancer (76–78), and widespread replication errors in microsatellite repeat loci are found in 10%–20% of sporadic tumors (79). These lesions still follow an adenomascarcinoma sequence, but there is a predilection for developing cancers in the proximal colon. In addition to inherited defects in mismatch repair genes, these cancers also often have somatic mutations in the APC and K-ras genes as well (80). The lifetime risk of colorectal cancer for gene carriers in these families is estimated at about 80%, and they also have high risks of endometrial, gastric, biliary tract, and ovarian cancers (81).

5. *Metabolic Polymorphisms*

Polymorphisms in the genes that produce metabolizing and detoxifying enzymes may be important in colorectal cancer, because they may influence the way in which different carcinogens are handled. Individual loci of the cytochrome P-450 (CYP), N-acetyltransferase (NAT), and glutathione S-transferase (GST) gene superfamilies have been implicated. No results are conclusive but there is some evidence for a role of the P-450 enzyme CYP1A2, the NAT1 and NAT2 genes, and the GSTM1 null genotype, all of which are involved in the metabolism of aromatic amines found in cooked meat and other foods (82,83).

III. SCREENING TESTS

A. Digital Rectal Examination

Digital palpation of the rectum is part of a routine physical examination in some parts of the world. It is more commonly used in men, because the prostate can also be palpated. The sensitivity of this test is severely limited by the short region (10 cm of distal rectum) that can be investigated and the crudeness of the method.

B. Guaiac-Based Occult Blood Tests

The concept of fecal occult blood detection (FOBT) by chemical means is usually credited to Van Dean, who in 1864 used gum guaiac. Benzidine and orthotolidins have also been evaluated, but have proven too sensitive for clinical use, because there is a normal amount of gastrointestinal blood loss. In 1901, Boas first emphasized the value of occult blood testing for the detection of bowel cancer, but little interest was shown in this until the mid-1960s. (See references 84–86 for historical reviews.) Greegor (87) developed guaiac-impregnated slides for home use and reported the detection of several asymptomatic cases of colorectal cancer by this means. This was the basis of the Hemoccult test, which has been widely marketed and evaluated since the early 1970s. Modification took place around 1977, which led to the current Hemoccult II test. This test consists of three cards, each of which contains two windows of guaiac-impregnated paper. For three successive bowel movements, a small amount of stool is applied to each of the two windows on a single slide using a supplied wooden applicator. It is usually recommended that the test be performed annually or biennially beginning around age 50 years.

This test depends on the oxidation of a phenolic compound to a quinone structure, resulting in a color change. Hydrogen peroxide facilitates the oxidation process, which is catalyzed by a number of naturally occurring peroxidases and catalases including hemoglobin. For the Hemoccult test, gum guaiac is impregnated in a test paper and hydrogen peroxide is provided in a developing solution. The resultant phenolic oxidation in the presence of blood yields a blue color. Guaiac tests detect heme in any form, provided that the iron has not been removed from the porphyrin ring and will react to any peroxidase. The heme in hemoglobin can be modified by the gut microflora, leading to a removal of iron and a loss of detectability. This can reduce the sensitivity of the test, especially for proximal cancers. Other products also catalyze the reaction, including animal blood, drugs, and certain fresh fruits and vegetables that contain peroxidases. This can lead to false-positive results, and to improve specificity, tests are often repeated after a positive result and following dietary restriction of red meat and sometimes dark fish (salmon, sardines), dark poultry (e.g., pheasant), and certain peroxidase-rich fresh fruits (e.g., melons) and vegetables or boiled turnips or horseradish. Oral iron preparations and aspirin can also lead to false-positive results, and the antioxidant activity of high-dose vitamin C supplementation may lead to a false-negative result by interfering with the peroxidase reaction. It is usually advised to stop these for at least 2 days before performing the test. A test is conventionally considered positive if at least one window turns blue. Test positivity has been noticed to decrease with storage time (88). Positivity can be greatly increased by rehydration of samples with a drop of water before processing (89). However, the added sensitivity appears to be outweighed by the loss of specificity (90,91) and this is usually no longer done.

A number of other commercial tests using guaiac have been marketed, but none are as popular as Hemoccult II. A more sensitive and easily read version of the test (HemoccultSENSA) has also been evaluated. A version for home use in which the reagents are impregnated in toilet paper and a color change is seen when placed in the toilet bowl is available in some countries including the United States. These tests have not been evaluated in terms of sensitivity and specificity and may provide false reassurance; they cannot be recommended until they have been properly evaluated.

C. Other Occult Blood Tests

Other chemical tests based on oxidation of phenolic compounds have been developed using orthotolidine (Hematest) and tetramethylbenzidine (Hemo-Fec), but these are not used for screening.

Newer tests have attempted to be more selective for hemoglobin and its degradation products. HemoQuant is a heme-porphyin test that detects not only intact heme but degraded de-ironed hemes (heme-derived porphyrins), which was predicted to increase sensitivity, especially for proximal cancers. The test is based on fluorescent detection of dicarboxylic porphyrins derived from hemes after iron removal and provides a quantitative output. The test is unaffected by stool dehydration and storage, dietary peroxidases, iron, and ascorbic acid, which offers potential advantages over guaiac-based tests. However, it is still affected by red meat ingestion and aspirin use.

Another approach has been to develop immunochemical tests that are immunoreactive with hemoglobin. These tests are the most specific for human blood and detect only human hemoglobin, globin, and perhaps some early degradation products. At least four versions of this approach have been reported. Most information is available for HemeSelect, which uses hemagglutination with formalin-fixed chicken erythrocytes coated with antihuman hemoglobin antibodies as a detection method.

D. Sigmoidoscopy

Endoscopic screening has a different goal than fecal occult blood tests (FOBT), which is primarily an early detection measure. For sigmoidoscopy, the primary aim is to detect and remove adenomas so that the adenoma-carcinoma sequence can be interrupted and cancer prevented. Such an approach, based on the detection of preinvasive precursor lesions, may allow for a much longer interval between screens and a nearly 100% cure rate for detected lesions.

Improvement in the equipment available has made endoscopic examination a much more attractive proposition for screening. Early work was carried out using a 25-cm rigid instrument, which only allowed visualization of the rectum. Development of flexible fiberoptic or video scopes has transformed the procedure and made it simpler, quicker, less painful, and more complete. Instruments of length 35 cm and 60 cm are available in addition to the full-length colonoscopes. The 60-cm scope is the instrument of choice for screening, permitting visualization of the rectum and whole sigmoid colon in approximately 80% of patients. Bowel preparation can be achieved by a single phosphate enema and most polypectomies can be performed at the same visit, unless the polyps are large or present specific difficulties (92). Typically, the examination takes 6–10 minutes. The approach is highly sensitive, allowing reliable detection of polyps of size down to 5 mm.

E. Barium Enema

Barium enema is often used for evaluation of symptomatic patients, especially when endoscopy is difficult. The lower sensitivity of FOBT for proximal lesions and the lack of coverage of the proximal bowel by flexible sigmoidscopy have led to a suggestion that it might be used as a screening test for this segment, the only alternative being colonoscopy, which is more expensive and more invasive. Eddy (93) has evaluated by simulation a range of possible screening protocols, some of which involve barium enema. As for endoscopy, bowel preparation with an enema is necessary, but it is possible for both procedures to be performed at the same visit. Both single and double contrast methods have been used, but the double contrast method is more common except for low-risk, seriously ill, or young patients. For the single-contrast method, the bowel is filled with a diluted barium suspension, allowing space-filling defects to be visualized. Good technique and colon compression are needed to obtain good films. In high-risk, occult blood positive, or symptomatic patients, the double-contrast procedure is more often used. It involves a high-density barium suspension and the bowel is then inflated with air to provide a contrast so that the lining can be visualized. Good mucosal coating with the suspension is critical. Typically, 4–5 large and 8–10 small films are obtained for each technique. In specialist centers, both techniques appear to have similar high sensitivity for cancer and polyps greater than 1 cm (70%–90%), but the double-contrast method has greater sensitivity for smaller adenomas. However these results may not translate to routine use, and good technique is essential if they are to be achieved. An additional limitation is that a further endoscopic procedure is needed for polypectomy if an adenoma is discovered. Further details can be found in references 94–96.

F. Colonoscopy

The limitation of sigmoidoscopy to the detection of distal bowel lesions has led some researchers to suggest complete colonoscopy of the large bowel as a screening test (97,98). This is a much more invasive procedure than flexible sigmoidoscopy with greater costs and higher complication rates, and is currently limited to symptomatic patients or those who have had an adenoma detected in the distal bowel. An experienced operator can visualize the entire large bowel in more than 80% of patients and all polyps can be removed at the same time. More complete bowel preparation is needed, usually involving overnight oral preparations. Sedation of the patient is also usually required for the procedure.

G. Future Approaches

There is much interest in developing less invasive methods of visualizing the bowel. Virtual colonoscopy based on helical computed tomography (CT) scanning and the use of a thin probe passed up the bowel is one possibility. Identification of the position of the endoscope within the colon to allow more accurate localization of visualized lesions and extent of coverage is also being developed (99). Undoubtedly endoscopic procedures will be much improved in the future.

H. Genetic Testing

In contrast to the aforementioned tests, genetic tests are designed to identify at-risk people rather than existing lesions. As a result, those who are positive generally require surveillance by a conventional method, usually endoscopy. There are three main areas where genetic testing is being evaluated. The germ-line mutation in the APC gene leads to familial polyposis coli. This is an autosomal dominant trait with high penetrance so that half of all offspring of carriers are affected, and regular endoscopic surveillance beginning at a young age is indicated. This syndrome is easily recognized, even in carriers of a new mutation, because the cancer is accompanied by many hundreds of adenomas. Laboratory testing for a mutation is time consuming. The gene itself produces an 8.9-kb RNA message, which is divided into 15 exons, the last of which contains most of the coding sequence. Complete sequencing requires 31 sets of PCR primers (100). The most common mutation is a 5 bp deletion at codon 1309 within exon 15, which is present in about 10% of patients with familial polyposis coli. Altogether, about one-fourth of all germ-line mutations are clustered in a hot spot of about 300 bp in exon 15 (Figure 10), and more than 90% of mutations lead to a premature stop codon. A number of simpler approaches to testing have been considered. One approach is to use a protein truncation test, which relies on the fact that most deletions result in a premature stop codon, so that a person with such a germ-line defect makes an abnormally short protein. Another approach is to amplify one or more short regions and

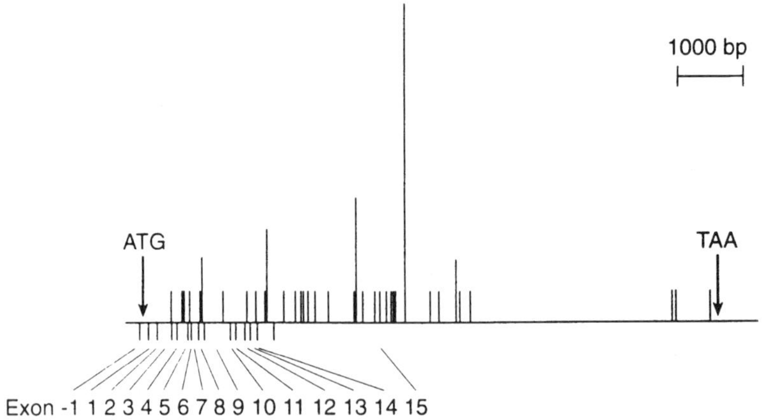

Figure 10 Distribution of germ-line mutations in APC gene. The length of the bar indicates the number of mutations at the indicated position. Positions translational initiation (ATG) and termination (TAA) codons are marked with arrows. (From Ref. 100.)

use single-strand conformation polymorphism analysis to identify a region containing a mutation (101). However, none of these methods detects all mutations and a complete analysis requires sequencing of all 15 exons and flanking regions. When a gene defect is identified in an individual, the children and siblings can be screened for the same mutation. This is much less labor intensive because only a single linked marker or specific mutation needs to be examined. Positive results in family members then leads to examination of their first degree blood relatives. The largest benefit accrues to those not affected, who can be spared intensive colonoscopic surveillance. However, even those who are affected have time to plan appropriate prophylactic measures.

New germ-line mutations in the APC gene are too rare in the general population to screen for them. However, most adenomas have a somatic mutation or deletion, and a test looking at exfoliated cells in stool samples may be a future possibility.

Defects in DNA repair genes are a much more common form of genetic colorectal cancer. All these genes leave tell-tale signals in DNA repeat regions and can be diagnosed replication error repair positive (RER+) by polymerase chain reaction (PCR) amplification of a few DNA repeat regions. Isolation of the specific gene involved is more difficult and specific genetic alterations are found in one of the four unknown replication error repair genes in about 75% of cases where a strong family history is present. The two most commonly implicated genes are MSH2 on chromosome 2p15-16, which contains 16 exons, and MLH1 on chronosome 3p21, which contains 19 exons. Nevertheless, this test can be useful in HNPCC families in whom exclusion of carrier status in offspring of those affected is particularly useful. Also, mutations in these genes are sufficiently common that screening for RER+ individuals in the general population may be an effective way of determining who needs intensive endoscopic screening. Screening is usually done by examining around 10 polymorphic microsatellite repeat markers. However, a precise definition of the criteria constituting the RER+ phenotype has not been agreed upon by investigators (e.g., the necessary number and sites of microsatellite mutation).

Another area of screening is for somatic K-RAS mutation in stool samples. About 40% of cancers and a similar proportion of adenomas contain such mutations, and almost all mutations affect either codon 12 or 13, making genetic testing much simpler. The testing procedures are able to pick up mutations from stool samples in about 50% of patients with RAS mutations in their cancer, but they are not yet sensitive enough to pick up a mutation arising in an adenoma, where the number of exfoliated cells is much lower (102–105).

IV. SCREENING STUDIES

A. Digital Rectal Examination

Digital rectal examination (DRE) was used in conjunction with flexible sigmoidoscopy in a study of multiphasic health check-ups at the Kaiser-Permanante centers in northern California (106). It was not possible to determine whether any effect was due to DRE. Herrinton et al. (107) have recently conducted a case-control study in which cases were in persons older than 45 years old who had died of distal rectal cancer. About one-fourth of controls had had such an examination in the past 10 years, but there was no evidence that the test had any value in preventing death (OR = 0.96, 95% CI 0.56–1.7).

B. Fecal Occult Blood Tests

1. *Detection Rates—Cross-Sectional Studies*

A major concern about FOBTs is their lack of sensitivity. The vast majority of results relate to Hemoccult. Even for patients with symptomatic bowel cancer, the test is positive only 60%–90% of the time (Table 4). In these patients, occult bleeding is more often detected in larger, grossly ulcerated, and advanced-stage cancers.

There have been a large number of studies of occult blood testing as a screening test. Lower sensitivities have been reported in these studies, typically on the order of 40%–70%, although the methods of assessing sensitivity are more variable here, and usually require follow-up for at least 6–12 months to identify interval cancers as falsely negative. Most of these studies are uncon-

Table 4 Sensitivity of Hemoccult for Bowel Cancer

Study	Reference	Sample size	Sensitivity (%)
Griffith *et al.*	137	28	82
Macrae and St. John	138	46	69
Doran and Hardcastle	139	50	70
Farrands and Hardcastle	140	61	72
McDonald and Goulston	141	36	53
St. John *et al.*	109	107	89
Ahlquist	142	11 symptomatic	36
		35 asymptomatic	23
Crowley	143	27	52
Songster	144	150	40
Kapparis	145	40	72
Reilly	177	51	61

trolled and thus cannot be used to assess effects on mortality rates, because of lead time bias and length bias (see Chapter 3). However, they do provide useful information on detection rates, specificity, and compliance. A selection of some of the larger studies is listed in Table 5.

Some of the studies have evaluated more than one method of occult blood testing. In a postresection group of 1217 patients with previous cancer, Ahlquist *et al.* (108) evaluated Hemoccult II and HemoQuant testing on three stool samples per subject. Patients also received annual endoscopic or radiological follow-up, which yielded 46 patients with new or recurrent cancers and 386 with polyps but not cancer. The sensitivity of Hemoccult was 26% and the specificity was 95.3%. When the cut-off level of HemoQuant was adjusted to produce the same specificity (2.1 mg Hb/g stool), the sensitivity for cancer for this test was also 26%. Sensitivity for polyps 1 cm or larger was 13% by Hemoccult and 11% by HemoQuant.

St. John *et al.* (109) compared Hemoccult II, HemoccultSENSA, HemoQuant, and HemeSelect on 107 patients with symptomatic colorectal cancer and 81 patients with primarily asymptomatic adenomas. HemeSelect and HemoccultSENSA had significantly higher sensitivities for colorectal cancer (97% and 94%, respectively) than either Hemoccult II (89%) or HemoQuant (71%). HemeSelect had the highest sensitivity for adenomas (58%) or large adenomas (76%) (Table 6). HemoccultSENSA and HemeSelect were also performed on 1355 screenees. In this screening population, HemoccultSENSA and HemeSelect had similar positive predictive values for neoplasia (32% versus 33%, respectively) but HemeSelect had a lower positivity rate overall (3.0% versus 5.0%) and better specificity (97.8% versus 96.1%, $P<0.05$).

Castiglione *et al.* (110) also compared Hemoccult II against HemeSelect in 24,282 screening subjects aged 40–70 years. Dietary restriction of red meat was advised for the day of the test and the preceding 2 days. A 37% compliance rate was achieved, leading to 8008 completed tests. The mean age was 54 years. The Hemoccult tests were rehydrated and produced a 6.0% positivity rate. A total of 15 cancers were detected, giving a positive predictive value (PPV) of 3.7%. HemeSelect was evaluated at two thresholds. The low threshold had a 8.2% positivity rate and detected 21 cancers (PPV = 3.8%), whereas the high threshold had a positivity rate of 3.1% and detected 17 cancers (PPV = 8.4%). Similar relative performances were seen for adenomas (see Table 5). Better results were also obtained when only individuals aged 50 years or older were considered, and the authors believed the tests should only be used in this group. Overall, HemeSelect gave better results than Hemoccult, but the appropriate threshold for positivity would depend on cost-benefit evaluations.

Petrelli *et al.* (111) evaluated a community screening program in which kits containing Hemoccult II, HemoccultSENSA, and HemeSelect were distributed to 39,000 people older than 40 years of age. No dietary restriction or drug

Table 5 Screening Studies of Occult Blood Tests

Study (first author)	Reference	No. tested (%)	Diet	No. positive (%)	No. cancer/polyps/ large adenomas	PPV cancer/polyps/ large adenomas	Comments
Uncontrolled studies							
Greegor	146	278	Y	24 (8.5)	2/3	8/14	
Hastings	147	2,625 (76)	Y	159 (6.1)	5/	3/	
Miller	148	2,332	N/Y	11 (0.5)	1/3	9/33	2 Stages/diet in 2nd stage
Goodman	149	1,701 (68)	Y	9 (0.5)	0/	0/	Women only
Helfrich	150	8,930	N	157 (1.8)	3/2	2/1	
Fruhmorgen	151	5,016 (84)	Y	136 (2.7)	13/83	10/61	
Heeb	152	3,956 (69)	Y	79 (2.0)	5/5	6/6	
Elwood	153	1,690 (15)	Y	58 (3.4)	2/	3/	
Bralow	154	3,008 (79)	N/Y	180 (6.0)	7/11	4/6	Rehydrated/2 stages
Winchester	155	14,074 (26)	Y	617 (4.4)	30/40	5/7	
Kurnick	156	5,420	N	120 (2.2)	9/	8/	
Gnauck	157	16,100	—	531 (3.3)	75/106	14/20	
Schwartz	158	3,480,000	—	38,030 (1.1)			
				4,262	322/	8/	Evaluable patients
Larkin	159	5,565 (95)	Y	88 (1.6)	4/6	1/1	Unrehydrated/2 stages
Chambers	160	1,160 (46)		68 (5.9)	3/10	4/15	
Stuart	161	4,498 (68)	Y	150 (3.3)	13/31	9/21	
Farrands	162	2,439 (27)	N	124 (5.1)	4/8	3/6	
Million	163	1,646 (28)	N	37 (2.3)	2/5	5/16	
Sontag	164	2,964 (22)	Y	135 (4.6)	14/44	10/33	
Hardcastle	165	3,613 (37)	Y	77 (2.1)	12/27	16/35	
Siba	166	3,791	N	97 (2.6)	6/18	6/19	
Habba	167	1,628 (76)	N	37 (2.3)	5/16	14/16	

Kapparis	145	200	Y	10 (5.0)	1/3	10/30	Not rehydrated
				21 (10.5)	1/5	5/24	Rehydrated
				15 (7.5)	2/4	13/27	Immunological
Armitage	168	1,304 (44)	N	40 (3.1)	3/25/20	8/62/50	Hemoccult
				106 (8.!)	5/22/22	5/21/10	Immunologic (Feca EIA)
Thomas	169	10,176 (58)	Y	131 (1.3)	20/76/27	15/58/21	3-Day test
		9,461 (54)		160 (1.7)	24/83/35	15/52/22	6-Day test
McGarrity	170	46,014 (63)		1,303 (2.8)			Follow-up only on subset
				799	67/174	8/22	
St. John	109	1,355	Y	68 (5.0)	1/15/9	1/22/13	HemoccultSENSA
				41 (3.0)	1/10/10	2/24/24	HemeSelect
Caffarey	171	5,934 (79)	?	287 (4.8)	44/38	15/13	High-risk population
Petrelli	111	8,933 (23)	N	456 (5.1)	13/44/18	5/18/7	Hemoccult
				849 (9.5)	16/85/24	4/19/5	HemoccultSENSA
				438 (4.9)	17/45/17	7/19/7	HemeSelect
Robinson	172	1,489 (38)		17 (1.1)	1/8/7	6/47/41	Hemoccult
				145 (9.7)	9/48/29	6/33/20	HemeSelect
Castiglione	173	8,008 (38)	Y	483 (6.0)	15/79/37	4/20/9	Rehydrated
				655 (8.2)	21/118/55	4/21/10	Low threshold—HemeSelect
				245 (3.1)	17/62/36	8/30/18	High threshold—HemeSelect
Controlled studies							
Kewenter	113	16,711 (65)	Y	835 (5.0)	38/159/95	5/19/11	Most samples rehydrated
Mandel	112						Most samples rehydrated
	Annual	14,051 (75)[a]	Y		174/	2/28	
				−(9.6)			
	Biennial	14,013 (78)[a]	Y		145/	3/29	
Kronberg	115	20,672 (67)	Y	215 (1.0)	37/ /68	17/ /32	
Hardcastle	114	40,214 (54)	Y	837 (2.1)	83/311	10/37	

[a] Includes follow-up tests

Table 6 Sensitivity of Four Occult Blood Tests (%)

Group	n	Hemoccult II	HemoccultSENSA	HemeSelect	HemoQuant (>2 mg/g)
Colorectal cancer	107	88.8	93.5	97.2	71.0
Cancer proximal to splenic flexure	28	89.3	92.9	100.0	89.3
Cancer distal to splenic flexure	79	88.6	93.7	96.2	64.6
Any adenoma	81	30.9	44.4	58.0	37.0
Adenoma >1 cm	45	42.2	60.0	75.6	42.2
Adenoma <1 cm	36	16.7	25.0	36.1	30.6

Source: Ref. 109.

cessation was required. A total of 8933 kits were returned (23%) and positivity rates were 5.1%, 9.4%, and 4.9%, respectively. The positivity rate for at least one test was 13%. Of the 1165 positive individuals, physician-verified results were obtained for 631. In this group, 20 cancers were found, of which, the respective tests found 13,16, and 17. Positive predictive values for cancer were slightly higher for HemeSelect, but the values were very similar for adenomas (see Table 5).

Two studies have used both hydrated and nonhydrated Hemoccult tests. Mandel *et al.* (112) found that positivity for the test increased from 2.4% to 9.8% and sensitivity for cancer increased from 80.8% to 92.2%, but positive predictive value decreased from 5.6% to 2.2%. Kewenter *et al.* (113) found the positivity rate increased from 1.9% for unrehydrated kits to 6.1% (and subsequently as high as 14.3%) for rehydrated tests. In this study, the high rate of positivity in rehydrated tests led to a policy of retesting positive subjects, which produced a positivity rate after the second test of 1.7%–4.6%, depending on the cohort. Detection rates at the prevalent screening were 0.9/1000 for unrehydrated tests and 2.7/1000 for rehydrated tests.

Rozen et al. (113a) have compared Hemoccult II against flexible 60-cm sigmoidoscopy in 1176 previously unscreened asymptomatic individuals of mean age 52 years. About one-third of these patients had a first degree relative with colorectal cancer, but otherwise they were at average risk. All abnormal results were followed by colonoscopy. FOBT detected only 18% (7/43) of the adenomas and 60% (3/5) of the cancers, compared to 95% (41/43) and 80% (4/5), respectively, for sigmoidoscopy. Only two of the cancers and five of the patients with an adenoma were positive by both tests, suggesting that FOBT might be a useful adjunct to sigmoidoscopy.

2. *Mortality—Follow-Up Studies*

Three randomized clinical trials have thus far reported on the effect on mortality of occult blood testing (Table 7). Mandel *et al.* (112) reported on 46,501 individuals aged 50–80 years in Minnesota who where randomized three ways to either control, annual FOBT, or biennial FOBT by Hemoccult II. Rehydrated tests were mostly used and positivity rates were very high (9.6%), leading to a 38% cumulative colonoscopy rate in the annually screened group and a 28% rate in the biennial group. After 13 years of follow-up, a 33% reduction (95% CI 18%–50%) in mortality rate was observed in the annual screening group, but thus far only a small and statistically nonsignificant 6% decrease (95% CI 32% decrease, 31% increase) in mortality rate has been seen in the group randomized to biennial screening.

Hardcastle *et al.* (114) reported on 152,850 individuals aged 45–74 years in Nottingham, United Kingdom, randomized to control or biennial Hemoccult tests. Overall compliance was just under 60%. Samples were not rehydrated and positivity rates were 2.1%. After a median follow-up of 7–8 years, a statistically significant 15% mortality rate reduction (95% CI 2%–26%) has been found. A very similar trial of 61,933 Danes has given similar results (115). After 10 years of follow-up, the mortality rate reduction was 18% (95% CI 1%–34%).

A further trial in Gothenberg, Sweden, involving approximately 50,000 subjects and initiated in 1982, has not yet reported on mortality effects, although a clear shift in stage distribution has been noted (113).

A fifth trial has studied the value of adding FOBT to rigid sigmoidoscopy (116). A total of 21,756 patients older than 40 years were entered from 1975 to 1979 and were assigned either to annual screening either by rigid sigmoidoscopy alone or in combination with Hemoccult. Initial compliance was high but decreased rapidly in follow-up visits. Follow-up was until 1984 when there were 199 colorectal cancers and 64 deaths. Randomization was not performed and patients were allocated to a treatment according to time of attendance at the clinic. A nonsignificant 26% reduction in colorectal cancer mortality rate was found.

These results have been generally supported by four published case-control studies (see Table 7). Selby *et al.* (117) conducted a case-control study of FOBT using Hemoccult II in northern California. Cases consisted of 485 persons in whom fatal colorectal cancer developed after age of 50 years, and there were 727 age-and sex-matched controls. Overall, a 31% (95% CI 9%–48%) reduction in colorectal cancer mortality rate was observed if at least one screening test had taken place in the past 5 years. Closer analysis indicated that the protection was confined to the 3 years following a screening test. The protection obtained was similar in men and women and was unaffected by age. Cases had fewer periodic health check-ups and sigmoidoscopies than controls, suggesting the

Table 7 Effect on Mortality of Occult Blood Testing

Study (first author)	Reference	Median FU (years)	Frequency (months)	Screened CRC deaths/pop.	Controls CRC deaths/pop.	OR (95% CI)	Compliance (%)	Age (years)
Clinical Trials								
Mandel	112	13	12	82/15520	121/15394	0.67 (0.50–0.82)	75	50–80
			24	117/15587	121/15394	0.94 (0.68–1.31)	78.4	50–80
Hardcastle	114	7.8	24	360/76466	420/76384	0.85 (0.74–0.98)	59.6	45–74
Kronberg	115	10	24	205/30967	249/30966	0.82 (0.68–0.99)	67	45–75
Winawer[a]	116	7	12	36/12974	28/8782	0.75 (0.45–1.21)	76 initially ~20 on FU	>40

[a]Rigid sigmoidoscopy with or without FOBT.

Study (first author)	Reference	Cases/controls	% Screened in last 5 years cases/controls	Screened in last 5 years OR (95% CI)	Screened in last 2 years OR (95% CI)	Duration of effect
Case control studies of FOBT						
Selby	117	486/727	31.5/42.6	0.69 (0.52–0.91)	0.76 (0.55–1.03)	2–3 years
Wahrendorf	118	372/684	14.6/22.3	Male 0.92[a] (0.54–1.57) Female 0.43[a] (0.27–0.68)	Male 0.83[b] (0.42–1.64) Female 0.30 (0.17–0.55)	Benefit only in years 1–3 after test
Saito	119	193/577	18.9/22.8	0.77 (0.34–1.74)	0.41 (0.20–0.82)	2–3 years
Lazovich	120	236/457	44/48	0.81 (0.62–1.19)	0.95 (0.66–1.35)	No trend

[a]6–36 months asymptomatic tests
[b]12–24 months asymptomatic tests

results may be at least partly due to a more healthy lifestyle in screenees. False-positive rates in controls were about 3%, in line with the clinical trials and series.

Wahrendorf *et al.* (118) conducted a case-control study based on a population screening program in Saarland, Germany. Screening by FOBT became available under a government program in 1977 and was offered to all men and women older than age 45 years, but participation was low and based solely on the individual's choice to attend. The area studied had a population of 1.1 million and was covered by the Saarland Cancer Registry. Cases were of individuals who died of colorectal cancer between 1983 and 1986 and were aged 55–75 years at death, and for whom screening histories could be retrieved (163 male, 209 female). Age-matched controls were obtained from general practitioner files of the cases and from gynecologist files for women. For men, 13% of cases and 14% of controls had at least one asymptomatic screen 6–36 months before diagnosis (OR = 0.92, 95% CI 0.61–1.75), whereas for women, 16% of cases and 29% of controls had such a test (OR = 0.43, 95% CI 0.27–0.68). Screening was more common in both male and female cases in the 6 months before diagnosis, suggesting that some tests were performed because of symptoms. The benefit in women was ascribed to a higher participation rate. However, the low overall participation rate (15%) and inability to reliably separate symptomatic from asymptomatic cases limit the interpretations that can be drawn from this study.

Saito *et al.* (119) have reported a study of similar design from Aomori Perfecture, Japan, which has a population of 1.5 million. The main difference was that the FOBT test routinely used was an immunochemical test for hemoglobin (HemeSelect). A total of 193 persons who died of colorectal cancer between 1986 and 1992 and who were diagnosed at ages 40–79 years after the screening program started in 1986 were considered. Three age-, sex-, and location-matched controls were selected for each case, and screening histories were abstracted for the 5 years before the date of diagnosis in the case from a central screening registry for the area. In the 5 years before diagnosis, 18.9% of case subjects and 22.8% of control subjects had been screened at least once (OR = 0.77, 95% CI 0.34–1.74). The benefit was strongest in the first 2 years after screening and was no longer apparent after 4 years. The odds ratios for the first 5 years since the most recent screening test were 0.40,0.39,0.58,0.90, and 1.20, respectively. Benefit did not differ by sex, age at diagnosis, or anatomical location of the cancer.

Lazovich *et al.* (120) conducted a case-control study based on a health maintenance organization in western Washington state. Case subjects were individuals aged 40–84 years who died of colorectal cancer between 1986 and 1991. A total of 248 eligible cases with available histories were matched on a two-to-one basis with 496 controls. Even though "definitely symptomatic" tests were excluded, an increased amount of screening was found in cases in the first year before diagnosis (20% versus 17%), leading to an odds ratio of 1.37, (95% CI 0.92–2.03), suggesting that there was some misclassification in this regard. The mortality rate was reduced in subsequent intervals and there was a reduced mortality rate for ever screening of 28% (95% CI 49% benefit, 2% harm) based on 50% of case subjects and 58% of control subjects having been screened at some time. The risk of fatal cancer was reduced almost 50% in the second year following a test (6% of cases versus 11% of controls), and benefit appeared to be confined to years 1–3 after a test. Benefit also appeared to be confined to patients younger than 75 years at diagnosis, and to tests performed at home compared to those performed in the physician's office.

C. Sigmoidoscopy

1. *Randomized Trials*

At least two large randomized trials are evaluating the efficacy of screening by flexible sigmoidoscopy. In the United States, the Prostate, Lung, Colorectal and Ovarian (PLCO) trial is examining the role of 3-yearly screening in 37,000 men and 37,000 women aged 55–74 years in a randomized trial with an equal number of controls subjects. In the United Kingdom, a trial of once-in-a-lifetime sigmoidoscopy for individuals aged 55–64 years is underway. This trial will involve 200,000 individuals in 14 centers, of which, 65,000 will be randomized for screening. It will be many years before either of these trials have definitive mortality results, but the United Kingdom trial has published baseline data for the first two centers (121). A two-stage entry procedure is used in which randomization is performed only among those who express an interest for screening in a prerandomization questionnaire. The positive response rate to this initial questionnaire has been 60%, and 75% of those invited to be screened have accepted, leading to screening being performed in 45% of the target population. In the first 1285 sigmoidoscopies, there have been 7 (0.5%) individuals with cancer, 6% with high-risk adenomas (>1 cm, villous, tubulovillous, severely dysplastic, or multiple), 10% with any adenoma, and a further 15% with metaplastic polyps only.

A pilot screening program has been reported on 3500 asymptomatic subjects aged 55–59 years in Freemantle, Western Australia (122). Initial compliance was 12% with a 3.5% increase after a telephone follow-up. Among 342 screened individuals, 35% had polyps, of which, just less than half (16% of those screened) were adenomas. Nineteen (5.6%) individuals had at least one adenoma greater than or equal to 1 cm. The procedure was well tolerated and 99% of the screenees said they would be tested again, if required.

Hoff *et al.* (123) have published results of a pilot trial of flexible sigmoidoscopy screening in Telemark County, Norway. Four hundred men and women aged 50–59 years were randomized to screening, and there were 399 unscreened controls. If a polyp was found at screening, colonoscopy was performed initially and at 2-and 6-year follow-up examinations. Attendance for screening was high at 81%. Polyps were found in 35% of the participants and adenomas greater than 5 mm were detected in 10%. After a 10-year follow-up, four cancers (one fatal) were found in the control group and one (fatal) cancer was found among the screened group, which was in a noncomplier. These numbers are too small to make any conclusions regarding efficacy. However, no significant pathology had occurred on the follow-up examinations, suggesting that follow-up could be much less frequent (or omitted altogether in some cases).

2. *Observational Studies*

Most of the information on the efficacy of endoscopic screening comes from clinical or screening series and case-control studies (Table 8). A number of studies have examined colorectal cancer rates in patients offered repeated sigmoidoscopy or colonoscopy. In an early study, Gilbertson and Nelms (124) reported an 85% reduction in rectal cancer in 21,000 subjects who had annual rigid sigmoidoscopy, but others have suggested that the reduction may not have been so large. Friedman *et al.* (106) reported a 60% reduction in distal colorectal cancer in 10,000 patients randomized to have a sigmoidoscopy every 3 years, but subsequent analysis showed that the use of this procedure was similar in

Table 8 Efficacy of Sigmoidoscopy

First author	Reference	Cases/no. of subjects	Reduction in colorectal cancer incidence in region examined (95% CI)	Type of study
Gilbertson	124	13/21,150	60–85%	Prospective uncontrolled
Friedman	106	110/10,713	60%	Prospective randomized
Selby	124	261/1,129	70% (52–81)	Case-control
Newcombe	125	66/290	79% (48–92)	Case-control
Atkin	65	3[b]/1,618	85%	Retrospective cohort
Winawer	116	5/1,418	90% (76–97)	Prospective cohort (colonoscopy)
Muller	126	4,358/16,351	59%[a] (50–67)	Case-control

[a] Any colorectal procedure
[b] Excluding cancers arising in incompletely excised adenomas

both arms of the study. However, a case-control study on the same population suggested a 70% reduction in mortality rate associated with a single sigmoidoscopy, which lasted up to 10 years (124a). This has also been confirmed in another smaller case-control study involving 66 cases and 196 controls in which a 79% reduction in mortality rate was found following a single sigmoidoscopy (125). A very large case-control study among American veterans has recently been reported (126). After excluding patients with familial polyposis coli or inflammatory bowel disease, the researchers examined the records of 4358 patients (98% males) who died of colorectal cancer between 1988 and 1992 and a set of 16,531 matched living controls and 16,199 matched dead controls. Overall, they found a 59% (95% CI 50%–67%) reduction in deaths associated with any prior colorectal procedure. The protection was 70% in the first year following the procedure but remained near 60% for the subsequent period of up to 10 years (Figure 11). Protection was most related to removal of tissue, suggesting a direct benefit of adenoma removal. Also, some of the procedures only examined part of the bowel (sigmoidoscopy), suggesting a larger benefit in the specific regions investigated.

Two prospective studies have also shown a marked reduction of the anticipated cancer rates in patients who have had adenomas completely removed. Atkin *et al.* (65) followed up a cohort of 1618 patients in whom rectosigmoid adenomas had been removed during rigid rectosigmoidoscopy up to 14 years in the past. Using the Stryker *et al.* (62) estimate of progression of large adenomas,

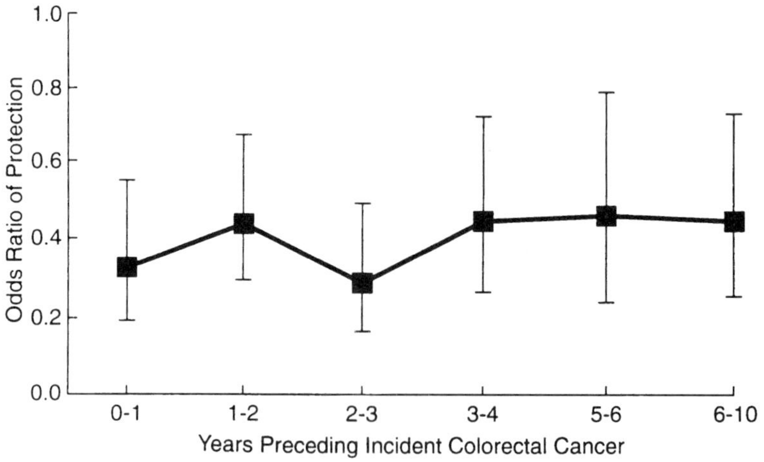

Figure 11 Relative protection against colorectal cancer death (and 95% confidence intervals) in years following an endoscopic procedure in the colon or rectum. (From Ref. 126.)

a predicted 85% reduction of distal colorectal cancer was associated with complete adenoma removal. However, rectal cancer rates remained high among individuals who had an incomplete removal of their adenoma.

In the National Polyp Study, 1418 patients with one or more adenomas removed at complete colonoscopy were followed up by periodic colonoscopy for an average of 5.9 years (127). A 76% (95% CI 44%, 92%) reduction in colorectal cancer was observed compared to the standard U.S. population and a 90% reduction was found if this was compared to other similar populations who did not receive follow-up surveillance. If the first 2 years of follow-up are omitted to partially adjust for a clear colon, the reductions were 66% and 85%, respectively.

A similar study from Australia (128) of 645 patients with adenomas followed up for a mean of 4.4 years found cancer rates similar to that of the general population (3 observed, 3.75 expected). The mean size of initial adenoma was 9 mm and 29% had a villous component; therefore, if a 2.5-fold increase in risk is assumed in this cohort compared to the general population, the cohort provides some evidence for a reduction of the order of 60%–70%.

One observational study has shown a mortality benefit associated with colonoscopic surveillance of patients with ulcerative colitis (129). The study consisted of 41 patients in whom carcinomas developed within a cohort of 2050 patients who had been seen for ulcerative colitis, 19 of whom were under colonoscopic surveillance. Although potentially subject to confounding and biases, both groups had long-standing ulcerative colitis before development of cancer (median 18 versus 15 years for surveillance versus not). An earlier stage distribution was associated with surveillance ($P=0.04$) and survival was significantly better (4 versus 11 deaths, $P=0.03$).

V. SAFETY AND COMPLICATIONS

The main complication of colorectal screening is perforation of the bowel. The risks are substantially greater for colonoscopy than for sigmoidoscopy. For colonoscopy perforation rates are about 1–3 per 1000 and mortality rate approximately 1 per 10,000 (130,131). Complications occur more frequently after polypectomy, with a 2% incidence of hemorrhage, 0.3% incidence of perforation, and 0.1% mortality rate. Rates are much lower for sigmoidoscopy, for which perforation rates are about 1–2 per 10,000 examinations (132,133), although larger series with the 60-cm fiberoptic scopes have not yet been reported. Somewhat higher rates occur after biopsy or polypectomy, but there are not yet mortality figures. Much lower perforation rates are reported for barium enema. A small risk of cardiac complication is associated with each of these procedures, relating possibly to the enema, sedation, or anxiety about the test.

Endoscopic procedures are uncomfortable and colonoscopy usually requires sedation. However, the discomfort is short-lived in that sigmoidoscopy usually requires 6–10 minutes and colonoscopy 15–30 minutes. Flatus and soiling are the most common complaints following these procedures.

Occult blood testing has virtually no direct complications, although many people find it an unacceptable procedure. However, because the test is performed frequently, the overall false-positivity rate is high, leading to a high proportion of screenees receiving colonoscopy with its attendant complications.

VI. COSTS AND COST EFFECTIVENESS

Several models exist for evaluating costs and benefits. They all are limited by a lack of detailed knowledge about the efficiency of the tests, the appropriate interval between tests, and the appropriate age for starting and stopping screening. The major costs relate to endoscopic examinations, which accrue directly for sigmoidoscopic screening or indirectly to evaluate positive tests for FOBT. A detailed analysis of the U.S. system involving strategies based on FOBT, sigmoidoscopy, barium enema, or colonoscopy found all strategies cost less than $20,000 per life-year saved (134). A simple analysis of once-in-a-lifetime sigmoidoscopy based in the United Kingdom suggested that the cost of screening would be largely offset by the savings in having fewer cancers to treat (92). The direct cost per cancer prevented was estimated to be £5500 and it was estimated that each life saved cost £8500. Based on data from the Nottingham trial, Walker *et al.* (135) have estimated that the cost per cancer detected by FOBT on the initial screening in persons aged 50–74 years is approximately £2700 and the total cost per person screened was £5.

VII. EDUCATION

There are deep-seated social taboos about bowel cancer. A comprehensive education program is necessary to break this down. The basic message is that screening tests do reduce mortality rates.

Underpinning this message is the fact that early Dukes stage A colorectal cancer has a very good prognosis and that cancer is preceded by a long preinvasive stage in which detectable and easily treatable polyps are present. Detection of these early lesions results in a cure rate approaching 100%. Simple dietary advice should also be a part of an education program.

VIII. RECOMMENDATIONS

A detailed set of recommendations has recently been published (136), which have been endorsed by the American Cancer Society and several other professional bodies. The recommendations are complex and overlapping. For average risk individuals aged 50 years, one or more of the following are recommended:

1. Annual fecal occult blood testing
2. Flexible sigmoidoscopy every 5 years
3. Double-contrast barium enema every 5–10 years
4. Colonoscopy every 10 years

It is acknowledged that strategies 3 and 4 are not supported by direct evidence (from randomized trial, nonrandomized trial, or case-control studies) that they reduce mortality rate from colorectal cancer.

A range of recommendations are made for individuals at increased risk. These include starting screening at age 40 years if a first degree blood relative had cancer before the age of 55 years and examination of the entire colon every 1–2 years beginning between ages 20 and 30 years for patients with a strong family history (e.g., Amsterdam criteria). Patients with large or multiple adenomatous polyps are advised to have a repeat examination of the colon after 3 years.

The 1995 American College of Physicians guidelines (136) recommend 10-yearly screening from age 50–70 years by sigmoidoscopy, double-contrast barium enema, or colonoscopy. Fecal occult blood testing is recommended for individuals who decline these tests.

IX. CONCLUSION

Although it is clear that lifestyle has a great deal to do with the development of colorectal cancer, the recommendations that can confidently be made are somewhat general. In terms of diet, the clearest message is a protective effect of fruit and vegetables. No food supplement can be confidently recommended and the British Health Education Authority's advice is to eat at least five servings of fruits and vegetables per day. This is the same as offered by the U.S. National Cancer Institute in its Five-a-Day Program. No specific items are emphasized. This recommendation is likely to reduce the risk of a number of cancers and supplies a good amount of fiber. The World Health Organization also recommends 400 g of fruits and vegetables daily.

The situation regarding animal fat and red meat is less certain for colon cancer, although there is clear evidence that reduced consumption of those is beneficial for heart disease and would be part of any general dietary advice.

Knowledge about specific micronutrients, food supplements, or drugs such as aspirin is not sufficiently advanced to offer advice on these substances.

Exercise may have a beneficial effect on colorectal cancer, but to be of value it must be regular and long term. Simple measures such as regular walking or bicycle riding that can be maintained for years are likely to be of more value than short-term high-energy exercise programs.

There is strong evidence that screening by either fecal occult blood testing or sigmoidoscopy are efficacious for reducing mortality rate and some form of screening is indicated. The potential benefits of FOBT are smaller but more clearly established, because they have been observed in all three randomized trials that have reported on mortality rate. Some of this benefit is due to the increased colonoscopic surveillance and serendipitous detection of polyps. The benefits of flexible sigmoidoscopy currently rest on observational studies, which are subject to biases. The mortality rate reductions appear to be quite large for the segment of the bowel that is examined, but unbiased estimates of the true magnitude must await the result of the current randomized trials. An advantage of sigmoidoscopy is that it probably only needs to be done infrequently (10-yearly) and may lead to fewer colonoscopies than biennial FOBT. Protocols for combining both tests have not been evaluated and there is much uncertainty about the best overall approach to screening. Use of double-contrast barium enema or colonoscopy has not been evaluated in a population screening context and I do not recommend them until evidence is forthcoming.

For the general population, there is little value in initiating screening much before the age of 55 years, but earlier testing may be appropriate when there is a strong family history. Familial polyposis coli is the extreme example, in which surveillance should begin by the age of 20 years. Here genetic testing offers the opportunity to avoid surveillance in half of the at-risk population and early planning for prophylactic measures can be begun for the other half.

Members of HNPCC (hereditary non-polyposis colon cancer) families require colonoscopic screening of the whole large bowel, because there is a tendency toward proximal cancers. Screening is usually recommended to commence 5 years before the youngest case in the family. Again, when genetic testing is informative, it can obviate the need for screening until later in life in family members found to not carry the relevant gene. First degree family members of anyone with colorectal cancer before the age of 50 years are at increased risk and may consider beginning endoscopic screening at an early age. One possible strategy is to begin 5 years before the age of diagnosis in the index case.

Individuals with inflammatory bowel disease are also at increased risk, and most organizations recommend endoscopic surveillance.

Genetic screening of the general population has not been properly evaluated and cannot be recommended at this stage, except in a research context.

REFERENCES

1. Parkin DM, Pisani P, Ferlay J. Estimates of the worldwide incidence of a eighteen major cancers in 1985. Int J Cancer 1993; 54:594–606.

2. Pisani P, Parkin DM, Ferlay J. Estimates of the world-wide mortality from eighteen major cancers in 1985. Implications for prevention and projections of future burden. Int J Cancer 1993; 55:891–903.
3. Armitage P, Doll R. The age distribution of cancer and a multi-stage theory of carcinogenesis. Br J Cancer 1954; 8:1–12.
4. Doll R. The age distribution of cancer: implications for models of carcinogenesis. Royal Statistical Society A 1971; 134:133–166.
5. Vukasin AP, Ballantyne GH, Flannery JT, et al. Increasing incidence of cecal and sigmoid carcinoma: data from the Connecticut Tumor Registry. Cancer 1990; 66: 2442–2449.
6. Levi F, Randimbison L, La Vecchia C. Trends in subsite distribution of colorectal cancers and polyps from the Vaud Cancer Registry. Cancer 1993; 72:46–50.
7. Slattery ML, Friedman GD, Potter JD, et al. A description of age, sex and site distributions of colon not rectum carcinoma in three geographic areas. Cancer 1996; 78:1666–1670.
8. Stower MJ, Hardcastle JD. The results of 1115 patients with colorectal cancer treated over an 8-year period in a single hospital. Eur J Surg Oncol 1985; 11: 119–123.
9. Miller BA, Ries LAG, Hankey BF, *et al*, eds. SEER Cancer Statistics Review: 1973–1990 [NIH Pub. No. 93–2789]. National Cancer Institute. 1993.
10. Chu KC, Tarone RE, Chow W-H, Hankey BF, Ries LA. Temporal patterns in colorectal cancer incidence, survival, and mortality from 1950 through 1990. J Natl Cancer Inst 1994; 86:997–1006.
11. International Agency for Research on Cancer. Trends in Cancer Incidence and Mortality. [IARC Scientific Publications No. 121]. Lyon: International Agency for Research on Cancer, 1993.
12. Haenszel W, Kurihara M. Studies of Japanese migrants. I. Mortality from cancer and other diseases among Japanese in the United States. J Natl Cancer Inst 1968; 40:43–68.
13. Steinitz R, Parkin DM, Young JL, Bieber CA, Katz L. Cancer Incidence in Jewish Migrants to Israel 1961–1981 [IARC Scienific Publications No. 98]. Lyon: International Agency for Research on Cancer, 1989.
14. McMichael AJ. Alimentary tract cancer in Australia in relation to diet and alcohol. Nutr Cancer 1979; 1:82–89.
15. Whittemore AS. Colorectal cancer incidence among Chinese in North America and the People's Republic of China: variation with sex, age and anatomical site. Int J Epidemiol 1989; 18:563–568.
16. Potter JD, Slattery ML, Bostick RM, Gapstur SM. Colon cancer: a review of the epidemiology. Epidemiol Rev 1993; 15:499–545.
17. Trock B, Lanza E, Greenwald P. Dietary fiber, vegetables, and colon cancer: critical review and meta-analyses of the epidemiologic evidence. J Natl Cancer Inst 1990; 82:650–661.
18. Block G, Patterson B, Subar A. Fruit, vegetables and cancer prevention: a review of the epidemiological evidence. Nutr Cancer 1992; 18:1–29.

19. Mehta RG, Liu J, Constantinou A, Thomas CF, Hawthorne M, You M, Gerhauser C, Pezzuto JM, Moon RC, Moriarty RM. Cancer chemopreventive activity of brassinin, a phytoalexin from cabbage. Carcinogenesis 1995;16:399–404.
20. Greenberg ER, Baron JA, Tosteson TD, et al. A clinical trial of antioxidant vitamins to prevent colorectal cancer. N Engl J Med 1994; 331:141–147.
21. MacLennan R, Macrae F, Bain C, Battistutta D, Chapuis P, Gratten H, Lambert J, Newland RC, Ngu M, Russell A, Ward M, Wahlqvist ML. Randomized trial of intake of fat, fiber, and beta carotene to prevent colorectal adenomas. J Natl Cancer Inst 1995; 87:1760–1766.
22. ATBC Cancer Prevention Study Group. The effect of vitamin E and beta carotene on the incidence of lung cancer and other cancers in male smokers. N Engl J Med 1994; 330:1029–1035.
23. Albanes D, Heinonen OP, Taylor PR, Virtamo J, Edwards BK, Rautalahti M, Hartman AM, Palmgren J, Freedman LS, Haapakoski J, Barrett MJ, Pietinen P, Malila N, Tala E, Liippo K, Salomaa E-R, Tangrea JA, Teppo L, Askin FB, Taskinen E, Erozan Y, Greenwald P, Huttunen JK. α-Tocopherol and β-carotene supplements and lung cancer incidence in the Alpha-tocopherol, Beta-Carotene Cancer Prevention Study: effects of base-line characteristics and study compliance. J Natl Cancer Inst 1996; 88:1560–1570.
24. Omenn GS, Goodman GE, Thornquist MD, Balmes J, Cullen MR, Glass A, Keogh JP, Meyskens FL, Valanis B, Williams JH, Barnhart S, Cherniack MG, Brodkin CA, Hammar S. Risk factors for lung cancer and for intervention effects in CARET, the Beta-Carotene and Retinol Efficacy Trial. J Natl Cancer Inst 1996, 88:1550–1559.
25. La Vecchia C, Ferraroni M, Mezzetti M, et al. Attributable risks for colorectal cancer in northern Italy. Int J Cancer 1996; 66:60–64.
26. Burkitt DP. Epidemiology of cancer of the colon and rectum. Cancer 1971; 28:3–13.
27. Howe GR, Benito E, Castellato R, et al. Dietary intake of fiber and decreased risk of cancers of the colon and rectum: evidence from the combined analysis of 13 case-control studies. J Natl Cancer Inst 1992; 84:1887–1896.
28. Macquart-Moulin G, Riboli E, Cornee J, et al. Case-control study on colorectal cancer and diet in Marseilles. Int J Cancer 1986; 38:183–191.
29. La Vecchia C, Negri E, Decarli A, D'Avanzo B, Gallotti L, Gentile A, Franceschi S. A case-control study of diet and colorectal cancer in northern Italy. Int J Cancer 1988; 41:492–498.
30. Wynder EL, Kajitani T, Ishikawa S, Dodo H, Takano A. Environmental factors of cancer of the colon and rectum. II. Japanese epidemiological data. Cancer 1969; 23:1210–1220.
31. Wasan HS, Goodlad RA. Fibre-supplemented foods may damage your health. Lancet 1996; 348:319–320.
32. Prentice RL, Sheppard L. Dietary fat and cancer: consistency of the epidemiologic data, and disease prevention that may follow from a practical reduction in fat consumption. Cancer Causes Control 1990; 1:81–97.

33. Thun MJ, Calle EE, Namboodiri MM, et al. Risk factors for fatal colon cancer in a large prospective study. J Natl Cancer Inst 1992; 84:1491–1500.
34. Goldbohm RA, van den Brandt PA, vant'Veer P, Dorant E, Sturmans F, Hermus RJ. Prospective study on alcohol-consumption and the risk of cancer of the colon and rectum in the Netherlands. Cancer Causes Control 1994;5:95–104.
35. Willett WC, Stampfer MJ, Colditz GA, Rosner BA, Speizer FE. Relation of meat, fat and fiber intake to the risk of colon cancer in a prospective study among women. N Engl J Med 1990; 323:1664–1672.
36. Giovannucci E, Rimm EB, Stampfer MJ, Colditz GA, Ascherio A, Willett WC. Aspirin use and the risk for colorectal cancer and adenoma in male health professionals. Ann Intern Med 1994; 121:241–246.
37. Key TJA, Thorogood M, Appleby PN, Burr ML. Dietary habits and mortality in 11000 vegetarians and health conscious people: results of a 17 year follow-up. BMJ 1996; 313:775–778.
38. International Agency for Research on Cancer. Coffee, tea, mate, methylxanthines and methylglyoxal. In: IARC Monographs on the Evaluation of Carcinogenic Risks to Humans. Vol. 51. Lyon: International Agency for Research on Cancer, 1991.
39. International Agency for Research on Cancer. Alcohol drinking. In: IARC Monographs on the Evaluation of Carcinogenic Risks to Humans. Vol. 44. Lyon: International Agency for Research on Cancer, 1988.
40. Longnecker MP, Orza MJ, Adams ME, Vioque J, Chalmers TC. A meta-analysis of alcoholic beverage consumption in relation to risk of colorectal cancer. Cancer Causes Control 1990; 1:59–68.
41. Giovannucci E, Egan KM, Hunter DJ, Stampfer MJ, Colditz GA, Willett WC, Speizer FE. Aspirin and the risk of colorectal cancer in women. N Engl J Med 1995; 333:609–614.
42. Jensen OM. The role of diet in colorectal cancer. International Congress Service 1985; 685:137–147.
43. Kune S, Kune GA, Watson LF. Case-control study of dietary etiological factors: the Melbourne colorectal cancer study. Nutr Cancer 1987; 9:21–42.
44. Giovannucci E, Stampfer MJ, Colditz GA, Rimm EB, Trichopoulos D, Rosner BA, Speizer FE, Willett WC. Folate, methionine, and alcohol intake and risk of colorectal adenoma. J Natl Cancer Inst 1993; 85:875–884.
45. White E, Jacobs EJ, Daling JR. Physical activity in relation to colon cancer in middle-aged men and women. Am J Epidemiol 1996; 144:42–50.
46. Albanes D, Taylor PR. International differences in body height and weight and their relationship to cancer incidence. Nutr Cancer 1990; 14:69–77.
47. Fraumeni JF, Lloyd JW, Smith EM, et al. Cancer mortality among nuns: role of marital status in etiology of neoplastic disease in women. J Natl Cancer Inst 1969; 42:455–468.
48. McMichael AJ, Potter JD. Reproduction, endogenous and exogenous sex hormones and colon cancer: a review and hypothesis. J Natl Cancer Inst 1980; 65: 1201–1207.

49. Cuzick J. Medicinal drugs and hormonal activity as chemopreventive agents. In: B.W. Stewart BW, McGregor D, Kleihues P, eds. Principles of Chemoprevention [IARC Scientific Publications No. 139]. Lyon: International Agency for Research on Cancer, 1996:99–114.
50. Calkins BM, Mandeloff AI. Epidemiology of inflammatory bowel disease. Epidemiol Rev 1986; 8:60–91.
51. Glickman RM. Inflammatory bowel disease. In: Braunwald E, Isselbacher K, Petersdorf RG, et al, eds. Harrison's Principles of Internal Medicine. New York: McGraw-Hill, 1987; 1277–1290.
52. Weeden DD, Shorter RG, Ilstrup DM, Huizenga KA, Taylor WF. Crohn's disease and cancer. N Engl J Med 1973; 289:1099–1103.
53. Greenstein AJ, Sachar DB, Smith H, Janowitz HD, Aufses AH Jr. Patterns of neoplasia in Crohn's disease and ulcerative colitis. Cancer 1980; 46:403–407.
54. Oshima M, Dinchuk JE, Kargman SL, Oshima H, Hancok B, Kwong E, Trzaskos JM, Evans JF, Taketo MM. Suppression of intestinal polyposis in apc^{716} knockout mice by inhibition of cyclo-oxygenase 2 (Cox-2). Cell 1966; 87:803–809.
55. Trujillo MA, Garewal HS, Sampliner RE. Nonsteroidal anti-inflammatory agents in chemoprevention of colorectal cancer at what cost? Dig Dis Sci 1994; 39:2260–2266.
56. Greenberg ER, Baron JA. Aspirin and other nonsteroid anti-inflammatory drugs as cancer-preventive agents. In: Stewart BW, McGregor D, Kleihues P, eds. Principles of Chemoprevention [IARC Scientific Publications No. 139]. Lyon: International Agency for Research on Cancer 1996:91–98.
57. Guillem JG, Neugut AI, Forde KA, Waye JD, Treat MR. Colonic neoplasms in asymptomatic first-degree relatives of colon cancer patients. Am J Gastroenterol 1988; 83:271–273.
58. Burt RW. Familial risk and colon cancer. Int J Cancer 1996; 69:44–46.
59. Morson BC, Bussey HJR, Day DW, Hill MJ. Adenomas of large bowel. Cancer Surv 1983; 2:451–477.
60. Morson BC. Factors influencing the prognosis of early cancer of the rectum. Proc R Soc Med 1966; 59:607–608.
61. Muto T, Bussey HJR, Morson BC. The evolution of cancer of the colon and rectum. Cancer 1975; 36:2251–2270.
62. Stryker SJ, Wolff BG, Culp CE, Libbe SD, Ilstrup DM, MacCarty RL. Natural history of untreated colonic polyps. Gastroenterology 1987; 93:1009–1013.
63. Eide TJ. Risk of colorectal cancer in adenoma-bearing individuals within a defined population. Int J Cancer 1986; 38:173–176.
64. Lotfi AM, Spencer RJ, Ilstrup DM, Melton LJ III. Colorectal polyps and risk of subsquent carcinoma. Mayo Clin Proc 1986; 61:337–343.
65. Atkin WS, Morson BC, Cuzick J. Long-term risk of colorectal cancer after excision of rectosigmoid adenomas. N Engl J Med 1992; 326:658–662.
66. Rex DK, Lehman GA, Hawes RH, Ulbright TM, Smith JJ. Screening colonoscopy in asymptomatic average-risk persons with negative fecal occult blood tests. Gastroenterology 1991; 100:64–67.

67. Williams AR, Balasooriya BAW, Day DW. Polyps and cancer of the large bowel: a necropsy study in Liverpool. Gut 1982; 123:835–842.
68. Clark JC, Collan Y, Eide TJ, Esteve J, Ewen S, Gibbs NM, Jensen OM, Koskela E, MacLennan R, Simpson JG, Stalsberg H, Zaridze DG. Prevalence of polyps in an autopsy series from areas with varying incidence of large-bowel cancer. Int J Cancer 1985; 36:179–186.
69. Johannsen LGK, Momsen O, Jacobsen NO. Polyps of the large intestine in Aarhus, Denmark. Scand J Gastroenterol 1989;24:799–806.
70. Vogelstein B, Fearon ER, Hamilton SR, Kern SE, Preisinger AC, Leppert M, Nakamura Y, White R, Smits AMM, Bos JL. Genetic alterations during colorectal tumour development. N Engl J Med 1988; 319:525–532.
71. Jessup JM, Gallick GE. The biology of colorectal carcinoma. Curr Probl Cancer 1992; 16:265–327.
72. Powell SM, Zilz N, Beazer-Barclay Y, Bryan TM, Hamilton SR, Thibodeau SN, Vogelstein B, Kinzler KW. APC mutations occur early during colorectal tumorigenesis. Nature 1992; 359:235–237.
73. Froggatt NJ, Joyce JA, Davies, R, Gareth D, Evans R, Ponder BA, Barton DE, Maher ER. A frequent hMSH2 mutation in hereditary non-polyposis colon cancer syndrome. Lancet 1995; 345:727.
74. Bos JL, Fearon ER, Hamilton SR, Verlaan-de Vries M, van Boom JH, Van der EB JA, Vogelstein B. Prevalence of ras gene mutations in human colorectal cancers. Nature 1987; 327:293–297.
75. Breivik J, Meling GI, Spurkland A, Rognum TO, Gaudernack G. K-ras mutation in colorectal cancer: relations to patient age, sex and tumour location. Br J Cancer 1994; 69:367–371.
76. Houlston RS, Collins A, Slack J, Morton NE. Ann Hum Genet 1992; 56:99–103.
77. Lynch HT, Smyrk TC, Watson P, Watson P, Lanspa SJ, Lynch JF, Lynch PM, Cavalieri RJ, Boland CR. Genetics, natural history, tumor spectrum, and pathology of hereditary nonpolyposis colorectal cancer: an updated review. Gastroenterology 1993; 104:1535–1549.
78. Deleon MP, Sassatelli R, Benatti P, Roncucci L. Identification of hereditary nonpolyposis colorectal cancer in the general-population—the 6-year experience of a population-based registry. Cancer 1993; 71:3493–3501.
79. Lothe RA, Peltomaki P, Meling GI, Aaltonen LA, Nystromlahti M, Pylkkanen L, Heimdal K, Andersen TI, Moller P, Rognum TO. Genomic instability in colorectal cancer—relationship to clinicopathological variables and family history. Cancer Res 1993; 53:5849–5852.
80. Aaltonen LA, Peltomäki P, Leach FS, Sistonen P, Pylkkanen L, Mecklin JP, Järvinen H, Powell SM, Jen J, Hamilton SR, Petersen GM, Kinzler KW, Vogelstein B, de la Chapelle A. Clues to the pathogenesis of familial colorectal cancer. Science 1993; 260:812–816.
81. Aarnio M, Mecklin JP, Aaltonen LA, Nyström-Lahti M, Järvinen HJ. Life-time risk of different cancers in hereditary non-polyposis colorectal cancer (HNPCC) syndrome. Int J Cancer 1995; 64:430–433.
82. Lang NP, Butler MA, Massengil J, et al. Rapid metabolic phenotypes for acetyl-

transferase and cytochrome P4501A2 and putative exposure to food-borne heterocyclic amines increase the risk of colorectal cancer of polyps. Cancer Epidemiol Biomarkers Prev 1994; 3:675–682.
83. Zhong S, Wyllie AH, Barnes D, et al. Relationship between GSTMI genetic polymorphism and susceptibility to bladder, breast and colon cancer. Carcinogenesis 1994; 14:1821–1824.
84. Simon JB. Occult blood screening for colorectal carcinoma: a critical review. Gastroenterology 1985; 88:820–837.
85. Lauridsen LH, Bohn L. Testing of feces for occult blood: a review. Dan Med Bull 1976; 23:230–235.
86. Irons GV, Kirsner JB. Routine chemical tests of the stool for occult blood: an evaluation. Am J Med Sci 1965; 249:247–260.
87. Greegor DH. Diagnosis of large-bowel cancer in the asymptomatic patient. JAMA 1967; 201:943–945.
88. Morris DW, Hansell JR, Ostrow JD, Lee CS. Reliability of chemical tests for fecal occult blood in hospitalized patients. Am J Dig Dis 1976; 21:845–852.
89. Wells HJ, Pagano JF. ''Hemoccult'' test-reversal of false-negative results due to storage (abstr). Gastroenterology 1977; 72:1148.
90. Winawer SJ, Andrews M, Fleichinger B, Sherlock P, Schottenfeld D, Miller DG. Progress report on controlled trial of fecal occult blood testing for the detection of colorectal neoplasia. Cancer 1980; 45:2959–2964.
91. Kewenter J, Björk S, Haglind E, Smith L, Svanvik J, Ahrén C. Screening and rescreening for colorectal cancer. A controlled trial of fecal occult blood testing in 27 700 subjects. Cancer 1988; 62:645–651.
92. Atkin WS, Cuzick J, Northover JMA, Whynes DK. Prevention of colorectal cancer by once-only sigmoidoscopy. Lancet 1993; 341:736–740.
93. Eddy DM. Screening for colorectal cancer. Ann Intern Med 1990; 113:373–384.
94. Ott DJ. Role of the barium enema in colorectal carcinoma. Radiol Clin North Am 1993; 31:1293–1313.
95. Ott DJ, Scharling ES, Chen YM, et al. Barium enema examination: sensitivity in detecting colonic polyps and carcinomas. South Med J 1989; 82:197–200.
96. Stevenson GW. Radiology and endoscopy in the pretreatment diagnostic management of colorectal cancer. Cancer 1993; 71:4198–4206.
97. Lieberman D. Cost-effectiveness of colon cancer screening. Am J Gastroenterol 1991; 86:1789–1794.
98. Lieberman DA, Smith FW. Screening for colon malignancy with colonoscopy. Am J Gastroenterol 1991; 86:946–951.
99. Williams C. Electronic 3-dimensional imaging of intestinal endoscopy. Lancet 1993;341:724–725.
100. Miyoshi Y, Ando H, Nagase H, Nishisho I, Horii A, Miki Y, Mori T, Utsunomiya J, Baba S, Petersen G, Hamilton SR, Kinzler KW, Vogelstein B, Nakamura Y. Germ-line mutations of the Apc gene in 53 familial adenomatous polyposis patients. Proc Natl Acad Sci USA 1992; 89:4452–4456.
101. Cottrell S, Bicknell D, Kaklamanis L, Bodmer WF. Molecular analysis of APC mutations in familial adenomatous polyposis and sporadic colon carcinomas. Lancet 1992; 340:626–630.

102. Sidransky D, Tokino T, Hamilton SR, Kinzler KW, Levin B, Frost P, Vogelstein B. Identification of ras oncogene mutations in the stool of patients with curable colorectal tumors. Science 1992;256:102–105.
103. Tobi M, Luo F-C, Ronai Z. Detection of K-ras mutation in colonic effluent samples from patients without evidence of colorectal carcinoma. J Natl Cancer Inst 1994;86:1007–1010.
104. Smith-Ravin J, England J, England J, Talbot IC, Bodmer W. Detection of c-Ki-ras mutations in fecal samples from sporadic colorectal cancer patients. Gut 1995; 36:81–86.
105. Villa E, Dugani A, Rebecchi AM, Vignoli A, Grottola A, Buttafoco P, Losi L, Perini M, Trande P, Merighi A, Lerose R, Manenti F. Identification of subjects at risk for colorectal carcinoma through a test based on K-ras determination in the stool. Gastroenterology 1996; 110:1346–1353.
106. Friedman GD, Collen MF, Fireman BH. Multiphasic health checkup evaluation: a 16-year follow-up. J Chron Dis 1986; 39:453–463.
107. Herrinton L, Selby JV, Friedman GD, Quesenberry CP, Weiss NS. A case-control study of digital-rectal screening in relation to mortality from cancer of the distal rectum. Am J Epidemiol 1995; 141:11.
108. Ahlquist DA, Wieand S, Moertel CG, McGill DB, Loprinzi CL, O'Connell MJ, Mailliard JA, Gerstner JB, Pandya K, Ellefson RD. Accuracy of fecal occult blood screening for colorectal neoplasia. JAMA 1993; 269:1262–1267.
109. St. John DJB, Young GP, Alexeyeff MA, Deacon MC, Cuthbertson A M, Macrae FA, Penfold JCB. Evaluation of new occult blood tests for detection of colorectal neoplasia. Gastroenterology 1993; 104:1661–1668.
110. Castiglione G, Zappa M, Grazzini G, Mazzotta A, Biagini M, Salvadori P, Ciatto S. Immunochemical vs guaiac faecal occult blood tests in a population-based screening programme for colorectal cancer. Br J Cancer 1996; 74:141–144.
111. Petrelli N, Michalek AM, Freedman A, Baroni M, Mink I, Rodriguez-Bigas M. Immunochemical versus guaiac occult blood stool tests: results of a community-based screening program. Surg Oncol 1994; 3:27–36.
112. Mandel JS, Bond JH, Church TR, Snover DC, Bradley GM, Schuman LM, Enderer F. Reducing mortality from colorectal cancer by screening for fecal occult blood. N Engl J Med 1993; 328:1365–1371.
113. Kewenter J, Asztely M, Engaras B, et al. A randomised trial of faecal occult blood testing for early detection of colorectal cancer. In: Miller AB, Chamberlain J, Day NE, et al, eds. Cancer Screening. Results of Screening and Rescreening of 51325 Subjects. Cambridge: Cambridge University Press 1991:116–125.
113a. Rozen P, Ron E, Fireman Z, Figer A, Legum C, Ron E. The relative value of fecal occult blood tests and flexible sigmoidoscopy in screening for large bowel neoplasia. Cancer 1987; 60:2553–2583.
114. Hardcastle JD, Chamberlain JO, Robinson MHE, Moss SM, Amar SS, Balfour TW, James PD, Mangham CM. Randomised controlled trial of faecal-occult-blood screening for colorectal cancer. Lancet 1996; 348:1472–1477.
115. Kronberg O, Fenger C, Olsen J, Jorgensen OD, Sondergaard. Randomised study of screening for colorectal cancer with faecal-occult-blood test. Lancet 1996; 348: 1467–1471.

116. Winawer SJ, Flehinger BJ, Schottenfeld D, Miller DG. Screening for colorectal cancer with fecal occult blood testing and sigmoidoscopy. J Natl Cancer Inst 1993; 85:1311–1318.
117. Selby JV, Friedman GD, Quesenberry CP, Weiss NS. Effect of fecal occult blood testing on mortality from colorectal cancer. A case-control study. Ann Intern Med 1993; 118:1–6.
118. Wahrendorf J, Robra BP, Wiebelt H, Oberhausen R, Weiland M, Dhom G. Effectiveness of colorectal cancer screening: results from a population-based case-control evaluation in Saarland, Germany. Eur J Cancer Prev 1993; 2:221–227.
119. Saito H, Soma Y, Koeda J, Wada Toyohito, Kawaguchi H, Sobue T, Aisawa T, Yoshida Y. Reduction in risk of mortality from colorectal cancer by fecal occult blood screening with immunochemical hemagglutination test. A case-control study. Int J Cancer 1995; 61:465–469.
120. Lazovich DA, Weiss NS, Stevens NG, White E, McKnight B, Wagner EH. A case-control study to evaluate efficacy of screening faecal occult blood. J Med Screening 1995, 2:84–89.
121. Atkin WS, Hart A, Edwards R, McIntyre P, Aubrey R, Wardle J, Sutton S, Cuzick J, Northover JMA. Uptake, yield of neoplasia and adverse effects of flexible sigmoidoscopy screening. Gut 1998; 42:560–565.
122. Olynyk JK, Aquilia S, Fletcher DR, Dickinson JA. Flexible sigmoidoscopy screening for colorectal cancer in average-risk subjects: a community-based pilot project. MJA 1996; 165:74–76.
123. Hoff G, Sauar J, Vatn MH, Larsen S, Langmark F, Moen IE, Foerser A, Thiis-Evensen E. Polypectomy of adenomas in the prevention of colorectal cancer: 10 years' follow-up of the Telemark Polyp Study I. Scand J Gastroenterol 1996; 31: 1006–1010.
124. Gilbertson VA, Nelms JM. The prevention of invasive cancer of the rectum. Cancer 1978; 41:1137–1139.
124a. Selby JV, Friedman GD, Quesenberry CP Jr, Weiss NS. A case-control study of screening sigmoidoscopy and mortality from colorectal cancer. N Engl J Med 1992; 326:653–657
125. Newcomb PA, Norfleet RG, Storer BE, Surawicz TS, Marcus PM. Screening sigmoidoscopy and colorectal cancer mortality. J Natl Cancer Inst 1992; 84:1572–1575.
126. Muller AD, Sonnenberg A. Protection by endoscopy against death from colorectal cancer: a case-control study among veterans. Arch Intern Med 1995; 155:1741–1748.
127. Winawer SJ, Zauber AG, Ho MN, O'Brien MJ, Gottleib LS, Sternberg SS, Waye JD, Schapiro M, Bond JH, Panish JF, Ackroyd F, Shike M, Kurtz RC, Hornsby-Lewis L, Gerdes H, Stewart ET, and the National Polyp Study Workgroup. Prevention of colorectal cancer by colonoscopic polypectomy. N Engl J Med 1993; 329:1977–1981.
128. Meagher AP, Stuart M. Does colonoscopic polypectomy reduce the incidence of colorectal carcinoma? Aust N Z J Surg 1994;64:400–404.

129. Choi PM, Nugent FW, Schoetz DJ, Silverman ML, Haggitt RC. Colonoscopic surveillance reduces mortality from colorectal cancer in ulcerative colitis. Gastroenterology 1993; 105:418–424.
130. Macrae FA, Tan KG, Williams CB. Towards safer colonoscopy: a report on the complications of 5000 diagnostic or therapeutic colonoscopies. Br Med Assoc 1983; 24:376–383.
131. Waye JD, Lewis BS, Yessayan S. Colonoscopy: a prospective report of complications. J Clin Gastroenterol 1992; 15:347–351.
132. Winnan G, Berci G, Panish J, Talbot TM, Overholt BF, McCallum RW. Superiority of the flexible to the rigid sigmoidoscope in routine proctosigmoidoscopy. N Engl J Med 1980; 302:1011–1012.
133. Portes C, Majarakis JD. Proctosigmoidoscopy: incidence of polyps in 50,000 examinations. JAMA 1957; 163:411–413.
134. Wagner JL, Tunis S, Brown M, Ching A, Almeida R. Cost-effectiveness of colorectal cancer screening in average-risk adults. In: Young G, Levin B, eds. Prevention and Early Detection of Colorectal Cancer. London: Saunders, 1996.
135. Walker A, Whynes DK, Chamberlain, JO, Hardcastle JD. The cost of screening for colorectal cancer. Epidemiol Community Health 1991; 45:220–224.
136. Winawer SJ, Fletcher RH, Miller L, et al. Colorectal cancer screening: clinical guidelines and rationale. Gastroenterology 1997; 112:594–642.
137. Griffith CDM, Turner DJ, Saunders JH. False negative results of Hemoccult test in colorectal screening. BMJ 1981; 283:472–473.
138. Macrae FA, St. John DJ. Relationship between patterns of bleeding and Haemoccult sensitivity in patients with colorectal cancers or adenomas. Gastroenterology 1982; 82:891.
139. Doran J, Hardcastle JD. Bleeding patterns in colorectal cancer: the effect of aspirin and the implications for faecal occult blood testing. Br J Surg 1982; 69:711–713.
140. Farrands PA, Hardcastle JD. Accuracy of occult blood tests over a six-day period. Clin Oncol 1983; 9:217–225.
141. McDonald C, Goulston K. Colorectal test for occult blood. Med J Aust 1984; 140:183.
142. Ahlquist DA, McGill DB, Schwartz S, Taylor WF, Owen RA. Fecal blood levels in health and disease. A study using HemoQuant. N Engl J Med 1985; 312:1422–1428.
143. Crowley ML, Freeman LD, Mottet MD, Strong RM, Sweeney BF, Brower RA, Sharma SP, Anderson DS. Sensitivity of guaiac-impregnated cards for the detection of colorectal neoplasia. J Clin Gastroenterol 1983; 5:127–130.
144. Songster CL, Barrows GH, Jarrett DD. Immunochemical detection of fecal occult blood—the fecal smear punch-disc test: a new non-invasive screening test for colorectal cancer. Cancer 1980; 45:1099–1102.
145. Kapparis A, Frommer D. Immunological detection of occult blood in bowel cancer patients. Br J Cancer 1985; 52:857–861.
146. Greegor DH. Detection of silent colon cancer in routine examination. CA 1969; 19:330–337.

147. Hastings JB. Mass screening for colorectal cancer. Am J Surg 1974; 127:228–233.
148. Miller SF, Knight AR. The early detection of colorectal cancer. Cancer 1977; 40: 945–949.
149. Goodman MJ. Mass screening for colorectal cancer—a negative report. JAMA 1977; 237:2380.
150. Helfrich GB, Petrucci P, Webb H. Mass screening for colorectal cancer. JAMA 1977; 238:1502–1503.
151. Fruhmorgen P, Demling L. Early detection of colorectal carcinoma with a modified guaiac test. A screening examination in 6000 humans. Acta Gastroenterol Belg 1978; 41:682–687.
152. Heeb MA, Ahlvin RC. Screening for colorectal carcinoma in a rural area. Surgery 1978; 83:540–541.
153. Elwood TW, Erickson A, Lieberman S. Comparative educational approaches to screening for colorectal cancer. Am J Public Health 1978; 68:135–138.
154. Bralow SP, Kopel J. Hemoccult screening for colorectal cancer. An impact study on Sarasota, Florida. J Florida Med Assoc 1979; 66:915–919.
155. Winchester DP, Shull JH, Scanlon EF, Murrel JV, Smeltzer C, Vrba P, Iden M, Streelman DH, Magpayo R, Dow JW, Sylvester J. A mass screening program for colorectal cancer using chemical testing for occult blood in the stool. Cancer 1980; 45:2955–2958.
156. Kurnick JE, Walley LB, Jacob HH, et al. Colorectal cancer detection in a community hospital screening program. JAMA 1980; 243:2056–2057.
157. Gnauck R, Thomas L. Haemoscreen in Vergleich mit Haemoccult als Suchtest auf kolorektalen. Krebs Dtsch Med Wschr 1980; 105:1642–1646.
158. Schwartz FW, Holstein H, Brecht JG. Preliminary report of fecal occult blood testing in Germany. In: Winawer SJ, Schottenfeld D, Sherlock P, eds. Colorectal Cancer: Prevention, Epidemiology, and Screening. New York: Raven, 1980:267–270.
159. Larkin KK. Mass screening in colorectal cancer. Aust NZ J Surg 1980; 50:467–469.
160. Chambers KJ, Morgan BP. Mass screening in colorectal cancer. Aust NZ J Surg 1980; 50:467–469.
161. Stuart M, Killingback MJ, Sakker S, et al. Hemoccult II test. Routine screening procedure for colorectal neoplasm? Med J Aust 1981; i:629–631.
162. Farrands PA, Griffiths RL, Britton DC. The Frome experiment—value of screening for colorectal cancer. Lancet 1981; i: 1231–1232.
163. Million R, Howarth J, Turnberg E, Turnberg LA. Faecal occult blood testing for colorectal cancer in general practice. Practitioner 1982; 226:659–663.
164. Sontag SJ, Durczak C, Aranha GV, Chejfec G, Frederick W, Greenlee HB. Fecal occult blood screening for colorectal cancer in a Veterans Administration hospital. Am J Surg 1983; 145:89–93.
165. Hardcastle JD, Farrands PA, Balfour TW, Chamberlain J, Amar SS, Sheldon MG. Controlled trial of faecal occult blood testing in the detection of colorectal cancer. Lancet 1983; ii:1–4.

166. Siba S. Experience with Haemoccult screening in Hungary. A multicenter trial. Hepatogastroenterol 1983; 30:27–29.
167. Habba SF, Doyle JS. Colorectal cancer screening of asymptomatic patients in Ireland. Ir J Med Sci 1983; 152:121–124.
168. Armitage N, Hardcastle JD, Amar SS, Balfour TW, Haynes J, James PD. A comparison of an immunological faecal occult blood test Fecatwin sensitive/FECA EIA with Haemoccult in population screening for colorectal cancer. Br J Cancer 1985; 51:799–804.
169. Thomas WM, Pye G, Hardcastle JD, Mangham CM. Faecal occult blood screening for colorectal neoplasia: a randomized trial of three days or six days of tests. Br J Surg 1990; 77:277–279.
170. McGarrity TJ, Long PA, Peiffer LP. Results of a repeat television-advertised mass screening program for colorectal cancer using fecal occult blood tests. Am J Gastroenterol 1990; 85:266–270.
171. Caffarey SM, Broughton CI, Marks CG. Faecal occult blood screening for colorectal neoplasia in a targeted high-risk population. Br J Surg 1993; 80:1399–1400.
172. Robinson MHE, Marks CG, Farrands PA, Thomas WM, Hardcastle JD. Population screening for colorectal cancer: comparison between guaiac and immunological faecal occult blood tests. Br J Surg 1994; 81:448–451.
173. Castiglione G, Biagini M, Barchielli A, Grazzini G, Mazzotta A, Salvadori P, Scillone L, Ciatto S. Effect of rehydration on guaiac-based faecal occult blood testing in colorectal cancer screening. Br J Cancer 1993; 67:1142–1144.
174. McCallum RW, Meyer CT, Marignani P, et al. Flexible sigmoidoscopy: diagnostic yield in 1015 patients. Am J Gastroenterol 1984; 79:433–437.
175. Rosevelt J, Frankl H. Colorectal cancer screening by nurse practitioner using 60-cm flexible fiberoptic sigmoidoscope. Dig Dis Sci 1984; 29:161–163.
176. Shida H, Yamamoto T. Fiberoptic sigmoidoscopy as the first screening procedure for colorectal neoplasms in an asymptomatic population. Dis Colon Rectum 1989; 32:404–408.
177. Reilly JM, Ballantyne GH, Fleming FX, Zucker KA, Modlin IM. Evalulation of the occult blood test in screening for colorectal neoplasms. A prospective study using flexible endoscopy. American Surgeon 1990; 56:119–123.

11

Head and Neck Cancer

Scott M. Lippman and Anita L. Sabichi
University of Texas M. D. Anderson Cancer Center, Houston, Texas

Jack L. Gluckman
University of Cincinnati Medical Center, Cincinnati, Ohio

I. INTRODUCTION

Interest in screening for head and neck cancer to facilitate early diagnosis has grown tremendously in recent years. This is a direct result of the absence over 2 decades of meaningful improvement in prognosis for these cancers. The two greatest causes of head and neck cancer are smoking and drinking alcohol (1,2). Although both habits appear to be declining overall in the United States, any direct impact on the incidence of this disease will be delayed 15–20 years. Even though these conditions provide an impetus for establishing a satisfactory screening program to increase early diagnosis, they do not mitigate the substantial difficulties and limitations encountered in cancer screening in the head and neck.

The general principles governing the appropriate use of cancer screening are divided into those involving the disease and those involving the nature of the screening test. The disease must have a high burden of morbidity, mortality, and costs, have effective treatment for early stages, have high prevalence and incidence, have known natural history and biology, and have a preclinical phase with high prevalence. The screening test must be able to detect disease in preclinical phase, be effective (sensitive and specific), be safe, simple, and inexpensive, and be acceptable to people without apparent disease, or signs and symptoms. In addition, effective screening tests impact on cause-specific mortality rate.

Screening for head and neck cancer generally meets the disease-specific principles of screening feasibility—for example, head and neck cancer has a very high burden of morbidity and mortality for the patient, has effective early stage treatment, and asymptomatic head and neck cancer frequently is detectable (1–4). In the mucosa of the head and neck, premalignant lesions usually present as leukoplakia (2). These certainly represent a preclinical, asymptomatic phase detectable by screening programs. Early frank invasive cancers may manifest as erythroplakia, small asymptomatic mucosal masses, or ulcers.

Principles governing the nature of the screening test, however, pose problems for screening in this region. Screening tests must be valid (as measured by sensitivity and specificity), reliable, reproducible, and not overly complex, expensive, or responsible for morbidity. Cancer screening for head and neck cancer has yet to live up to this last group of requirements.

In addition to discussing the tests themselves, this chapter presents several other key issues involved with screening for potentially deadly cancers in this location—tumor burden, epidemiology, natural history and biology of this disease, and future considerations for cancer screening approaches. Clinical cancer screening lags behind and may become productive only in company with molecular risk assessment and screening in the exceedingly complex set of diseases called head and neck cancer.

II. DISEASE BURDEN

It was estimated that head and neck cancers would comprise 3% of new cancer cases (41,650, of which 30,750 are oral and pharynx cancers) and 2.3% of cancer deaths (12,670–8440 from oral and pharynx cancers) in the United States and more than 600,000 incident cancer cases worldwide in 1997 (3). Since the mid-1960s, overall survival rate for patients with head and neck cancer has improved only marginally. The current overall U.S. 5-year survival rates for laryngeal cancer are 68% for whites and 52% for blacks. The respective figures for oral cavity and pharyngeal cancer are 55% and 33%.

These cancers are cosmetically, functionally, and economically devastating and have yet to be controlled by standard approaches of prevention, screening and early detection, or therapy (2). In addition to having the substantial biological and clinical heterogeneity inherent in all epithelial cancers, squamous cell carcinoma of the head and neck includes many distinct primary sites and subsites, extending from the lips to the cervical esophagus. Not only are the sites heterogeneous, but each subsite is biologically and clinically heterogeneous as well. The broad category of head and neck cancer also includes entirely unrelated cancers, such as adenocarcinomas, nasopharyngeal carcinomas, paranasal sinus tumors, and salivary gland tumors. This great anatomical, clinical, and biological diversity is responsible for the difficulty in controlling these cancers.

Standard treatment modalities for cancer of the head and neck are surgery and radiotherapy, either one or the other (single modality) for early stage disease, or combined for locally advanced disease (2). The two major types of failure after standard therapy of head and neck cancer are primary disease recurrence (local recurrence or regional [cervical lymph nodes] or distant metastasis) and development of second primary tumors. The major cause of death after therapy of locally advanced disease is primary disease recurrence (local-regional in 50%–60% of deaths and distant in 20%–30%), occurring usually within the first 2 years after initial therapy (2,4–6). Second primary tumors are the major cancer-related cause of death after primary therapy of early-stage disease and in long-term survivors, regardless of initial disease stage (5). Four prospective studies indicate that second primary tumors develop at a constant annual rate of from 3% to 7% (4–8), roughly a two-fold higher rate than was indicated in earlier retrospective studies. There is no current standard approach to control these second primary tumors. All these troubling head and neck data offer proof that new approaches are needed to enhance control of this disease.

III. EPIDEMIOLOGY

Head and neck and the other major aerodigestive tract cancers (lung and esophageal) are linked by a common causative agent—tobacco (Table 1) (1,2). Tobacco-use figures are staggering—there are approximately 1 billion smokers and 600 million tobacco chewers worldwide. In the United States, there are approx-

Table 1 Epidemiology and Risk Factors

Tobacco
Smoked, smokeless
Field defect
Molecular data (p53)
Ethanol
Marijuana
Occupational/chemical exposures
Age
Diet
Radiation
Oral hygiene
Viral
Genetic
Human leukocyte antigen
Polymorphisms
DNA repair defect

imately 44 million smokers, 44 million former smokers, and 12 million chewers. Prospective epidemiological studies show that smokers have high relative risks of death from head and neck cancers. Recent randomized trials indicate that intensive behavioral smoking-cessation efforts have had only minimal success in highest risk heavy smokers (9), and despite the wide availability of tobacco-cessation programs, the incidence of and deaths resulting from head and neck cancers have remained unchanged for more than 2 decades. These data parallel those of standard local therapy in supporting the urgent need for new prevention and screening strategies to control head and neck cancers.

Alcohol, especially in combination with tobacco use, is the other major risk factor for head and neck cancers. In only a small fraction of people exposed to these carcinogens, however, does cancer develop. This suggests wide differences in inherent (genetic) susceptibility to carcinogen exposure (10). Mutagen-sensitivity assays have been developed to assess sensitivity to environmental carcinogens. One such assay quantitates bleomycin-induced chromosome breaks in cultured peripheral blood lymphocytes and is thought to reflect sensitivity of cells to clastogenic carcinogens (i.e., those that cause physical disruption or breakage of chromosomes) and ability to repair resultant DNA damage. The importance of DNA repair in cancer risk is well illustrated by xeroderma pigmentosa. This rare genetic disorder is characterized by a profound DNA repair defect and more than 100-fold increased risk of ultraviolet radiation-induced skin cancers.

There have been two case-control studies of the Hsu mutagen sensitivity assay for risk of primary head and neck cancer. The first study included 75 cancer subjects and 108 control subjects, the second, 108 cancer subjects and 108 control subjects. (11). In both studies, mutagen sensitivity (defined as >0.8 chromosomal break per cell after in vitro bleomycin exposure) was a significant independent risk factor for head and neck cancer. Both studies also found that risk associated with mutagen sensitivity varies with smoking status. In the larger of the two studies, the odds ratios of development of head and neck cancer for mutagen sensitivity, smoking, and both combined were 3.2 (0.5–18.7), 8.1 (1.7–37.8), and 23.0 (5.0–106.0), respectively (95% confidence interval) (12).

More recent data suggest that mutagen sensitivity in head and neck-cancer patients may involve specific chromosomes. Chromosomes 3 and 7 have the highest frequencies of bleomycin-induced breaks (13). In studies of locations and frequencies of mutagen-induced chromatid breaks in Q-banded cytogenetic preparations, breaks occur in specific sites (''hot spots'') on chromosomes 3 (3p21, 3q21) and 7 (7q22,7q32). Continued work in this area may lead to identifying individuals at highest risk and the precise molecular genetic basis of this risk.

As in the development of primary head and neck cancer, mutagen sensitivity appears to be important in the development of second primary tumors. A recent prospective study involving 278 head and neck-cancer patients showed that mutagen hypersensitivity (>1.0 break per cell) occurred in 44% of patients and was associated with a 2.67 relative risk of development of a second primary tumor (14). The mean number of chromatid breaks per cell in patients in whom second primary tumors developed was 1.17, compared with 0.98 break per cell in patients in whom second primary tumors did not develop (two sided $P=0.04$). These findings were confirmed by another recent study in which a higher mean number of breaks per cell occurred in patients in whom second primary tumors developed (1.20 breaks) than in patients with no second primary tumors (0.96 break) ($P=0.025$, one-sided Student's t-test) (15).

Genetic susceptibility factors include human leukocyte antigen (HLA) phenotypes, DNA-repair defects, and polymorphisms involving carcinogen-metabolizing enzymes (e.g., glutathione S-transferases and acetyltransferases) (10). The importance of DNA repair defects is supported by their presence in certain family cancer syndromes, such as those associated with xeroderma pigmentosum, Bloom syndrome, ataxia-telangiectasia, and Fanconi's anemia, and by case-control studies of mutagen sensitivity (discussed previously), the association of which with cancer proclivity may reflect in part an altered DNA repair process.

The mutagen-sensitivity phenotype also may involve an inherent chromatin alteration that increases the efficiency of translating DNA damage into chromosome damage after mutagen exposure (10). Recently a host-cell reactivation assay was developed in parallel with the mutagen-sensitivity assay to examine DNA repair capacity in 16 established lymphoblastoid cell lines, including three of the head and neck (16). Using ultraviolet radiation and nitro-quinoline-4-oxide (4-NQO) as the test mutagens for both assays, reduced cellular DNA repair capacity was found to correlate significantly with increased frequency of mutagen-induced chromatid breaks. This finding strongly suggests that repair fidelity may be impaired in persons with mutagen hypersensitivity.

IV. NATURAL HISTORY AND BIOLOGY OF HEAD AND NECK CANCER

A. Natural History

Because they are malignancies of the epithelial mucosa, squamous cell carcinomas of the head and neck frequently are directly or indirectly apparent (e.g., in the oral cavity and larynx, respectively) and palpable, depending primarily on their location. Therefore, these lesions frequently can be diagnosed in premalignant stages, such as leukoplakia and erythroplakia, which occur most often in individuals exposed to known risk factors and are associated with a high rate of transformation (2). Clinically apparent premalignant lesions usually require many years to progress to invasive carcinoma. Head and neck squamous cell carcinomas have metastatic potential, frequently involve regional lymph nodes, and commonly, in advanced cases, spread to distant organs such as the lungs.

Lesions that may be diagnosed within relatively early stages include visually apparent lesions, such as those within the oral cavity, and visually nonapparent lesions that produce distinct early symptoms, such as those of the larynx producing hoarseness. Other nonapparent lesions, such as those of the pyriform sinus or nasopharynx, which do not cause specific early symptoms, may be diagnosed only after much greater advancement (e.g., regional metastases).

Head and neck lesions are most frequent in individuals with known carcinogen exposure, they usually progress through a years-long process of multistep carcinogenesis to invasive disease, and they frequently are clinically, historically, or molecularly detectable in premalignant stages. Although these natural history characteristics lend themselves to screening approaches, no system that would reduce head and neck cancer mortality and morbidity rates through early detection has yet been established.

The two seminal biological concepts behind screening as well as chemoprevention in this region are field carcinogenesis and multistep carcinogenesis (2,4,8,10,17–25). Field carcinogenesis links carcinogenesis of the head and neck to that of the rest of the aerodigestive tract. Multifocal lesions develop at various rates diffusely through the region as a result of repeated field-wide carcinogenic assault. Tobacco is the major agent for this effect (24,25). In 1953, Slaughter et al. (26) were the first to report this phenomenon, observed in their study of tissue surrounding primary cancer in resection specimens of the oral cavity. They found frequent and widespread carcinogenic changes—from hyperplasia and dysplasia to areas of in situ and invasive cancer.

This field phenomenon of multiple, developmentally different neoplastic lesions supports the concept of multistep carcinogenesis, which is that neoplasia progresses sequentially over a period of time from normal, to hyperplastic, to dysplastic, to fully malignant phenotypes. The multistep model has been well worked out in head and neck carcinogenesis and the molecular biology of this process is discussed in the following section.

B. Molecular Biology

Several types of genes are involved in carcinogenesis. Recently developed molecular techniques allow primary-tumor DNA to be isolated and analyzed for the direct identification of genes involved in the pathogenesis of cancer (oncogenes). Regulation of normal cell proliferation results from a balance between the function of growth-promoting genes (proto-oncogenes) and growth-suppressing genes. Growth-promoting genes, when altered through "activating" mutations such as overexpression or amplification, are called dominant oncogenes and lead to deregulated cellular proliferation and cellular transformation.

These genes usually encode proteins that function as growth factors or growth factor receptors. Tumor suppressor genes encode proteins involved in growth suppression, and consequently, when mutated, result in uncontrolled cell proliferation. Mutations in both alleles of ''recessive'' oncogenes must occur before transformation occurs. Sophisticated methodologies have led to the identification of critical genetic events that drive the carcinogenic process and have provided a preliminary genetic progression model of multistep head and neck carcinogenesis that is beginning to indicate which genetic targets are suitable for novel molecular screening strategies.

The concept of clonality is the basis of the genetic model of carcinogenesis (27). Carcinogenesis develops and progresses through the accumulation of specific genetic changes (multistep carcinogenesis) that activate proto-oncogenes and inactivate tumor suppressor genes. Proto-oncogenes are ''progenitors'' of oncogenes, in that, when altered by mutagenesis, translocation, or amplification, they are converted to a dominant oncogene with ability to transform cells to neoplastic growth. Certain genetic alterations provide cells with a growth advantage, allowing unregulated cell growth (clonal expansion) to occur. Statistical models suggest that 6 to 10 genetic alterations are required to develop clinically apparent cancer of the head and neck. These models are supported by laboratory transformation studies suggesting that several genes interact in the process of carcinogenesis.

1. *Dominant Oncogenes.*

Dominant oncogenes are genes that exert their effects by amplification or overexpression. Only a few altered proto-oncogenes have been identified in primary cancers of the head and neck or premalignant lesions (2,4,). Proto-oncogenes often amplified in head and neck carcinogenesis are *c-erb*B-1 and several others on chromosome 11q13—*cyclin D1, int*-2, *hst*-1, and *ems*-1 (28). Other genes variably reported to appear altered in head and neck cancer are *c-myc, ras, her-2/neu*, and E-*cadherin*.

2. *Tumor Suppressor Genes*

Tumor suppressor genes are genes whose loss of function can lead to cancer. Whereas apparently only a few oncogenes are involved in head and neck carcinogenesis, many tumor suppressor genes are involved in the process (27). The loss or inactivation of many tumor suppressor gene loci has been identified from preinvasive and invasive head and neck disease. Regions on chromosomes 9p, 3p, and 17p are frequently lost early (often before histological dysplasia), followed by losses on chromosomes 11q, 13q, and 14q and p53 mutations in severe dysplasia/carcinoma in situ, followed by losses on 9p, 8, and 4q (usually seen only in invasive cancer) (27–33). It is the accumulation and not necessarily the precise order of these genetic events that determine histopathological progression. This is illustrated by the presence of a solitary ''late'' event as the only genetic alteration in some histopathologically early lesions.

V. SCREENING

A. Overview

Screening for cancer in the mucosa of the upper aerodigestive tract to decrease cause-specific mortality rate is a relatively poorly understood approach, even though groups at high risk of this disease have been recognized for many years (34–36). As discussed, prognosis for head and neck cancer patients, often diagnosed at relatively late stages, has improved little over the past 20 years, so the concept of screening in this region is becoming more attractive to clinicians.

Cancer screening in the head and neck can be broadly divided into two categories: 1. screening for first primary cancers (e.g., in smokers and/or alcohol users) and 2. screening for second primary tumors (every patient with a primary cancer is at high risk of second primary tumors). Screening for first primary tumors has met with highly questionable results. Many problems with this approach include poor compliance, lack of mortality-benefit data, high cost, and substantial morbidity.

Perhaps the only group in which a screening program can truly be justified is patients in whom an overt carcinoma of the upper aerodigestive tract mucosa already has developed. These patients have an extremely high incidence of second primary tumors (3%–7% per year), approximately 80% of which occur within tobacco-exposed tissue, including the esophagus, lung, and remaining upper aerodigestive tract (4–8). In the discussion that follows, screening in the setting of second primary tumor detection precedes the more problematic area of screening for first primary cancers. This chapter concludes with a discussion of molecular screening approaches (Future Directions).

B. Clinical Screening Techniques

Mucosa of the oral cavity, oropharynx, larynx, hypopharynx, and even nasopharynx, is easily visualized during routine clinical examination using contemporary diagnostic techniques, including mirrors and fiberoptic endoscopy (Table 2). The inferior hypopharynx, esophagus, trachea, and bronchi, however, are not easily evaluated and usually require more formal endoscopy for visualization. Even though second primary tumors at these inaccessible sites represent more than 50% of postindex malignancies (37,38), follow-up endoscopy of these lower areas is not routinely performed. This procedure is generally thought to be logistically difficult and expensive to perform and to compromise other rec-

Table 2 Head and Neck Cancer Screening Techniques

Head and neck examination
Direct: observation and palpation
Indirect: mirror
Radiographic evaluation
Chest radiograph
Computed tomography
Magnetic resonance imaging
Contrast studies (e.g., esophagram)
Cytological evaluation
Head and neck
Esophagus
Lungs
Endoscopic evaluation
Panendoscopy (laryngoscopy, bronchoscopy, esophagoscopy)
Nasopharyngoscopy
Newer adjuvants
Vital dyes, fluorescence spectroscopy
Detection of clonal genetic alterations (p53 mutations)

ommended screening procedures, such as chest radiograph and esophagogram (even though the latter two have proven to be less sensitive screening techniques).

Endoscopy also is thought to be of value only in detecting overt, clinically apparent lesions (e.g., ulcers or masses and erythroplakia and leukoplakia). Detection of premalignant lesions poses a dilemma because of the risks involved with field carcinogenesis (39). High-grade premalignancy, such as carcinoma in situ, is usually resected, if well localized. But how should clinicians address the problem of multifocal, diffuse premalignant lesions common in the head and neck resulting from field carcinogenic effects? Chemoprevention is attempting to address control of these lesions, for which there is no standard of care (17–23,39).

In vivo staining techniques have been proposed to facilitate the diagnosis of early cancers. Toluidine blue is the most frequently used stain in the oral cavity, oropharynx, larynx, and esophagus. This is a basic metachromatic stain for the nuclear material of malignant lesions and spares normal mucosa. It is quick, inexpensive, and noninvasive, and it has an acceptable specificity rate for malignancy if used correctly. It has not gained popular acceptance, however, because of a high false-positive rate caused by staining of inflammatory lesions. Poor technique may be the major cause of this problem. The key is for the solution to be carefully constituted as recommended and only applied after all debris and saliva are removed. The false-positive rate declines from approximately 50% to 6%–10% when these guidelines are followed (40,41). Perhaps toluidine blue will be more useful in selecting biopsy sites than in helping to diagnose early cancer.

Devices to detect tumor autofluorescence and drug-induced fluorescence (e.g., hematoporphyrin) may provide useful screening tools, particularly when combined with image enhancers and endoscopy, such as with fiberoptic bronchoscopy for the detection of early bronchial cancer (42,43). These devices, however, are expensive, cumbersome, and experimental. Cytologic evaluation and DNA analysis of suspicious areas also are potentially useful, but ways to incorporate these techniques in a screening program have yet to be worked out. For these reasons, contemporary screening techniques remain somewhat unsatisfactory, particularly for cancer of the esophagus and bronchi.

C. Screening for Second Primary Tumors

The discovery that field carcinogenesis (leading to the development of second primary tumors) poses a significant risk to long-term survival, even in successfully treated early stage disease, has stimulated great interest in chemoprevention and screening in primary cancer patients. These patients are at a 3% to 7% annual risk of second primary tumors for as long as they live. Second primary tumors are classified temporally into three types: 1. simultaneous tumors, which are diagnosed along with the index cancer; 2. synchronous tumors, which although not seen initially, are diagnosed within 6 months and presumably were present initially; and 3. metachronous tumors, which are diagnosed 6 months or more after the index cancer, possibly developing many years later (5,6).

The hypothesis for the development of field-wide, multicentric mucosal cancers became stronger with the emergence of tobacco as the prime etiological agent for head and neck cancers (36,44–46). Once this etiological relationship was established, it became obvious that tobacco users were at risk for the development of multicentric cancers and would need careful monitoring of the mucosa of the upper aerodigestive tract and other organ systems subject to tobacco carcinogens.

In 1943, even before the dangers of tobacco use and field carcinogenesis were understood in respect to this region, close surveillance of head and neck cancer patients was recommended. These patients were prescribed radiographic evaluation of the total aerodigestive tract to detect second primary tumors (47). Frequent lifelong follow-up examinations with routine semiannual chest radiographs to detect metachronous lung cancers (48) were recommended (49). The clear rationale behind these early screening programs was an attempt to improve mortality rates by early detection.

Until field carcinogenesis and its association with multiple primary cancers became an accepted concept, limited screening was the rule (50,51). This consisted of a thorough initial history and physical examination, including visualization of as much of the upper aerodigestive tract as possible and a chest radiograph. Oncologists also performed standard close follow-up at regular intervals after treatment of the index primary lesion to screen for recurrence and second primary tumors. Since the establishment of field carcinogenic risk, this approach, plus fiberoptic endoscopy for inaccessible areas if indicated, has continued to be standard post-therapy screening. Beyond this standard, direct laryngoscopy, esophagoscopy, and bronchoscopy under general anesthesia have been added to the evaluation at the initial examination in many centers, but the subsequent screening value of these procedures remains inconclusive. Endorsements have been based on the prevalence rates of concurrent second primary tumors (52–56), with concomitant improved mortality rates usually being presumed but not proven (55–57). The following aspects need to be considered when evaluating screening programs for cancer of the upper aerodigestive tract.

Many screening strategies for second primary tumors have been described. These include laryngoscopy, esophagoscopy, bronchoscopy, panendoscopy (all three endoscopic procedures), contrast studies of the esophagus, chest radiograph, and sputum cytology. The newest modality, which is purely investigative, involves molecular analyses of cells within saliva and sputum (see Future Directions). No agreement on optimal second primary tumor screening has been reached with respect to use or timing of these screening modalities or the necessary duration of follow-up.

Perhaps the greatest unresolved issue is whether to perform panendoscopy at the time of treatment of the index cancer. The argument for this procedure is based on the high rate of second primary tumor detection, reported in at least one prospective study, to occur in up to 16% of patients (53). Disease found in this setting is presumed to be of an earlier (curable) stage. The argument against this procedure is based on disagreement with this high rate of second primary tumor detection and the belief that panendoscopy is too costly and of questionable value in altering disease course and patient survival.

Two subgroups of patients, tobacco and alcohol abusers after initial treatment and possibly those with index cancers at certain sites, are at highest risk of development of second primary tumors. Moore (58) determined that the incidence of second primary tumors was 6% in patients who quit smoking, compared with 40% in continuing smokers. No significant history of alcohol and/or tobacco use conferred the lowest overall (independent of site of index cancer) risk of development of second primary tumors.

Index cancers in the soft palate (59) and, especially, the larynx (60–62) confer a higher risk of second primary tumors, occurring most commonly in the lung, oropharynx, and esophagus and almost never in the sinonasal passage or nasopharynx (62,63). Many oncologists believe that screening of these latter two sites would be futile because of the low incidence of second primary tumors.

Although the second primary tumor incidence data apparently support screening in all index patients, several issues need to be worked out before a cohesive screening program can be recommended. For example, what modalities should be used? What screening frequency and duration should be used? Should patients with advanced index cancers and little chance of cure or improved survival be screened?

Synchronous esophageal or pulmonary second primary tumors are reported to occur in 0.3% to 7.4% of these patients. These tumors can adversely affect survival and change treatment management. Even so, there are differing opinions on how to search for second primary tumors in newly diagnosed head and neck cancer patients. Some investigators recommend routine universal esophagoscopy and bronchoscopy in this population. Based on cost-effectiveness concerns, others recommend only symptom-directed, selective endoscopy. It is unclear whether early detection of asymptomatic synchronous esophageal or pulmonary second primary tumors improves the mortality rate or quality of life compared to later detection of symptomatic second primary tumors. The effects of screening on life expectancy and quality of life have not been examined in a randomized controlled trial.

Clinical practice guidelines recently published by the American Society for Head and Neck Surgery and the Society for Head and Neck Surgery do not recommend routine esophagoscopy or bronchoscopy in asymptomatic patients with normal chest radiographs. Nevertheless, recent studies have demonstrated that the combination of mucosal iodine staining with routine esophagoscopy can yield a high rate of detection of esophageal second primary tumors (between 3.3% and 6.3%) in high-risk asymptomatic populations. The majority of esophageal second primary tumors detected in these studies were confined to the mucosa (without submucosal invasion) and were potentially curable and thought to be undetectable with routine esophagoscopy. These results provide some rationale for the use of iodine staining in patients receiving esophagoscopy and perhaps also for screening in asymptomatic patients. Screening esophagoscopy and bronchoscopy are generally performed more frequently in patients with index tumors of the pharynx or supraglottis than in patients with index tumors of the oral cavity and oral pharynx. This increased frequency of screening probably results from the need of initial direct laryngoscopy in these patients and so greater ease of esophagoscopy and bronchoscopy. This scenario contrasts with that of patients not needing direct laryngoscopy for evaluation of the index tumor. An example of the latter is patients with primary cancer of the oral cavity who do not require direct laryngoscopy for initial evaluation of the primary lesion. Therefore, these patients would require an additional procedure that would include operating room personnel and time, and possibly a general anesthetic to screen for second primary tumors.

Esophagoscopy and bronchoscopy entail many costs, including fees for the operating room, surgeon, and anesthesia and costs in time of medical personnel taken from other medical services. Potential morbidity of these procedures includes airway compromise, esophageal perforation, cardiac arrhythmias, pneumothorax, and aspiration. Controversy continues regarding whether fiberoptic endoscopy, that is, bronchoscopy and esophagoscopy (which can be performed under local anesthesia) or rigid endoscopy (requiring general anesthesia) is better. In the final analysis, it is the physician's experience that dictates preference.

Controlled randomized study is necessary to resolve whether potential improvements in mortality rates and quality of life associated with routine endoscopic screening outweigh the morbidity and costs from these procedures. Until such confirmation is achieved, uncertainty and disagreement about the efficacy and appropriate use of second primary tumor screening will continue.

Patients given radiation therapy may be at additional risk of development of second primary tumors, particularly cutaneous, mucosal, salivary gland, bone, and soft tissue malignancies (64–68). Radiation-associated risk of second primary tumors was first noted for primary cancers (see Section VD). Although thyroid second primary tumors are the most studied, second primary tumors of the salivary glands, parathyroid, central nervous system, and mucosa of the upper aerodigestive tract also have been attributed to childhood irradiation (69–73). Some data do not support the concept of high-dose, therapeutic radiation-induced head and neck squamous cell carcinoma. Several retrospective investigations of head and neck cancer patients given irradiation failed to demonstrate a statistically significant increased risk of second primary tumors in the same region (59,73–76).

Synchronous or metachronous second primary tumors associated with head and neck primaries frequently develop in the esophagus (38,55,77,78), just as second primary tumors associated with esophageal primaries frequently occur in other head and neck sites. The true incidence of the latter set of second primary tumors is difficult to determine because the long-term survival rate after treatment for esophageal cancer is less than 10% (79–81).

The survival rate after treatment of esophageal second primary tumors is extremely poor, reflecting typically late diagnosis. Earlier diagnosis through screening potentially could improve mortality rate.

The possibility of diagnosing esophageal second primary tumors early has been proved in northern China, where mass screening over the past 20 years has been associated with the 5-year survival rate from 10% prescreening to 30% for advanced cancer and 90% for early cancers (82). Cytological specimens are collected from a mesh-covered balloon pulled through the length of the esophagus. Positive tests are followed up with endoscopy and biopsy to confirm the diagnosis. This technique has an extremely high detection rate, and 75% of detected cancers are early second primary tumors.

The Chinese screening program is an example of successful screening, but it is not specifically applicable to the United States population because of the relatively low overall incidence of esophageal cancer in the United States. Similar cytological screening programs, however, have been implemented in other countries suffering high rates of esophageal cancer. The success of this screening may be due to the prolonged preclinical phase (83) of esophageal cancer, yielding a high potential for early detection and cure.

Esophagoscopy is preferred, but timing and cost of this test remain difficult issues. Esophagogram has been recommended as an alternative to esophagoscopy but has little screening value because it is not very effective in diagnosing early cancer (53,80,84,85–88).

Currently, esophagoscopy is used in evaluating the index cancer and only on a symptom-directed basis in follow-up. Sensitivity and specificity (with biopsy) of esophagoscopy make this a good screening tool. Even so, up to 20% of patients with early cancer have normal-appearing mucosa under endoscopy (82). Toluidine blue has been suggested as a means to enhance visualization of early lesions, but, as discussed, its value is marginal (89). Although useful in initial evaluations of index cancers, toluidine blue for routine follow-up remains controversial and subject to questions of cost effectiveness and patient convenience, among others (54,55,90,91), and few oncologists follow this regimen.

Two methods of esophagoscopy are available. Head and neck surgeons consider rigid esophagoscopy the gold standard, but this procedure requires general anesthesia and so has concomitant increased costs. Fiberoptic esophagoscopy does not require general anesthesia and is equally accurate in expert hands. Complication rates of both techniques are low, but considerations of logistics and convenience make both unacceptable for primary or second primary tumor screening. Use of this procedure is therefore recommended only for evaluation of the index cancer and on a symptom-directed basis rather than as a routine screening procedure.

The incidence of second primary tumors in the bronchi is high and associated with poor prognosis (38,92). Reflecting relatively advanced stage at diagnosis, survival rates of lung second primary tumors are slightly lower than those of index lung cancers (93).

Annual chest radiographs have been the traditional screen for early detection of lung cancer. This has failed to improve lung cancer mortality rates, however, and so has failed as a screening technique. Sputum cytology to complement chest radiography has been tried under the rationale that cytology can better detect endobronchial cancers than can chest radiography, which better detects peripheral lesions. Three large National Cancer Institute (NCI)-sponsored long-term, randomized controlled trials examined the use of this dual screening for early stage lung cancer in smokers. The Mayo Lung Project found no difference in the lung cancer mortality rate between those receiving dual screening compared to a control group (93). Similarly, two other studies found that the addition of standard sputum cytology assessments to chest radiography did not improve resectability or mortality rates (94,95).

Therefore, many investigators recommend that routine bronchoscopy be incorporated into screening for early stage second primary tumors, even though others doubt its value because of a low detection rate with bronchoscopy in the presence of normal chest radiographs 90. It has been demonstrated, however, that bronchoscopy combined with bronchial lavage and cytology of the washings can detect early lung cancer (96). False-positive cytology results can occur, however, because malignant cells from head and neck cancers can shed into the lungs (97).

Whereas rigid bronchoscopy is valuable in the initial examination because of the ability to secure the airway with this scope, fiberoptic bronchoscopy may be preferable in follow-up screening for second primary tumors because this procedure, as discussed earlier, has several logistical advantages over rigid bronchoscopy, including the fact that it does not require general anesthesia. As with esophagoscopy, the recommendation for use of fiberoptic bronchoscopy for detection of second primary tumors might be considered following index cancer diagnosis. There are no randomized studies addressing the role of bronchoscopy as a screening test in this setting and it is unclear whether such a trial will ever be done. Although there are no definitive data available to show benefit or harm, the use of this procedure is common practice based on a high rate of second primary tumors and anecdotal experience.

EUROSCAN was a European screening program designed to evaluate bronchoscopy, sputum cytology, and chest radiography for detecting lung second primary tumors in patients treated for oral cavity, laryngeal, or lung cancer. This ambitious program, which included 4 years of screening and 10 years of follow-up, experienced feasibility problems and was stopped early before it could provide the first meaningful information on the effectiveness of bronchoscopic screening for lung second primary tumors.

D. Screening for First Primary Cancers

Cure rates are high for relatively small primary squamous cell carcinomas of the head and neck that have no nodal or distant metastases. As the primary lesions increase in size or as nodal metastases develop, the success of therapy decreases dramatically. Therefore, as with many other solid tumors, detection of primary tumors when small may theoretically be beneficial. Notwithstanding these general conditions and the documented ability of head and neck cancer screening to detect prevalent cases, no effective screening for this disease has been established in the United States (i.e., there is no evidence of reduced mortality and morbidity rates) and no large-scale routine screening programs in asymptomatic individuals are recommended at the present time.

Problems with screening in this region include the distinct natural histories of different head and neck cancers and differing intensities of examination required to diagnose early primary disease of different sites. Because of the very low incidence of head and neck primary cancer in the general U.S. population, national screening could not be cost effective, even if a specific disease mortality rate possibly could be improved.

Because resources are limited for cancer screening, some organizations require proof of mortality or morbidity benefit before screening tests are recommended for the general public. The U.S. Preventive Health Services Task Force does not recommend screening for oral cancer because there is little evidence to show that it can reduce mortality rates (98). Some data suggest that dentists are more effective than physicians in detecting early stage oral cancer. This suggests that screening in the head and neck could be improved by educating primary care providers about high-risk profiles and effective clinical methods. Education programs for this have been implemented. This approach has been limited, however, by time constraints for prevention activities imposed by other health care demands within the practice setting.

Studies of screening for oral cancer in Sri Lanka and India have demonstrated the ability to detect premalignant lesions and early cancers in these populations at high risk because of habits of tobacco and betel nut chewing and reverse smoking (placing the lit end of the cigarette in the mouth). A prospective study of educational intervention in primary oral cancer prevention was conducted in 36,000 high-risk tobacco users in three districts of India—Ernakulam, Srikakulam, and Bhavnagar (99) (Table 3). The population of selected villages in each district was screened completely and all available tobacco users aged 15 years and older were chosen as part of the intervention cohort. This study included a house-to-house interview and examination of all subjects, and all oral lesions were documented and photographed. Annual follow-up with examination was provided over a 5-year period. Subjects entered onto the study were provided health education programs and encouragement to give up their tobacco habits. Results of this group were compared with results in a control cohort, which included other people from the same districts, who received no educational intervention. The follow-up rate was 97%. In two of the three districts (Ernakulam and Srikakulam), substantially more people in the intervention group stopped their tobacco habit and reduced their frequency of tobacco use than did in the control group. However, in the third district, only a slightly higher proportion in the intervention group ceased their tobacco habits, and there was no difference between the groups in decreased tobacco usage. The rate ratio

Table 3 Prospective Trial of Primary Oral Cancer Prevention in 36,000 Indian Tobacco Users: *Results*

	Compliance (1 F/U [%]; lost to F/U [%])		Tobacco cessation % (decrease in use)		5-Year age-adjusted OPL rate ratio per 1000 (intervention/control)	
District	Intervention	Control	Intervention	Control	Men	Women
Ernakulam	98.2; 1	94.8; 2.4	9 (28)	3 (9)	0.24 (11.4/47.8)	0.18 (5.8/33)
Srikakulam	97.6; 1.3	91.5; 5.6	17 (49)	5 (8)	0.22 (59.8/260.8)	0.59 (289.5/489.5)
Bhavnagar	96.5; 2.6	83.5; 13.9	13 (20)	9 (21)	1.31 (59.7/45.4)	NA

Abbreviations: F/U, follow-up; OPL, oral premalignant lesion; NA, not applicable.
Source: Ref. 101.

of oral premalignant lesion detection (intervention/control) in the three districts correlated with the changes in tobacco-use patterns (99).

Three large screening programs for oral cancer in Sri Lanka and India have demonstrated the ability to detect premalignant lesions and early cancers in these populations at high risk due to habits of tobacco use, reverse smoking and betel nut chewing (100–102). These studies assessed the performance of primary health care workers in detecting oral cancer/precancerous lesions. More than 250,000 people in Sri Lanka and India were included in these projects. General findings revealed that, of those screened, 12%–26% were found to have oral lesions that did not require referral to a specialist. The prevalence of referable oral mucosal lesions ranged from 1.3% to 4.2%. Of this latter group, 45%–80% were correctly referred. The false-positive rate ranged from 9% to 29%. A total of 44 new oral cancers were detected by primary health care workers.

Feasibility of these types of studies remains uncertain because difficulties in compliance were documented following the initial screening. Compliance rates varied from 54% in one study to 72% in another, and it was determined that compliance was only marginally improved by institution of a health education program.

Less than optimal compliance in the Indian and Sri Lankan studies is relevant to screening in the United States and other countries, where data also indicate compliance problems in high-risk tobacco and alcohol users (103). The opposite recommendations of the U.S. Preventive Health Services Task Force (against) and the American Dental Association (for) screening for oral cancers highlight the complexity of screening in this region. Oral screening with oral exfoliative cytology is the most extensively studied screening procedure in the United States. This screen has yielded a high proportion of false-negative examinations, and highest risk individuals (heavy tobacco and alcohol users) are least likely to voluntarily participate.

Whereas screening for cancer of the oral cavity may be performed by visual inspection and palpation, screening for cancers in other head and neck sites is more complex. Screening methods such as panendoscopy (triple endoscopy), laryngoscopy, and esophagoscopy are both expensive and time consuming. Even in high-risk patients, the screening effectiveness of these procedures is uncertain. An arm of the EUROSCAN trial began to test intensive semiannual screening with flexible bronchoscopy, which proved to be unfeasible and so was stopped after the pilot phase.

In general, identifying high-risk populations can lead to improved screening strategies by allowing intensive follow-up and invasive and costly procedures to be limited to the highest risk individuals. On the other hand, many highest risk populations, such as Chinese or Aleuts who are at a very high risk of nasopharyngeal cancer, live in developing countries or regions where cost and logistical considerations make screening programs prohibitive. Also, this general principle has yet to be proven in the United States. Screening studies in heavy tobacco and alcohol users do not indicate improved outcomes in these high-risk groups, probably in large part because of lack of compliance. Screening has not been well studied in other high-risk groups such as workers with occupational exposure to asbestos, coal soot, petroleum products, chromium, and iron oxide (104,105).

Screening may be beneficial in groups at high risk because of known medical disorders. Medical conditions that predispose toward development of esophageal cancer include achalasia (106), chronic esophageal strictures following lye ingestion (107), Plummer-Vinson syndrome (108), celiac disease (109), Barrett's esophagus resulting from chronic reflux (110), and tylosis, a rare autosomal dominant disorder that confers a 95% risk of development of esophageal cancer by age 65 years (111). Overt oral lichen planus, submucosal fibrosis, and established leukoplakia also dispose toward head and neck cancer, and affected patients should be screened along with close follow-up of the established lesion.

The risk of radiation-induced head and neck cancer was first recognized in radium watch-dial painters, who had a high incidence of bone tumors in the paranasal sinuses, mastoid, and other sites (112). The full risk of cancer from external beam irradiation was not realized until 1950 when an unexpectedly high incidence of thyroid cancer was observed (and subsequently confirmed) in adults who had been irradiated as children (113). This incidence was particularly high after low-dose radiation treatment for benign conditions of the head and neck such as thymic or tonsillar hyperplasia, acne, or tinea capitis [69–71].

Although many screening programs have been designed for thyroid cancers in patients previously irradiated for benign conditions (114,115), no screening for mucosal cancers in these patients has been conducted. Despite the lack of definitive data, routine screening is often undertaken in patients who have a history of radiation to the head and neck to detect cancers of the mucosa of the upper aerodigestive tract, thyroid, and salivary glands.

Perhaps the most promising and specific means of identifying high-risk individuals will be through validation of molecular genetic markers (see Section III); these include mutagen sensitivity assays and testing for genetic polymorphisms and DNA repair defects.

E. Nasopharyngeal Cancer

Nasopharyngeal cancer is common in many parts of the world, such as southeast China, where nasopharyngeal cancer develops in 1%–8% of the population. Nonspecific symptoms and signs and difficulty in visualizing the nasopharynx frequently impede diagnosis. This leads to late detection in many cases and poor prognosis. Early detection and accurate diagnosis could have a major impact on controlling this disease.

Nasopharyngeal cancer is rarely associated with other head and neck cancers, and most investigators do not recommend routine nasopharyngoscopy in patients with other index cancers of this region (38,53,60,90,116) because the etiology of nasopharyngeal cancer differs from that of other cancers in the head and neck area. Although there are no data from randomized trials, many head and neck surgeons would recommend regular screening with mirror, direct, or fiberoptic endoscopy if the patient falls into a high-risk group for nasopharyngeal cancer, such as the Aleuts.

An alternative to direct visualization is laboratory tests for Epstein-Barr virus (EBV) in the plasma or nasopharyngeal tissue. Epstein-Barr virus is closely, perhaps casually, linked to nasopharyngeal cancer (117,118). The association of EBV with nasopharyngeal cancer is independent of histopathological subtype (WHO I-III), geographic or ethnic setting, sporadic or endemic pattern, or preinvasive or invasive disease status. The EBV-specific antigens in infected cells can be grouped into latent phase antigens, early replicative antigens, and late antigens (118). Latent phase antigens include the EBV-associated nuclear antigen (EBNA) and latent membrane proteins (LMPs). Early antigens include the early membrane antigen (EMA) and early intracellular antigen (EA). Late antigens include the viral capsid antigen (VCA) and late membrane antigen (LMA). The IgA response to VCA and EA in nasopharyngeal cancer patients is high, which can be used as a diagnostic indicator. Two antibodies have been used for screening in endemic areas: IgA anti-VCA (the more sensitive) and IgA anti-EA (the more specific). Endoscopy and biopsy of the nasopharynx were conducted in patients with high IgA levels to detect early nasopharyngeal cancer.

A large prospective seroepidemiological study of the accuracy of endoscopy versus biopsy for diagnosing subclinical nasopharyngeal cancer was conducted in southeast China (119). A total of 67,891 healthy persons between 30 and 59 years old were tested for IgA levels against VCA—6102, or 9%, of which had elevated serum antibody titer. Clinical examinations of these seropositive, but asymptomatic, subjects found histologically confirmed nasopharyngeal cancer in 48 (0.8%). One hundred thirty of the remaining 6054 seropositive patients were randomly recruited for endoscopic examination and biopsy of the nasopharynx. They received complete clinical examinations to rule out obvious tumor in the nasopharynx, followed by fiberoptic endoscopy and biopsy of the nasopharynx under local anesthesia.

A total of 780 biopsy specimens were obtained from the 130 subjects. Nasopharyngeal cancer was detected in 11 biopsy specimens from seven patients. Seven of the 11 positive biopsy specimens were from the pharyngeal recess. Overall, subclinical nasopharyngeal cancer was diagnosed in 5.4% of patients with elevated IgA titer against EBV. This relatively high rate of detection supports the use of endoscopic screening of the nasopharynx in individuals with elevated antibody levels in high-risk geographical regions. Even though the gross appearance of the mucosa may be unremarkable, biopsy is advisable, especially of the pharyngeal recess. This screening for early nasopharyngeal cancer could increase the chance of cure.

Recent investigations suggest that polymerase chain reaction (PCR) based screening for EBV-DNA may have value as a molecular marker of nasopharyngeal cancer. Virtually all tumors and preinvasive lesions of the nasopharynx carry clonal EBV genomes (117). The practicality of diagnostic EBV detection remains unclear, however, because more than 90% of human adults carry the virus, and nasopharyngeal tissue may represent a primary reservoir for latent EBV.

F. Major Problems and Limitations of Clinical Screening

The major limitations of morbidity of endoscopic screening and the undocumented ability of this or any other head and neck cancer screening method to reduce cancer morbidity and mortality rates are previously discussed in some detail. Indeed, these methods plus time and financial limitations may offset the benefits of endoscopy for cancer screening in the head and neck region (120–122).

Even if endoscopy is the most effective method of detecting preclinical cancer, its cost-benefit ratio still requires careful evaluation, which is difficult. Parker and Hill (122) are among several investigators who have studied this issue. They concluded that esophagoscopy and bronchoscopy have low yields and high additional costs and that laryngoscopy is the only method among these three head and neck cancer screening procedures that rationally could be implemented routinely (122). Whereas Shaha et al. (120) concluded that high costs disqualify routine triple endoscopy for all head and neck cancer patients, others have argued that the cost of routine panendoscopy in this setting is justified (123).

Although it may be that panendoscopy to detect early cancer is common practice in many settings and is often used on an individual patient basis, there is no evidence that these expensive, invasive procedures can be used in a successful broad-based screening program to effect a reduction in mortality rate. As health care dollars become scarcer, logistical and monetary issues make it difficult to conduct a definitive study to address this issue, and less sensitive and specific noninvasive radiographical or cytological procedures may have to be used for screening practices.

VI. FUTURE DIRECTIONS

It is not clear whether generalized screening for head and neck cancer will improve mortality rates or ever become cost effective. Nevertheless, individual patients with documented high risk, such as all definitively treated primary cancer patients, receive routine endoscopy, radiographical evaluation, and frequent clinical examinations as part of routine clinical follow-up.

It is to be hoped that more sensitive, lower cost screening methods for subjects/patients at risk ultimately may lead to improved mortality rates and less morbidity. Perhaps further identification of particular subgroups among head and neck cancer patients with a higher risk of second primary tumors (e.g., those with immunological anomalies or genetic influences) will improve screening efficiency in this setting (124,125).

The future of head and neck cancer screening may lie in the field of molecular biology, where exciting and novel research in cancer detection is being conducted. This work has major implications for screening of primary and second primary tumors (126,127). This technology has been used successfully to detect clonal genetic alterations in clinical samples, including in body fluids. These clonal alterations are generally considered to represent specific markers for the presence of neoplastic cells, which standard cytological methods previously may have found to be nonmalignant. Often in clinical samples the number of neoplastic cells are far outnumbered by normal cells within the same specimen. Therefore, very sensitive and specific techniques have been developed to detect these rare clonal genetic alterations among many normal, or wild-type, DNA molecules. Gene mutations that have been detected include clonal *ras* mutations in the stool of colon cancer patients, p53 mutations in the urine of bladder cancer patients, and *ras* and p53 mutations in the sputum of lung cancer patients (126).

These molecular detection techniques are just beginning to be studied in the head and neck, and Table 4 indicates the current status of this field (1,118,127–133). Mutations of p53 have been shown to mark tumor cells in glove or tongue-blade swabbings of the oral cavity mucosa and in the saliva of head and neck cancer patients.

Based on the premise that clonal tumor cells would exhibit the same p53 mutation as occurs in primary tumor cells (134), a study using polymerase chain reaction and oligomer-specific hybridization techniques was conducted to determine whether a clonal population of tumor cells with p53 mutations could be detected in histologically negative (nonmalignant) lymph nodes and surgical margins (135). Clonal tumor cells with identical p53 mutations were detected in 52% (13/25) of surgical margins and 21% (6/28) of lymph nodes found to be normal by standard light microscopy.

Future large-scale translational study with long-term follow-up will be required to establish whether this molecular detection of preclinical asymptomatic disease will lead to improved clinical outcomes. Initial results of the aforementioned study are especially encouraging (135). Recurrence occurred in five of the 13 cases with positive (mutant) p53 margins, compared with no recurrences in the 12 patients with negative (wild type) p53 margins ($P=0.02$). This work and related follow-up studies may lead to standards for assessing the appropriateness and type of adjuvant therapy.

The greatly increased specificity and sensitivity of molecular techniques could make this diagnostic tool a valuable adjunct to standard histopathological study for early detection of head and neck cancer (135). Because only 50% of head and neck cancers have a p53 mutation or protein accumulation (136,137),

Table 4 Molecular Screening for Head and Neck Cancer Risk and Cancer

Screen for	Abnormality
Risk	
Familial medullary thyroid cancer	Germline *ret* mutation (1, 133)
Squamous cell carcinoma	Mutagen sensitivity (11–15, 128)
Cancers	
Nasopharyngeal cancer	EBV (serology) (118)
	IgA anti-VCA
	IgA anti-EA
Squamous cell carcinoma	p53 Mutations (cytology) (129)
	p53 Antibodies (serum) (130)
	Telomerase (oral rinses) (131)
	Microsatellite alterations (127,132)

Abbreviations: EBV, Epstein Barr virus; VCA, viral capsid antigen; EA, early antigen.

molecular detection studies could potentially be expanded to include other markers as they are validated. Using other documented markers may alleviate the problem of the large sample size currently required to determine the mutated p53 sequence and to synthesize the probe. A future panel of probes for the most common head and neck cancer gene mutations may allow screening of at-risk subjects (137).

Recently, investigators have shown that it is possible to detect antibodies to p53 (130) and microsatellite DNA alterations (132) in serum from patients with advanced head and neck squamous cell carcinoma. In another study, DNA of normal and tumor tissue from each of 35 head and neck cancer patients was tested with trinucleotide and tetranucleotide repeat markers for microsatellite alterations (138). Microsatellite alterations in at least one locus were identified in 29% of tumors. The same clonal alterations detected in the primary tumors were identified in normal (histopathologically negative) tissue adjacent to the tumor. In the future, it might be possible to apply this technique to detection of earlier stage cancers.

It has recently been shown that telomerase activation is a frequent and early event in head and neck carcinogenesis (139). A novel and exciting study by Califano et al. (131), based on this work, reported that a modified PCR-based telomerase assay was positive in oral rinses from 80% of patients with head and neck squamous cell cancer and in 50% of patients with dysplastic lesions. The false-positive rate was only 5%. This assay has tremendous potential for detection of early lesions.

Molecular studies of risk and multistep carcinogenesis of the head and neck are advancing current understanding of the natural history and biology of this disease (27,126,137,140). As molecular biological techniques become more refined and their results more clinically applicable, molecular screening techniques will be incorporated into clinical practice. Notwithstanding the futuristic dimension of molecular technology, aspect such as p53 mutation studies are immediately relevant to early detection of recurrence and should be tested for application in this setting.

Head and neck cancers are devastating diseases with many known etiological factors. Nasopharyngeal cancer, with its strong relationship to Epstein-Barr virus, is a prime example of future screening possibilities in the head and neck (118). Disease of this site offers the unique opportunity to develop effective screening with viral markers in high-risk geographical areas. Molecular detection of thyroid cancer risk associated with the multiple endocrine neoplaisa 2 syndrome has also been developed with the detection of loss of the ret gene on chromosome 10q11 (1,133).

Working together, molecular and classical epidemiology and etiology facilitate screening research that offers strong hope for future reductions in rates of head and neck cancer mortality.

REFERENCES

1. Schottenfeld H, Fraumeni JF Jr, *Cancer Epidemiology and Prevention.* 2d ed. New York: Oxford Press, 1996.
2. Vokes EE, Weichselbaum RR, Lippman SM, Hong WK. Head and neck cancer. N Engl J Med 1993; 328:184–194.
3. Parker SL, Tong T, Bolden S, Wingo PA. Cancer statistics, 1997. Ca Cancer J Clin 1997; 47:5–27.
4. Lippman SM, Hong WK. Chemotherapy and chemoprevention. *In*: Myer EN, Suen JY, eds. Cancer of the Head and Neck. 3rd ed. Philadelphia: WB Saunders, 1996: 782–804.
5. Lippman SM, Hong WK. Second malignant tumors in head and neck squamous cell carcinoma: the overshadowing threat for patients with early-stage disease (editorial). Int J Radiat Oncol Biol Phys 1989; 17:691–694.
6. Lippman SM, Hong WK. Not yet standard: retinoids versus second primary tumors (editorial). J Clin Oncol 1993; 11:1204–1207.
7. Vikram B. Changing patterns of failure in advanced head and neck cancer. Arch Otolaryngol 1984; 110:564–565.
8. Hong WK, Lippman SM, Itri LM, et al. Prevention of second primary tumors with isotretinoin in squamous-cell carcinoma of the head and neck. N Engl J Med 1990; 323:795–801.
9. The COMMIT Research Group. Community intervention trial for smoking cessation (COMMIT): I. Cohort results from a four-year community intervention. Am J Public Health 1995; 85:183–192.
10. Khuri FR, Lippman SM, Spitz MR, et al. Molecular epidemiology and retinoid chemoprevention of head and neck cancer. J Natl Cancer Inst 1997; 89:199–211.
11. Spitz MR, Fueger JJ, Beddingfield NA, et al. Chromosome sensitivity to bleomycin-induced mutagenesis, and independent risk factor for upper aerodigestive tract cancers. Cancer Res 1989; 49:4626–4628.
12. Spitz MR, Fueger JJ, Halabi S, et al. Mutagen sensitivity in upper aerodigestive tract cancer: a case-control analysis. Cancer Epidemiol Biomarkers Prev 1993; 2: 329–333.
13. Dave BJ, Hsu TC, Hong WK, et al. Non-random distribution of mutagen-induced chromosome breaks in lymphocyte of patients with different malignancies. Int J Oncol 1994; 5:733–740.
14. Spitz MR, Hoque A, Trizna Z, et al. Mutagen sensitivity as a risk factor for second malignant tumors following malignancies of the upper aerodigestive tract. J Natl Cancer Inst 1994; 86:1681–1684.
15. Cloos J, Braakhuis BJ, Steen I, et al. Increased mutagen sensitivity in head-and-neck squamous-cell carcinoma patients, particularly those with multiple primary tumors. Int J Cancer 1994; 56:816–819.
16. Wei Q, Spitz MR, Gu J, et al. DNA repair correlates with mutagen sensitivity in lymphoblastoid cell lines. Cancer Epidemiol Biomarkers Prev 1996; 5:199–204.
17. Lippman SM, Benner SE, Hong WK. Cancer chemoprevention. J Clin Oncol 1994; 12:851–873.

18. Sporn MB, Lippman SM. Chemoprevention of cancer. In: Holland JF, Frei E, Bast RC, Jr., Kufe DW, Morton DL, Weichselbaum R, eds. *Cancer Medicine*. 4th ed. Baltimore: Williams & Wilkins, 1997:495–508.
19. Mayne ST, Lippman SM. Retinoids and carotenoids. In: DeVita VT, Hellman S, Rosenberg SA, eds. *Cancer: Principles and Practice of Oncology*. 5th ed. New York: Lippincott-Raven, 1997: 585–599.
20. Lippman SM, Batsakis JG, Toth BB, et al. Comparison of low-dose isotretinoin with beta-carotene to prevent oral carcinogenesis. N Engl J Med 1993; 328:15–20.
21. Hong WK, Lippman SM, Hittelman WN, et al. Retinoid chemoprevention of aerodigestive cancer: from basic research to the clinic. Clin Cancer Res 1995; 1:677–686.
22. Lippman SM, Lee JS, Lotan R, et al. Biomarkers as intermediate end points in chemoprevention trials. J Natl Cancer Inst 1990; 82:555–560.
23. Lotan R, Xu C, Lippman SM, et al. Suppression of retinoic acid receptor (in oral premalignant lesions and its upregulation by isotretinoin. N Engl J Med 1995; 332:1405–1410.
24. Mao L, Lee JS, Kurie JM, et al. Clonal genetic alterations in the lungs of current and former smokers. J Natl Cancer Inst 1997; 89:857–662.
25. Brennan JA, Jay OB, Koch WM, et al. Association between cigarette smoking and mutation of the p53 gene in squamous-cell carcinoma of the head and neck. N Engl J Med 1995; 332:712–717.
26. Slaughter DP, Southwick HW, Smejkal W. 'Field cancerization' in oral stratified squamous epithelium. Cancer 1953; 6:963–968.
27. Califano J, van der Riet P, Westra W, et al. Genetic progression model for head and neck cancer: implications for field cancerization. Cancer Res 1996; 56:2488–2492.
28. Shin DM, Hittleman WN, Hong WK. Biomarkers in upper aerodigestive tract tumorigenesis: a review. Cancer Epidemiol Biomarkers Prev 1994; 3:697–709.
29. Nawroz H, van der Riet P, Hruban RH, et al. Allelotype of head and neck squamous cell carcinoma. Cancer Res 1996; 54:1152–1155.
30. Bedi GC, Westra WH, Gabrielson E, et al. Multiple head and neck tumors: evidence for a common clonal origin. Cancer Res 1996; 56:2484–2487.
31. Reed AL, Califano J, Cairns P, et al. High frequency of p16 (CDKN2/MTS-1/INK4A) inactivation in head and neck squamous cell carcinoma. Cancer Res 1996; 56:3630–3633.
32. Papadimitrakopoulou V, Izzo J, Lippman SM, et al. Frequent inactivation of *p*16 in oral premalignant lesions. Oncogene 1997; 14:1799–1803.
33. Mao L, Lee JS, Fan YH, Ro JY, et al. Frequent microsatellite alterations at chromosome 9p21 and 3p14 in oral premalignant lesions and its value in cancer risk assessment. Nature Med 1996; 2:682–685.
34. Warren S, Gates O. Multiple primary malignant tumors: a survey of the literature and a statistical study. Am J Cancer 1932; 51:1358–1403.
35. Lung CC. Second primary cancer in cases of cancer of the buccal mucosa: a mathematical study of susceptibility to cancer. N Engl J Med 1933; 209:1144–1152.

36. Lombard HL, Doering CR. Cancer studies in Massachusetts. Habits, characteristics and environment of individuals with and without cancer. N Engl J Med 1928; 198:481–487.
37. Gluckman JL, Cissman JD, Donegan JO. Multicentric squamous cell carcinoma of the upper aerodigestive tract. Head Neck Surg 1980; 3:90–96.
38. Shons AR, McQuarrie DG. Multiple primary epidermoid carcinomas of the upper aerodigestive tract. Arch Surg 1985; 120:1007–1009.
39. Toth BB, Martin JW, Lippman SM, Hong WK. Chemoprevention as a form of cancer control. Journal of the American Dental Association 1993; 124:243–246.
40. Mashberg A. Re-evaluation of toluidine blue application as a diagnostic adjunct in the detection of asymptomatic oral squamous cell carcinoma: a continuing prospective study of oral cancer III. Cancer 1980; 46:758–763.
41. Strong MS, Vaughan CW, Incze JS. Toluindine blue in the management of carcinoma of the oral cavity. Arch Otolaryngol 1968; 87:101–105.
42. Hayat Y, Kato H, Onco J. Fluorescence fiberoptic bronchoscopy in the diagnosis of early stage lung cancer. Recent Results Cancer Res 1982; 82:121.
43. Balchum OF, Doiron DR, Profia AE, et al. Fluorescence bronchoscopy for localizing early bronchial cancer and carcinoma in situ. Recent Results Cancer Res 1982; 82:97–120.
44. Levin ML, Goldstein H, Gerhardt PR. Cancer and tobacco smoking: a preliminary report. JAMA 1950; 143:336–338.
45. Mills CA, Porter MM. Tobacco smoking habits and cancer of the mouth and respiratory system. Cancer Res 1950; 10:539–542.
46. Wynder EL, Graham EA. Tobacco smoking as a possible etiologic factor in bronchogenic carcinoma. A study of 684 proved cases. JAMA 1950; 143:329–336.
47. Hellendall H. Multiple carcinoma: the clinical picture, diagnosis and prognosis. Am J Surg 1943; 60:22–35.
48. Cahan WG. Lung cancer associated with primary cancer in other sites. Am J Surg 1955; 89:494–514.
49. Byars LT, Anderson R. Multiple cancers of the oral cavity. Am Surg 1952; 18: 386–391.
50. Shedd DP, Kligerman MM, Gowen GF. Multifocal carcinogenesis in the oral cavity. Am J Surg 1962; 104:692–696.
51. Gluckman JL, Zitsch RP. Early detection and screening for head and neck cancer. In: *Head and Neck Cancer*. Hong WK, Weber RS, eds. Kluwer Academic Publishers, 1995; pp. 141–157.
52. Atkinson D, Fleming S, Weaver A. Triple endoscopy: a valuable procedure in head and neck surgery. Am J Surg 1982; 144:416–419.
53. Leipzig B, Zellmer JE, Klug D. The panendoscopy study group: the role of endoscopy in evaluating patients with head and neck cancer: a multi-institutional prospective study. Arch Otolaryngol 1985; 111:589–594.
54. McGuirt WF. Panendoscopy as a screening examination for simultaneous primary tumors in head and neck cancer: a prospective sequential study and review of the literature. Laryngoscope 1982; 92:569–576.

55. Shapshay SM, Hong WK, Fired MP, et al. Simultaneous carcinomas of the esophagus and upper aerodigestive tract. Otolaryngol Head Neck Surg 1980; 88:373–377.
56. Weaver A, Fleming SM, Knechtges TC, et al. Triple endoscopy: a neglected essential in head and neck cancer. Surgery 1979; 86:493–496.
57. Atkins JP, Keane WM, Young KA, et al. Value of panendoscopy in determination of second primary cancer: A study of 451 cases of head and neck cancer. Arch Otolaryngol 1984; 110:533–534.
58. Moore C. Cigarette smoking and cancer of the mouth, pharynx and larynx: a continuing study. JAMA 1971; 218:553–558.
59. Kogelnik HD, Fletcher GH, Jesse RH. Clinical course of patients with squamous cell carcinoma of the upper respiratory and digestive tracts with no evidence of disease 5 years after initial treatment. Radiology 1975; 115:423–427.
60. Cohn AM, Peppard SM. Multiple primary malignant tumors of the head and neck. Am J Otolaryngol 1980; 1:411–417.
61. DeVries N, Snow GB. Multiple primary tumors in laryngeal cancer. J Laryngol Otol 1986; 100:915–918.
62. Vrabec DP. Multiple primary malignancies of the upper aerodigestive system. Ann Otol Rhinol Laryngol 1979; 88:846–854.
63. Berg JW, Schottenfeld D, Ritter F. Incidence of multiple primary cancers. III. Cancers of the respiratory and upper digestive system as multiple primary cancers. J Natl Cancer Inst 1976; 44:263–274.
64. Cronin J. Radiation-induced cancer of the larynx and pharynx. J Laryngol Otol 1971; 25:621–622.
65. Goolden AWG. Radiation cancer: a review with special reference to radiation tumors in the pharynx, larynx and thyroid. Br J Radiol 1957; 30:626–640.
66. Sakamoto A, Sakamoto G, Sugano H. History of cervical radiation and incidence of carcinoma of the pharynx, larynx and thyroid. Cancer 1979; 44:718–723.
67. Schindel J, Castoriano IM. Late-appearing (radiation-induced) carcinoma. Arch Otolaryngol 1972; 95:205–210.
68. Southwick HW. Radiation-associated head and neck tumors. Am J Surg 1977; 134:438–443.
69. De Groot L, Palyan E. Thyroid carcinoma and radiation. JAMA 1973; 225:487–491.
70. Schneider AB, Shore-Fredman E. Radiation-induced tumors of the head and neck following childhood irradiation. Medicine 1985; 64:1–5.
71. Scanlon EF. Head and neck tumors associated with irradiation for benign conditions. CA Cancer J Clin 1981; 31:177–182.
72. Schneider AB, Favus MJ, Stachura ME, et al. Salivary gland neoplasms as a late consequence of head and neck irradiation. Ann Intern Med 1977; 87:160–164.
73. Modan B, Baidaz D, Mart H, et al. Radiation-induced head and neck tumors. Lancet 1974; 1:277–279.
74. Jy DMC. Salivary gland tumors occurring after radiation of the head and neck area. Am J Surg 1968; 116:518–523.
75. Parker RG, Enstrom JE. Second primary cancers of the head and neck following

treatment of initial primary head and neck cancer. Int J Radiat Oncol Biol Phys 1988; 14:561–564.
76. Seydel HG. The risk of tumor induction in man following medical irradiation for malignant neoplasm. Cancer 1975; 35:1641–1645.
77. Vikram B, Strong EW, Shah JP, et al. Second malignant neoplasms in patients successfully treated with multimodality treatment for advanced head and neck cancer. Head Neck Surg 1984; 6:734–737.
78. Odette J, Szymanowski RT, Nichols RD. Multiple head and neck malignancies. Trans Am Acad Ophthalmol Otol 1977; 84:805–813.
79. Fogel TD, Harrison LB, Son YN. Subsequent upper aerodigestive tract malignancies following treatment of esophageal cancer. Cancer 1985; 55:1882–1885.
80. Goldstein HM, Zornoza J. Association of squamous cell carcinoma of the head and neck with cancer of the esophagus. Am J Roentgenol 1978; 131:791–794.
81. Mantravadi RVP, Lad T, Briele H, et al. Carcinoma of the esophagus. Sites of failure. Int J Radiat Oncol Biol Phys 1982; 8:1897–1901.
82. Shu YJ. Cytopathology of the esophagus. An overview of cytopathology in China. Acta Cytol 1983; 27:7–16.
83. Guanrei Y, He H, Sungliang Q. Endoscopic diagnosis of 115 cases of early esophageal carcinoma. Endoscopy 1982; 14:157–161.
84. Schuller DE, Fritsch MH. An assessment of the value of triple endoscopy in the evaluation of head and neck cancer patients. J Surg Oncol 1986; 32:156–158.
85. Bundrick TJ, Cho SR. Evaluation of the esophagus in patients with head and neck cancer. Am J Roentgenol 1984; 142:1082–1083.
86. Moss AA, Koehler RE, Margulis AR. Initial accuracy of esophagrams in detection of small esophageal carcinoma. Am J Roentgenol 1976; 127:909–913.
87. Zornoza J, Lindell MM Jr. Radiographic evaluation of small esophageal carcinoma. Gastrointest Radiol 1980; 5:107–111.
88. Grossman TW, Kita MS, Toohill RJ. The diagnostic accuracy of pharyngoesophargram compared to esophagoscopy in patients with head and neck cancer. Laryngoscope 1987; 97: 1030–1032.
89. Jessen K, Paolucci P, Classen M. Endoscopic vital staining of the esophagus in high risk patients: detection of dysplasia and early carcinoma. In : Sherlock P, Morson BC, Barbara L, Veronesi V, eds. Precancerous Lesions of the Gastrointestinal Tract. New York: Raven Press, 1983: 65–70.
90. Maisel RH, Vermeersch H. Panendoscopy of second primaries in head and neck cancer. Ann Otol Rhinol Laryngol 1981; 90:460–464.
91. Abemayor E, Moore DM, Hanson DG. Identification of synchronous esophageal tumors in patients with head and neck cancer. J Surg Oncol 1988; 38:94–96.
92. Atabek U, Mirseyed AM, Suresh R, et al. Lung cancer in patients with head and neck cancer: incidence and long-term survival. Am J Surg 1987; 154:434–437.
93. Fontana RS, Sanderson DR, Woolner LB, et al. Lung cancer screening: the Mayo program. J Occup Med 1986; 28:746–750.
94. Melmed MR, Flehinger BJ, Zaman MB, et al. Screening for early lung cancer: results of the Memorial-Sloan Kettering study in New York. Chest 1984; 86:44–53.

95. Tochman MS. Survival and mortality from lung cancer in a screened population: The Johns Hopkins Study. Chest 1986; 89:325S–326S.
96. Leipzig B. Bronchoscopy in the staging and evaluation of head and neck carcinoma. Ann Otol Rhinol Laryngol 1983; 92:373–376.
97. Johnson JT, Turner J, Dekker A. The significance of positive bronchial cytology in the presence of squamous cell carcinoma of the upper aerodigestive tract. Ann Otol Rhinol Laryngol 1981; 90:454–456.
98. Screening for oral cancer. In: Fisher M, ed. Guide to Clinical Preventive Services: An Assessment of the Effectiveness of 169 Interventions. Report of the U.S. Preventive Services Task Force. Baltimore, MD: Williams & Wilkins, 1989:91–94.
99. Gupta PC, Pindborg JJ, Bhonsie RB, et al. Intervention study for primary prevention of oral cancer among 36,000 Indian tobacco users. Lancet 1986; i:1235–1239.
100. Warnakulasuriya KA, Nanayakkara BG. Reproducibility of an oral cancer and precancer detection program using a primary health care model in Sri Lanka. Cancer Detect Prev 1991; 15:331–334.
101. Mehta FS, Gupta PC, Bhonsie RB, et al. Detection of oral cancer using basic health workers in an area of high oral cancer incidence in India. Cancer Detect Prev 1986; 9:219–225.
102. Warnakulasuriya KAAS, Ekanayake ANI, Sivayoham S, et al. Utilization of primary health care workers for early detection of oral cancer and precancer cases in Sri Lanka. Bull WHO 1984; 62:243–250.
103. Prout MN, Sidari JN, Witzburg, RA, et al. Head and neck cancer screening among 4611 tobacco users older than forty years. Otolaryngol Head Neck Surg 1997; 116:201–208.
104. Cole P, Goldman MB. Occupation. Persons at high risk of cancer. In: Fraumeni JF, ed. An Approach to Cancer Etiology and Control. New York: Academic Press, 1975:167–184.
105. Benner SE, Lippman SM, Hong WK. Cancer of the head and neck. In: Greenwald P, Kramer BS, Weed DL, eds. Cancer Prevention and Control. New York: Marcel Dekker, 1995:641–657.
106. Carter R, Brewer LA III. Achalasia and esophageal carcinoma. Am J Surg 1975; 130:114–120.
107. Hopkins RA, Postlewait RW. Caustic burns and carcinoma of the esophagus. Ann Surg 1981; 194:146–148.
108. Ahlbom HE. Simple achlorydric anemia, Plummer-Vinson syndrome and carcinoma of the mouth, pharynx and esophagus in women. BMJ 1936; 2:331–333.
109. Swinson CM, Slavin G, Coles EC, et al. Coeliac disease and malignancy. Lancet 1983; i:111–115.
110. Naef AP, Savary M, Ozello L. Columnar-lined lower esophagus: an acquired lesion with malignant predisposition. Report on 140 cases of Barrett's esophagus with 12 adenocarcinomas. J Thorac Cardiovasc Surg 1975; 70:826–835.
111. Harper PS, Garper RMJ, Howel-Evans AW. Carcinoma of the esophagus with tylosis. QJ Med 1970; 39:317–333.
112. Martland HS. The occurrence of malignancy in radioactive persons. Am J Cancer 1931; 15:2435–3118.

113. Duffy BJ, Fitzgerald PJ. Cancer of the thyroid in children: a report of 28 cases. J Clin Endocrinol Metab 1950; 10:1296–1308.
114. Royce PC, MacKay BR, DiSabella PM. Value of postirradiation screening for thyroid nodules: a controlled study of recalled patients. JAMA 1979; 242:2675–2678.
115. Schneider AB, Pinsky S, Bekerman C, et al. Characteristics of 108 thyroid cancers detected by screening in a population with a history of head and neck irradiation. Cancer 1980; 46:1218–1227.
116. Tepperman BS, Fitzpatrick PJ. Second respiratory and upper digestive tract cancers after oral cancer. Lancet 1981; 2:547–549.
117. Pathmanathan R, Prasad U, Sadler R, et al. Clonal proliferations of cells infected with Epstein-Barr virus in preinvasive lesions related to nasopharyngeal carcinoma. N Engl J Med 1995; 333:693–698.
118. Wei WI, Sham JS. Cancer of the nasopharynx. *In*: Myer EN, Suen JY, eds. Cancer of the Head and Neck. 3d ed. Philadelphia: WB Saunders, 1996:277–293.
119. Sham JS T, Wei WI, Zong Y, et al. Detection of subclinical nasopharyngeal carcinoma by fiberoptic endoscopy and multiple biopsy. Lancet 1990; 335:371–374.
120. Shaha AR, Hoover EL, Mitrani M, et al. Synchronicity, multicentricity and metachronicity of head and neck cancer. Head Neck Surg 1988; 10:225–228.
121. Neew HB III. Routing panendoscopy—is it necessary every time? (commentary). Arch Otolaryngol 1984; 110:531–532.
122. Parker JT, Hill JH. Panendoscopy in screening for synchronous primary malignancies. Laryngoscope 1988; 98:147–149.
123. McGuirt WF. Panendoscopy (letter). Laryngoscope 1988; 98:688.
124. DeVries N, Delange G, Drexhage H, et al. Immunoglobin allotypes in head and neck cancer patients with multiple primary cancers. Acta Otolaryngol 1987; 104: 187–191.
125. DeVries N, Drexhage HA, Dewaal LP, et al. Human leukocyte antigens and immunoglobulin allotypes in head and neck cancer patients with and without multiple primary tumors. Cancer 1987; 60:957–961.
126. Sidransky D. Cancer of the head and neck. *In*: DeVita VT Jr, Hellman S, Rosenberg SA, eds. Cancer: Principles and Practice of Oncology. 5th ed. New York: Lippincott-Raven, 1996:735–847.
127. Leong PP, Rezai B, Koch WM, et al. Distinguishing second primary tumors from lung metastases in patients with head and neck squamous cell carcinoma. J Natl Cancer Inst 1998; 90:972–977.
128. Cloos J, Spitz MR, Schantz SP, et al. Genetic susceptibility to head and neck squamous cell carcinoma. J Natl Cancer Inst 1996; 88:530–535.
129. Boyle JO, Mao L, Brennan JA, et al. Gene mutations in saliva as molecular markers for head and neck squamous cell carcinomas. Am J Surg 1994; 168:429–432.
130. Bourhis J, Lubin R, Roche B, et al. Analysis of p53 serum antibodies in patients with head and neck squamous cell carcinoma. J Natl Cancer Inst 1996; 88:1228–1233.

131. Califano J, Ahrendt SA, Meininger G, et al. Detection of telomerase activity in oral rinses from head and neck squamous cell carcinoma patients. Cancer Res 1996; 56:5720–5722.
132. Nawroz H, Koch W, Anker P, et al. Microsatellite alterations in serum DNA of head and neck cancer patients. Nature Med 1996; 2:1035–1037.
133. Heshmati HM, Gharib H, Khosla S, et al. Genetic testing in medullary thyroid carcinoma syndromes: mutation types and clinical significance. Mayo Clin Proc 1997; 72:430–436.
134. Boyle JO, Hakim J, Koch W, et al. The incidence of p53 mutations increases with progression of head and neck cancer. Cancer Res 1993; 53:4477–4480.
135. Brennan JA, Mao L, Hruban RH, et al. Molecular assessment of histopathological staging in squamous-cell carcinoma of the head and neck. N Engl J Med 1995; 332:429–435.
136. Shin DM, Lee JS, Lippman SM, et al. p53 expression: predicting recurrence and second primary tumors in head and neck squamous cell carcinoma. J Natl Cancer Inst 1996; 88:519–529.
137. Yun O, Mao L. Biomarkers in head and neck carcinoma. Curr Opin Oncol 1997; 9:247–256.
138. Mao L, Lee DJ, Tockman MS, et al. Microsatellite alterations as clonal markers for the detection of human cancer. Proc Natl Acad Sci U S A 1994; 91:9871–9875.
139. Mao L, El-Naggar AK, Fan YH, et al. Telomerase activity in the pathogenesis of head and neck squamous cell carcinoma. Cancer Res 1996; 56:5600–5604.
140. Spitz MR, McPherson RS, Jiang H. et al. Correlates of mutagen sensitivity in patients with upper aerodigestive tract cancer. Cancer Epidemiol Biomarkers Prev 1997; 6:687–692.

12

Liver Cancer

Robert S. Brown, Jr.
Center for Liver Disease and Transplantation, Columbia Presbyterian Medical Center, New York, New York

Bruce F. Scharschmidt
University of California, San Francisco, San Francisco, California

I. INTRODUCTION

Hepatocellular carcinoma (HCC) is one of the most common cancers worldwide with an estimated incidence of 1,000,000 cases per year (1). Because of its high morbidity and mortality rates, there is extensive interest in the possibility of screening for HCC, particularly in high-risk groups. In 1988, a National Institutes of Health (NIH) consensus statement recommended routine screening for HCC in patients with hepatitis B virus (HBV) infection (Table 1) (2). However, the utility of screening for HCC has not yet been clearly established. One year after the consensus statement, the 1989 report of the U.S. Preventive Services Task Force did not even address screening patients for HCC. Successful cancer screening programs are based on four principles: 1. presence of an easily identifiable high-risk population or high prevalence of the disease in the general population; 2. the existence of a preclinical stage of the disease; 3. the availability of sensitive and specific screening tests that are reasonably low cost and safe; 4. increased efficacy of treatment during the preclinical phase of the disease process (3). Hepatocellular carcinoma clearly meets the first two criteria, the latter two are less clear. This chapter, reviews the epidemiological data regarding screening for HCC in high-risk groups according to the aforementioned paradigm and assesses both the costs and the benefits of screening to the persons at risk and to society.

Table 1 Current Recommendations for Screening for HCC Based on the 1986 NIH Consensus Conference

Risk	HBV	Onset	Race	Sex	Cirrhosis	AFP	US
High[a]	Yes	Infant	Non-white	Male	Yes	3–4 months	4–6 months
Moderate[a]	Yes	Adult	White	Female	No	3–4 months	12 months[b]
"Low"	No					Currently not recommended	

[a] High and moderate risk patients have all of these characteristics.
[b] Some authors have not recommended US in moderate risk individuals unless AFP is elevated.
Abbreviations: AFP, alpha fetoprotein; US, ultrasound; HBV, hepatitis B virus; HCC, hepatocellular carcinoma.
Source: Adapted from references 2, 3, 5.

II. EPIDEMIOLOGY AND BURDEN OF DISEASE

The prevalence of HCC has a wide geographical variation but a constant male predominance with a male-to-female ratio of approximately 4:1. In Africa, China, and much of the Far East, HCC is among the top four malignancies, with incidence rates of up to 530 cases/100,000 persons. In the United States, the incidence is less than 4/100,000 (e.g., there were 2500 new cases in 1985 according to the National Cancer Institute) and HCC ranks 22nd among cancers (2). The geographical variation in the incidence is most likely due to differences in the rate of exposure to putative causative agents, particularly hepatitis B and C viruses and chemical carcinogens. Incidence rates can be difficult to estimate accurately because they are frequently based on clinical diagnoses and lack histological confirmation. Therefore, the statistics likely contain some non-HCC liver cancers and nonmalignant nodules (particularly regenerating nodules in a cirrhotic liver) and miss many unsuspected HCCs in the setting of death from liver disease when autopsy was not obtained.

In the United States, there are no studies investigating the efficacy of population-based screening for HCC. Because the incidence of HCC and its risk factors are much lower in the United States, these studies would be difficult to perform. However, based on data from other countries, principally Asia, there is some evidence to support screening in certain high-risk groups. The major risk factor for HCC is cirrhosis of the liver. Although HCC can occur with other forms of liver disease, cirrhosis is present in 80%–90% of cases (4). This is particularly true in low incidence areas such as the United States. Among patients with cirrhosis, certain etiologies carry a higher risk for HCC. Hepatitis B (HBV), hepatitis C (HCV), and hereditary hemochromatosis carry the largest risk, whereas alcoholic cirrhosis, $alpha_1$-antitrypsin deficiency, and other etiologies are lower risk.

A. Hepatitis B

The data linking HBV to HCC are the best developed and are based largely on epidemiological and animal studies. First, HCC tends to cluster in areas of high incidence of HBV. For example, in China where the prevalence of chronic HBV carriage is 15%, the incidence of HCC is 150/100,000 compared to a U.S. incidence of 1–5/100,000 and an HBV prevalence of less than 1% (5). In addition, in high prevalence areas, case-control studies have shown an attributable risk for HBV positivity of 42%–79% (6). A large prospective study in Taiwan of 3454 chronic HBV carriers and 19,253 age-matched control subjects found an annual incidence of HCC of 495/100,000 and 5/100,000, respectively, for a relative risk of 98 (7). This finding, combined with a large body of data collected in animals, particularly in woodchucks, which are frequently infected with HBV and in which HCC develops (2), is strong evidence for a role of hepatitis B virus infection in HCC.

B. Hepatitis C

Hepatitis C virus (HCV), which was isolated and cloned in 1989, represents the major cause of transfusion-related and non-A, non-B hepatitis (8). Whether patients with HCV were at increased risk of HCC was initially debated, but there is a growing body of evidence establishing HCV as a risk factor for HCC. Ikeda et al. (9) prospectively followed 795 Japanese patients with alcoholic or viral (HBV, HCV) cirrhosis for 2 to 17 years (median 5.8 years). There were 568 men and 227 women with a mean age of 53 years. Tests for HBV were positive in 262 patients and for HCV in 364. Although all had documented cirrhosis, only 65 had advanced or decompensated liver disease with a history of complications (e.g., encephalopathy or confusion and/or ascites). Screening for HCC was performed with blood tests for a tumor marker (alpha fetoprotein [AFP]) bimonthly and an ultrasound examination of the liver at least yearly. The overall incidence of HCC was 27.8% with a calculated 50% occurrence at 11.8 years. Of the patients with HCC, 21% were positive for HBV and 83% were positive for HCV. Cox proportional hazard analysis found HCV to be a strong predictor of HCC ($P=0.001$) and the risk to be higher than in patients with HBV ($P=0.003$).

A second study analyzed 917 Japanese patients from 40 to 69 years of age with chronic hepatitis or early cirrhosis for a mean of 36 months (10). During this time period, HCC developed in 54 patients. The incidence of HCC was low, 5.8% for the duration of the study, and this might be because many patients did not have cirrhosis at entry. To analyze potential risk factors for HCC, Cox proportional hazard analysis was performed and revealed age, presence of cirrhosis, HBV, or HCV to be significant predictors of HCC. The adjusted rate ratio for HCV was 4.1 (95% CI 1.3 to 12.4); that for HBV was 6.9 (95% CI 2.9 to 16.4). Sex, alcohol use, and smoking were adjusted for as potential confounders and none of these were significant risk factors in this analysis. Thus, presence of HCV appears to be associated with an increased risk of HCC.

There is increasing evidence that different genotypes of HCV exhibit different biological behavior, with some evidence that genotype 1 may lead to more aggressive disease. A recent study from Italy raised the possibility that certain genotypes of HCV, specifically type 1b, may have a higher risk of development of HCC, making this group of patients more attractive as candidates for screening (11). In this study, genotype 1b was the single largest independent risk factor for the development of HCC with an odds ratio of 6.14 (95% CI 1.77 to 21.37). Other risk factors analyzed included age, sex, and alcohol use. This raises the question as to whether screening would be more effective if only applied to patients with HCV genotype 1. There were not enough patients with genotype 1a to address whether there are differences in risk between patients with genotype 1a and 1b. If this represents an even higher prevalence risk group, the cost effectiveness of screening would increase. This interesting preliminary result needs further confirmation before recommendations can be made.

C. Alcoholic Liver Disease

Whether alcohol is a risk factor for HCC is controversial. Some studies have shown increased risk (9) and others have not (10). The question is clouded by the fact that a high percentage of alcoholics (~50%) are also infected with HCV. Alcohol may act as a cocarcinogen or hasten the process of carcinogenesis by causing cirrhosis. Currently, there is no strong evidence linking alcoholic liver disease to HCC, and screening cannot be recommended for patients with alcoholic cirrhosis in the absence of other risk factors.

D. Other Liver Diseases

Other forms of chronic liver disease do have increased risks of HCC, but because they are less prevalent, they have not been major targets for study or for screening protocols. Hereditary hemochromatosis, a disease of iron metabolism in which excess iron accumulates in the liver, presents an increased risk of HCC. An autopsy study found the prevalence of HCC in patients with hemochromatosis to be 7.3%–8.2% compared with 3.4%–4.5% in those with other causes of cirrhosis (12). The ratio of observed to expected cases was 240. Other causes of liver disease including alpha$_1$-antitrypsin deficiency (13) and aflatoxin exposure (14) have been linked to HCC, but, because of their rarity (alpha$_1$-antitrypsin deficiency) and difficulty identifying the at-risk group (aflatoxin exposure), they are not good targets for population-based screening.

E. Demographic Factors

Because incidence of HCC increases with age (Figure 1) and is rarely seen before age 40, many investigators believe age is a risk factor for development of HCC. This relationship is confounded by the fact that increased age is associated with an increased risk of cirrhosis and with increased duration of the predisposing conditions (viral hepatitis or other liver disease), both of which increase the risk of HCC. The importance of duration of disease is supported by the fact that the risk of HCC is highest in patients with perinatally acquired HBV who have a relative risk of more than 200 when compared to age-matched HBV negative control subjects, a relative risk much higher than for HBV acquired as an adult (15). In the group with perinatal HBV, the incidence of HCC is nearly zero until age 30 years and increases progressively thereafter (15). In contrast, one study has shown an increased risk of HCC in patients older than 60 years (adjusted rate ratio 6.5) after adjusting for potential confounders, including presence of cirrhosis (10). Although these results could also be confounded by duration of disease, the degree of confounding should be minimized by adjusting for the presence of cirrhosis. Therefore, age may be an independent risk factor.

Hepatocellular carcinoma is more common in males than females and some investigators have postulated that sex is an independent risk factor, but it is

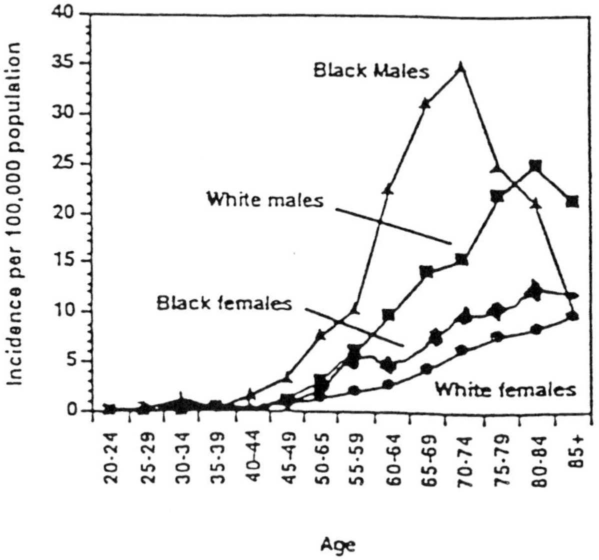

Figure 1 Average annual incidence of HCC in the United States by age, sex, and race.

difficult to control for other exposures that tend to be more common in men. In the Tsukuma study, when adjusted for potential confounders such as HBV or HCV infection, alcohol use, and presence of cirrhosis, male sex carried a relative risk of 1.3, which did not reach statistical significance (10). Similarly, non-white race is frequently cited as a risk factor, but most likely reflects higher rates of HBV, HCV, and chemical exposure among people of Asian and African descent.

F. Severity/Duration of Liver Disease

The stage of liver disease is thought to influence the risk of HCC. Cirrhosis is staged clinically using the Child's classification scheme, which measures the degree of liver impairment. Child's A denotes cirrhosis or scarring of the liver with minimal impairment of liver function; Child's C denotes advanced or decompensated liver cirrhosis in which the liver is not making necessary proteins, and complications such as encephalopathy (confusion), jaundice, and ascites (fluid accumulation in the abdomen) are present; and Child's B is an intermediate condition. The risk of HCC is higher in patients with Child's C cirrhosis than in those with Child's A or B cirrhosis (16,17). Once again it is difficult to separate out the effects of more advanced cirrhosis from increased duration of the underlying cause of the cirrhosis, because all forms of liver disease that increase the risk of HCC are progressive over time. Cirrhosis may also be part of the causal pathway through which HBV and HCV (as well as other risk factors for HCC) lead to malignant change. This is supported by the low incidence of HCC in some forms of cirrhosis. Thus, it is unclear whether cirrhosis *per se* is a risk independent of the underlying condition and its duration. However, the increased frequency of advanced cirrhosis at the time of HCC discovery has profound implications for screening, because it decreases the likelihood of the patient being able to survive surgery.

G. Summary

The fact that the prevalence of HCC increases with the severity of liver disease complicates the decision of when to start screening. If patients most likely to tolerate surgery and thus benefit from screening are chosen, then the incidence of HCC is lower. Conversely, in the highest risk group, those with Child's C cirrhosis, the proportion of patients whose treatment is successful decreases. Balancing these two issues in terms of cost and potential benefit is a challenge to any screening strategy. For the risk groups with annual incidences that range from 3% to 5% (e.g., chronic hepatitis B and C in Asian patients, [Table 2]), most screening efforts have been started at or after age 40 years.

Table 2 Annual Incidence of HCC According to Risk Factor

Risk factor	Annual incidence
Hepatitis B	3%
Hepatitis C	5%
Hemochromatosis	5%
Cirrhosis	3%–5%
Alcoholic liver disease	?
African, Asian origin	0.01%–0.1%

III. NATURAL HISTORY AND TUMOR BIOLOGY

Hepatocellular carcinoma exhibits wide biological variation with tumor size doubling times that range from 1 to 19 months. The average doubling time of 4 months makes HCC a relatively slow-growing malignancy, which would support a potential role for screening (3). Thus, a 0.5-cm HCC lesion could take 8 months to reach 2 cm, a size that can be detected with ultrasound screening. Furthermore, because these tumors are usually asymptomatic for 2 years or more, screening represents the best strategy for early diagnosis.

Tang et al. (18) analyzed 391 subclinical HCC tumors defined as those discovered on routine screening in the absence of symptoms. These were then compared to 1251 clinical HCC lesions found during investigation of symptomatic patients. Subclinical cases had smaller tumors (median 5 cm compared with 9 cm) with a greater proportion being less than 5 cm (61% versus 11%). The rate of resectable tumors was almost two-fold higher (80% versus 47%) and the operative mortality rate was lower in the subclinical group. More importantly, patients with subclinical HCC 5-year survival rate was two-to three-fold higher both for the entire series (51% versus 21%) and for the subgroup of those resected. These data, summarized in Table 3, point toward a clear subclinical stage that can be identified by screening and that has improved outcome. However, there are two major potential sources of bias in the study. The first is lead time bias. The patients in the subclinical group were relatively young (median age 50 years); the demographics of the other group is not known. The fact that 89% of the patients in the subclinical group already had established cirrhosis might minimize the effect of lead time bias, although it is difficult to exclude in this nonrandomized, nonprospective study. The second is length bias, which is difficult to eliminate in any retrospective study of screening. A potential bias against screening could stem from the use of alpha fetoprotein (AFP) screening only in the early portion of the study. The low sensitivity of this test might bias toward later stages of HCC in the screened group compared with

Table 3 Characteristics of HCC Discovered by Screening (Preclinical HCC) Compared with Those Found Based on Symptoms (Clinical HCC)

	Preclinical (n=391)	Clinical (n=1251)
Mean size of tumor	5 cm	9 cm
HCC tumor <5 cm	61%	11%
Resectable tumors	80%	47%
Operative mortality rate	1.9%	6.0%
5-year survival		
Overall	51%	21%
Resected patients	60%	37%

Source: Ref. 18.

current screening strategies. In summary, because both the percentage of patients with HCC eligible for resection as well as the prognosis following resection appear improved for small (<5 cm) tumors, the discovery of HCC at a preclinical stage could potentially benefit screened patients.

IV. AVAILABLE SCREENING TESTS

A. Blood Tests

There has been extensive interest in developing serum markers for HCC. Most have focused on abnormal variants of normal proteins produced by the liver (e.g., des-gamma-carboxyprothrombin) or fetal antigens (AFP and carcinoembryonic antigen [CEA]). Alpha fetoprotein has been used the longest and, despite its limitations, none of the newer tests have shown any advantage over AFP, largely because of poor specificity; that is, these markers are elevated in many patients with chronic liver disease without cancer.

Alpha fetoprotein is a normal serum protein synthesized primarily by fetal liver cells and yolk sac cells. Healthy nonpregnant adults have AFP levels less than 20 ng/mL. Elevated levels are seen in the setting of normal pregnancy, acute and chronic hepatitis, germ cell tumors, and HCC. Patients with consistently normal AFP over long periods of follow-up have the lowest rates of HCC and those with increases of more than 100 ng/mL have the highest rates (17). However, the sensitivity and specificity of AFP for HCC depends largely on what level is classified as a positive test. Because there is no clear level that distinguishes benign from malignant disease, where to define the cut-off point is controversial. Lower levels (e.g.,>20 ng/mL) have higher sensitivity but low specificity, resulting in a high percentage of false-positive results, whereas higher levels (e.g., >200 ng/mL) have lower sensitivity but high specificity (Table 4). In addition, there are questions as to whether AFP can detect small, subclinical HCC tumors. Because up to 50% of large tumors secrete AFP, the use of this test to accurately detect small HCC tumors is still undetermined.

Several studies have investigated the usefulness of AFP at identifying small HCC tumors. Nomura et al. (19) analyzed 606 Japanese patients with HCC identified either by AFP level or radiological tests. There was no routine screening in this population. They found elevated AFP at the time of diagnosis of HCC in 79%. However, the percentage of patients with elevated AFP was much lower in those with tumors less than 5 cm (60%–72% depending on the cut-off used) than those with tumors greater than 5 cm (84%, see Table 4). Median survival was lower for persons with larger tumors and surprisingly was also lower for those with elevated AFP even when matched for size of tumor.

Lok et al. (20) prospectively screened 290 Chinese patients with chronic HBV infection with AFP every 3–6 months. Patients with elevated levels had the test repeated and if the level was still elevated underwent ultrasound study. Forty-four patients had increased AFP at entry or during the 12 to 50 months (mean 27 months) of follow-up. Of these 44 patients, seven (16%) had HCC. All the patients with HCC had progressively increasing or persistently elevated AFP. Four of seven had potentially resectable tumors. Hepatocellular carcinoma did not develop in any patient with normal AFP, for an overall sensitivity of 100%. However, the sensitivity for small resectable HCC tumors was low because only about half of the tumors were found at a resectable stage. In addition, using a cut-off of 20 ng/mL produced an 84% false-positive rate.

A study by Maringhini et al. (21) analyzed the sensitivity and specificity of AFP and ultrasound in an Italian patient population where there was a clinical suspicion of HCC based on symptoms (pain, fever, weight loss) or elevated liver function tests (alkaline phosphatase). There were 363 patients (75% men), the majority of whom had cirrhosis secondary to HBV infection. All patients received AFP measurements and ultrasound every 4 months. A total of 145 patients were found to have HCC. Of these, 128 had lesions on ultrasound and 114 had AFP level greater than 20 ng/mL. The overall sensitivity and specificity were estimated to be 90% and 93% respectively, for ultrasound and 78% and 78% for AFP. However, the sensitivity of AFP for small HCC tumors (<5 cm, n=56) was only 62%. In patients with a normal ultrasound, the value of AFP was extremely limited, with only 6% being positive. Levels greater than 100 ng/mL (95%–99%) have high specificity, which may justify its use in conjunction with ultrasound. Even though, in this study, the sensitivity of the protocol for detecting HCC was more than 90%, there are several limitations to the data. First, because the patients were symptomatic, this is a study of a diagnostic strategy rather than a screening protocol and hence has limited external validity for the target population of asymptomatic at-risk individuals. Second, this group

Table 4 Sensitivity, Specificity, and Positive Predictive Value of AFP and Ultrasound from Screening Studies

	Italy (28)	Italy (29)	France (30)	Canada (31)	China (18)	Taiwan (25)	Japan (26)	Japan (27)	Overall
Alpha fetoprotein									
Sensitivity									
>15–20	41%	40%	50%	64%	77%	75%	75%	68%	60%
>200		0%	<20%			50%	32%	40%	28%
>1000		0%				38%	15%	13%	16%
Positive predictive value		~42%		9%				71%	
Ultrasound									
Sensitivity	~46%	100%	78%	67%	95%		82%		>80%
	50%		93%	95%					
Specificity									
Positive predictive value			73%	15%					

Blank values indicate the data were not provided. ~ values are estimated based on the data provided.

of symptomatic patients had a predictably higher prevalence of HCC (40%) but less than half of the patients with HCC (38%) were identified at a stage at which they could benefit from therapy (i.e., lesions were smaller than 5 cm). Third, most of the patients were cirrhotic; the positive predictive value of AFP is far lower in noncirrhotic patients because of its frequent elevation in chronic active hepatitis.

In summary, AFP is not a very sensitive test for small preclinical HCC tumors, but high levels are very specific. Given the lower rate of elevated AFP in non-Asian populations and with small HCC lesions, AFP would need to be used together with ultrasound in a group of patients before development of symptoms for screening to be successful. Clearly, improved serum markers for HCC are needed. Until that time, however, the ease, safety, and relatively low cost of the blood test, the lack of need for technical expertise (on which the sensitivity of ultrasound is strongly dependent), and the specificity of high values can justify the inclusion of AFP in a screening study protocol.

B. Ultrasound

Many radiological techniques have been investigated to improve the diagnostic accuracy of AFP screening. Of these, ultrasound has been most frequently used because of its ease of use, low cost, and lack of radiation. Of the noninvasive techniques (ultrasound, computed tomography [CT], and magnetic resonance imaging [MRI]), ultrasound is also the most sensitive (22). Even though invasive techniques (e.g., lipiodol-CT in which contrast is injected directly into the hepatic artery) have higher positive and negative predictive values, they are unfeasible for screening and thus reserved as confirmatory tests.

The best results for ultrasound screening come from Japan where physicians or other specialized technicians perform these studies frequently and where 80%–90% sensitivities have been reported. Shinagawa et al. (23) retrospectively reviewed approximately 400 patients followed up by ultrasound and AFP every 2–3 months (patients with cirrhosis) or every 3–5 months (patients with hepatitis). Over a 5-year period, there were 51 small (<5 cm) HCC lesions found, approximately half of which were tumors smaller than 3 cm. The patients with HCC had a mean age of 56 years. Tumors were detected based on elevated AFP in 22% and by ultrasound in 73%. The sensitivity of ultrasound was 92%, and there were 21 false-positive studies (the specificity cannot be calculated because the number of true-negative results is not stated, but would be approximately 94% if none of the other patients truly had cancer). Twenty-five percent of patients with HCC had normal AFP, and only 33% had levels greater than 200 ng/mL.

A second study looked prospectively at 528 Chinese patients with cirrhosis or chronic HBV infection using AFP and ultrasound screening every 3–6 months

(24). The prevalence of HCC at entry was 3.2% and the incidence during 348 person-years of follow-up was 2 per 100 person-years. Overall, 27 HCC were found, 20 of which were less than 5 cm. Only half of the tumors were thought to be resectable. Of the patients in whom tumors were resected, the survival rate was 67% after limited follow-up (2 months to 2 years). All the patients with lesions larger than 5 cm and two-thirds of the patients who refused surgery died. There were three false-positive ultrasound studies, for a specificity of 99%. Alpha fetoprotein level was normal in 46% of patients with HCC tumors smaller than 5 cm. This study confirmed the sensitivity and specificity of ultrasound and the limitations of AFP screening. It also demonstrated both the promise and the problems with screening. Even though the number of patients was small, those with small HCC tumors tended to have better outcomes. However, tumors were resectable in only half of the patients identified with HCC, and the number of patients who actually benefited from screening was small (at most six of 528).

All of the prior studies have been criticized because small HCC lesions may be missed by ultrasound and thus the sensitivity could be overstated because of short length of follow-up. To truly address the sensitivity and specificity of ultrasound, Dodd et al. prospectively evaluated all patients evaluated for liver transplantation at one center. Two hundred consecutive patients who subsequently underwent transplantation had their explants systematically reviewed for the presence of malignancy. The underlying cause of liver disease is not reported. Pathological study revealed HCC in 28 patients. Twelve of the 28 had preoperative ultrasound revealing tumor for a sensitivity of 43%. The majority of patients with HCC had normal AFP levels. Of the missed lesions, 60% were 1 or smaller cm and only 20% were larger than 3 cm. This study cannot be extrapolated to a screening program because it is based on one examination at one time point. In addition, the sensitivity may have been lowered by the fact that none of the patients had received transplants, hence the advanced cirrhosis made ultrasound detection more difficult. Also, unlike many studies in Asia, screening tests were frequently performed by technicians rather than physicians with extensive experience in this technique. Because the study sample consisted of candidates for transplantation who would not be able to tolerate resection because of advanced cirrhosis, they are not comparable to the population that would be selected for routine screening. Last, it is unclear whether the low sensitivity is clinically relevant in that most of the missed tumors were less than or equal to 1 cm and possibly could have been picked up on the following screening examination while still treatable. However, the lower sensitivity of ultrasound in this and other non-Asian studies (see Table 4) raises questions about the usefulness of screening programs in the United States and Europe.

IV. STUDIES OF SCREENING PROGRAMS

There have been several studies of screening for HCC yielding conflicting results (summarized in Tables 4 and 6). All studies of screening are difficult because of the possibilities of lead time bias, in which screening identifies a disease at an earlier stage without affecting overall outcome, and length bias, in which the patients with disease identified by screening have a fundamentally different and more indolent course than those who are not found by screening. Both of these biases tend to overestimate the utility of a screening program and are difficult to evaluate without control groups. In addition, interpretation of the published studies of HCC screening is limited by differing populations, nonrandomized design, short follow-up, and inconsistent therapies.

A. Screening Studies in Asia

Liaw et al. (25) prospectively followed up 432 Chinese patients with chronic hepatitis resulting from HBV infection and 105 patients with non-A, non-B hepatitis every 3–6 months with AFP and ultrasound. During follow-up, new asymptomatic cases of HCC were identified in eight HBV patients and one non-A, non-B hepatitis patient. This represents an incidence of 2.8% per year in patients in the HBV cohort who were older than 35 years old at entry. The relative risk for HBV-positive patients was 5 for all patients older than 35 years. Six of eight cases of HCC were identified initially by an elevated AFP and two by ultrasound with normal AFP. Six patients underwent successful resection and pathological study revealed HCC lesions smaller than 3 cm in 67% and liver cirrhosis in 50%. Seven of 509 ultrasound studies had false-positive results and there was one false-negative result, yielding a sensitivity and specificity for ultrasound of 89% and 99%. The positive predictive value was 53%. One-fourth of patients with HCC had a normal AFP level. The authors calculate a sensitivity of their overall screening strategy was 100% but the false-negative rate could not be accurately assessed because of a lack of follow-up. The lack of follow-up also does not allow calculation of either the recurrence or cure rate of HCC found on screening. The overall success and high rate of respectability in the screening program may be because the patients were, for the most part, young and noncirrhotic. This increases the sensitivity of ultrasound and the ability of patients to undergo resection, but it increases the cost of screening because there is a longer time between initiation of screening and detection of HCC. Overall, only 1.4% of all ultrasound examinations were truly positive. The fact that the patients were screened early in their disease course is also supported by the low incidence (2.8% per year) compared with other studies and the fact that the overall incidence in the study (0.8% per year) was much lower than that seen in the subgroup of patients older than 35 years. However, the sensitivity may not be as high as stated in that the denominator of all patients with HCC cannot be determined without longer follow-up because there is no gold standard for HCC in the short term.

Oka et al. (26) followed up 140 Japanese patients with cirrhosis prospectively for a mean of 41 months with AFP measurements made every 2 months and ultrasound every 3 months. The mean age of the patients was 55 years and the patients were evenly divided among HBV infection, transfusion-related cirrhosis (presumably HCV), and alcohol consumption as the cause of their liver disease. Hepatocellular carcinoma was diagnosed in 40 patients (29%) over the 6-year period, with annual incidences between 5% and 11%. Sixty-two percent of HCC lesions were smaller than 2 cm. The sensitivity of ultrasound was 82% and of AFP was 75%. The incidence of HCC was highest in patients with HBV infection and lowest in patients with alcoholic liver disease. Most patients (83%) underwent treatment, seven with resection and the remainder with embolization or ethanol injection (see Section V). Overall survival was 13 of 40 at the end of the study, with 1-and 3-year survival rates of 72% and 41%, respectively. Only 43% of patients with HCC were Child's class A and thus eligible for resection. Among patients with Child's class C cirrhosis, the survival was predictably lower, 56% and 11%, respectively, at 1 and 3 years. Doubling time for untreated tumors was highly variable, ranging from 15.6 to 313 days. In their discussion, Oka et al. stress the high survival rate and attribute it to the effects of early diagnosis and therapy. Given the high percentage of Child's class C cirrhosis and HCC in their study population, the survival rates appear excellent; however, survival rate in the 100 patients without HCC is not available for comparison. It is also difficult to exclude lead time or length bias as a contributor to the survival rates. In addition, the very aggressive screening protocol (every 2–3 months) greatly increases the costs associated with screening.

Kobayashi et al. (27) prospectively screened 95 Japanese patients with documented cirrhosis with AFP every 2 months, with ultrasound every 3 months, and with angiography and CT every year. All patients were older than 40 years and were followed up for a mean of 4.2 years. Patients had liver disease based on HBV infection in 35% and alcohol in consumption in 22%. The remainder of cases were most likely secondary to HCV infection but antibody testing was not yet available at the time of the study. Ultrasound discovered eight lesions of which six were HCC. There were two false-negative results for ultrasound, yielding a sensitivity of 75% and a specificity of 98%. The AFP level was normal in three of eight patients with HCC. All tumors were less than 5 cm. Only two patients underwent successful resection and the length of follow-up was only 3–6 months. The mortality rate among unresected patients was 50% at approximately 1 year of follow-up. Thus, despite a fairly high prevalence of HCC (8.4%) and early detection (all tumors less than 5 cm), only 25% of HCC detected and 2% of all patients screened benefited from the screening program and were able to undergo successful resection.

B. Screening Studies in Europe and North America

Colombo et al. (28) followed up a cohort of 447 patients with early cirrhosis (Child's class C patients were excluded because they were unlikely to benefit from therapy) for up to 4 years to determine the prevalence, incidence, and natural history of HCC as well as the outcomes of screening. Patients were all older than 35 years with a mean age of 55 years. Forty-five percent had cirrhosis from hepatitis C, 16% from hepatitis B, 14% from alcohol consumption, and 24% from more than one cause. At entry, 86 patients had a positive screening test (AFP and/or ultrasound) of which 30 were found to have HCC for an overall prevalence of 6.7%. Tumors had a median size of 4 cm. Screening was performed annually with ultrasound and AFP. During follow-up, 29 additional HCC lesions were found for an annual incidence of 2–3 per 100 patients. The risk was lower in patients with normal AFP level at baseline; the relative risk of HCC in patients with persistently elevated AFP level was 14 and with fluctuating AFP levels was 6 when compared with persistently normal AFP level. However, 21% of patients with HCC had persistently normal AFP levels. Sixty percent of the tumors were smaller than 3 cm, but only 30% were potentially resectable. Resectable tumors were more likely to be found on entry than on follow-up screening (43% of tumors found versus 14%, $P=0.027$). Among the patients who underwent surgery, 1-year survival rate was 67%; however, recurrence was seen in 60%, and long-term survival data are not available. Among patients who refused resection, tumor size doubling time was 6 months, and one of seven patients was alive at 2 years. Thus, even though the screening program was successful at identifying small cancers, only half of those small cancers were resectable and the recurrence rate following surgery was high, making it unlikely that any long-term benefits from screening would be realized. One weakness of the study was that it used a 1-year interval between screening examinations, which would allow tumors to double two to three times between examinations. This could bias screening toward finding more unresectable tumors without decreasing the sensitivity of the screening program. Another unusual finding is that, even though tumors found on follow-up (''incident'' HCC) were of smaller size than those found at entry (''prevalent'' HCC), fewer were resectable, which is at odds with most of the available data.

Cottone et al. (29) prospectively followed up 147 Italian patients with Child's class A cirrhosis and no evidence of HCC at entry with biannual ultrasound and AFP for 8 years. All patients were older than age 40 and the cause of liver disease was hepatitis C infection in 71%. Follow-up was longer than in the other studies (mean 65 months). Survival and cancer-free survival rates were calculated by the Kaplan-Meier method and adjusted for age, sex, hepatitis B and C status, and alcohol use. The researchers found 30 cases of HCC in 8 years. The annual incidence increased from 1.5%–3% in the beginning of the study to 10% per year at the end. Small tumors (<5 cm) were found in 83% of patients. Although the overall sensitivity of AFP was only 40%, the odds ratio of having HCC for patients with AFP between 50 and 400 at entry was 3.6 when compared with values less than 40. Overall survival rate tended to be lower in patients with HCC than in those without cancer, 46% compared with 62%, but the difference was not statistically significant. Of 10 patients who underwent surgery or ethanol injection, eight were alive at the end of the study, and no deaths were attributable to therapy. Therefore, even though the screening program identified a large number of tumors at an early stage, only 20% of those with cancer underwent successful treatment. Cottone et al. point out that ethanol injection was introduced late in the study and that potentially another 12 patients might have benefited. This latter point is a shortcoming of all these studies; rarely do they include treatment advances that might influence the efficacy of treatment of tumors found by screening. For example, in none of these studies have any patients undergone liver transplantation, which could theoretically have improved survival rate in both the patients with and without HCC. The availability of treatment advances, although it would not affect the sensitivity and specificity of screening, could change the proportion of patients who could benefit from therapy and thus further clouds interpretation, because it is increased survival among those screened which represents the true efficacy of any screening program.

Pateron et al. (30) prospectively followed up 118 French patients with predominantly alcoholic cirrhosis of Child's class A or B with ultrasound and AFP every 6 months for a median of 3 years. Patients with abnormal screening tests at baseline were excluded. The mean age was 55 years. Hepatocellular carcinoma was found in 14 patients with an annual incidence of 5%–7%. The patients with HCC had a higher mean age and tended to have a greater prevalence of hepatitis C infection. Only 25% had tumors less than or equal to 3 cm. False-positive results were seen with ultrasound in four cases and with AFP in 14. The sensitivity and specificity was 78%, and 93%, respectively, for ultrasound and 50% and 86% for AFP (see Table 4). Surgery was attempted in only one patient, with tumor recurrence in 15 months. Overall survival rate was 14% at approximately 2 years. Thus, in this population, screening was relatively ineffective at detecting small, treatable tumors.

The only prospective study in a North American population was carried out by Sherman et al. (31) who followed up 1069 chronic carriers of hepatitis B virus. Most of their patients (71%) were of Asian descent, and the mean age was only 39 years. As might have been expected from the young age of the patients, only a minority (4%) had cirrhosis. Patients were randomized to be screened at 6-month intervals by AFP or by AFP plus ultrasound. There were 254 withdrawals during the course of the study. The mean follow-up period was 26 months, for a total of 2340 person-years of observation. At entry, the prevalence of HCC was 281/100,000 (95% CI 36 to 598/100,000). The annual incidence for the remainder of the study was 470/100,000 (95% CI 192 to 747/100,000). The racial distribution of patients diagnosed with HCC was comparable to the overall study population. Half of the patients had resectable lesions. The overall survival rate among those resected was 83% at the end of the study. However, the range of follow-up following diagnosis was only 2 to 27 months and there was one recurrent case of HCC. There were 11 deaths in the overall cohort, five from HCC, for a disease-specific mortality rate of 214/100,000 (95% CI 27 to 401/100,000).

In their study, Sherman et al. (31) attempted to determine the sensitivity and specificity of the screening tests. The sensitivity and specificity of AFP was 64% and 91%; however, 91% of patients with elevated AFP levels did not have HCC. The negative and positive predictive values were 99.5% and 9%, respectively. In the 538 patients randomized to ultrasound in addition to AFP, 2150 examinations discovered 24 new focal lesions, of which four were HCC. In these patients, three of four had normal AFP level. There were two patients with HCC and normal ultrasound for an overall sensitivity and specificity of 67% and 95%, respectively, and a positive predictive value of 15%.

The strengths of this study were the large sample size and the fact that subjects were recruited through publicity about the program as well as through referrals. In addition, it allowed comparison of two strategies of screening in a randomized manner. However, there was not a control group that received no screening. The authors do not separate whether survival or detection of HCC differed in the two screening strategies, although any differences would be difficult to assess because of the small number of cases. Only five patients (0.5%) potentially benefited from the screening strategy, raising questions about cost effectiveness. In addition, the success rate of surgery may be overstated because of the short length of follow-up. Given that the patients did not have advanced disease at entry, it is surprising that the incidence rates were as high as they were, comparable to rates seen in other countries, 387 to 527/100,000 (28,29,32–34). However, despite the early stage of liver disease, very few patients could undergo resection for cure. The rate of withdrawals was very high, with almost one-fourth of the patients' outcomes being censored during the study. This is a potential source of bias whose impact cannot be assessed.

A recent study from Italy raised the possibility that certain genotypes of HCV, specifically type 1b, may have a higher risk of development of HCC, making this group of patients more attractive as candidates for screening (11). This study represented a retrospective analysis of a prospective program of screening in patients with cirrhosis. Of 240 patients who were HCV positive in the study, 163 without hepatitis B, other causes of liver disease, or suspicion of HCC at entry were followed up with AFP and ultrasound every 6 months. Ninety percent of the patients were Child's class A, 62% had genotype 1b, and the remainder predominantly were genotype 2 a/c (30%). Fifty-one percent were given interferon. The authors state that there were no differences in genotype among those given interferon, but these patients were more likely to be Child's class A. Over a follow-up period of 5–7 years (median 68 months), HCC developed in 22 patients (13.5%) for an annual incidence of 2.5%. Of these, 16 had unifocal HCC with a mean diameter of 2 cm. The AFP level was normal in 21 patients and was greater than 400 ng/mL in only one patient, once again highlighting the limitation of the AFP assay. Fourteen patients, or 64%, were eligible for resection, although only four underwent resection with one death, the cause of which was not reported. Two patients underwent transplantation, and five received ethanol injection. Overall, only seven of 22 patients survived to the end of study. In the analysis of risk factors for HCC, genotype 1b, male sex, and older age were associated with the development of HCC in both the univariate and multivariate analysis. Alcohol use was significantly associated with HCC in the univariate analysis and had an odds ratio of 2.31 in the multivariate analysis, but the confidence interval included unity (95% CI 0.93 to 5.74). Genotype 1b was the single largest independent risk factor with an odds ratio of 6.14 (95% CI 1.77 to 21.37). Interferon therapy was not associated with a decreased rate of HCC in the multivariate analysis but was in the univariate analysis. This study raises the question as to whether screening should be more focused in higher risk groups to improve its cost effectiveness. Evidence that such groups may exist is exciting yet preliminary. There are several possible alternative explanations. First, genotype may be a reflection of duration of infection or other cohort effects in that the epidemiology of HCV may have changed over time. Although there was no difference in mean age between groups of patients of different genotypes, this does not exclude secular effects over time. Second, genotype is clearly related to route of infection, and this may be related to risk of HCC. The poor response of HCV 1b to interferon may also confound the relationship, if interferon is truly a protective factor for the development of HCC. Third, the small number of patients of genotypes other than 2, for the most part, limits the analysis to genotype 1b and 2 a/c. Nevertheless, it does appear that genotype 1b may represent a higher risk group that might benefit from screening if this is confirmed in other studies. However, the small number of patients who benefited (at most seven of the 22 with HCC or 4.2% of the entire study population) continues to raise questions about the cost effectiveness of the current screening strategies.

In summary, eight prospective studies have been carried out yielding data that are difficult to synthesize in any meaningful way. The study populations, screening protocols, and treatment strategies were very different. In addition, the short length of follow-up made it difficult to assess outcome, because it is tumor cure rather than detection that is the goal of screening. As with all non-randomized screening studies, it is difficult to exclude lead time and length bias as factors in the results. Several features from these studies do stand out, however. Ultrasound is consistently the most sensitive and specific screening test and should be included in any proposed screening program. The AFP measurement is less sensitive, particularly in non-Asian populations and for small HCC tumors (see Tables 4 and 5), but it can probably be justified in a screening program because of its lower cost (approximately 10% of the cost of ultrasound), high specificity (if a high cut-off is used), and the fact that the combination of AFP and ultrasound approaches 100% sensitivity. What is less clear from these studies is what proportion of the people screened will benefit. In these studies, from 7%–80% of HCC tumors were deemed resectable, and both recurrence and survival rates were highly variable (Table 6). Most of the high rates of success come from Asia where very aggressive surgical and nonsurgical approaches to HCC are the standard. With new treatment advances and increased experience in the United States, the proportion of patients who benefit may change. Thus, it is unclear whether the lack of success is due to insensitivity of the current screening strategy for small, treatable HCC tumors or to inadequate therapies for the preclinical stage of HCC.

V. TREATMENT OPTIONS

For any screening strategy to be effective, the available therapies need to be better when applied in the preclinical stage. Although advanced HCC is uniformly untreatable with poor prognosis, there is no evidence that early diagnosis improves outcome.

Table 5 Correlation Between AFP and Tumor Size in 606 Japanese Patients

	AFP Level			
Tumor size	<20	21–200	201–1000	>1000
≤2 cm	40%	32%	21%	6%
2–3 cm	24%	50%	18%	7%
3–5 cm	28%	35%	20%	17%
>5 cm	16%	20%	14%	49%

Source: Adapted from Ref. 19.

Table 6 Outcomes of Screening for Hepatocellular Carcinoma in Selected Studies

	Italy (26)	Italy (29)	France (30)	Canada (31)	China (18)	Taiwan (25)	Japan (26)	Japan (27)
Total with HCC	56	30	14	14	391	8	40	8
Lesion <3 cm	23%	53%	21%			75%	70%	50%
Lesion <5 cm	30%	83%			61%	88%	85%	100%
Multifocal/metastases	48%	13%	35%	50%	~20%		13%	13%
Resectable tumor	21%	7%	8%	43%	80%	75%	18%	25%
Recurrence rate	60%		100%	16%	>14%			
5-year survival rate			0%		60%			

Blank values indicate the data are not provided; ~ values are estimated based on the data provided.

A. Surgical Resection

There are limited data from Asia to suggest that postoperative screening and early and aggressive surgical management improve survival rate among patients who have undergone resection for HCC. Izumi et al. (35) retrospectively reviewed 104 patients undergoing surgical resection over a 14-year period in Japan. The mean age was 61 years. All patients were followed up postoperatively with ultrasound, CT, and AFP every 2–3 months. Even though most (81%) of the patients had cirrhosis at the time of surgery, perioperative mortality rate was reasonably low (7.1%). Five-and 7-year survival was 59% and 41%, although disease-free survival rate was significantly less, 20% and 13%, respectively. Cox regression analysis revealed tumor size to be a significant predictor of survival, with patients with tumors larger than 5 cm having significantly lower survival rate than patients with tumors smaller than 5 cm ($P<0.05$). There was no difference between patients with tumors smaller than 2 cm and those with lesions between 2 and 5 cm, supporting a possibility that screening does not need to identify very small tumors in order to be successful. Recurrence rates did not differ significantly among patients with and without cirrhosis or based on size of the tumor among those patients whose tumors could be resected. The lack of impact of size of tumor on recurrence rate may be due to insufficient power given the high overall recurrence rate of 80%. The 5-year survival in this study is higher than in many prior studies, which have ranged from 25%–39% (36–38). This may be due to the fact that eight patients who died within 30 days were excluded from analysis and that aggressive therapy for recurrence with repeat surgery, embolization, or ethanol injection was used. However, these results and the 60% 5-year survival rate seen in the Tang et al. (18) study with small tumors found on screening (see Table 4) is somewhat encouraging compared to the 25% 3-year survival rate seen in patients with tumors smaller than 3 cm who do not receive treatment. A final study of 62 asymptomatic cases found on screening yielded 2-and 4-year survival rates of 44% and 34%, respectively, compared with 4-year survival of 8% in a group of 47 patients with symptomatic HCC (39). Most studies, however, have been limited by short follow-up and inability to assess lead time and length biases. In addition, studies in Europe have, in general, yielded much less encouraging results with recurrence rates as high as 100% and survival rates as low as 0% at 5 years (see Table 4) (30).

B. Tumor Ablation

In addition to resection, there are several new therapies that may have a role in adjuvant or palliative therapy for HCC, particularly in patients with small lesions. Percutaneous intratumor injection of ethanol under ultrasound guidance causes immediate tissue necrosis and can be used for lesions up to 3 cm in size. Small, uncontrolled studies have suggested up to 60% survival rate, although follow-up has been short (40). Whereas systemic chemotherapy has limited efficacy in HCC, directed chemotherapy via a catheter in the hepatic artery has some efficacy. This strategy is predicated on the fact that HCC lesions derive their blood flow via branches of the hepatic artery and the remainder of the liver is supplied predominantly by the portal vein. Targeted chemotherapy can either use lipiodol, an ether ester of poppy seed oil, which is selectively retained in the tumor, or embolization with Gelfoam (chemoembolization) to retain chemotherapeutic agents within the tumor. Similar to resection, both strategies require adequate hepatic reserve. Lipiodol chemotherapy is reported to result in 5-year survival rates of 70% when combined with subsequent resection in small HCC (41) and 1-year survival rates of 78% have been reported with chemoembolization (40).

C. Liver Transplantation

Lastly, there is one therapy, liver transplantation, which can be applied to patients with advanced cirrhosis who are at the highest risk of HCC. Liver transplantation has over the past decade become the treatment of choice for most forms of decompensated cirrhosis. Although initial results with liver transplantation for HCC were disappointing, recent reports of transplantation for carefully selected patients with small HCC tumors have been encouraging. One center prospectively evaluated liver transplantation in conjunction with preoperative chemoembolization for patients without hepatitis B and HCC less than 5 cm in diameter. Of 15 patients with HCC, 11 patients were able to undergo transplantation (metastases developed in three who died before an organ became available and one was able to have the tumor resected after chemoembolization) (42). Survival rate among those transplanted was 91% at a median of 40 months; overall survival rate (including those not transplanted) was 73%. This is an encouraging treatment for HCC, because it does not require adequate liver function. However, this study excluded patients with HBV cirrhosis, so it cannot necessarily be generalized to a large portion of the population at risk. In addition, liver transplantation is limited by the number of donor organs available and the long waiting times. Therefore, HCC screening can only be used to exclude patients whose tumors are larger than 5 cm before transplantation, because there is no added priority or mechanism to expedite transplantation in patients who are found to have HCC on screening. Appearance of HCC in a previously well-compensated patient with cirrhosis can be viewed as an indication for transplantation.

Thus, data do not support a clear survival benefit for patients with HCC discovered at a preclinical stage, particularly in North America. However, there is some evidence that smaller tumors can do well with aggressive therapy and resection. In addition, new treatment options including chemoembolization and liver transplantation may further improve survival, particularly in small tumors, and increase the utility of HCC screening. Given the improved therapies for many of the other complications of cirrhosis, including gastrointestinal bleeding and ascites, the life expectancy of patients with cirrhosis is increasing. As a result, an increased incidence of HCC can be expected in the United States. The efficacy of therapies for HCC must be critically evaluated as an important step in determining whether screening for HCC is a beneficial use of limited health care resources.

VI. RISKS OF SCREENING

Positive screening tests using ultrasound or AFP are followed with a CT or MRI, and a fine needle aspiration (FNA) if a mass is identified. The risks of FNA are small (<1% risk of bleeding). Negative or equivocal FNA findings may be followed by a biopsy, either percutaneously or laparoscopically, which carries higher risk. In addition, there is the possible risk of dye-induced renal failure from CT contrast as well as the anxiety caused by false-positive results. There are psychological costs to all patients with testing every 3 months and the uncertainty associated with each test. Finally, there is the psychological cost of finding a tumor that may not be amenable to therapy, which is not infrequent. However, the largest barrier to a screening program is the high cost of screening, which escalates markedly with the follow-up of positive screening tests.

VII. COSTS OF SCREENING

If this screening strategy is adopted, what will be the costs involved? Given a moderately high estimate of the incidence of HCC of 5% per year, there would be 50 new cases per 1000 people screened per year (Figure 2). If the overall sensitivity of screening is 80%, 40 cases would be found. With three AFP determinations (~$70) and two ultrasound examinations (~$600), the costs for screening would be approximately $1,401,000 for 1 year of screening 1000 at-risk individuals. With the current rate of resectable tumors being ~50% and with a 5-year survival rate of 60%, the cost per patient "cured" would be $117,000. These figures reflect the cost of screening only; the costs of CT scans (~$800) or fine needle aspirates (~$1800) to confirm the diagnosis, and the costs of investigating false-positive test results would greatly increase the cost. Even in the United States, where the prevalence of HBV infection is less than 1%, a screening program would need to consider between 200,000 and 1,000,000 people, and this number would markedly increase if patients with HCV (U.S. prevalence ~1%) or other causes of cirrhosis were screened. It seems

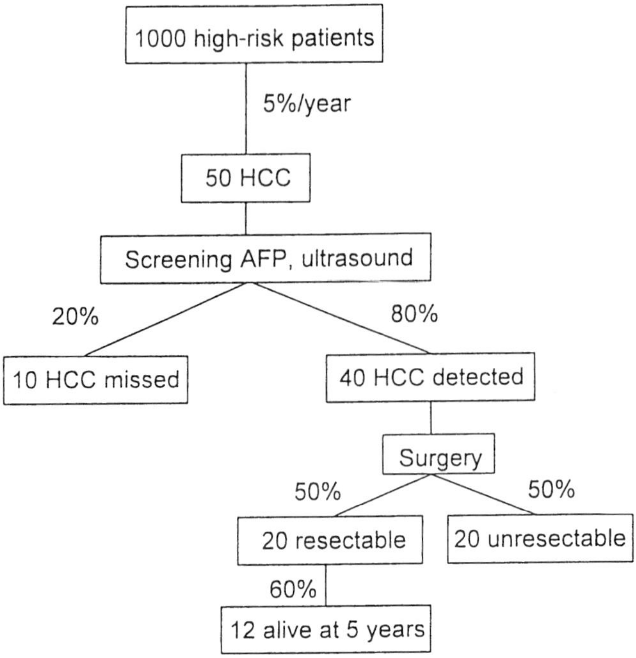

Figure 2 Anticipated results of HCC screening.

likely that, for a screening strategy to be cost effective and feasible, it needs to be limited to those at highest risk (e.g., patients with established cirrhosis or perinatally acquired disease) and to exclude those unable to undergo therapy (e.g., Child's C cirrhosis in a patient who is not a candidate for transplantation).

VIII. PATIENT EDUCATION

Patients with liver disease should be informed about the risk of HCC, the risks and benefits of screening, and the lack of demonstrated efficacy at improving survival. The screening process, in particular the need for invasive tests to follow-up lesions seen on ultrasound, needs to be explained. The treatment options and the risks of those treatments, including resection and chemoembolization, should be discussed along with the absence of data that therapy can prolong survival. If after extensive discussion highly motivated patients, particularly those of Asian or African descent, those with perinatal HBV or HCV, or those with hemochromatosis desire screening, they should be offered it (Table 7).

Table 7 Patient Education Regarding Utility of Screening

Factor	Favors screening	Favors no screening
Demographic/clinical factors		
Male	√	
Age older than 45 years	√	
Motivation/compliance	√	
Cirrhosis	√	
Etiologic factors		
HBV infection	√	
Perinatally acquired	√	
Acquired as adult		√
Asian, African	√	
HCV infection	√	
Genotype 1a/1b	√	
Other genotypes		√
Hereditary hemochromatosis	√	
Alcohol abuse		√
Other causes of liver disease		√

IX. RECOMMENDATIONS FROM PROFESSIONAL ORGANIZATIONS

Despite the limitations in the data on screening, a National Institutes of Health (NIH) consensus conference recommended screening for HCC in patients with chronic HBV infection (2,3). Patients at high risk (particularly those with cirrhosis, men, non-whites, and those with perinatally acquired infection) are recommended to have AFP screening every 3–4 months and ultrasound every 4–6 months, and, for those at moderate risk, a similar strategy except ultrasound is done annually. No recommendations have been made for those with HCV infection or other causes of cirrhosis (Table 1). The American Association for the Study of Liver Diseases, a professional organization of liver researchers and clinicians, are formulating a position statement on the topic, which likely will take a more moderate stance given the limitations of the data. All parties agree that further research is needed to define the best approach to screening for this deadly complication of chronic liver disease.

X. CONCLUSION

Hepatocellular carcinoma presents a difficult clinical problem because of its aggressive nature, absence of early symptoms, and poor prognosis. In addition, studies of HCC have been frustrated by different study populations with widely different prevalences and outcomes, different screening strategies and therapies used, and short periods of follow-up. Clearly, HCC is a tumor that might benefit from an accurate screening strategy if there were successful therapies available. However, difficulties with early diagnosis and inability to tolerate surgery in patients with more advanced liver disease when HCC is more common have frustrated most attempts.

Based on current data, it is difficult to accept recommendations for routine population-based screening in the United States until better screening tests and therapies are developed and tested. Although there are limited data that support a role for screening and early therapy in prolonging survival, the studies are flawed and thus a high level of certainty cannot be placed on the conclusions. However, given the recommendations of professional organizations and the possibility of survival benefit, patients with liver disease should be informed about the risk of HCC, the risks and benefits of screening, and the lack of demonstrated efficacy at improving survival. Highly motivated patients, particularly those of Asian or African descent, those with perinatal HBV or HCV infection, or those with hemochromatosis who desire screening could be offered it. Finally, further studies should endeavor to determine the long-term effects of screening on survival in different populations using newer and improved treatment modalities, including liver transplantation.

REFERENCES

1. London WT. Primary hepatocellular carcinoma—etiology, pathogenesis, and prevention. Hum Pathol 1981; 12:1085–97.
2. Di Bisceglie AM, Rustgi VK, Hoofnagle JH, Dusheiko GM, Lotze MT. NIH conference. Hepatocellular carcinoma. Ann Intern Med 1988; 108:390–401.
3. McMahon BJ, London T. Workshop on screening for hepatocellular carcinoma. J Natl Cancer Inst 1991; 83:916–919.
4. Okuda K, Nakashima T, Kojiro M, Kondo Y, Wada K. Hepatocellular carcinoma without cirrhosis in Japanese patients. Gastroenterology 1989; 97:140–146.
5. Regan LS. Screening for hepatocellular carcinoma in high-risk individuals. A clinical review. Arch Intern Med 1989; 149:1741–1744.
6. Croce LS, Bercich L, Sasso F, Masutti F, Molino R, Tiribelli C. Strategies in the study of hepatocellular carcinoma worldwide. J Surg Oncol Suppl 1993;3:37–39.
7. Beasley RP. Hepatitis B virus. The major etiology of hepatocellular carcinoma. Cancer 1988;61:1942–1956.
8. Alter HJ. To C or not to C: these are the questions. Blood 1995; 85:1681–1695.
9. Ikeda K, Saitoh S, Koida I, et al. A multivariate analysis of risk factors for hepatocellular carcinogenesis: a prospective observation of 795 patients with viral and alcoholic cirrhosis. Hepatology 1993; 18:47–53.

10. Tsukuma H, Hiyama T, Tanaka S, et al. Risk factors for hepatocellular carcinoma among patients with chronic liver disease [see comments]. N Engl J Med 1993; 328:1797–1801.
11. Bruno S, Silini E, Crosignani A, et al. Hepatitis C virus genotypes and risk of hepatocellular carcinoma in cirrhosis. Hepatology 1997; 25:754–8.
12. Bradbear RA, Bain C, Siskind V, et al. Cohort study of internal malignancy in genetic hemochromatosis and other chronic nonalcoholic liver diseases. J Natl Cancer Inst 1985; 75:81–84.
13. Eriksson S, Carlson J, Velez R. Risk of cirrhosis and primary liver cancer in alpha 1-antitrypsin deficiency. N Engl J Med 1986; 314:736–739.
14. Van Rensburg SJ, Cook-Mozaffari P, Van Schalkwyk DJ, Van der Watt JJ, Vincent TJ, Purchase IF. Hepatocellular carcinoma and dietary aflatoxin in Mozambique and Transkei. Br J Cancer 1985; 51:713–72b.
15. Beasley RP, Hwang LY, Lin CC, Chien CS. Hepatocellular carcinoma and hepatitis B virus. A prospective study of 22 707 men in Taiwan. Lancet 1981; 2:1129–1133.
16. Imberti D, Fornari F, Sbolli G, Buscarini E, Squassante L, Buscarini L. Hepatocellular carcinoma in liver cirrhosis. A prospective study. Scand J Gastroenterol 1993; 28:540–544.
17. Oka H, Tamori A, Kuroki T, Kobayashi K, Yamamoto S. Prospective study of alpha-fetoprotein in cirrhotic patients monitored for development of hepatocellular carcinoma. Hepatology 1994; 19:61–66.
18. Tang ZY, Yu YQ, Zhou XD, Yang BH, Ma ZC, Lin ZY. Subclinical hepatocellular carcinoma: an analysis of 391 patients. J Surg Oncol Suppl 1993; 3:55–58.
19. Nomura F, Ohnishi K, Tanabe Y. Clinical features and prognosis of hepatocellular carcinoma with reference to serum alpha-fetoprotein levels. Analysis of 606 patients. Cancer 1989;64:1700–1707.
20. Lok AS, Lai CL. Alpha-fetoprotein monitoring in Chinese patients with chronic hepatitis B virus infection: role in the early detection of hepatocellular carcinoma [see comments]. Hepatology 1989;9:110–115.
21. Maringhini A, Cottone M, Sciarrino E, et al. Ultrasonography and alpha-fetoprotein in diagnosis of hepatocellular carcinoma in cirrhosis. Dig Dis Sci 1988; 33:47–51.
22. De Santis M, Romagnoli R, Christani A, et al. MRI of small hepatocellular carcinoma: comparison with US, CT, DSA, and lipiodol-CT. J Comput Assist Tomogr 1992; 16:189–197.
23. Shinagawa T, Ohto M, Kimura K, et al. Diagnosis and clinical features of small hepatocellular carcinoma with emphasis on the utility of real-time ultrasonography. A study in 51 patients. Gastroenterology 1984; 86:495–502.
24. Sheu JC, Sung JL, Chen DS, et al. Early detection of hepatocellular carcinoma by real-time ultrasonography. A prospective study. Cancer 1985; 56:660–666.
25. Liaw YF, Tai DI, Chu CM, et al. Early detection of hepatocellular carcinoma in patients with chronic type B hepatitis. A prospective study. Gastroenterology 1986; 90:263–267.
26. Oka H, Kurioka N, Kim K, et al. Prospective study of early detection of hepatocellular carcinoma in patients with cirrhosis. Hepatology 1990; 12:680–687.

27. Kobayashi K, Sugimoto T, Makino H, et al. Screening methods for early detection of hepatocellular carcinoma. Hepatology 1985; 5:1100–1105.
28. Colombo M, de Franchis R, Del Ninno E, et al. Hepatocellular carcinoma in Italian patients with cirrhosis [see comments]. N Engl J Med 1991; 325:675–680.
29. Cottone M, Turri M, Caltagirone M, et al. Screening for hepatocellular carcinoma in patients with Child's A cirrhosis: an 8-year prospective study by ultrasound and alphafetoprotein. J Hepatol 1994; 21:1029–1034.
30. Pateron D, Ganne N, Trinchet JC, et al. Prospective study of screening for hepatocellular carcinoma in Caucasian patients with cirrhosis. J Hepatol 1994; 20:65–71.
31. Sherman M, Peltekian KM, Lee C. Screening for hepatocellular carcinoma in chronic carriers of hepatitis B virus: incidence and prevalence of hepatocellular carcinoma in a North American urban population. Hepatology 1995; 22:432–438.
32. Heyward WL, Lanier AP, McMahon BJ, Fitzgerald MA, Kilkenny S, Paprocki TR. Early detection of primary hepatocellular carcinoma. Screening for primary hepatocellular carcinoma among persons infected with hepatitis B virus. JAMA 1985; 254:3052–3054.
33. Hsu HC, Wu TT, Wu MZ, Sheu JC, Lee CS, Chen DS. Tumor invasiveness and prognosis in resected hepatocellular carcinoma. Clinical and pathogenetic implications. Cancer 1988; 61:2095–2099.
34. Liver Cancer Study Group of Japan. Primary liver cancer in Japan. Clinicopathologic features and results of surgical treatment. Ann Surg 1990; 211:277–287.
35. Izumi R, Shimizu K, Ii T, et al. Prognostic factors of hepatocellular carcinoma in patients undergoing hepatic resection. Gastroenterology 1994; 106:720–727.
36. Tsuzuki T, Sugioka A, Ueda M, et al. Hepatic resection for hepatocellular carcinoma. Surgery 1990; 107:511–520.
37. Nagao T, Inoue S, Goto S, et al. Hepatic resection for hepatocellular carcinoma. Clinical features and long-term prognosis. Ann Surg 1987; 205:33–40.
38. Chen MF, Hwang TL, Jeng LB, Jan YY, Wang CS, Chou FF. Hepatic resection in 120 patients with hepatocellular carcinoma. Arch Surg 1989; 124:1025–1028.
39. Lee CS, Sung JL, Hwang LY, et al. Surgical treatment of 109 patients with symptomatic and asymptomatic hepatocellular carcinoma. Surgery 1986; 99:481–490.
40. Dusheiko GM, Hobbs KE, Dick R, Burroughs AK. Treatment of small hepatocellular carcinomas. Lancet 1992; 340:285–288.
41. Kanematsu T, Furuta T, Takenaka K, et al. A 5-year experience of lipiodolization: selective regional chemotherapy for 200 patients with hepatocellular carcinoma. Hepatology 1989; 10:98–102.
42. Venook AP, Ferrell LD, Roberts JP, et al. Liver transplantation for hepatocellular carcinoma: results with preoperative chemoembolization. Liver Transplantation and Surgery 1995; 1:242–248.

13

Lung Cancer

William C. Black
Dartmouth-Hitchcock Medical Center, Lebanon, and Dartmouth Medical School, Hanover, New Hampshire

I. BACKGROUND

Despite early optimism for screening with chest radiography and sputum cytology, several randomized controlled trials of screening have failed to demonstrate a reduction in lung cancer mortality rate. Consequently, no medical organization in the United States recommends lung cancer screening for the general population of smokers (1).

However, there are several reasons why it may be useful to critically review previous studies of lung cancer screening and to seriously consider its prospects for the future. First, lung cancer is the leading cause of cancer-related death in both men and women (2). Second, previous studies do not exclude the possibility that screening with chest radiography can reduce lung cancer mortality rate by 10% to 20%. Third, a large proportion of community physicians recommend screening with chest radiography every 1 to 2 years for patients without known risk factors for lung cancer even though no medical organization does so (3). Fourth, there have been dramatic advances in diagnostic imaging and the molecular detection of cancer that could play a future role in screening. Finally, a critical review of the past and serious consideration of new technology may help determine which future screening strategies, if any, are worthwhile.

This chapter gives an overview of lung cancer, then describes chest radiography and sputum cytology (the major tests that have been used for lung cancer screening), analyzes their effectiveness in screening, and discusses new technologies and future prospects for screening.

A. Disease Burden

In the United States, carcinoma of the lung causes about three-fold more cancer deaths and person-years of life lost than any other cancer (2,4). In 1996, there were an estimated 177,000 newly diagnosed cases of lung cancer and 158,700 deaths resulting from this disease (2). The high ratio of mortality to incidence in lung cancer is a reflection of its poor prognosis. Combining all stages, the 5-year survival rate from diagnosis is only 13% (4).

In men, lung cancer has been the leading cause of cancer-related death since the early 1950s (2) and currently causes about one-third (32%) of all cancer deaths (Figure 1a). In women, lung cancer surpassed breast cancer as the leading cause of cancer-related death in the 1987 (5), and currently it causes one-fourth (25%) of all cancer deaths (see Figure 1b) (2).

Historically, lung cancer mortality rate has increased with the prevalence of cigarette smoking, lagging behind by about 20 years (6). Although lung cancer was rare before 1930, thereafter, the mortality rate in men increased rapidly and peaked at 75.3 per 100,000 in 1990 (4). The mortality rate in women has lagged behind that in men, reflecting the fact that women began to smoke later in the century than did men. In 1992, the mortality rate in women was 33.1 per 100,000 (4), and it is projected to increase until at least 2010 (6). Although the prevalence of cigarette smoking has decreased steadily among adults since the mid-1960s (4), the relative risk of dying of lung cancer remains elevated in former smokers for decades after they quit (7). In some medical centers, about half of the newly diagnosed lung cancers are in former smokers (8). In addition, the prevalence of cigarette smoking has been increasing recently among adolescents (9).

B. Cellular Classification

The most widely used classification system for lung cancer is that proposed by the World Health Organization (WHO) in 1981 (10). Four cell types account for the vast majority of all lung cancer cases. In the past several decades, the incidence of adenocarcinoma has increased relative to that of the other three major cell types (11) and this increase has been attributed to changes in cigarette smoking (12) and histopathological typing criteria (13). Adenocarcinoma currently comprises about 40% of all newly diagnosed cases, squamous cell carcinoma, about 30%, and large cell and small cell carcinoma, each about 15% (14). The first three cell types are often categorized together as non-small cell lung cancer (NSCLC) because, when localized, they have a similar treatment and prognosis, because these three cell types commonly coexist within a single tumor (15) and because pathologists have difficulty distinguishing large cell carcinoma from poorly differentiated forms of squamous cell carcinoma and adenocarcinoma (16,17).

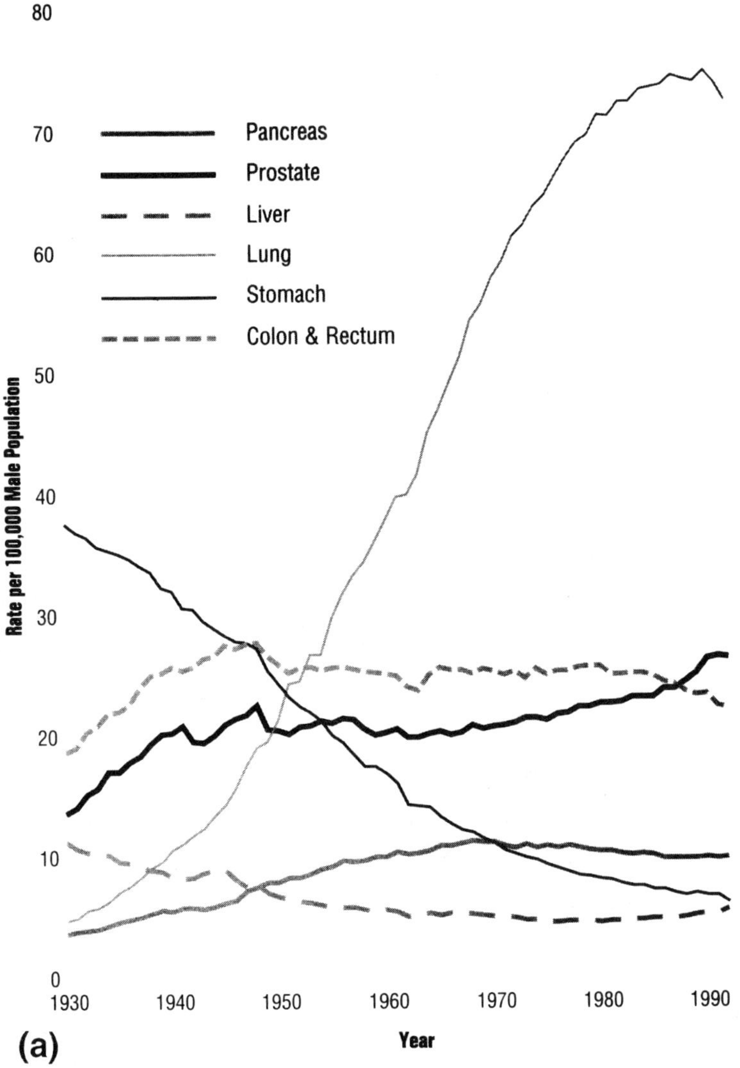

Figure 1 The age-adjusted cancer death rates for selected sites in the United States from 1930 to 1992. (a) Males, (b) females.

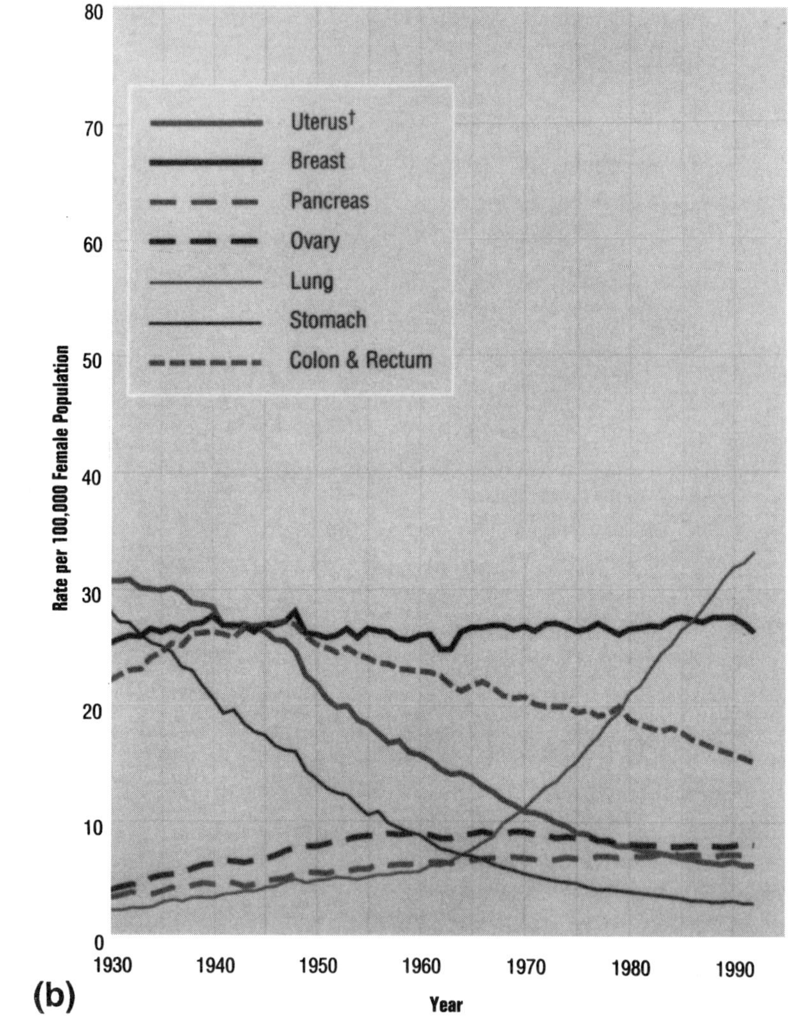

Figure 1 Continued
†Cervix and corpus combined

C. Etiology

1. *Smoking*

Tobacco smoke is the single most important etiological factor in the development of lung cancer (18). It is estimated that about 90% of all lung cancers are attributable to smoking (19). Cigarette smokers, overall, have about 10 times the risk of nonsmokers and the risk increases with the intensity and duration of smoking (20). Cigar and pipe smokers and those exposed to the cigarette smoke from others have risks in between those of cigarette smokers and nonsmokers (21). Smoking has been most strongly associated with squamous cell and small cell carcinoma (22), but there is recent evidence that smoking is also associated with adenocarcinoma, including the bronchioloalveolar subtype (23).

2. *Radon*

Radon gas accounts for more than one-half of the total radiation dose to the U.S. population (24). The gas, which is naturally present in soil and rocks, accumulates in poorly ventilated indoor environments and decays into daughter products that emit heavily ionizing alpha particles. When these alpha particles are inhaled, they irradiate the epithelium lining of the airways. Epidemiological studies of miners have clearly shown an excess lung cancer risk with exposure to radon (25). Using epidemiological evidence from miners and a linear non-threshold model to extrapolate risk to lower levels of radon exposure, Lubin et al. (25) estimated that the lifetime relative risk for residing in a home at the Environmental Protection Agency action level of 4 picocuries per liter is about 1.4 for smokers and 2.0 for nonsmokers. The authors also estimated that 10% of all lung cancer deaths and 30% of lung cancer deaths in never-smokers are attributable to indoor radon. Although the validity of this model-based extrapolation from miners is uncertain, a recent meta-analysis of case-control studies of lung cancer and indoor radon exposure provides similar estimations of risk (26,27).

3. *Inhaled Substances*

Several inhaled substances have been implicated in development of lung cancer (20). Asbestos (insulation and shipyard workers) increases the risk of lung cancer development in both smokers and nonsmokers. As there is with cigarette smoking, there is a strong dose-response relationship between the asbestos and the development of lung cancer (20). In addition, smoking and asbestos have synergistic effects (20). Several other particulates related to occupation have been implicated, including arsenic (copper smelters), chromium (chromate workers), nickel (refinery workers), and silica (foundry workers and miners). There is also evidence that air pollution from polycyclic hydrocarbons and other products of the combustion of fossil fuels may cause a substantial proportion of lung cancer cases among nonsmokers (28).

4. *Heredity*

The observations that lung cancer does not develop in most smokers and that the disease is sometimes clustered in families suggest that genetics may play an etiological role. The search for genetic markers of susceptibility has focused on genetically determined enzymes of metabolism and DNA repair. The activity of cytochrome P-450 enzymes, which activate certain carcinogens and catabolize retinoic acid, has been associated with an increased risk of lung cancer development. A recent study suggests that the rapid catabolizers of retinoic acid have an increased risk of squamous and large cell lung cancers (29). An inability to repair damaged DNA, which can be indirectly measured by a mutagen sensitivity assay using peripheral lymphocytes, has been shown to be an independent risk factor for lung cancer development (30). The overall relative risk for mutagen sensitivity after adjustment for age, sex, and smoking history has been shown to be about 3 to 4 (31,32).

5. *Diet*

There is much evidence that diet plays a role in the development of lung cancer. Increased consumption of fresh fruits and vegetables rich in beta carotene and reduced consumption of fat are associated with a lower risk of lung cancer (33). However, the reduction in risk is at most two-fold and the interaction between diet and smoking is unclear. In fact, two recently published trials of beta carotene supplementation (34,35) have demonstrated a higher risk of lung cancer in heavy smokers randomized to receive the supplement than those randomized to receive the placebo. It has been postulated that beta carotene is oxidized in the lungs of heavy smokers and that the by-products of beta carotene oxidation promote the development of lung cancer (36).

6. *Pulmonary Fibrosis*

Lung cancer has been observed to develop in focal areas of pulmonary fibrosis (37), which may be caused by infection, other inflammatory conditions, or infarction. This association has given rise to the term ''scar carcinoma.'' However, it is not clear whether these focal areas of fibrosis cause or result from lung cancer (20).

D. Clinical and Radiological Manifestations

In clinical practice, most patients with lung cancer present with symptoms arising from the lungs or adjacent intrathoracic structures (14). The most common symptoms related to the lungs are cough, hemoptysis, shortness of breath, and fever from obstructive pneumonitis. Infiltration of the pleura commonly causes chest pain that is exacerbated by breathing. Mediastinal lymph node metastases in the aortopulmonary window may cause hoarseness by compressing the left recurrent laryngeal nerve. Patients presenting with constitutional symptoms of weakness, malaise, and weight loss usually have distant metastases. Occasionally, patients with lung cancer are seen with extrathoracic metastases or one of a variety of paraneoplastic syndromes. Only 10% to 25% percent of lung cancer patients are asymptomatic at the time of diagnosis.

Almost all patients with clinical signs or symptoms of lung cancers have chest radiographical findings (14). One of the most common findings is postobstructive atelectasis, which is most typical of squamous cell carcinoma arising in a segmental or lobar bronchus (Figure 2). Another common finding is a hilar mass and mediastinal widening resulting from lymph node metastases, which is most typical of small cell carcinoma (Figure 3). A variety of other radiographical abnormalities are seen in patients with advanced disease. In asymptomatic patients, lung cancer is usually first suspected on the basis of incidental findings on a chest radiograph obtained for another purpose, such as evaluation for surgery or hospital admission. The most common finding is a solitary pulmonary nodule or mass, which is typical of adenocarcinoma (Figure 4). (By convention, nodules measure less than 3 cm in diameter and masses measure more than 3 cm in diameter [38].)

Among those participating in lung cancer screening with chest radiography, 50% to 70% of cases are detected by screening, the most common finding being

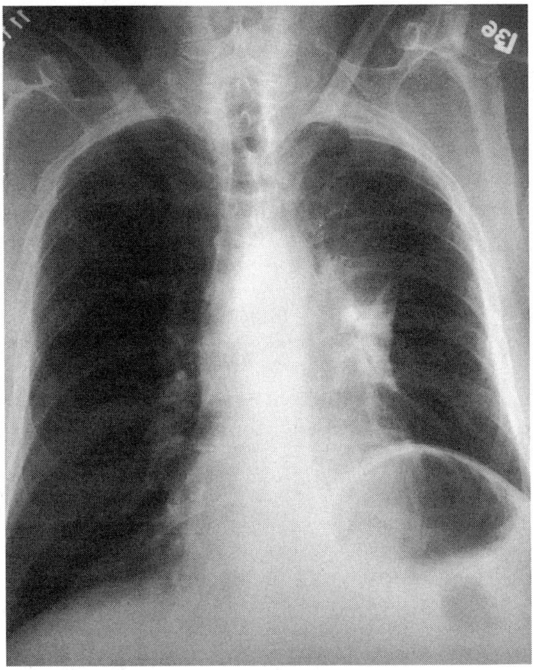

Figure 2 Non-small cell carcinoma (cell type not further specified) in smoker with cough and shortness of breath. Chest radiograph demonstrates left upper lobe collapse and resulting elevation of the left hemidiaphragm. Bronchoscopy revealed tumor completely obstructing left upper lobe bronchus.

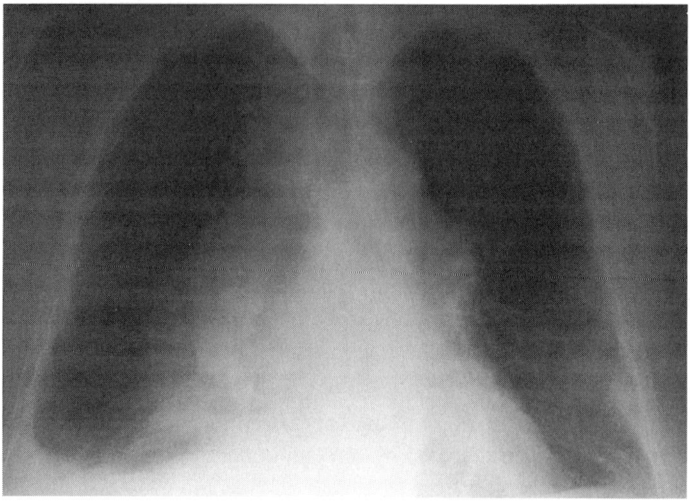

Figure 3 Small cell carcinoma in smoker with signs and symptoms of superior vena cava obstruction. Chest radiograph demonstrates widening of the mediastinum and small left pleural effusion.

a solitary pulmonary nodule (39,40). Screen-detected lung cancers tend to progress relatively slowly and are often retrospectively visible on chest radiographs for more than 1 year (39–41). In contrast, interval cases, those diagnosed between screening examinations, tend to progress relatively rapidly and present with the clinical and radiographical findings that are more commonly seen in patients who are not screened. This contrast between screen-detected and interval cases is common to all cancer screening, and is known as length bias favoring survival of screen-detected cases (42).

E. Diagnosis

Although some clinical and radiological findings may be strongly suggestive of lung cancer, the final diagnosis is usually based on cytology or histology. This may be obtained from expectorated sputum, percutaneous needle biopsy under radiological guidance, fiberoptic bronchoscopy, or thoracotomy. The diagnostic evaluation of suspected lung cancer has been modified since most of the screening studies were carried out, largely because of the development of computed tomography (CT). However, the evaluation is variable and depends on clinical and radiological presentation, and the availability of CT and other diagnostic procedures.

Most relevant to screening is the evaluation of a suspicious radiographical finding in an asymptomatic patient. A repeat radiograph using a different pro-

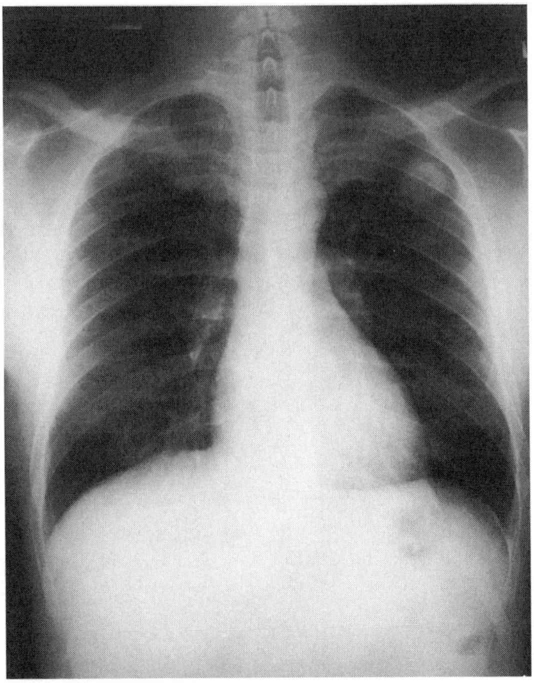

Figure 4 Adenocarcinoma in asymptomatic smoker. Chest radiograph shows left upper lobe nodule, which was surgically resected.

jection or markers may be helpful in excluding an artifact, such as a nipple shadow. Comparison with older or follow-up radiographs may be helpful in excluding an anatomical variant or benign solitary pulmonary nodule, both of which should be stable over time. Historically, 2 years of stability has been accepted as a criterion for confirming that a solitary pulmonary nodule is benign. However, this criterion has recently been challenged, especially for the small nodules detectable by CT, in which the measurement error may be large relative to the diameter of the nodule (43).

Chest CT may also be useful in the evaluation of a suspicious radiographical finding. Computed tomography eliminates superimposition of normal anatomical structures in the lung, hila, mediastinum, or chest wall that can raise suspicion on the chest radiograph. In addition, because CT has much greater contrast resolution than radiography, CT can more reliably detect certain patterns of calcification and fat that exclude malignancy (Figure 5) (44,45). Lack of enhancement with contrast has also recently been reported as a potentially reliable sign for excluding malignancy (46).

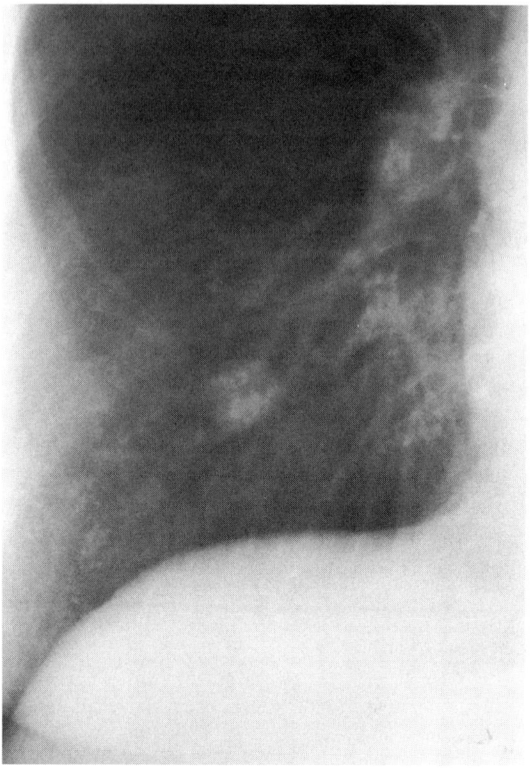

A.

Figure 5 Granuloma. A, Chest radiograph shows 1.5-cm nodule without definite signs of calcification in right middle lobe. B, Computed tomography reveals dense homogeneous calcification characteristic of a granuloma. Patient required no further evaluation.

Often it is helpful to analyze the radiological findings in conjunction with the patient's risk factors. Gurney has developed and validated a method for quantifying the probability of malignancy in solitary pulmonary nodules with Bayesian analysis using likelihood ratios for age, smoking history, rate of growth, and several radiographical and CT findings (47,48). However, when these noninvasive procedures fail to exclude malignancy or sufficiently lower its probability, percutaneous needle biopsy or thoracotomy is usually performed (sputum cytology may be diagnostic for central, endobronchial tumors, but is rarely diagnostic for solitary pulmonary nodules, which usually arise peripherally [14].) Although the positive predictive value of a percutaneous needle biopsy is nearly 100%, a negative result is less reliable for excluding malignancy (14,44). Consequently, many patients proceed directly to thoracotomy, especially when the probability of malignancy is high.

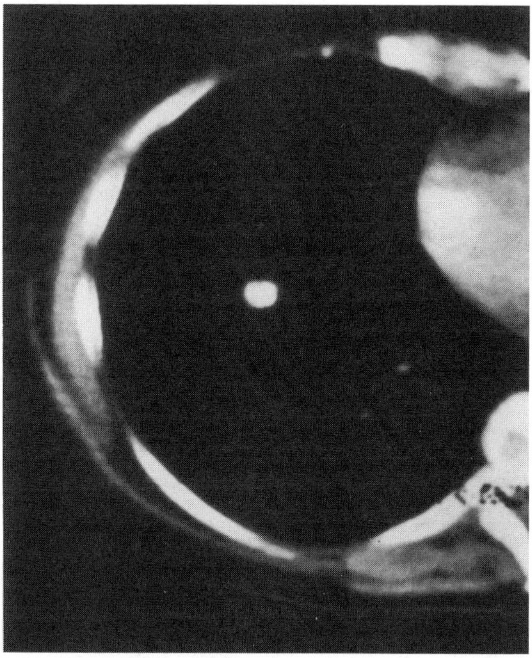

B.

F. Staging

The primary tumor may invade adjacent intrathoracic structures, metastasize to hilar or mediastinal lymph nodes via the lymphatics, or metastasize to distant sites via the blood stream. Non-small cell lung carcinoma (NSCLC) is staged according to the International Staging System, which was introduced in 1986 (49). In this system, staging is based on the extent of TNM involvement, where T refers to the primary tumor, N refers to regional lymph nodes, and M refers to distant metastasis (Tables 1 and 2). Small cell carcinoma is usually staged more simply as either limited (stages I–III) or extensive (stage IV) (50).

For NSCLC, the primary tumor extent (T) can usually be based on the combination of clinical, chest radiographical, CT, and bronchoscopic findings. Magnetic resonance imaging (MRI) can be very helpful in determining extent of chest wall involvement, particularly for superior sulcus tumors, in which invasion of the branchial plexus and subclavian vessels are best visualized in the coronal and sagittal planes. At most institutions, CT is routinely performed to assess lymph node involvement (N) in patients who are surgical candidates (14), although there is ongoing controversy about the cost effectiveness of CT

Table 1 TNM Staging System for Lung Cancer

T Descriptor	Definition
TX	Positive malignant cells; no lesion seen
Tis	Carcinoma in situ
T1	≤3 cm diameter
T2	>3 cm diameter or distal atelectasis or pneumonitis extending to the hilar region
T3	Extension to pleura, chest wall, diaphragm, or pericardium or <2 cm from carina or atelectasis/pneumonia involving an entire lung
T4	Invasion of mediastinal organs or malignant pleural effusion
N Descriptor	**Node involvement**
N0	No involvement
N1	Ipsilateral bronchopulmonary or hilar
N2	Ipsilateral or subcarinal mediastinal
N3	Contralateral mediastinal, hilar, or supraclavicular
M Descriptor	**Metastatic involvement**
M0	None
M1	Metastases present

Source: Ref. 14.

Table 2 Stage Grouping of TNM Subsets

Stage	T	N	N
0	Tis	N0	M0
IA	T1	N0	M0
IB	T2	N0	M0
IIA	T1	N1	M0
IIB	T2	N1	M0
	T3	N0	M0
IIIA	T3	N1	M0
	T1–3	N2	M0
IIIB	T4	Any N	M0
	Any T	N3	M0
IV	Any T	Any N	M1

Source: Ref. 49.

in the subset of patients staged T1N0M0 by clinical and chest radiographical criteria (51–54). Lymph nodes in the hila and mediastinum measuring greater than 1 cm in diameter are considered suspicious. However, because lymph nodes can be enlarged by nonneoplastic conditions, such as previous histoplasmosis or concurrent pneumonitis, enlarged nodes are usually sampled by mediatinoscopy, mediastinotomy, or transbronchial needle biopsy before a patient is excluded from potentially curative resection. Distant metastases to the liver or adrenal glands are occasionally detected on the staging chest CT. Head CT or MRI and radionuclide bone scan are also useful for detecting distant metastases in patients with signs or symptoms of extrathoracic disease, but they are not recommended for routine staging (55–58).

G. Treatment

Surgical resection is the major potentially curative therapeutic option for NSCLC and is the treatment of choice for localized disease (stages I and II) and selected cases of locally (T3) or regionally advanced (N2) stage IIIA disease (14). Lobectomy is the most common form of surgical treatment. However, a smaller amount of lung may be resected (''wedge resection'') from patients with high operative risks and the entire lung is resected from patients with hilar involvement. The immediate postoperative mortality rate ranges from 1% to 8%, depending on the type of resection and age of the patients (14). Patients with localized NSCLC and medical contraindications to surgery may be considered for potentially curative radiation therapy.

Most patients with locally (T3-T4) or regionally advanced (N2-N3) NSCLC are given radiotherapy or some combination of radiation therapy, chemotherapy, and surgical resection (14). Patients with distant metastases (M1) are usually given radiation therapy or chemotherapy for palliation. Some patients with a solitary distant metastasis may undergo resection of both primary tumor and metastasis.

The primary treatment for small cell carcinoma is combination chemotherapy (50). Radiotherapy to the chest and other sites is often used with chemotherapy in patients with limited stage disease. A small minority of patients with limited small cell carcinoma undergo surgical resection with or without adjuvant chemotherapy. Various regimens of combined modality therapy are under clinical investigation.

H. Prognosis

Stage at diagnosis is the single most important predictor of survival for lung cancer (14). In completely resected T1N0M0 lung cancer, relative 5-year survival rate exceeds 70% (14). According to the most recent Surveillance, Epidemiology, and End Results (SEER) data, relative 5-year survival rate for

NSCLC is as follows: localized (stage I), 50%; regional (stages II and III), 19%; distant (stage IV), 2% (4). These stage-specific survival statistics are almost identical to those reported by Mountain in 1986 (59). The relative 5-year survival rate for all stages combined has not changed significantly since 1977 and is only 15%, reflecting the fact that less than 20% of patients have localized disease at the time of diagnosis. The prognosis for small cell carcinoma is even worse (4).

The strong relationship between stage and survival suggests that early diagnosis through screening may be beneficial. However, this strong relationship does not necessarily imply that early diagnosis is effective in reducing lung cancer mortality rate. Survival statistics can be misleading and the effectiveness of early detection can only be reliably determined in properly designed prospective studies that directly demonstrate a reduction in lung cancer mortality rate (42,60–62).

II. SCREENING TESTS

A. Chest Radiography

Chest radiography has been the principal test used in lung cancer screening. It is safe, acceptable to screenees, widely available, and relatively inexpensive (see section II. D. 3). The estimated effective dose equivalent is about 6 mrem (63). By comparison, in the United States, the average annual effective dose equivalent from natural sources is about 300 mrem per person, two-thirds of which comes from inhalation of indoor radon (24).

1. *Technique*

The standard chest radiographical examination consists of upright posteroanterior (PA) and lateral views obtained during full inspiration (64). The standard radiograph is 14 × 17 inches and includes the lung apices and costophrenic angles. Technical factors for the PA view include a mean exposure not exceeding 30 mR per examination (skin surface) and exposure time not exceeding 40 msec. A high kilovoltage technique (110–140 kVp) and wide latitude screen-film combination are used with an antiscatter technique, producing scatter reduction equivalent to that of at least a 10:1 grid. A dedicated chest unit with a focal spot size not exceeding 2 mm is used and, to minimize magnification, the source-image distance is at least 6 feet.

Most of the aforementioned standards were met in randomized controlled trials that were conducted in the United States (65), although several did not use the lateral view. However, most of the studies that were conducted outside the United States evaluated screening with chest photofluorography instead of standard chest radiography. Chest photofluorography consists of miniature (70–110 mm) photographic recordings of PA chest fluorographic images. This technique has not been used in the United States since the 1970s because of its relatively poor spatial resolution and high radiation dose (66).

2. *Sensitivity*

In screening for lung cancer, the sensitivity of chest radiography depends on several factors, including the size and anatomical location of the lung cancer, the radiographical technique, and the radiologist's skill and interpretation threshold. The threshold size for visibility of a nodule is about 6 mm when it is projected over an intercostal space (38). However, nodules measuring greater than 1 cm in diameter are often hidden by ribs or clavicle, hilar structures, mediastinum, or diaphragm (38,67). In a recent study of missed lung cancers, the mean diameter was 1.6 cm and the largest was 3.4 cm. Most of the lesions were in the upper lobes and obscured by overlying ribs, clavicles, and blood vessels (68). In addition to these anatomical problems, there are many other perceptual errors that hinder detection and reporting (67,69). In the Mayo Clinic Study, only one of the 50 peripheral screen-detected lung cancer lesions was less than 1 cm in diameter (8 mm), 17 were 1 to 2 cm, and 19 were 2 to 3 cm (39).

It has been estimated that 75% of the natural history of cancer transpires before it is detectable by chest radiography (38). A single cancer cell 10 μm in diameter must go through 30 doubling times to become a 1-cm nodule. Assuming a constant doubling time between 30 and 300 days, a lung cancer lesion grows for 2.5 to 25 years before its diameter reaches 1 cm (lung cancer growth rates probably do vary over time).

It is difficult to determine the true sensitivity of a screening test because of the lack of a definitive gold standard that can be applied to subjects with negative test results (70). For this reason, sensitivity is usually defined with regard to a screening program rather than a screening test per se. Screen-detected cases are considered true-positive results and cases that present between screening intervals are considered false-negative results. This definition of sensitivity depends not only on the intrinsic characteristics of the test, the tumor growth rate, and the screening interval, but also on the threshold of the medical system for recognizing and recording signs and symptoms of lung cancer.

The largest, most systematic, and most recent evaluation of chest radiographical sensitivity and specificity in screening for lung cancer was performed in Nagasaki (40). (Screening was performed with a 100-mm indirect PA radiograph of the chest, which has been shown to be almost equivalent to full-sized direct PA chest radiograph [40]). In this population-based study of annual screening, the overall program sensitivity was 70% (140/199). The sensitivities for the four major cell types were as follows: adenocarcinoma, 85% (91/107); squamous cell, 52% (26/50); small cell, 50% (12/24); and large cell, 75% (9/12). Of the screen detected cases, 50% were stage I at diagnosis. In retrospect, 85% of the stage I adenocarcinomas were radiographically visible 1 year earlier

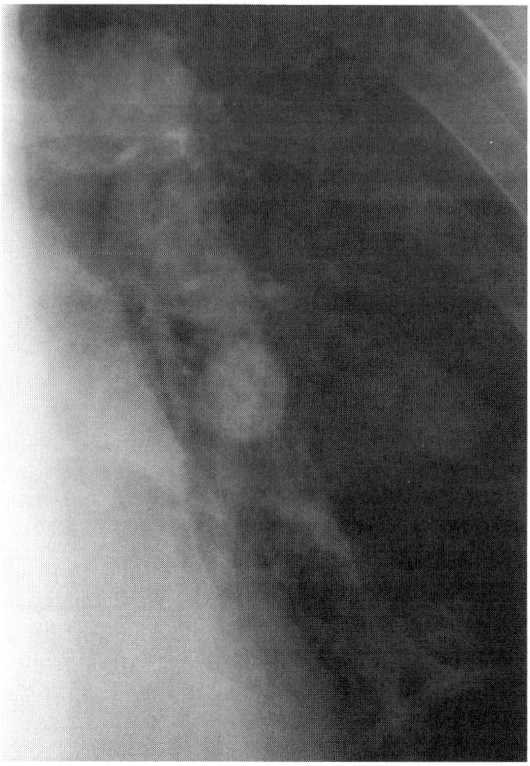

A.

Figure 6 Hamartoma. A, Chest radiograph shows 1.0-cm nodule in left mid lung region. B, computed tomography confirms presence of nodule in left lower lobe but is otherwise nondiagnostic. Nodule was surgically resected.

and 58% were visible 2 years earlier. In contrast, only 12% (7/59) of the interval cases were stage I at diagnosis and less than half of the interval cases were radiographically visible retrospectively.

Similar findings were observed in an earlier Japanese screening program (71) and the NCI cooperative trials (66,72–74), in which the sensitivity of chest radiography ranged from 50% to 70%, and about 50% of the screen-detected lung cancers were stage I. In the Mayo Clinic Study, 90% of the peripheral cancers were retrospectively visible on radiographs obtained 4 months earlier and 36% were retrospectively visible 1 year earlier (39).

3. *Specificity*

The specificity of chest radiography also depends on many factors, including the prevalence of benign lesions, the radiographical technique, and the radiol-

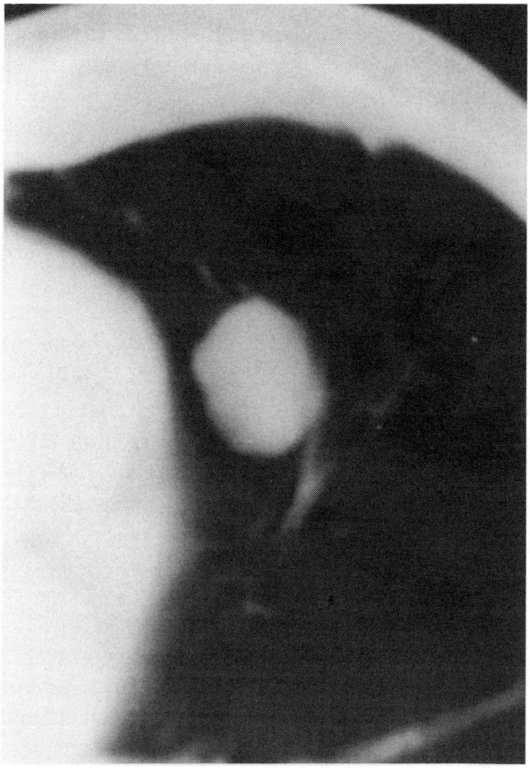

B

ogist's skill and interpretation threshold. Although the most common radiographical finding of screen-detected lung cancer is a solitary pulmonary nodule, only about 5% of radiographically detected solitary pulmonary nodules in the general population are malignant (38). However, about 50% of solitary pulmonary nodules that are surgically resected in men older than 50 years of age are primary lung cancer (38,75). Most of the remaining cases are granulomas, hamartomas, and other benign lesions (Figure 6). In their large population-based study of annual screening, Soda et al. (40) reported a specificity of 99.6% (919 false-positive and 204,283 true-negative screening examinations). However, the causes for the false-positive results and their management were not reported. In the NCI cooperative trials (76), the specificity of chest radiography during the initial screen depended strongly on how ''indeterminate'' was categorized: when categorized as positive, the specificity was 90%; when categorized as negative, the specificity was 99%. Most of the false-positive results were resolved by comparison with old or follow-up chest radiographs. However, many patients did require invasive diagnostic procedures and at least two died after thoracotomy for a benign tumor (77).

4. *Positive Predictive Value*

The positive predictive value depends on the accuracy of the test and the prevalence or incidence of disease. In the Nagasaki study, the incidence of lung cancer was very low, about 0.7 per 1000 person-years, because the study included residents 40 years of age and older, regardless of smoking history. Consequently, despite the very high specificity, the positive predictive value of a positive chest radiograph was only 13% (140/1059). In the NCI cooperative trials (76), the prevalence at the initial screen was much higher, 7.6 per 1000 persons. As with specificity, the positive predictive value depended strongly on how ''indeterminate'' was categorized: when ''indeterminate'' was categorized as positive, the positive predictive value was only 7% (127/1892); when ''indeterminate'' was categorized as negative, the positive predictive value was 46% (92/202).

B. Sputum Cytology

Sputum cytology has also been evaluated as a screening test for lung cancer. However, this test has not been used for screening since the completion of the National Cancer Institute (NCI) cooperative trials, which demonstrated that screening with sputum cytology had no effect on the lung cancer mortality rate. However, other potential uses of sputum cells for lung cancer screening are under investigation (78).

1. *Technique*

In the NCI cooperative trials, sputum specimens were induced on the first visit by an aerosol spray procedure that required about 20 minutes (65). Subsequently, each specimen consisted of 3-day pooled collection of sputum obtained spontaneously from the subject at home on three consecutive mornings. Induced and spontaneous specimens were processed by the Saccomanno technique and four Papanicolaou stained slides were prepared for each specimen. The slides were interpreted by cytotechnologists and cytopathologists and categorized as negative, slight atypia, moderate atypia, marked atypia, carcinoma, or ''unsatisfactory.''

2. *Sensitivity*

The sensitivity of sputum cytology depends on several factors, including the proximity of the tumors to the major airways, the tumor cell type, and the ability of the subject to produce an adequate specimen (14,76). In the NCI cooperative trials, 160 lung cancers were detected during the prevalence screen of subjects who received both sputum cytology and chest radiography (76). Of these 160 prevalent cases, 37 (23%) were detected by sputum cytology (reported as marked atypia or carcinoma) alone and 67 (42%) were detected by sputum cytology and chest radiography. Of the 37 cases detected by sputum cytology alone, 35 (95%) were squamous cell carcinoma and the other two were large cell carcinoma with squamous components. Thirty (81%) were stage I.

3. *Specificity*

The specificity of sputum cytology depends on the definition of a "positive" examination. Of the 21,127 subjects who received both sputum cytology and chest radiography during the prevalence screen of the NCI cooperative trials, 300 (1.4%) had sputum examinations reported as moderate atypia (221), marked atypia (27), or carcinoma (52). When reports of moderate atypia, marked atypia and carcinoma were considered positive, the specificity was 98.9% (233/20,967). When reports of only marked atypia and carcinoma were considered positive, the specificity was 99.94%.

4. *Positive Predictive Value*

Similarly, the positive predictive value of sputum cytology depends on the definition of a "positive" examination. When reports of moderate atypia, marked atypia, and carcinoma were considered positive, the positive predictive value was 23% (70/300). When reports of only marked atypia and carcinoma were considered positive, the positive predictive value was 85% (67/79). One major disadvantage of sputum cytology is that it does not localize the anatomical site of the tumor. "Positive" examinations must be followed by a detailed evaluation of the upper respiratory tract and systematic bronchoscopic visualization of all lobar and segmental bronchi up to and including the fifth or sixth general branches (72).

III. STUDIES OF SCREENING EFFECTIVENESS

A. Demonstration Projects

During the 1940s, several countries established tuberculosis screening programs that used chest photofluorography. The subsequent improvement in the prognosis for tuberculosis was attributed partly to earlier detection (and antimicrobial therapy) and created optimism that earlier detection might also improve the prognosis for lung cancer, which was also detectable by chest photofluorography. Screening for lung cancer was first investigated in uncontrolled demonstration projects that were based on tuberculosis screening programs. The three largest projects were the Philadelphia Pulmonary Neoplasm Research Project

(79,80), American Cancer Society and Veterans Administration (VA) study (81), and South London Lung Cancer Study (82).

1. *Philadelphia Pulmonary Neoplasm Research Project*

Beginning in 1951, men volunteers 45 years and older were recruited from the Philadelphia area to participate in the lung cancer screening project (79,80). Screening consisted of a 70-mm PA chest photofluorogram and questionnaire for lung cancer symptoms every 6 months for 10 years. Subjects with suspicious photofluorographic findings or symptoms were initially evaluated by a 14 × 17-inch PA chest radiograph and a visit to the project physician, who recommended further evaluation when appropriate. Over a period of 4 years, 6136 volunteers were enrolled and entered into follow-up for the 10-year screening period. In addition, lung cancer cases diagnosed during the screening period were followed up for a minimum of 5 years from time of diagnosis. There was no control group.

During the 10-year screening period, 121 incident cases of lung cancer were diagnosed in men who had no evidence of lung cancer on the initial photofluorogram; the annual incidence was about 2 per 1000 person-years. Seventy-four (62%) of these cases were first detected at the time of screening whereas the remainder were diagnosed because of symptoms that developed during the interval between screenings. A higher percentage of lung cancer cases were detected by screening in those who complied with screening than in those who did not, 72% (48/67) versus 48% (26/54). Only 33 (27%) of the incident cases were resectable.

The overall 5-year survival in men diagnosed with lung cancer was only 8%. One-year survival was much higher in the screen-detected than in the interval cases, 65% versus 9%, but the 5-year survival rate was very low for both groups, 11% versus 5%. Similarly, the 1-year survival rate was much higher in the resected cases than in the unresected cases, 58% versus 38%, but the 5-year survival rate was very low for both groups, 18% versus 5%. The operative mortality rate within 6 weeks of resection was 30%.

The investigators attributed the large difference in short-term survival rate to lead time bias and the similarity in long-term survival rate to a low potential for curing screen-detected lung cancer. They concluded that the small benefit of screening was not worth the morbidity associated with working up suspicious findings, the mortality of surgical resection, and the monetary costs.

2. *American Cancer Society and Veterans Administration Study*

The study population consisted of the 14,607 male residents of six VA domiciles, which were established to provide for disabled veterans (heart disease and alcoholism were the two most common disabilities) (81). The median age was 62.8 years and the range was from younger than 45 to older than 75 years.

During the 3-year period from 1958 to 1961, the study population underwent six semiannual screens for lung cancer. Screening consisted of a PA radiograph (two stereo views) and a spontaneous or aerosol induced sputum specimen for cytology. All radiographs were initially read at the domicile and mailed to the radiology center to be read independently by two qualified radiologists. Smears of the sputum were made at the domicile and mailed to a cytology center to be stained and read independently by two cytotechnicians. Suspicious findings were followed up by additional radiographical views, tomography, bronchoscopy, and exploratory thoracotomy, depending on the nature of the findings. There was no control group.

During the 3-year period of screening, 73 cases of lung cancer were detected by the screening program. The sensitivity and specificity of chest radiography alone were reported as 42% and 98%, respectively. The sensitivity and specificity of sputum cytology alone were reported as 33% and 98%, respectively. The two detection methods were complementary and the sensitivity and specificity of the two methods combined were 63% and 96%, respectively.

Twenty-nine (40%) of the screen-detected cases were surgically explored and 26 (36%) underwent resection (lobectomy or pneumonectomy). The cumulative 32-month survival rate was 17% for all newly detected cases and 12% for the subset of patients who underwent resection. Although there were only two immediate postoperative deaths (7.7%), more than 40% of the resected patients died within 4 months of surgery.

Lilienfeld et al. (81) attributed the low rates of survival to the poor health of the disabled veterans. The authors concluded that screening with chest radiography and sputum cytology was feasible and might benefit a healthier population.

3. *South London Lung Cancer Study*

In the South London study, 67,400 male volunteers older than 45 years of age were recruited from those who attended the London Mass X-ray Services at least once during the period from January 1959 to June 1963 (82). More than 75% of the volunteers participated at 197,500 subsequent semiannual screenings, which consisted of chest photofluorography and a health questionnaire. During these screenings, 147 new cases of lung cancer were discovered by radiography, 38 (27%) of which were symptomatic. Based on these cases, the average annual incidence was about 1.4 per 1000 person-years and increased with age and with the number of cigarettes smoked per day. The cell type distribution was as follows: squamous cell, 42%; small cell, 14%; adenocarcinoma, 12%; other, 32%.

Eighty-three (56%) of the incident cases were resected. The 4-year survival rate was much higher for asymptomatic than symptomatic lung cancer, 33% versus 16%; much higher for the resected than unresected cases, 47% versus 1%; much higher for patients older than 55 years than younger than 55 years,

32% versus 9%; and much higher for squamous cell and adenocarcinoma than for the other cell types, 46% versus 6%.

Nash et al. (82) concluded that screening with semiannual chest radiography had the potential to improve survival in male smokers older than 55 years of age.

B. Controlled Trials

The relatively high rates of survival in the South London study suggested that screening could substantially advance the time of diagnosis. However, to the extent that screening does advance the time of diagnosis, survival rate is a biased measure of the effectiveness (42,60,83). The appropriate endpoint for screening effectiveness is the reduction in disease-specific mortality rate. Randomized clinical trials are considered the best method of determining screening effectiveness because they eliminate selection bias and the three biases that affect comparisons of survival: lead time bias, length bias, and overdiagnosis bias.

1. *North West London Mass Radiology*

The first controlled trial of lung cancer screening was begun in 1960 by the Mass Radiography Service of the North West Metropolitan Area of London (84,85). The trial was designed to determine the effectiveness of screening for lung cancer with chest radiography. A total of 55,034 men aged 40 years and older from 119 industrial firms volunteered for the study and completed the initial radiographical examination and questionnaire. These firms, which were mainly factories, were categorized according to type of work and locality. Within each category, each firm was assigned to the screened group or control group by the use of random sampling numbers. Thus, randomization was carried out at the level of the industrial firms that employed the volunteers rather than at the level of the individual volunteer. After randomization, the screened group consisted of 29,723 men from 75 firms and the control group consisted of 25,311 men from 44 firms (Table 3). The baseline characteristics of the two groups with respect to smoking and age were virtually identical, suggesting that the method of randomization was highly effective. The screened group was scheduled to have a total of seven chest radiographs, one at the beginning of the study and one every 6 months for the next 3 years, while the control group was scheduled to have only two chest radiographs, one at the beginning of the study and one 3 years later at the completion of the study.

On the initial chest radiograph, 31 cases of lung cancer were detected in the screened group versus 20 in control group (Table 3). Even though the numbers of resectable cases in the screened and control groups were similar, 16 versus 15, there was a marked discrepancy in the numbers of unresectable cases, 15 versus 5. There was also a discrepancy in the numbers of cases with poorly differentiated histology in the screened and control groups, 12 versus 7.

Table 3 North West London Mass Radiology Summary Statistics

	Screened group	Control group
Subjects	29,723	25,311
Prevalent cases[a] (per 1000)	31 (1.0)	20 (0.8)
Incident cases (per 1000)	101 (3.4)	76 (3.4)
% resectable	44	29
% 5-year survival[b]	15	6
Lung cancer deaths (per 1000)	62 (2.1)	59 (2.3)
Relative risk[c] (95% confidence interval)	0.89 (0.63–1.28)	

[a] Excluded from analysis of resectability, survival, and mortality.
[b] Based on extended follow-up of cases diagnosed during the 3-year study period (85).
[c] Based on cumulative lung cancer mortality rate at 3 years (84).

After the initial screen, about 76% of the planned screening chest radiographs were completed in the screened group. In the control group, no chest radiographs were obtained between the beginning and end of the study. At the completion of the study, 101 incident cases were diagnosed in the screened group (3.4 per 1000 subjects) and 76 incident cases were diagnosed in the control group (3.4 per 1000 subjects) (Table 3). In the screened group, 65 cases (64%) were detected by screening whereas the remainder were diagnosed because of symptoms that developed during the intervals between screenings. In the screened group, 65% of screen-detected cases and 44% of all cases were resected, whereas only 29% of the cases in the control group were resected.

For the calculation of lung cancer mortality rate, only deaths that occurred during the 3-year period were included. In addition, the 29 deaths from the 51 prevalent cases detected at the initial screening were excluded to avoid underestimating the effectiveness of screening. After 3 years, there were 62 lung cancer deaths in the screened group and 59 lung cancer deaths in the control group (Table 3). Comparing the cumulative lung cancer mortality rate in the screened group to that in the control group, the relative risk was 0.89. However, most of the deaths from lung cancer detected at the initial screening, 20 of the 29, were in the screened group. If these deaths had been included, the lung cancer mortality rate would have been slightly higher in the screened group than in the control group and the relative risk would have been 1.03.

For the purpose of comparing survival from time of diagnosis, the trial was reanalyzed 5 years later, when each case of lung cancer diagnosed during the study period had at least 5 years of follow up (84). In the test group, the 5-year survival rate was 23% for screen-detected cases and 15% for all incident cases (Table 3). In the control group, the 5-year survival rate was only 6%. Mortality rate was not reported in the reanalysis.

Brett concluded that a ‘‘modest improvement’’ in lung cancer prognosis can be achieved through 6-monthly chest radiographical screening. He speculated that the lack of a statistically significant difference in lung cancer mortality rate was due to either a more accurate ascertainment of lung cancer deaths in the screened group than in the control group (see Controversies) or to the short duration over which mortality was observed (84).

2. *National Cancer Institute Cooperative Trials*

During the 1960s radiographical technique and cytological evaluation of the sputum improved and flexible fiberoptic bronchoscopy was developed. This procedure made possible the localization of radiographically occult, cytologically positive lung cancer. These technological advances and the confusion generated by previous studies provided a rationale for new randomized controlled trials (RCTs) of lung cancer screening.

In the early 1970s, the National Cancer Institute initiated three large randomized clinical trials of lung cancer screening at the Mayo Clinic, the Johns Hopkins Medical Center, and the Memorial Sloan-Kettering Cancer Center (65). The purpose of these cooperative trials was to determine whether frequent screening with some combination of chest radiography and sputum cytology could decrease the lung cancer mortality rate. The study population consisted of men who were 45 years and older, smoked at least one pack of cigarettes per day, and had no history or clinical suspicion of respiratory tract cancer or medical contraindications to an attempt at curative surgical resection. To ensure sufficient statistical power to detect a modest reduction in lung cancer mortality rate, about 10,000 subjects were enrolled at each center, screened regularly for 5 years, and followed up for at least 5 years thereafter. Mortality rate from lung cancer was the major outcome by which the effectiveness of screening was evaluated.

a. Mayo Clinic Study. The Mayo Clinic study was designed to determine the effectiveness of screening with the combination of 4-monthly chest radiography and 4-monthly sputum cytology versus usual care at the Mayo Clinic (73,86). The Mayo Clinic study was different in design than the other two studies in two important respects. First, it was the only study designed to determine the effectiveness of screening with chest radiography. Second, subjects with prevalent lung cancers detected at the first screening were excluded from the study.

From late 1971 to mid-1976, 10,933 Mayo Clinic outpatients aged 45 years and older were recruited. Of these, 91 (8.3 per 1000) were excluded because of unsuspected lung cancer detected by the initial PA chest radiograph or 3-day pooled sputum cytology (Table 4). More than 1000 additional patients either refused to participate or were excluded because of medical conditions that contraindicated surgery or decreased their life expectancy to less than 5 years. The

Table 4 Mayo Clinic Study Summary Statistics

	Screened group (CXR + SC)	Control group (usual care)
Subjects[a]	4618	4593
Incident cases	206	160
% resectable	46	31
% 5-year survival	31	13
Lung cancer deaths	122	115
Relative risk[b] (95% confidence interval)	1.06 (0.82–1.36)	

Abbreviations: CXR, chest radiography; SC, sputum cytology.
[a] After the initial screening of 10,933 recruits, 91 prevalent cases and 1631 outpatients with contraindications were excluded before randomization.
[b] Based on cumulative lung cancer mortality rate at 11 years (73).

remaining 9211 men were randomized to two groups. The 4618 men randomized to the screening group were asked and reminded to have chest radiography and sputum cytology every 4 months for 6 years. The 4593 men randomized to the control group received only usual care, which was a recommendation for annual chest radiography and sputum cytology.

The subjects in the intensively screened group completed about 75% of their tests during the 6-year screening period (73). Fifty-five percent of the control subjects had a chest radiograph during the final year of the screening period and 73% had a chest radiograph during the final 2 years (73), although it is not clear why these radiographs were obtained. Throughout the 11-year period of observation, there were more incidence cases in the intensively screened group than in the control group. At the end of the 6-year screening period, there were 143 cases in the intensively screened group versus 87 cases in the control group. Of the cases in the intensively screened group, 75% were detected by screening or nonstudy radiographs, and the remainder were diagnosed by symptoms that developed between screenings. Of the cases in the control group, 26% were detected by nonstudy radiographs. At the end of the 11-year period, 5 years after screening stopped, there were 206 cases in the intensively screened group versus 160 cases in the control group (see Table 4). The excess 46 cases in the intensively screened group was comprised mainly of early stage squamous cell carcinomas and adenocarcinomas.

In the intensively screened group, 46% (94/206) were resectable versus 32% (51/160) in the control group (see Table 4). Considering only deaths from lung cancer, the 5-year survival in the intensively screened group was 33% versus 15% in the control group. Despite the markedly increased rates of lung cancer detection, resectability, and survival in the intensively screened group, there was no significant difference in the number of advanced cases or lung cancer deaths, the ultimate measure of screening effectiveness (see Table 4). At the completion of the study, there were 107 advanced stage (stages III-IV) cases in the intensively screened group versus 109 in the control group. Similarly, there were 122 deaths from lung cancer in the intensively screened group versus 115 in the control group and the relative risk was 1.06.

As discussed, prevalent cases of lung cancer were excluded from the Mayo Clinic study and did not contribute to the estimation of screening effectiveness. Nevertheless, the prevalent cases are relevant because their exclusion decreased the power of the study to detect a mortality rate reduction from screening and could have biased the estimation of relative risk against screening if it had been more effective on the first than subsequent rounds. However, the 5-year survival rate for the prevalent cases was the same as that for the incident cases detected by screening or nonstudy radiographs, about 40% (73). This similarity suggests that the exclusion of the prevalent cases did not substantially bias the Mayo Clinic study against screening.

Sanderson and Fontana concluded that mass screening for lung cancer is not justified on the basis of the Mayo Clinic study (73,87). They also acknowledged that lung cancer detection, resectability, and survival statistics can be highly misleading and that a decrease in lung cancer mortality rate is the ultimate measure of screening effectiveness. However, they raised also two important points. First, the site of origin of adenocarcinoma is often in doubt. It is likely that some of the excess deaths from adenocarcinoma were due to a combination of overdiagnosis of lung cancer in the intensively screened group and underdiagnosis in the control group, which biases the comparison of lung cancer mortality rate against screening. Second, the Mayo Clinic study did not have sufficient power to reliably exclude a modest 20% mortality rate reduction, especially considering the degree of contamination in the control group. These and other controversies pertaining to the Mayo Clinic study are discussed further later in this chapter.

b. Memorial Sloan-Kettering Study. Unlike the Mayo Clinic Study, the Memorial Sloan-Kettering Study was designed to determine the effectiveness of adding 4-monthly sputum cytology to annual chest radiography (87,88). From 1974 to 1978, 10,040 volunteers aged 45 years and older from the New York metropolitan area were enrolled (Table 5). Volunteers were not excluded on the basis of a screening chest radiograph or sputum cytology exam, as they were in the Mayo study. Instead, the volunteers in the Memorial study were randomized before the first screening. The 5,072 subjects in the dual screening group received 4-monthly sputum cytology in the addition to annual chest radiography. The 4968 in the control group received only the annual chest radiography. The two groups were very similar in age distribution, smoking history, residence,

Table 5 Memorial Sloan Kettering Study Summary Statistics

	Dual screened group (CXR + SC)	Control group (CXR alone)
Subjects	5072	4968
Prevalent cases	30	23
Incident cases	114	121
% resectable	51	53
% 5-year survival	36	33
Lung cancer deaths	78	87
Relative risk[a] (95% confidence interval)	0.88 (0.65–1.19)	

Abbreviations: CXR, chest radiography; SC, sputum cytology.
[a] Based on cumulative lung cancer mortality at 8 years (72).

and exposure to chemical carcinogens. Screening continued until 1982, by which time the subjects had been screened for five to eight years, depending on their time of enrollment.

At the first screening, 30 prevalent cases were detected in the dual screen group and 23 were detected in the control group (see Table 5) (88,89) In the dual screen group, nine cases were detected by sputum cytology alone, all of which were stage 0 or I squamous cell carcinomas. There were a total of 18 squamous cell carcinomas detected in the dual screen group versus 12 in the control group. For the three other major cell types, the number of cases in the two groups were very similar.

After the first screening, 114 incident cases were detected in the dual screen group and 121 were detected in the control group (see Table 5) (88). Of the cases in the dual screen group, 61% were detected by screening, and the remainder were diagnosed by symptoms that developed between screenings. Of the cases in the control group, 54% were detected by screening (chest radiography). In the dual screen group, nine additional squamous cell carcinomas were detected by sputum cytology alone. However, there were nine more incident cases of squamous cell carcinoma in the dual screen group than in the control group, 37 versus 28. For the three other major cell types, the number of incident cases in the two groups were very similar.

When screening was completed in 1982, the total number of prevalent and incident cases was 144 in each group (see Table 5). The distribution of stage and cell type and the percentage of cases undergoing complete resection were almost identical, 51% in the dual screen group versus 53% in the control group. There were 78 deaths resulting from lung cancer in the dual screen group versus 87 in the control group and the relative risk was 0.88, but the total number of deaths from all causes were nearly identical, 493 versus 491.

It was concluded that the addition of sputum cytology to annual chest radiography did not decrease lung cancer mortality rate in high-risk individuals (72,89). The addition of sputum cytology did appear to advance the time of diagnosis for some squamous cell carcinomas. However, the similarity in lung cancer resectability and mortality rate in the two groups suggested that the cases that had been detected by sputum cytology alone were slow growing and would have still been localized if they had been detected slightly later by annual chest radiography.

c. *Johns Hopkins Study.* The Johns Hopkins study was identical to the Memorial Sloan-Kettering study in purpose and design. Volunteers were solicited through radio, television, newspaper advertisements, and direct mailings to a list of licensed drivers in the Baltimore area (90). From 1973 to 1978, 10,387 eligible adult males 45 years of age and older volunteered for the study. Of these, 5226 were randomized to the dual screen group and 5161 to the control group (Table 6). Screening continued until 1982, by which time the subjects had been screened for 5 to 8 years, depending on their time of enrollment.

At the first screen, there were a similar number of prevalent cases in the two groups, 39 in the dual screen group and 40 in the control group (90). However, 74% of the cases in the dual screen group were stage I compared with 40% in the control group. In the dual screen group, 11 cases were detected by sputum cytology alone and all of these were stage I squamous cell carcinomas. In the dual screen group, 69% of patients underwent complete resection versus 42% in the control group. Five-year survival rate was about 60% in the dual screen group versus 35% in the control group.

After the first screen, 194 additional cases were diagnosed in the dual screen group versus 202 in the control group (see Table 6) (91). In the dual screen group, 22 of the cases were detected by sputum cytology alone. Age-adjusted lung cancer mortality rate was not significantly lower in the dual screen group than in the control group, 3.4 versus 3.8 per 1000 person-years. Some of this difference might have been attributable to chance misrandomization, which would explain the excess number of lung cancer cases in the control group. Regardless, it was concluded that the addition of sputum cytology to annual chest radiography did not significantly decrease lung cancer mortality rate in high-risk individuals (91).

3. *Erfurt County Trial*

A controlled, prospective, nonrandomized trail of lung cancer screening was carried out in the former German Democratic Republic (GDR) (92). The study population consisted of 143,880 men in Erfurt County who were between 40 and 65 years of age on December 31, 1971. Assignment to the semiannual screened group and control group was based on district of residence. The 41,532 subjects in the screened group were scheduled to undergo semiannual chest

Table 6 Johns Hopkins Study Summary Statistics

	Dual screened group (CXR + SC)	Control group (CXR alone)
Subjects	5226	5161
Prevalent cases	39	40
Incident cases	194	202
% resectable	NA	NA
% 5-year survival	NA	NA
Lung cancer deaths[a]	141	173
Relative risk[b] (95% confidence interval)	0.80 (0.65–1.00)	

Abbreviations: CXR, chest radiography; SC, sputum cytology; NA, not available from published reports.
[a]Based on extended follow-up (74).
[b]Based on cumulative lung cancer mortality rate reported by Flehinger et al. (74).

fluorography during the 6-year period from 1972 to 1977. During the same time period, the 102,348 subjects in the control group were scheduled to undergo chest fluorography at 1-to 2-year intervals (average interval 18 months). A retrospective study revealed similar incidence rates of lung cancer in the two groups before the study, but no baseline characteristics were reported.

During the 6-year period of the study, the rates of lung cancer incidence, resectability, and survival were higher in the screened group than in the control group (Table 7). The cumulative incidence of lung cancer was 9.0 per 1000 in the screened group versus 6.5 per 1000 in the control. In the screened group, 28% of cases were resected versus 19% in the control group. Five-year survival rate was 52% in the screened group versus 39% in the control group (based on

Table 7 Erfurt County Study Summary Statistics

	Screened group (CPF q 6 mos)	Control group (CPF q 2 yrs)
Subjects	41,532	102,348
Cumulative incidence per 1000	9.0	6.5
% resectable	28	19
% 5-year survival	52	39
Cumulative lung cancer deaths	8	6
Relative risk[a] (95% confidence interval)	1.34 (0.94–1.98)	

Abbreviations: CPF, miniature chest photofluoroscopy.
[a]Based on cumulative lung cancer mortality rate at 6 years (92).

an additional 4 years of observation of lung cancer cases). In addition, 47% of cases in the screened group were first detected by chest fluorography versus 27% in the control group. However, cumulative lung cancer mortality and all cause mortality rates were slightly higher in the screened group than in the control group, 8 versus 6 per 1000 and 13.1% versus 12.6%, respectively.

Although Wilde (92) reported no mortality reduction from semiannual screening, he did speculate that it may prolong survival rate for patients with peripheral lung cancers, particularly adenocarcinomas. Wilde attributed the excess incidence of lung cancer in the screened group and apparent lack of mortality rate reduction to a combination of overdiagnosis in the screened group and ascertainment bias in the cause of death.

4. *Czechoslovakian Trial*

The randomized clinical trial in Czechoslovakia was similar to the Mayo Clinic study in that both were designed to determine the effectiveness of dual screening with chest radiography and sputum cytology in a high-risk population (93,94). Screening consisted of semiannual PA chest photofluorography and sputum cytology. Suspicious photofluorographic findings were evaluated with a full-size PA and lateral chest radiograph, and suspicious cytological findings were evaluated by fiberoptic bronchoscopy.

Between June 1976 and July 1977, 6364 male smokers aged 40 to 64 years (150,000 lifetime consumption of cigarettes) were drawn from the general population of six districts in Czechoslovakia. Previously unsuspected lung cancer was detected by chest radiology or sputum cytology in 19 of these potential subjects (prevalence 3.0 per 1000), who were excluded from the study (Table 8). The remaining 6345 subjects were randomized to either the screening group or the control group after being stratified by age, lifetime number of cigarettes,

Table 8 Czechoslovakian Trial Summary Statistics

	Screened group	Control group
Subjects[a]	3171	3174
Incident cases	108	82
% resectable[b]	25	16
% 5-year survival[b]	25	0
Lung cancer deaths	64	47
Relative risk[c] (95% confidence interval)	1.36 (0.94–1.98)	

[a] After the initial screening of 6364 recruits, 19 prevalent cases were excluded before randomization.
[b] Based on cases diagnosed during the first 3 years of the study.
[c] Based on cumulative lung cancer mortality rate at 6 years (94).

socioeconomic group, occupational exposure to pollutants, and district of residence. During the first 3-year period of the study, the 3171 subjects in screening group were scheduled to receive six semiannual dual screenings while the 3174 subjects in the control group were scheduled to receive one dual screening at the end of the period. During the second 3-year period of the study, both groups were scheduled to receive three annual screenings with chest photofluorography alone to allow for the accrual of the lung cancer cases in the control group that corresponded to the cases that had already been detected in the screened group. All subjects were followed up for at least 6 years from entry and mortality rate from lung cancer and other causes was reported at the end of the 6th year. In addition, all lung cancer cases were followed up for at least 5 years from diagnosis.

After randomization, the two groups were nearly identical with respect to age distribution, smoking history, and other risk factors. During the first 3 years, the compliance in the screening group was 90%–95% at each screening. Compliance in the control group was nearly 100%. Immediately before the end of the 3rd year, 36 cases had been diagnosed in the screened group and 19 in the control group. In the screening group, 26 of the cases (67%) were detected at screening. Ten cases (33%) presented with symptoms that developed between screenings. In the control group, 15 (79%) cases presented with symptoms, three were detected at autopsy, and one was detected by chest radiography outside of the study. In the screening group, nine cases (25%) were resectable versus three (16%) in the control group (see Table 8). Five-year relative survival rate was 25% in the screened group versus 0% in the control group. During the seven episodes of dual screening, only two of the 55 detected cancers were detected by cytology alone.

During the remainder of the study, 72 additional cases were diagnosed in the screened group (35 screen detected and 37 interval cases) and 63 additional cases were diagnosed in the control group (38 screen detected and 25 interval cases). By the end of 6th year, 26 more cases of lung cancer had been diagnosed in the screened group than the control group, 108 versus 82, and there were 17 more lung cancer deaths in the screened group than in the control group, 64 versus 47 (see Table 8). The 30-day postoperative mortality rate was 9.1%.

Kubic et al. concluded that screening did not decrease lung cancer mortality rate. However, they could not explain the 16 excess lung cancer cases in the screened group. By their own reasoning, there were no more than six cases of pseudodisease detected by semiannual screening (20 of the 26 patients with screen-detected disease eventually died from lung cancer). In addition, they could not explain the 17 excess lung cancer deaths in the screened group. Ascertainment bias in the cause of death alone could not account for the excess lung cancer deaths because there were more total deaths (341 versus 293), cancer deaths (128 versus 99), and smoking-related deaths (202 versus 162) in the screened group than in the control group. These differences suggest that the screened group had a higher lung cancer risk (owing to some unknown confounder), incomplete ascertainment of death resulting from lung cancer and other causes, or some combination of both. Under all of theses conditions, the comparison of lung cancer mortality rate would have been biased against screening.

5. *PLCO Trial*

In November 1993, the National Cancer Institute initiated a multicenter randomized control trial of screening for cancer of the prostate, lung, colorectal, and ovary (PLCO) (95,96). The NCI plans to recruit a total of 148,000 subjects 55–74 years of age, half of them women, without a history of any of these four cancers. Half of the subjects will be randomized to the screened group, which is scheduled to undergo annual multiphasic screening for these cancers for four consecutive years (an initial screen plus three annual screens). The remaining half will be randomized to the control group, which will receive usual medical care. All subjects will be followed up for at least 10 years from entry. Cancer-specific mortality rate is the major endpoint for determining the effectiveness of screening, but incidence, stage at detection, survival, quality of life, complications from workup and treatment, and other data for each of the four cancer types will also be obtained.

Screening for lung cancer is performed with a full-size PA chest radiograph. Technical specifications include a high-energy (110–140 kVp) dedicated chest unit, wide latitude film, a 12 to 1 or higher grid, and source-to-film distance of 6 to 10 feet. At each participating center, each film is checked for quality by a radiological technologist and read by the study radiologist. A sample of the radiographs is read twice for quality control.

Because the eligibility criteria do not include a history of smoking, the subjects will have a lower risk of lung cancer development than the subjects in previous randomized trials of lung cancer screening. However, because of the large sample size, the study will have the power to detect a 10% lung cancer mortality rate reduction with a probability of 89% (assuming 100% compliance in both groups).

6. *Overview of Trials*

Despite early optimism for screening with chest radiography and sputum cytology, the controlled trials that have been completed have failed to demonstrate a reduction in lung cancer mortality rate (Table 9). However, all of the trials have limitations and the interpretation of the Mayo Clinic study, which is the study most relevant to screening with chest radiography, remains controversial.

a. Excess Cases in Screened Groups. The major source of controversy in this study is the excess of 46 lung cancer cases in the intensively screened group. Three major explanations for this excess have been provided, each of

Table 9 Overview of Prospective Controlled Trials of Lung Cancer Screening

Study (reference)	Start date	Study population	Study group	Control group	RR (95% CI)
London Mass Radiology[a] (84,85)	1960	Male factory workers	CXR q 6 mos	Usual care	0.89 (0.63–1.28)
Mayo Clinic study (73)	1971	Volunteer male smokers >45 years	CXR, SC q 4 mos	Usual care	1.06 (0.82–1.36)
Johns Hopkins study (74,90,91)	1973	Volunteer male smokers >45 years	SC q 4 months, CXR q 12 months	CXR q 12 months	0.80 (0.65–1.00)
Memorial Sloan-Kettering study (72,89)	1974	Volunteer male smokers >45 years	SC q 4 months, CXR q 12 months	CXR q 12 months	0.88 (0.65–1.19)
Erfurt County trial[b] (92)	1972	All males 40–65 years	CPF q 6 months	CFP q 18 months	1.34 (1.13–1.59)
Czechoslovakian trial (93,94)	1976	Male smokers 40–64 years	CPF, SC q 6 months	CPF, SC at 3 years	1.36 (0.94–1.98)
PLCO trial (95)	1993	Volunteers 60–74 years	CXR q 12 months	Usual care	In progress

Abbreviations: CPF, miniature chest photofluoroscopy; CXR, full-size chest radiography; SC, sputum cytology.

[a] Randomized by industrial firm

[b] Nonrandomized

which has a different implication for estimating the effectiveness of screening. Strauss et al. have attributed the excess cases to chance misrandomization (97,98) and population heterogeneity (99). They estimated that there would have been 154 lung cancer deaths in the control group instead of the observed 115 if the lung cancer incidence had been equal in both groups. Given the 122 lung cancer deaths in the intensively screened group, the relative risk would have been 0.79 (95% confidence interval 0.62 to 1.00). However, the probability that there would have been an excess of 46 or more in either group is only about 1% (100). In addition, chance misrandomization in the Mayo Clinic study would not explain the excess lung cancer cases observed in the other controlled trials that were designed to determine the effectiveness of screening with chest radiography (North West London, Erfurt County and Czechoslovakian trials).

A second explanation is that lung cancer was underdiagnosed in the control group (73). Some of the undiagnosed cases may have died of lung cancer and had their deaths attributed to other causes. This explanation is most relevant to the adenocarcinoma cell type, the origin of which is often unknown, and could account for some of the 15 excess deaths from adenocarcinoma in the intensively screened group. This explanation could also account for the excess lung cancer mortality rate observed in the screened groups of the other controlled trials. Underdiagnosis has more recently been observed in a large necropsy study of lung cancer (101). About one in six cases observed at necropsy had been falsely diagnosed premortem as another primary tumor. Symptoms may have eventually developed in some of the other undiagnosed cases in the control group of the Mayo Clinic study, after the observation period.

A third explanation is overdiagnosis of lung cancer survivors in the screened group, that is, the detection of pseudodisease by chest radiography or sputum cytology (77,100,102,103). Two types of pseudodisease have been distinguished (104). Type I is nonprogressive and could be due to false-positive errors in pathological diagnosis. Type II is preclinical disease that progresses too slowly to cause symptoms before the individual dies of other causes, such as coronary artery disease or chronic obstructive pulmonary disease, which are also related to smoking. Overdiagnosis may explain some of the 16 squamous cell carcinomas in the Mayo Clinic study that were detected only by sputum cytology and had a 5-year survival rate of 83%. Overdiagnosis may also explain some of the excess cases of adenocarcinoma and large cell carcinoma.

Autopsy studies provide compelling evidence that pseudodisease is abundant for cancer in general and that its reported prevalence increases with the intensity of the detection method (61). In the Chan necropsy study, about one in six lung cancers observed at necropsy had not been recognized at all premortem, and, in 65 of these 68 ''surprise'' cases, death was due to another cause (101). During the latter part of the study, from 1973 to 1982, the prevalence of ''surprise'' cases in men was nearly 1%, and 70% of these cases had been resectable and, presumably, asymptomatic. In other words, about 0.7% of the men died with some combination of type I and type II pseudodisease. In a recent study of emphysematous patients being considered for lung reduction surgery, unsuspected primary lung cancer was found by preoperative chest radiography in 2%. Thus, it is not unrealistic to expect 6 years of intensive screening to detect lung cancer pseudodisease in 1% of heavy smokers 45 years of age and older, or 46 cases of pseudodisease among the 4618 subjects in the intensively screened group of the Mayo Clinic study.

Sobue et al. (105) from Japan have argued that overdiagnosis by chest radiography is minimal on the basis of their follow-up study of 42 nonsurgically treated screen-detected stage I lung cancers. In their study, median survival was only 2 years and all patients died within 10 years. However, 17% of the deaths were attributed to other diseases, and causes of death other than lung cancer were used only when they were clearly identified. Thus, 17% or more of the cases may have been type II pseudodisease. In addition, five cases from the nationwide registry of lung cancer screening had been excluded from the study because of inappropriate evidence of malignancy. Some of these cases may have been type I pseudodisease. However, much overdiagnosis actually did occur in the Japanese screening program, the detection of types I and II pseudodisease would be higher in programs with higher rates of radiographical detection and less stringent pathological criteria for malignancy. In a recent study on the reproducibility of histological typing of lung cancer, Japanese pathologists classified more of the slide sets as benign or inadequate than did American pathologists, 39% versus 25% (17), suggesting that type I pseudodisease may be more common in the United States than in Japan. Nevertheless, the study by Sobue et al. strongly suggests that overdiagnosis does not fully account for the excess lung cancer cases in the Mayo Clinic study and other trials of screening done with chest radiography.

Perhaps the best explanation for the excess lung cancer cases is a combination of underdiagnosis in the control group and overdiagnosis in the screened group. In other words, intensive screening for asymptomatic lung cancer detects more cases that will eventually become symptomatic and more that will not. Accordingly, comparisons of lung cancer mortality rates underestimate benefit because of an underdiagnosis of lung cancer deaths in the control group, whereas comparisons of lung cancer survival rates overestimate benefit because of an overdiagnosis of lung cancer survivors (and lead time and length bias) in the screened group. In the lung cancer necropsy study, the number of wrongly diagnosed and ''surprise'' cases were about equal, which suggests that the problems of underdiagnosis and overdiagnosis in screening trials may be of about the same magnitude. If this had been true in the Mayo Clinic study, then 23 of the 46 excess cases would have resulted from underdiagnosis of lung cancer deaths in the control group and the remaining 23 excess cases would have resulted from overdiagnosis of lung cancer survivors in the intensively screened group. The relative risk would have been 0.88 (122 lung cancer deaths in the intensively screened group versus 138 in the control group) rather than the 1.06, as reported.

b. Statistical Power. Another problem with the Mayo Clinic study and other studies of lung cancer screening is a lack of statistical power. Even if it is assumed that there had not been chance misrandomization or an underdiagnosis in the control group, the Mayo Clinic study did not have sufficient power to detect a medically important benefit. Considering the burden of lung cancer, even modest risk reduction would not be trivial. In addition, to the extent that the control group was contaminated by screening outside of the study, any mortality rate reduction that might have existed would have been underestimated (83,106). Finally, the power of the Mayo Clinic study could have been further decreased by the exclusion of prevalent cases. These statistical considerations should not be a major problem in the ongoing PLCO trial, which has been designed to detect a 10% mortality rate reduction in men and women with a power of nearly 90% (96). In addition, this trial does not exclude prevalent cases from the first radiographical screening.

C. Case-Control Studies

Although it is widely agreed that the randomized controlled clinical trial is the most valid method of determining the effectiveness of screening, there are certain advantages to the case-control method. First, the case-control method can be used to evaluate screening interventions that are actually occurring in the population (107,108). In an RCT, subjects may not be willing to comply with their assignments to screened and control groups. Second, a case-control study requires many fewer subjects than an RCT (107,109). Consequently, it is possible to explore a multitude of hypotheses about a screening intervention, such as the relative efficacy of screening at different intensity levels. Finally, a case-control study can be completed in a much shorter period of time than can an RCT (months versus years), assuming that screening had actually taken place long enough ago for its effects to have become manifest.

1. *German Democratic Republic*

The first case-control study of lung cancer screening was performed in the former German Democratic Republic (GDR) (Table 10) (110), which had a screening program for tuberculosis during the early 1950s. Screening included a miniature, 70 × 70 mm, PA photofluorogram of the chest, which was initially obtained annually in all persons 16 years of age and older. In 1974, the interval

Table 10 Case-Control Studies of Lung Cancer Screening

Study (reference)	Study population	Intervention	Case group	Control group	RR (95% CI)
GDR study (110)	Male residents of Berlin district	At least one screening CPF within 2 years before time of diagnosis	130 lung cancer deaths 1980–1985	A: age-matched from district	0.88 (0.53–1.45)
				B: age-matched from hospital	1.09 (0.67–1.78)
GDR study (111)	Residents of Berlin <60 years with at least one CPF within 10 years before diagnosis	At least one screening CPF within 2 years before time of diagnosis	278 lung cancer deaths 1979–1987	Age-sex-matched from district	0.93 (0.65–1.33)
Japanese study (113)	Residents >16 years in 50 municipalities	At least one screening CPF within 1 year before time of diagnosis	273 lung cancer deaths 1981–1988	Matched by age, sex, smoking history, and health insurance	0.72 (0.50–1.03)

Abbreviations: CPF, miniature chest photofluorogram.

was prolonged to 2 years and, in 1984, the lower age limit was increased to 40 years. Although the importance of tuberculosis had since decreased, the screening program had continued. Its effect on lung cancer mortality rate was investigated by the case-control method and reported in 1987.

The cases included all 130 lung cancer deaths that occurred between January 1980 and December 1985 in one city district of Berlin in men who were younger than 70 years of age at time of diagnosis. All cases were histologically confirmed and classified as follows: squamous cell carcinoma, 48%; small cell carcinoma, 19%; large cell carcinoma, 18%; adenocarcinoma, 11%; other or unknown, 5%. At the time of diagnosis, 22% of the cases were stage I or II. There were two control groups selected during the beginning of 1986. For control group A, two living males matched for year of birth were randomly selected from the residents of the same district. For control group B, two living males matched for year of birth were randomly selected from the outpatient records of a general hospital in the area.

Information on the radiographical screening for cases and controls was obtained from the district records. For the cases, screening examinations were counted up to the time of the lung cancer diagnosis and included the examination on which the diagnosis was based if the disease was detected by screening. For the control subjects, screening examinations were counted up to the time of diagnosis in the matched case. Because smoking could potentially confound the relationship between screening and lung cancer death, smoking histories were obtained for the case group and control group B from the district screening records and hospital records, respectively. As expected, smoking was strongly related to lung cancer death. The relative risk for heavy smokers ($\geq$ 20 cigarettes per day) compared to nonsmokers was 12.4. However, smoking was not correlated with participation in screening for either the cases or controls. Therefore, smoking history was not considered in the analysis of screening effectiveness.

No statistically significant relationship between screening and lung cancer mortality rate could be found (see Table 10). In addition, there was no trend in the relative risk with regard to the number of tests performed in the 10 years before diagnosis or with regard to the time between the last test and diagnosis.

A few years later, the screening program was evaluated by a larger case-control study that included men and women from three city districts of Berlin (111). The study was restricted to subjects younger than 60 years of age, who are most likely to benefit. To minimize the healthy screenee effect (107,112), whereby prior screening would have selected control subjects with lower risks of eventual development of clinically overt lung cancer, the cases and control subjects were restricted to those who had been screened at least once during the 10 years preceding diagnosis. Again, no statistically significant relationship between screening and lung cancer mortality rate could be found (see Table 10). However, there were slight, nonstatistically significant trends related to the intensity of screening and time since last screening.

2. *Japan*

A case-control study was also undertaken in Japan where there is a national program for lung cancer screening (113). The program targets all residents aged 16 years or older from 50 local municipalities. Screening consists of annual chest photofluorography, using 90–140 kVp and 100-× 100-mm film. High-risk subjects (those who smoked more than 60 pack-years) also received annual 3-day pooled sputum cytology.

The cases consisted of 273 lung cancer deaths that occurred between 1981 and 1988. The cancers were diagnosed after the screening program was started and the patients died of lung cancer between the ages of 40 and 74. The distribution of cell type was as follows: squamous cell carcinoma, 33%; small cell carcinoma, 12%; large cell carcinoma, 7%; adenocarcinoma, 33%; unknown, 14%. At the time of diagnosis, only 12% of the cases were stage I or II. For each case, two to five control subjects were selected from those who were alive at the time of diagnosis of the corresponding case and matched by sex, age, smoking history, and type of health insurance. The total number of control subjects was 1269.

Information on the screening histories for cases and control subjects was obtained from the program records in each municipality. For the cases, screening examinations were counted up to the time of the lung cancer diagnosis and included the examination on which the diagnosis was based if the disease was detected by screening. For the control subjects, screening examinations were counted up to the time of diagnosis in the matched case.

The relationship between screening and lung cancer mortality rate approached statistical significance (see Table 10). The odds ratio of dying of lung cancer for persons screened within 12 months versus those not screened was 0.72. In addition, the odds ratio increased toward 1 as the length of time for which screening histories were compared was increased to 60 months. Furthermore, adjustment for other variables associated with opportunities for radiographical screening did not change the odds ratio.

Although there are several advantages to the case-control method, there are also several disadvantages. Case-control studies in general are susceptible to confounding (114). One major source of confounding in studies of screening is self-selection bias, in which subjects who are screened may be more or less likely to die of the disease of interest for reasons other than screening (108). In addition, there are several problems in design and analysis that are unique to case-control studies of screening (107,112).

D. Summary of Benefits, Harms, and Costs

1. *Benefit*

The bulk of the evidence to date suggests that screening with chest radiography does not substantively reduce lung cancer mortality rate, but a modest, 10%–20%, relative risk reduction has not been excluded. Given the enormous burden of disease, such a relative risk reduction might make radiographical screening worthwhile, especially for high-risk populations. In addition, earlier detection through chest radiographical screening might decrease the morbidity from lung cancer and the morbidity and mortality from treatment by providing less invasive options, such as lobectomy instead of pneumonectomy.

2. *Harms*

Chest radiography has no directly harmful effects. However, 1% to 10% of screenees have false-positive results, depending on how indeterminate results are classified. Although most of these cases in the NCI cooperative trials were resolved by follow-up chest radiography, many others required invasive procedures and at least two patients with benign lesions died of diagnostic thoracotomies. In addition, some screenees with screen-detected lung cancers are harmed by earlier detection. In the NCI cooperative trials, at least six patients died of treatment in the immediate postoperative period. Many of those who survived treatment eventually died of lung cancer and suffered the psychological (115) and physical effects of diagnosis and treatment for a longer duration than they would have without early detection. Those with pseudodisease suffer the greatest harm: diagnosis of and treatment for a condition that would not have become symptomatic during their lifetime. Reich has argued that screening for lung cancer ''does no good and may do much harm'' (116).

3. *Costs*

Screening for lung cancer incurs costs related to the screening tests, evaluation of positive results, and care for the disease. Rigorous cost-effectiveness analyses have not been published, probably because screening is not recommended. However, the cost can be estimated roughly from Medicare payments for physician services and hospitalizations related to lung cancer screening (Table 11).

For example, annual screening with a PA chest radiograph would cost about \$25 per screenee per year for the screening test alone. The evaluation of false-positive results would add considerably to the cost. If, for example, each year 5% of screenees required a CT scan and 1% a surgical resection to exclude malignancy, the cost of the screening would increase to about \$100 per screenee per year (\$25 + .05 × \$300 + .01 × \$6000). Whether or not screening would be cost effective would depend greatly on its effectiveness and the incidence of disease in the targeted population. If, for example, screening did actually reduce

Table 11 Medicare Payments[a] for Procedures Related to Lung Cancer Screening

Procedure (billing codes)	Amount
PA chest radiograph (CPT 71010)	$25
PA and lateral chest radiograph (CPT 71020)	$32
Chest CT with contrast (CPT 71260)	$302
Percutaneous biopsy under CT guidance (CPT 76360 + CPT 32405)	$476
Bronchoscopy and biopsy[b] (CPT 31625)	$246
Surgical resection for benign pulmonary nodule[b] (CPT 32140 + DRG 102)	$4,884
Lobectomy for lung cancer[b] (CPT 32480 + DRG 075)	$21,168

Abbreviations: CPT, current procedural terminology; DRG, diagnostic related group.
[a] Based on 1997 national fee schedule.
[b] Does not include payments for anesthesia and pathology.

lung cancer mortality rate by 10% and was targeted on the general adult population older than 50 years of age, which has an annual incidence of about 2 per 1000, then screening would cost about $500,000 per patient cured, excluding treatment costs ($\$10/(.002\times0.1)$). Assuming that cure prolongs survival by 10 years, the cost per year of life saved would be about $50,000, which is near the limit of acceptability. The cost of screening could be increased or decreased depending on how the population was selected, how chest radiographs were interpreted, and how patients were managed.

Although it is commonly assumed that earlier diagnosis reduces the cost of cancer care, in a recent study of Medicare payments for five common cancers (lung, breast, prostate, colon, and bladder), patients diagnosed at earlier stages incurred higher total payments between diagnosis and death than those diagnosed at later stages (117). For lung cancer in particular, total payments were $47,779 for patients diagnosed with localized disease versus $21,705 for those diagnosed with distant disease. Some of this difference would have to be added to the cost of lung cancer screening.

4. *New Screening Methods*

Even if chest radiography and sputum cytology are completely ineffective in screening for lung cancer, there have been dramatic advances in chest imaging and the molecular detection of cancer that provide opportunities for screening in the future. One advance in chest imaging is spiral (helical) CT (118), which can scan the entire thorax during a single breath-hold. The rapid and continuous scanning eliminates the problem of respiratory misregistration in conventional CT scanning (which may cause a nodule to be missed) and makes possible multiplanar and three-dimensional display of the anatomy. In a recent study (119) of 1369 Japanese smokers who were screened with both chest radiography

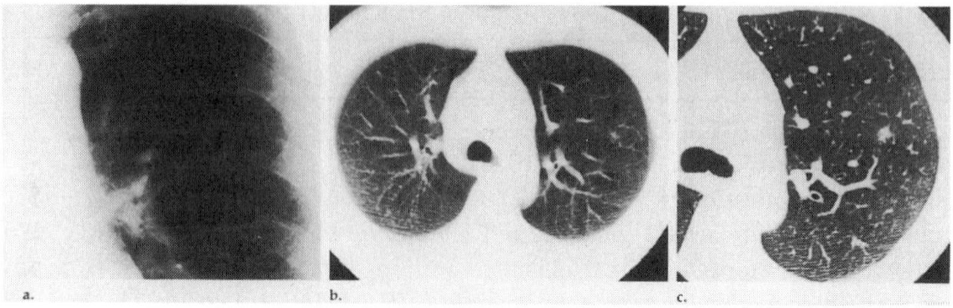

Figure 7 Well-differentiated adenocarcinoma detected by screening. a. Chest radiograph shows no nodule. b. Low-dose spiral computed tomography (CT) scan shows vague density in the left mid lung. c. Thin-section CT scan shows 7-× 7-mm spiculated nodule, which was surgically resected. (From Ref. 119.)

and low-dose spiral CT, only four peripheral lung cancers were detected by the former test while 15 were detected by the latter (Figure 7). The mean diameter of tumors detected by spiral CT alone was only 12 mm. Spiral CT also has potential for detecting endobronchial lesions. With perspective volume rendering, a new three-dimensional image processing technique, the tracheobronchial tree can be visualized from an internal perspective that simulates fiberoptic bronchoscopy and is termed ''virtual bronchoscopy'' (Figure 8) (119,120).

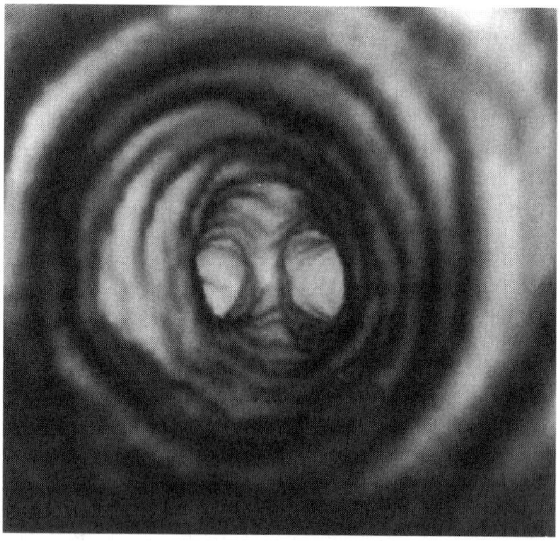

Figure 8 Virtual bronchoscopy. Perspective-volume-rendered spiral CT image of the trachea and main stem bronchi. (From Ref. 120.)

In addition to these advances in imaging, new molecular techniques can detect neoplastic transformation in the bronchial epithelial cells in sputum before morphological changes appear (78,122). One such technique is immunostaining with monoclonal antibodies directed at cell surface markers of neoplasia. Using archived sputum samples from a subset of participants in the John Hopkins study with atypia, Tockman et al. have demonstrated that this immunostaining technique can predict the development of lung cancer 2 years before clinical diagnosis with a 91% sensitivity and 88% specificity (123). This immunostaining technique is currently being evaluated in a prospective trial of patients with previously resected stage I lung cancer, in whom the incidence of new primary lung cancer is 1%–3% per year (123). In addition, bronchoalveolar lavage samples from a subset of these patients is being analyzed for tumor growth factors, oncogene mutations in exfoliated cells, and other possible tumor markers. Another new staining technique that could play a role in screening is fluorescence bronchoscopy using porphyrin derivatives that preferentially accumulate in neoplastic epithelium (124).

Undoubtedly, these and other advances in imaging and biomarker technology enable lung cancer and its precursors to be detected earlier than ever before. However, earlier diagnosis is a double-edged sword (42). On the positive side, it reduces morbidity or mortality from lung cancer in the subset for whom earlier treatment is more effective. In addition, earlier diagnosis reduces the morbidity or mortality from treatment in the subset for whom milder treatment becomes possible. On the negative side, however, earlier diagnosis increases the duration of psychological (115) and physical effects of diagnosis and treatment in the subset for whom earlier diagnosis is ineffective or unnecessary. Because the process of malignant transformation may evolve over many decades and at numerous sites throughout the lung, screening with biomarkers has the potential to detect an enormous amount of pseudodisease. To some extent, these benefits and harms of early diagnosis are inseparable and the screening tests that have great potential for benefit also have great potential for harm. For this reason, the effectiveness of new screening strategies should be proven in well-designed clinical trials that demonstrate a reduction in lung cancer mortality rate. In addition to clinical trials, quality of life assessments (125,126) and decision analysis (127) should play an increasing role in weighing the benefits, harms, and costs of the numerous screening alternatives that will become available.

E. Conclusion

No screening strategy has been shown to reduce lung cancer mortality rate. However, the evidence to date does not exclude a small mortality rate reduction from screening with chest radiography that could justify this intervention in high-risk populations. Eventually, the ongoing PLCO trial should help to resolve the uncertainty about the effectiveness of screening with chest radiography.

Regardless, there have been dramatic advances in chest imaging and the molecular detection of cancer that provide opportunities for lung cancer screening in the future. However, earlier diagnosis is a double-edged sword. New clinical trials and rigorous analysis will be required to ensure that the benefits of new screening strategies outweigh their harms and costs.

ACKNOWLEDGMENT

The author appreciates the thoughtful advice on this chapter by H. Gilbert Welch, MD.

REFERENCES

1. United States Preventive Services Task Force. Screening for lung cancer. In: Guide to Clinical Preventive Services. 2d ed. Baltimore: Williams & Wilkins, 1996:135–139.
2. Parker S L, Bolden S, Tong T, Wingo W A. Cancer statistics, 1996. CA Cancer J Clin 1996; 46:5–27.
3. Czaja R, McFall S L, Warnecke R B, Ford L, Kaluzny A D. Preferences of community physicians for cancer screening guidelines. Ann Intern Med 1994; 120: 602–608.
4. Ries L A G, Kosary C L, Hankey B F, Miller B A, Harras A, Edwards B K (eds). SEER Cancer Statistics Review, 1973–1994 [NIH Pub. No. 97–2789]. Bethesda, MD: National Cancer Institute, 1997:285.
5. Ernster V L. The epidemiology of lung cancer in women. Ann Epidemiol 1994; 4:102–110.
6. Shopland D R. Effect of smoking on the incidence and mortality of lung cancer. In: Johnson B E, Johnson D H, eds. Lung Cancer. New York: Wiley-Liss, 1995: 1–14.
7. Halpern M T, Gillespie B W, Warner K E. Patterns of absolute risk of lung cancer mortality in former smokers. J Natl Cancer Inst 1993; 85:457–464.
8. Tong L, Spitz M R, Fueger J J, Amos C A. Lung carcinoma in former smokers. Cancer 1996; 78:1004–1010.
9. Tobacco use and usual source of cigarettes among high school students—United States, 1995. MMWR Morb Mortal Wkly Rep 1996; 45:413–418.
10. Kreyberg L. International histological classification of tumours. In: Histological Typing of Lung Tumours. Geneva: World Health Organization, 1981.
11. Devesa S S, Shaw G L, Blot W J. Changing patterns of lung cancer incidence by histological type. Cancer Epidemiol Biomarkers Prev 1991; 1:29–34.
12. Wynder E L, Muscat J E. The changing epidemiology of smoking and lung cancer histology. Environ Health Perspect 1995; 103(suppl 8):143–148.

13. Campobasso O, Andrion A, Ribotta M, Ronco G. The value of the 1981 WHO histological classification in inter-observer reproducibility and changing pattern of lung cancer. Int J Cancer 1993; 53:205–208.
14. Ginsberg R J, Vokes E E, Raben A. Non-small lung cancer. In: Devita V T, Hellman S, Rosenberg S A, eds. Cancer: Principles and Practice of Oncology. 5th ed. Philadelphia: Lippincott-Raven, 1997: 858–911.
15. Linnoila R I, Aisner S C. Pathology of lung cancer: an exercise in classification. In: Johnson B E, Johnson D H, eds. Lung cancer. New York: Wiley-Liss, 1995: 73–95.
16. Feinstein A R, Gelfman N A, Yesner R. Observer variability in the histolopathological classification of lung cancer. Am Rev Respir Dis 1970; 101:671–684.
17. Keehn R, Auerbach O, Nambu S, et al. Reproducibility of major diagnoses in a binational study of lung cancer in uranium miners and atomic bomb survivors. Am J Clin Pathol 1994; 101:478–482.
18. U.S. Department of Health and Human Services. Reducing the health consequences of smoking: 25 years of progress. A report of the Surgeon General. [DHHS Publication No. (CDC) 89–8411]. Rockville, MD: U.S. Department of Health and Human Services, 1989.
19. Shopland D R, Eyre H J, Pechacek T F. Smoking-attributable cancer mortality in 1991: is lung cancer now the leading cause of death among smokers in the United States? J Natl Cancer Inst 1991; 83:1142–1148.
20. Blot W J, Fraumeni J F. Cancers of the lung and pleura. In: Schottenfeld D, Fraumeni J F, eds. Cancer epidemiology and prevention. New York: Oxford University Press, 1996:637–665.
21. Chow W H, Schuman L M, McLaughlin J K, et al. A cohort study of tobacco use, diet, occupation, and lung cancer mortality. Cancer Causes Control 1992; 3:247–254.
22. Morabia A, Wynder E L. Cigarette smoking and lung cancer cell types. Cancer 1991; 68:2074–2078.
23. Morabia A, Wynder E L. Relation of bronchioloalveolar carcinoma to tobacco. BMJ 1992; 304:541–543.
24. National Research Council. Committee on the Biological Effects of Ionizing Radiation. Health effects of exposure to low levels of ionizing radiation (BEIR V). Washington, DC: National Academy Press, 1990.
25. Lubin J H, Boice J D J, Edling C, et al. Lung cancer in radon-exposed miners and estimation of risk from indoor exposure. J Natl Cancer Inst 1995; 87:817–827.
26. Lubin J H, Boice J D J. Lung cancer risk from residential radon: meta-analysis from eight epidemiologic studies. J Natl Cancer Inst 1997; 89:49–57.
27. Samet J M. Indoor radon and lung cancer: risky or not?—all over again. J Natl Cancer Inst 1997; 89:4–6.
28. Cohen A J, Pope CAr. Lung cancer and air pollution. Environ Health Perspect 1995; 103(suppl 8): 219–224.
29. Rigas J R, Miller V A, Zhang Z F, et al. Metabolic phenotypes of retinoic acid and the risk of lung cancer. Cancer Res 1996; 56:2692–2696.

30. Spitz M R, Hsu T C. Mutagen sensitivity as a marker of cancer risk. Cancer Detect Prev 1994; 18:299–303.
31. Spitz M R, Hsu T C, Wu X, Fueger J J, Amos C I, Roth J A. Mutagen sensitivity as a biological marker of lung cancer risk in African Americans. Cancer Epidemiol Biomarkers Prev 1995; 4:99–103.
32. Strom S S, Wu S, Sigurdson A J, et al. Lung cancer, smoking patterns, and mutagen sensitivity in Mexican-Americans. J Natl Cancer Inst Monogr 1995; 18:29–33.
33. Ziegler R G, Mayne S T, Swanson C A. Nutrition and lung cancer. Cancer Causes Control 1996; 7:157–177.
34. The Alpha-Tocopherol BCCPSG. The effect of vitamin E and beta carotene on the incidence of lung cancer and other cancers in male smokers. N Engl J Med 1994; 330: 1029–1035.
35. Omenn G S, Goodman G E, Thornquist M D, et al. Effects of a combination of beta carotene and vitamin A on lung cancer and cardiovascular disease. N Engl J Med 1996; 334:1150–1155.
36. Mayne S T, Handelman G J, Beecher G. Beta carotene and lung cancer promotion in heavy smokers—a plausible relationship? J Natl Cancer Inst 1996; 88:1513–1515.
37. Auerbach O, Garfinkle L, Parks V R. Scar carcinoma of the lung: increase over a 21-year period. Cancer 1979; 43:636–642.
38. Fraser R S, Pare J A P, Fraser R G, Pare P D. Neoplastic diseases of the lungs. In: Synopsis of Diseases of the Chest. Philadelphia: W B Saunders, 1994: 445–538.
39. Muhn J R, Miller W E, Fontana R S, Sanderson D R, Uhlenhopp M A. Lung cancer detected during a screening program using four-month chest radiographs. Radiology 1983; 148:609–615.
40. Soda H, Tomita H, Kohno S, et al. Limitation of annual screening chest radiography for the diagnosis of lung cancer. A retrospective study. Cancer 1993; 72: 2341–2346.
41. Hakama M, Holli K, Visakorpi T, Pekola M, Kallioniemi O. Low biological aggressiveness of screen-detected lung cancers may indicate over-diagnosis. Int J Cancer 1996; 66:6–10.
42. Cole P, Morrison A S. Basic issues in population screening for patients. J Natl Cancer Inst 1980; 64:1263–1272.
43. Yankelevitz D F, Henschke C I. Does 2-year stability imply that pulmonary nodules are benign? Am J Roentgenol 1997; 168:325–328.
44. Armstrong P. Basic patterns in lung disease. In: Armstrong P, Wilson A G, Dee P, Hansell D M, eds. Imaging of Diseases of the Chest. 2d ed. St. Louis: Mosby, 1995:58–124.
45. Armstrong P. Neoplasms of the lungs, airways, and pleura. In: Armstrong P, Wilson A G, Dee P, Hansell D M, eds. Imaging of Diseases of the Chest. 2d ed. St. Louis: Mosby, 1995:272–368.
46. Swensen S J, Brown L R, Colby T V, Weaver A L, Midthun D E. Lung nodule enhancement at CT: prospective findings. Radiology 1996; 201:447–455.

47. Gurney J W. Determining the likelihood of malignancy in solitary pulmonary nodules with Bayesian analysis. Part I. Theory. Radiology 1993; 186:405–413.
48. Gurney J W, Lyddon D M, McKay J A. Determining the likelihood of malignancy in solitary pulmonary nodules with Bayesian analysis. Part II. Application. Radiology 1993; 186:415–422.
49. Mountain C F. Revisions in the International System for Staging Lung Cancer [see comments]. Chest 1997; 111(6):1710–1717.
50. Idhe D C, Pass H I, Glatstein E J. Small cell lung cancer. In: Devita V T, Hellman S, Rosenberg S A, ed. Cancer: Principles and Practice of Oncology. 5th ed. Philadelphia: Lippincott-Raven, 1997:911–949.
51. Black W C, Armstrong P, Daniels T M. Cost-effectiveness of CT in T1N0M0 lung cancer. Radiology 1988; 167:373–378.
52. Conces D J, Klink J F, Tarver R D, Moak G D. T1N0M0 lung cancer: evaluation with CT. Radiology 1989; 170:643–646.
53. Becker G L, Whitlock W L, Schaefer P S, Tenholder M F. The impact of thoracic computed tomography in clinically staged T1, N0, M0 chest lesions. Arch Intern Med 1990; 150:557–559.
54. Duncan K A, Gomersall L N, Weir J. Computed tomography of the chest in T1NoMo non-small cell bronchial carcinoma. Br J Radiol 1993; 66:20–22.
55. Michel F, Soler M, Imhof E, Perruchoud A. Initial staging of non-small cell lung cancer: value of routine radioisotope bone scanning. Thorax 1991; 46:469–473.
56. Hatter J, Kohman L J, Mosca R S, Graziano S L, Veit L J, Coleman M. Preoperative evaluation of stage I and stage II non-small cell lung cancer. Ann Thorac Surg 1994; 58: 1738–1741.
57. Silvestri G A, Littenberg B, Colice G L. The clinical evaluation for detecting metastatic lung cancer. A meta-analysis. Am J Respir Crit Care Med 1995; 152:225–230.
58. Colice G L, Birkmeyer J D, Black W C, Littenberg B, Silvestri G. Cost-effectiveness of head CT in patients with lung cancer without clinical evidence of metastases. Chest 1995; 108:1264–1271.
59. Mountain C F. A new international staging system for lung cancer. Chest 1986; 89(suppl):225–233.
60. Prorok P C, Hankey B F, Bundy B M. Concepts and problems in the evaluation of screening programs. J Chronic Dis 1981; 34:159–171.
61. Black W C, Welch H G. Advances in diagnostic imaging and overestimations of disease prevalence and the benefits of therapy. N Engl J Med 1993; 328:1237–1243.
62. Kramer B S, Gohagan J, Prorok P C. NIH Consensus 1994: screening. Gynecol Oncol 1994; 55(3 pt 2):S20–S21.
63. National Council on Radiation Protection and Measurements (NRCP). Ionizing radiation exposures of the population of the United States. Report No. 93. Washington, DC: National Council on Radiation Protection and Measurements, 1987.
64. American College of Radiology Standards. The performance of adult chest radiography. 1995:41–45.

65. Berlin N I, Buncher C R, Fontana R S, Frost J K, Melamed M R. The National Cancer Institute Cooperative Early Lung Cancer Detection Program: results of the initial screen (prevalence). Am Rev Respir Dis 1984; 130:545–549.
66. Stitik F P, Tockman M S, Khouri N G. Chest radiology. In: Miller A B, ed. Screening for cancer. Orlando, FL: Academic Press, 1985: 163–191.
67. Woodring J H. Pitfalls in the radiologic diagnosis of lung cancer. Am J Roentgenol 1990; 154:1165–1175.
68. Austin J H, Romney B M, Goldsmith L S. Missed bronchogenic carcinoma: radiographic findings in 27 patients with a potentially resectable lesion evident in rctrospect. Radiology 1992; 182:115–122.
69. Kundel H L, Nodine C F, Carmody D. Visual scanning, pattern recognition, and decision making in pulmonary nodule detection. Invest Radiol 1978; 13:175–181.
70. Morrison A S. Early detection: sensitivity and lead time. In: Screening in Chronic Disease. 2d ed. New York: Oxford University Press, 1992: 43–73.
71. Sobue T, Suzuki T, Matsuda M, Horai T, Kajita A, Kuriyama K. Sensitivity and specificity of lung cancer screening in Osaka, Japan. Jpn J Cancer Res 1991; 82: 1069–1076.
72. Melamed M R, Flehinger B J, Zaman M B, Heelan R T, Perchick W A, Martini N. Screening for early lung cancer: results of the Memorial Sloan-Kettering Study in New York. Chest 1984; 86:44–53.
73. Fontana R S, Sanderson D R, Woolner L B, et al. Screening for lung cancer. Cancer 1991; 67(suppl):1155–1164.
74. Flehinger B J, Melamed M R. Current status of screening for lung cancer. Chest Surg Clin North Am 1994; 4:1–14.
75. Steele J D. The solitary pulmonary nodule: report of a cooperative study of asymptomatic solitary pulmonary nodules in males. J Thorac Cardiovasc Surg 1963; 46: 21–39.
76. Early lung cancer detection. Summary and conclusions. Am Rev Respir Dis 1984; 130:565–567.
77. Bailar J C. Screening for lung cancer: where are we now? Am Rev Respir Dis 1984; 130:541–542.
78. Mulshine J L, Linnoila R I, Jensen S M, et al. Rational targets for the early detection of lung cancer. Monogr Natl Cancer Inst 1992; 13:183–190.
79. Weiss W. Survivorship among men with bronchogenic carcinoma: three studies in populations screened every six months. Arch Environ Health 1971; 22:168–173.
80. Boucot K R, Weiss W. Is curable lung cancer detected by semiannual screening? JAMA 1973; 224: 1361–1365.
81. Lilienfield A, Archer P G, Burnett C H, et al. An evaluation of radiologic and cytologic screening for early detection of lung cancer: a cooperative pilot study of the American Cancer Study and the Veterans Administration. Cancer Res 1966; 26:2083–2121.
82. Nash F A, Morgan J M, Tomkins J G. South London lung cancer study. BMJ 1968; 2:715–721.

83. Black W C, Welch H G. Screening for disease. Am J Roentgenol 1997; 168:3–11.
84. Brett G Z. The value of lung cancer detection by six-monthly chest radiographs. Thorax 1968; 23:414–420.
85. Brett G Z. Earlier diagnosis and survival in lung cancer. BMJ 1969; 4:260–262.
86. Fontana R S, Sanderson D R, Taylor W F, et al. Early lung cancer detection: results of the initial (prevalence) radiologic and cytologic screening in the Mayo Clinic study. Am Rev Respir Dis 1984; 130:561–565.
87. Sanderson D R. Lung cancer screening: the Mayo study. Chest 1986; 89(suppl): 324.
88. Flehinger B J, Melamed M R, Zaman M B, Heelan R T, Perchick W A, Martini N. Early lung cancer detection: results of the initial (prevalence) radiologic and cytologic screening in the Memorial Sloan-Kettering Study. Am Rev Respir Dis 1984; 130:555–560.
89. Martini N. Results of the Memorial Sloan-Kettering study in screening for lung cancer. 1986; 89(suppl):325.
90. Frost J K, Ball W C, Levin M L, et al. Early lung cancer detection: results of the initial (prevalence) radiologic and cytologic screening in the Johns Hopkins study. Am Rev Respir Dis 1984; 130:549–554.
91. Tockman M S. Lung cancer screening: the Johns Hopkins study. Chest 1986; 89(suppl):324.
92. Wilde J. A 10-year follow-up of semi-annual screening for early lung cancer in the Erfurt County (GDR). Eur Respir J 1989; 2:656–662.
93. Kubic A, Polak J. Lung cancer detection. Results of a randomised prospective study in Czechoslovakia. 1986; 57:2427–2437.
94. Kubic A, Parkin D M, Khlat M, Erban J, Polak J, Adamec M. Lack of benefit from semi-annual screening for cancer of the lung: follow-up report of a randomized controlled trial on a population of high-risk males in Czechoslovakia. Int J Cancer 1990; 45:26–33.
95. Kramer B S, Gohagan J, Prorok P C, Smart C. A National Cancer Institute sponsored screening trial for prostatic, lung, colorectal, and ovarian cancers. Cancer 1993; 71:589–593.
96. National Cancer Institute. Prostate, Lung, Colorectal, and Ovarian Cancer Screening Trial. Manual of Operations and Procedures. Rockville, MD: Westat, Inc, 1993.
97. Strauss G M, Gleason R E, Sugarbaker D J. Screening for lung cancer reexamined: a reinterpretation of the Mayo Lung Project randomized trial on lung cancer screening. Chest 1993; 103(suppl):337–341.
98. Strauss. Chest X-ray screening for lung cancer (letter). Chest 1995; 108:1770–1772.
99. Strauss G M, Gleason R E, Sugarbaker D J. Screening for lung cancer. Another look; a different view [see comments]. Chest 1997; 111:754–768.
100. Parkin D M, Pisani P. Lung cancer screening (letter). Chest 1994; 106:977.
101. Chan C K, Wells C K, McFarlane M J, Feinstein A R. More lung cancer but better

survival: implications of secular trends in ''necropsy surprise'' rates. Chest 1989; 96:291–296.
102. Eddy D. Screening for lung cancer. Ann Intern Med 1989; 111:232–237.
103. Weiss W. Chest X-ray screening for lung cancer (letter). Chest 1995; 108:1770.
104. Morrison A S. The natural history of disease in relation to measures of disease frequency. In: Screening in Chronic Disease. 2d ed. New York: Oxford University Press, 1992:25–42.
105. Sobue T, Suzuki T, Matsuda M, Kuroishi T, Ikeda S, Naruke T. Survival for clinical stage I lung cancer not surgically treated. Comparison between screen-detected and symptom-detected cases. The Japanese Lung Cancer Screening Research Group. Cancer 1992;69:685–692.
106. Lachin J M. Introduction to sample size determination and power analysis for clinical trials. Controlled Clin Trials 1981;2:93–113.
107. Weiss N S, McKnight B, Stevens N G. Approaches to the analysis of case-control studies of the efficacy of screening for cancer. Am J Epidemiol 1992; 135:817–823.
108. Connor R J, Prorok P C, Weed D L. The case-control design and the assessment of the efficacy of cancer screening. J Clin Epidemiol 1991;44:1215–1221.
109. Sasco A J, Day N E, Walter S D. Case-control studies for the evaluation of screening. J Chron Dis 1986; 39:399–405.
110. Ebeling K, Nischan P. Screening for lung cancer—results from a case control study. Int J Cancer 1987; 40:141–144.
111. Berndt R, Nischan P, Ebeling K. Screening for lung cancer in the middle-aged. Int J Cancer 1990; 45:229–230.
112. Morrison A S. Assessing the value of early treatment: non-experimental studies. In: Screening in Chronic Disease. 2d ed. New York: Oxford University Press, 1992:98–128.
113. Sobue T, Suzuki T, Naruke T, Group J L-C-S R. A case-control study for evaluating lung-cancer screening in Japan. Int J Cancer 1992; 50:230–237.
114. Hennekens C H, Buring J E. Case-control studies. In: Epidemiology in Medicine. Boston: Little, Brown, 1987:132–152.
115. Wardle J, Pope R. The psychological costs of screening for cancer. J Psychosom Res 1992; 36:609–624.
116. Reich J M. More on lung cancer screening (letter). Chest 1995; 108:294.
117. Riley G F, Potosky A L, Lubitz J D, Kessler L G. Medicare payments from diagnosis to death for elderly cancer patients by stage at diagnosis. Med Care 1995; 33:828–841.
118. Heiken J P, Brink J A, Vannier M W. Spiral (helical) CT. Radiology 1993; 189: 647–656.
119. Kaneko M, Eguchi K, Ohmatsu H, et al. Peripheral lung cancer: screening and detection with low-dose spiral CT versus radiography. Radiology 1996; 201:798–802.
120. Rubin G D, Beaulieu C F, Argiro V, et al. Perspective volume rendering of CT and MR images: applications for endoscopic imaging. Radiology 1996; 199:321–330.

121. Vining D J, Liu K, Choplin R H, Haponik E F. Virtual bronchoscopy. Relationships of virtual reality endobronchial simulations to actual bronchoscopic findings. Chest 1996; 109:549–553.
122. Mulshine J L, Scott F. Molecular markers in early cancer detection. New screening tools. Chest 1995; 107(suppl):280–286.
123. Tockman M S. Monoclonal antibody detection of premalignant lesions of the lung. In: Fortner J G S P, ed. Accomplishments in Cancer Research, 1995. Philadelphia: Lippincott-Raven, 1996:169–177.
124. Wagner H, Ruckdeschel J C. Lung Cancer. In: Reintgen D S, Clark R A, eds. Cancer Screening. St. Louis: Mosby, 1996:118–149.
125. Kaplan R M. Quality of life assessment for cost/utility studies in cancer. Cancer Treat Rev 1993; 19(suppl A): 85–96.
126. Nease R F J, Tsai R, Hynes L M, Littenberg B. Automated utility assessment of global health. Qual Life Res 1996; 5:175–182.
127. Russell L B, Gold M R, Siegel J E, Daniels N, Weinstein M C. The role of cost-effectiveness analysis in health and medicine. Panel on Cost-Effectiveness in Health and Medicine. JAMA 1996; 276:1172–1177.

14

Melanoma

Howard K. Koh
Massachusetts Department of Public Health; Cancer Prevention and Control Center, Boston University School of Medicine; and Boston University School of Public Health, Boston, Massachusetts

Alan C. Geller
Cancer Prevention and Control Center, Boston University School of Medicine, and Boston University School of Public Health, Boston, Massachusetts

Robert A. Lew
Cancer Prevention and Control Center, Boston University School of Medicine, and Multipurpose Arthritis Center, Brigham and Women's Hospital, Boston, Massachusetts

I. INTRODUCTION

In theory, screening and early detection programs could save lives otherwise lost to melanoma. This external cancer is uniquely visible, risk factors are well documented, and screening tests are safe and acceptable to the public (1). Furthermore, early melanoma can be cured by simple surgical excision.

Numerous melanoma control programs around the world have begun over the past decade. However, no randomized trials have been conducted to prove screening efficacy for melanoma, leaving much controversy in current guidelines (2). In addition, precise distinctions between screening and other forms of early detection or even education can be challenging in this visible tumor (3,4). This chapter reviews the current status of melanoma screening with attention to unresolved scientific questions.

Melanoma incidence and death rates have been increasing in most industrialized nations. In the United States (1997), melanoma incidence rates increased faster than for all other cancers (except prostate cancer), leading to more than 40,000 new cases (5,6). Mortality is increasing at rates surpassed only by lung cancer and non-Hodgkin's lymphoma, causing 7300 deaths per year (6). However, several recent studies using the Connecticut Tumor Registry, and Surveillance, Epidemiology, and End Results (SEER) data project a future downward trend in melanoma death rates in the United States (7,8). Similarly in Australia, melanoma mortality rates have appeared to plateau (9). International attention to early detection is one key explanation; cohort effects are another possibility (1,7,9–11).

II. THEORETICAL AND PRACTICAL ASPECTS OF MELANOMA SCREENING

According to the United States Preventive Services Task Force (USPSTF), effective screening must satisfy two major requirements (2):

1. The test must be able to detect the target condition earlier than without screening and with sufficient accuracy to avoid producing large numbers of false-positive and false-negative results (accuracy of screening test).
2. Screening for and treatment of persons with early disease should improve the likelihood of favorable health outcomes (e.g., reduced disease-specific morbidity or mortality) compared to treatment of patients with signs or symptoms of the disease (effectiveness of early detection) (2).

By these criteria, many researchers consider melanoma screening a concept that deserves scientific attention. The 5-year survival rates of melanoma decline steadily as the tumor thickness (in millimeters) and stage increase (1,10,11). Most people with thin stage I lesions can expect prolonged disease-free survival and even cure, whereas persons with thicker, later stage lesions (e.g., >4.0 mm) are more likely to die of metastatic disease (Table 1) (11). In 1986–1991, 82% of melanomas in the United States were diagnosed at a localized stage, 12% regional or distant, and 6% unstaged (6). Moreover, a substantial proportion of melanomas may arise from melanocytic nevi or atypical nevi (1,10,12,13). The screening examination (a visual examination by a qualified observer) can take only a few minutes, is safe, acceptable to the public, and regarded by many as reliable in diagnostic situations (1,12,13). Such examinations could uncover melanomas on the back and posterior legs (which cannot be easily viewed by the affected person) (1,10,12).

However, not all biological and clinical aspects of melanoma are conducive to screening, most notably, 1. some fraction of melanomas may be clinically unrecognizable, reducing the sensitivity of skin examinations (14); 2. the detectable preclinical phase in some melanomas may be virtually absent (e.g., nodular melanoma) or of too short a duration for detection by periodic screening (15); 3. screening often finds nonmelanoma skin cancers (basal cell carcinoma,

Table 1 The 1988 American Joint Committee on Cancer/International Union Against Cancer pTNM Staging System[a]

Stage	Criteria
IA	Primary melanoma ≤0.75 mm thick and/or Clark's level II (pT1); no nodal or systemic metastases (N0, M0)
IB	Primary melanoma 0.76 to 1.50 mm thick and/or Clark's level III (pT2); N0, M0
IIA	Primary melanoma 1.51 to 4.00 mm thick and/or Clark's level IV (pT3); N0, M0
IIB	Primary melanoma >4.0 mm thick and/or Clark's level V (pT4); N0, M0
III	Regional lymph node and/or in-transit metastases (any pT, N1 or N2, M0)
IV	Systemic metastases (any pT, any N, M1)

[a] The authors and the AJCC Melanoma Committee recommend that when there is a discordance between thickness and level, the measured tumor thickness shall take precedence and be used for pT staging.
Source: Balch C, Houghton A, Milton GW, et al eds. Cutaneous Melanoma. 2d ed. Philadelphia: JB Lippincott, 1992:217.

squamous cell cancer), which uncommonly cause death (16); 4. screening can introduce a host of biases (e.g., lead time bias, length bias [16]) or uncover ''pseudodisease'' (thin nonmetastasizing form of melanoma) (17).

III. APPLICATIONS OF SCREENING DEFINITIONS TO MELANOMA

Some public health experts try to distinguish melanoma screening from other forms of early detection, such as ''surveillance,'' ''opportunistic surveillance,'' ''case-finding'' (when searching for cancer within a routine physical examination by a physician), and other (4,16,18–24). Precise distinctions between melanoma screening and other forms of early detection and education are extraordinarily challenging for this visible cancer (3,4,10,24).

Broad definitions of cancer screening include the following:

1. ''Testing . . . of apparently healthy volunteers from the general population for the purpose of separating them into groups with high and low probabilities for a given disorder'' (19)

2. ''. . . the early detection of those diseases whose treatment is either easier or more effective when undertaken at an earlier point in time'' (19)
3. ''a process which allows detection of disease at an early or preinvasive stage before symptoms prompt the diagnosis'' (18,19)

Such broad definitions when applied to melanoma could refer to any process of visually examining the skin for early detection, ranging from an average-risk person conducting casual self-examination to high-risk persons under-going systematic monthly examination of the entire skin by a skin cancer specialist (10).

However other, more narrow cancer screening definitions apply less readily to melanoma. Sackett defines screening as an ''organized attempt to detect among apparently healthy people in the country, disease or risk factors of which they are *unaware*'' (19). Miller defines screening as ''the presumptive identification of *unrecognized* disease by the application of tests, examinations or other procedures that can be applied rapidly'' (18). The International Union Against Cancer (UICC) defined screening as an ''organized program in which *unsuspected* lesions in asymptomatic subjects in a defined population are found'' (20). Morrison defined screening as a ''purposeful extension of the process by which people *perceive symptoms* of their own illnesses and then consult physicians for diagnosis and treatment'' (21).

These definitions cannot fully capture the layers of subtlety involved in melanoma screening, especially in attempting to apply the terms ''asymptomatic,'' ''unsuspected,'' ''unaware,'' and ''unrecognized'' to skin lesions (3,4,22). Melanoma can be considered either asymptomatic (in that it uncommonly causes itching, discomfort, or pain) or symptomatic (in that persons may be able to see their own skin cancer) (3,4,22).

Furthermore, persons with melanoma/skin cancer may be one of the following (3,4,22):

1. *Unaware* of the lesion, because it is hard to see (e.g., on the back of the body)
2. *Aware* of the lesion, but *not suspicious* that it is cancer (even though it may be large, irregular in border, or changing)
3. *Aware* and *suspicious* of the lesion, but not concerned enough to have it checked promptly
4. *Aware* and *suspicious* of a lesion, but with real or perceived barriers to having it checked (such as lack of access to appropriate health care) (3,4,22).

Because these subtleties are unique to melanoma screening, scientists at the Second Epidemiology of Melanoma Conference in Vancouver, Canada (1991)

(4) first wrestled with these issues and generally concluded that the unique external and visible nature of skin cancer made screening and early detection indistinguishable. They gave less importance to rigid distinctions between them and more weight to whether a program improved early detection and found cases that would not otherwise be detectable at such an early stage by other prevailing means.

In this chapter, melanoma screening is discussed in such a broad sense. In addition, this chapter reviews the emerging melanoma control programs that attempt to exploit the unique visibility of this skin cancer by aligning screening with education to enhance early detection (3).

IV. SCREENING STUDY DESIGN AND OUTCOME MEASURES

A. Randomized Trials

The optimal screening study is a prospective randomized trial demonstrating sustained reductions in the melanoma mortality rates in a defined population (compared to a control population) (3,10,16,18).

Implementing a randomized controlled trial for melanoma screening presents many challenges, including logistical barriers, maintenance of a control population, and inherent difficulties arising from the unique nature of this visible cancer (22).

Logistical barriers include the challenges of randomizing at least 500,000 individuals, costs of follow-up of the cohort (especially the estimated 1.5% of persons with suspected melanoma), the training of nondermatologist physicians or other screeners to recognize melanoma and take appropriate action, establishment of systems to route persons with suspected melanoma to appropriate treatment, and the time burden on physician screeners (22). Ideally, randomized trials should consider blinding of physician screeners but, this, too, would be difficult.

Building and sustaining a control population would be hampered by the challenges of finding a population who neither examines their skin nor is seen by a physician for routine medical care. As well, a control population could be easily contaminated by growing media publicity regarding early detection of melanoma and skin cancer (22).

A randomized study design for melanoma screening in a medical setting may be hampered for the following reasons: physician reluctance to place persons with obviously apparent and visible atypical moles into a control group or restrictions on the physician's ability to provide skin cancer education or perform routine examinations.

Randomized prospective trials that compare cancer screening at regular intervals to no screening are very expensive, but feasible when a large group that has a substantial annual incidence rate can be followed up for a period of 10 years or more (25,26). But, the vast resources needed to mount a cancer study need sufficient power to minimize anticipated costs (27). Despite increasing incidence, melanoma remains a relatively uncommon cancer, making affordable prospective studies a challenge. Even affordable studies need careful design and strict assumptions need to be taken into account for statistical procedures used in analyzing prospective data (28–30).

Hence, a prospective study with the theoretically ideal endpoint of death from melanoma may lie beyond practical limits. No randomized trial has been conducted for melanoma to date, hence, policy makers must examine data from other types of studies.

B. Case Control Studies

Case-control studies can offer initial data to evaluate screening programs (10,16,18,24,31–35). Such a study could examine the rates of screening exposure in persons with advanced/metastatic melanoma versus control subjects selected at random from the population in which screening is offered (13). One immediate challenge is in determining exactly what constitutes a screening examination for melanoma. Unlike the memorable, tangible screening involved in sigmoidoscopy, mammography, or Papanicolaou smears, the process of visual screening for melanoma may not be readily recalled by persons screened. Similarly, physicians may perform only a cursory skin examination or not accurately document such an examination in the medical record even if one was done.

The major case-control study that has been published is a population-based Connecticut study of skin self-examination (SSE) for melanoma that included 650 cases and 549 controls (31). The study design allowed for separate estimation of the impact of SSE on reduced melanoma incidence and survival rate among incident cases. While practiced by only 15% of all subjects, SSE was associated with a reduced risk of melanoma incidence and a reduced risk of advanced disease among melanoma patients (unadjusted risk ratio = 0.58; 95% CI = 0.31–1.11). Berwick et al. speculated that SSE might reduce mortality rate from melanoma by 63%, but cautioned that longer follow-up is required to confirm this estimate (31). Further analyses of this study led Begg et al. to argue that the results of SSE (removal of precancerous moles) may aid primary prevention by reducing cancer incidence (32). Others note that selection bias, information bias, and confounding may explain some of the association between SSE and reduced incidence (33). In addition, the relatively short follow-up and the likelihood of lead time bias limited the association between SSE and improved survival rate after diagnosis (31).

C. Descriptive Studies

With no randomized trials and only a single major case-control study, the bulk of the published melanoma screening investigations fall in the category of descriptive studies. Such studies should be interpreted with caution, given the many possible biases involved. However, they can offer initial data relevant for future trials.

Because reaching mortality outcomes consumes much time, effort, and cost, investigators have searched for appropriate intermediate outcome measures to assess melanoma screening efficacy (4,10,16). Elwood stresses that a reduction in absolute incidence rates of advanced lesions in a defined population after screening could presage a future decrease in melanoma mortality rate because thin lesions are associated with improved survival rate (16). Other intermediate endpoints of interest that have been pursued include the following (3):

1. Demonstrating decreasing rates of thick melanomas in screened populations (including high-risk persons) compared with unscreened populations
2. Demonstrating greater proportions of high-risk persons attending screening sessions
3. Increasing access to medical care through screening, as proposed by Cole and Morrison (34)
4. Documenting the educational effects of screening programs.
5. Reversing years of potential life lost (YPLL) per death for melanoma (36)
6. Improving quality of life through early detection (reducing levels of physical and psychosocial morbidity seen with recurrent melanoma) (37)
7. Cost-effectiveness outcomes

The remainder of this chapter reviews the status of these descriptive studies, with attention to comparison groups, strengths and weaknesses of the outcomes proposed previously, initial data generated, and possible biases.

V. THE SCREENING EXAMINATION FOR MELANOMA

Cutaneous melanoma is perhaps more easily discovered than other types of cancer because it ''writes its message in the skin with its own ink, and it is there for all to see'' (38). Ideally, the physician examination for skin cancer (and ascertainment of risk factors) can be carried out in tandem, taking only a few minutes. Examiners need only a source of bright light and a magnifying lens. Some also use a blow dryer to better examine scalp lesions (Figure 1) (12).

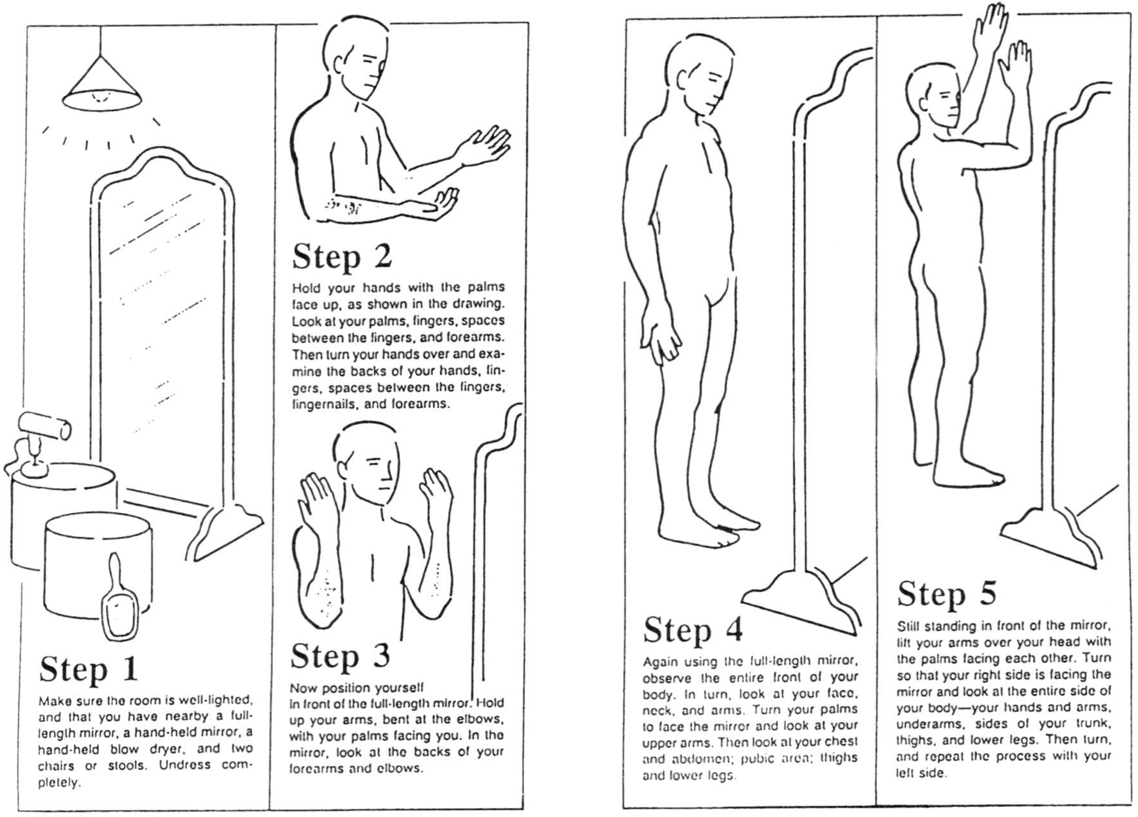
Step 1
Make sure the room is well-lighted, and that you have nearby a full-length mirror, a hand-held mirror, a hand-held blow dryer, and two chairs or stools. Undress completely.
Step 2
Hold your hands with the palms face up, as shown in the drawing. Look at your palms, fingers, spaces between the fingers, and forearms. Then turn your hands over and examine the backs of your hands, fingers, spaces between the fingers, fingernails, and forearms.
Step 3
Now position yourself
in front of the full-length mirror. Hold up your arms, bent at the elbows, with your palms facing you. In the mirror, look at the backs of your forearms and elbows.
Step 4
Again using the full-length mirror, observe the entire front of your body. In turn, look at your face, neck, and arms. Turn your palms to face the mirror and look at your upper arms. Then look at your chest and abdomen; pubic area; thighs and lower legs.
Step 5
Still standing in front of the mirror, lift your arms over your head with the palms facing each other. Turn so that your right side is facing the mirror and look at the entire side of your body—your hands and arms, underarms, sides of your trunk, thighs, and lower legs. Then turn, and repeat the process with your left side.

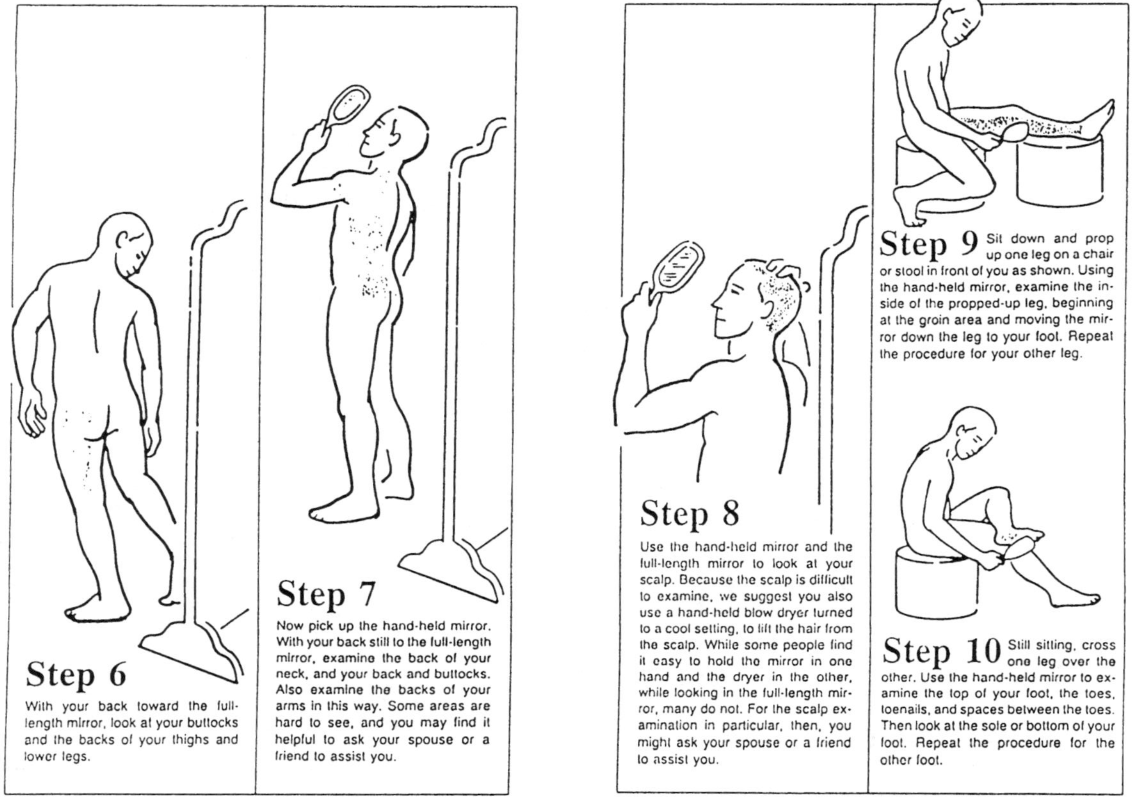

Figure 1 Skin self-examination. (From Ref. 12.)

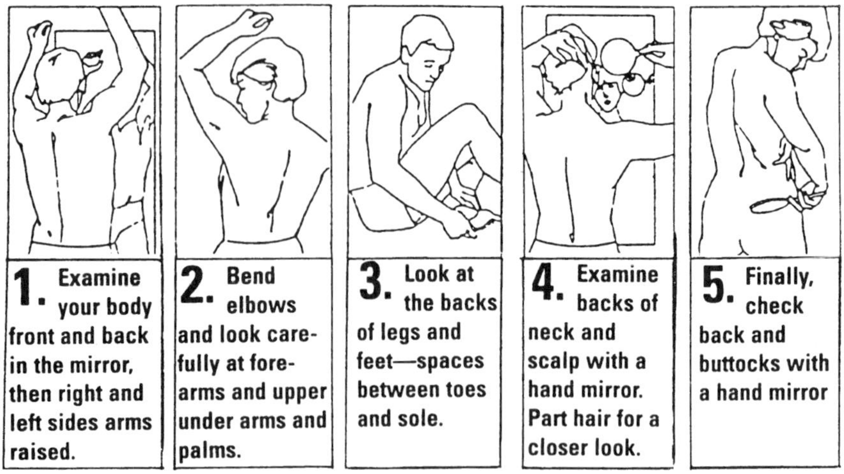

Figure 2 ABCD bookmark for self-examination for melanoma. (From Koh HK, Geller AC, Grossbart T, Miller DR, Lew RA. Prevention and early detection strategies for melanoma/skin cancer. Arch Dermatol 1996; 132:436–443.)

Unlike benign pigmented lesions, which are generally symmetrical, evenly bordered, and uniform in color, the characteristic clinical features of early melanoma follow the **ABCD** rule: **A**symmetry, **B**order irregularity, **C**olor variegation (or dark black color), and **D**iameter greater than 0.6 cm (the size of a pencil eraser) (Figure 2) (1,12).

Examiners should screen the total skin surface (except possibly the undergarment area), which may increase by six times the likelihood of detecting melanoma than partial examinations (39). In a systematic fashion, the screener examines every pigmented lesion looking for ABCD qualities or features of a skin lesion clearly different from its neighbors (''marching to a different drummer'') (1,12). In addition, screeners should be aware that men have more lesions on the back and women have more lesions on the lower legs (12,40). Because patients may be unaware of melanoma on the back and other areas that are difficult for self-inspection, screening of those sites could particularly aid early detection (Figure 3) (40).

Dermoscopy (or epiluminescence microscopy [ELM], which uses a handheld lens in combination with oil immersion to better view a pigmented lesion, could be of future potential value for improving melanoma detection. In addition, digital ELM offers the capacity to retrieve computerized images for skin lesions followed up in the clinical setting (41).

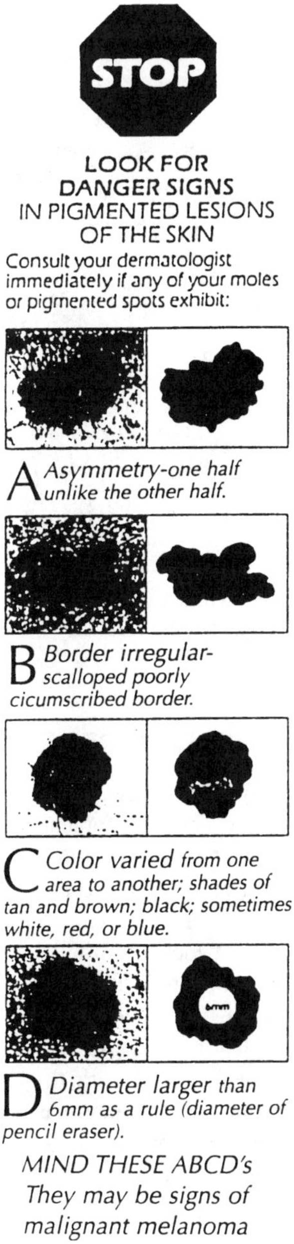

Figure 2 Continued

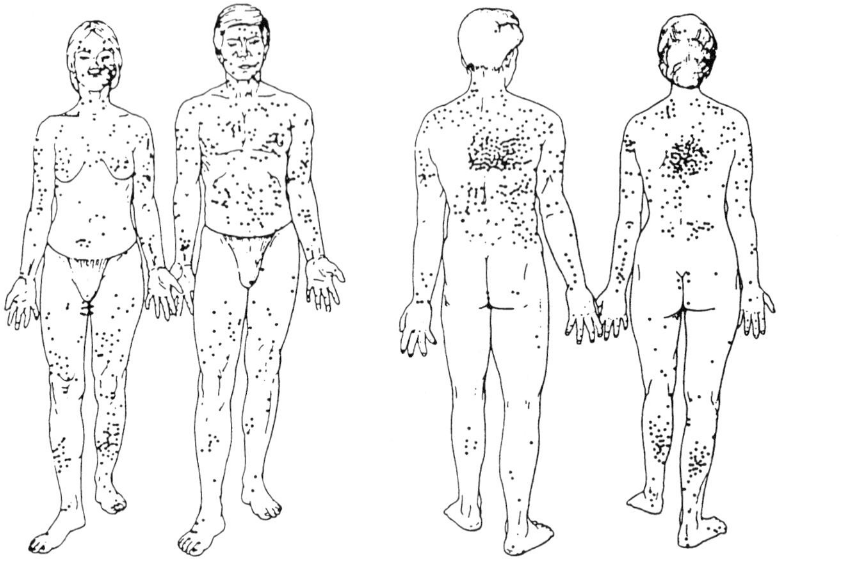

Figure 3 Sites of melanoma in men and women. (From Greenwald et al., eds. Cancer Prevention and Control. New York: Marcel Dekker, 1995.)

VI. SCREENING OF HIGH-RISK PERSONS

A precedent for screening high-risk persons has evolved by monitoring and educating persons with multiple nevi or atypical moles/dysplastic nevi within the familial melanoma setting (42,43). Even though such persons comprise only a small fraction of the population, they represent a model population for integrating all possible modalities to enhance early detection. All such studies to date have monitored endpoints of melanoma thickness, not mortality rates.

Surveillance and education of family members of melanoma patients in the Netherlands and in Philadelphia led to the detection of thinner melanomas. In the Leiden area of the Netherlands, Vasen et al. (42) classified 50 primary familial melanomas (among nine kindreds with the dysplastic nevus syndrome) into three groups. Comparing invasive melanomas at three time periods (before the start of screening, at initial examination, and during routine follow-up) found the average thickness decreasing from 1.75 mm to 0.80 mm to 0.54 mm, respectively ($P=0.001$), suggesting a benefit (42).

Masri et al. (43) identified family members of 264 familial melanoma kindreds at the University of Pennsylvania Pigmented Lesion Group. In a surveillance program offered to 555 persons, the mean thickness of detected melanomas was 1.44 mm for index lesions compared with a mean thickness of 0.52 mm for surveillance incident melanoma and 0.55 mm for nonsurveillance incident melanomas (for differences across the three groups). The authors concluded that surveillance and education within such a high-risk population could have a beneficial effect (43).

Investigators in Scotland followed up 116 patients with three or more clinically atypical nevi (without a personal or family history of melanoma) with education, regular skin examinations (3–6 months apart), and photographs (44). After 583 person-years of follow-up, the investigators found five invasive melanomas in 85 patients, compared with an expected number of invasive melanomas of .054 (RR 92, 95% CI 30–216) (44). Another study involving regular examination of 452 patients with dysplastic nevi found 18 newly diagnosed melanomas during a 27-month follow-up (12 in situ, six invasive melanomas 0.10 to 0.88 mm thick), comparing favorably with SEER populations (45). In a recent case-control study, Tucker et al. (46) noted that family members of unselected persons with dysplastic nevi could be appropriate targets for monitoring because they are more likely to have dysplastic nevi (RR 7.2 95% CI 2.1–24) and may be at increased risk for melanoma (46). Several other studies concluded that careful monitoring of patients with atypical nevi yielded thin, early melanomas (47,48). Similarly, patients with classic atypical-mole syndrome are more likely than control subjects to be diagnosed with invasive melanoma during a 10-year follow-up (48).

VII. SKIN SELF-EXAMINATION AND SELF-SCREENING

The case-control study by Berwick et al. (31) remains the major published investigation on skin self-examination in the United States.

Two Australian studies have published data on prevalence and accuracy of SSE. Girgis et al. (49) found that about half (48%) of 1344 individuals (from 1075 randomly selected households) self-screened or had their skin checked by another lay person. Australian men and persons of lower occupational and educational status, particularly the unemployed and infirmed, and persons with only basic medical insurance had poor quality in self-screening and medical screening (49). Also, the frequency of screening decreased sharply with increasing age. Among the same cohort, 12% had discovered one or more signs of melanoma during the past year, but only 32% sought medical opinion about these signs within the recommended period. Among the 1344 individuals, five were diagnosed with melanoma (50).

Studies on self-screening vary—one report found that 46% of all Americans conduct SSE, while another study found only one in five persons with melanoma reporting self-screening before diagnosis (40,51). Recently one study found that 61% of self-selected screenees performed SSE at least once during the past year, but only 20% reported monthly SSE (52). Only a minority of the screenees were knowledgeable about most of the recommended SSE steps (52). Multivariate analysis showed that psychological factors, such as optimism and perceived self-efficacy correlated positively with frequency of SSE (52). Yet another study found that only 6% of 874 patients seen in dermatology settings followed all recommendations for monthly self-examination, yearly professional examination, and sun protection (53). The authors concluded that physicians should perform total skin examination for patients, because most patients did not follow recommendations for at-home screening (53).

VIII. OTHER SCREENING PROGRAMS

A variety of screening programs have begun around the world. None are randomized and many combine education with screening.

A. United States

The American Academy of Dermatology (AAD) has provided the largest screening experience to date, with more than one million free skin screenings (1985–1997) (54). Each spring, the schedule of screening sessions is publicized and the public is invited to attend. The AAD physicians perform visual examinations and, if possible skin cancer is found, they ask the screening participants to see their own doctor for diagnosis and treatment (39,54–56). This effort represents one type of screening—a nationally organized effort of episodic screening by dermatologists on self-selected individuals (39,54–56). Initial evaluation has focused on a number of areas.

1. *Screening Yield and Stage Distribution of Melanoma*

In a first national attempt to evaluate this program, Koh et al. (54) contacted 96% of the 4458 persons with the screening designation of melanoma (from more than 282,000 screened, 1992–1994) to determine ultimate clinical outcome. Yield of pathology-confirmed melanomas was 129 per 100,000 persons screened. Even though earlier reports in state screenings showed biopsy-confirmed melanoma yields of 14/2239 in New York (39) and 9/2560 in Massachusetts (55) those yields contained mostly ''prevalent'' cases with data lacking on ''incident'' cases or interval cancers.

Almost all melanomas (98%) were localized (stages I and II), with a median thickness of 0.30 mm. Ninety percent of the confirmed melanomas were either in situ or less than 1.50 mm thick. Notably, 39% of persons with pathology-confirmed melanomas reported that they would not have otherwise seen a physician in the absence of screening. A total of 8.3% of cases were advanced melanoma (metastatic disease, regional disease, or lesions equal to or greater than 1.51 mm)—a lower proportion than that reported by the 1990 SEER registry (54).

2. *Potential Biases*

Because a number of biases may apply to these data, further research must determine exactly how these AAD findings apply to ongoing melanoma screening policy (4,16,54). Early detection and screening can detect the least aggressive tumors (length bias) (13,18,21). Some investigators even speculate that a proportion of melanomas are nonmetastasizing and perhaps more amenable to screening (17). Overdiagnosis bias may also apply (i.e., at least some of the in situ lesions may not have progressed to invasive cancer if left untreated) (54). All attendees in AAD screenings were self-selected and theoretically could be more health conscious than the general population; however, older men, who are at highest risk for late-stage melanoma and death and who are generally considered to be less health conscious, comprised a greater proportion of screened AAD cases than SEER cases (54,57).

Some patients with melanoma found in the screening program might die of their disease despite ''early'' detection (lead time bias) (4,13). The projected 5-year survival rates for persons found in these programs need to be confirmed through extended observation and comparison with case survival in the general (nonparticipating) population (3,4,58).

3. *Risk Factor Profiles of Attendees*

Even without specific targeting approaches, persons at high risk for skin cancer appear to be appropriately selecting themselves to be screened in these programs (59,60). Data from AAD programs in Massachusetts show that more than 86% had at least one risk factor and 78% had at least two risk factors (1986–1987) (59).

4. *Combining Education with Screening*

National surveys (1995) document that only one-third of Americans even recognize the term melanoma as a type of skin cancer, with approximately 50% of men and 35% of women being unaware (51). Levels of awareness are lowest among youngest respondents (ages 18–24 years of age) (51). Few Americans (26%) know the early signs of melanoma (51). In older persons (aged 50 years and older), skin cancer knowledge and prevention practices are significantly worse among men and persons with no education beyond high school (61).

Such data point to the need for more national skin cancer educational efforts. The AAD early detection initiative combines screening and education, blurring the distinction between the two (3,4,22). In this effort, national and local media (newspaper, radio, and television) have publicized skin cancer information each spring since 1985 (3). In May 1992, 238 television stations in more than 150 U.S. cities (with populations greater than 50,000) carried skin cancer messages during prime-time news coverage and publicized the availability of screening sessions (3). These messages penetrated all 30 of the top television markets in the United States, reaching an estimated 50 million Americans (3).

Further studies are required to measure the ''ripple effects'' or diffusion of this information (62,63), of not only the prescreening publicity but also the personalized health education delivered by the physician at the ''teachable moment'' of screening (22,58). In this way, as Morrison has noted, ''publicity efforts for cancer control have now given large numbers of well people advice as to when and how they should be examined for various diseases during their early, asymptomatic phases'' (21).

Only preliminary data document the educational effects of screening programs. Leffell et al. (64) found that participants receiving educational literature before the screening felt less anxiety about a full body examination than those who did not receive any information (64).

5. *Access to Care*

The program also provides an opportunity for health access studies (54,65). Based on 1992 and 1993 data (n=195,660), 80% of AAD screening participants reported not having a regular dermatologist, 52% would not have seen their physician without the screen, and 9% had no health insurance (65). Eighty-three percent of persons were attending their first program, suggesting screenings also offer an opportunity for many persons to receive their first 'expert' complete skin examination (3,59,65).

6. *Unresolved Questions*

More studies are needed to assess many scientific questions relevant to the AAD programs and melanoma screening in general. The impact of screening on melanoma mortality rates remains unclear. Analyses should determine yields of melanoma in participants with negative findings (false-negative rates). In addition, further investigations could determine the yield of repeat examinations, the impact of complete or partial skin examination on the yield, the frequency of interval cancers, and the precise impact of the biases on the results. Monitoring melanoma thickness over time would be best achieved by requiring that all U.S. cancer registries routinely record this important variable (66,67). Studies should try to link publicity messages that arouse concern with those at highest risk and determine the level of concern that convinces them to attend. Further research should also attempt to clarify the long-term educational effects of the screening program, for example, greater awareness and recognition of suspicious lesions, and vigilant self and family screening of suspected lesions (post-screening), as well as cost and benefits of screening.

B. New Zealand and Australia

The New Zealand Cancer Society sponsors ''skin check'' days, attended by more than 12,000 people between 1988 and 1989 (68). Published yields from one of these clinics found melanoma in three of 746 attendees. Evaluation of the effects of the New Zealand Spotcheck program in one large regional center found a significant increase in proportions of thin (less than 0.76 mm) invasive melanomas treated, (from 30% [1985] to 70% [1991] (69). High-risk individuals seem to attend these skin cancer screening programs (70).

In Australia, screening caravans, referred to as ''battle stations'' examine persons at beaches, city squares, and other sites (68). Screening procedures vary by state. Patients with suspicious lesions are encouraged to visit their own doctor after screening (68). Results from clinics in Western Australia (1991–1993) found four confirmed melanomas and two in situ melanomas among 3397 persons screened (71).

In Western Australia, a population-based study of 4176 persons (76% of all adults aged 40–64 years found on the electoral roll) were examined by local dermatologists (72). Among the 920 presumptively diagnosed skin cancers and suspicious pigmented lesions, there were 39 suspected melanomas, of which 12 were confirmed pathologically. An additional eight melanomas were found among persons clinically categorized as basal cell cancers (BCC) or suspicious pigmented lesions (72).

C. Netherlands

Melanoma control activities include public education campaigns throughout the Netherlands, and screening programs in two areas (73–75). In 1989, skin cancer screening programs used a mobile trailer along the beach in the western region of the Netherlands. Among the 3069 individuals screened, 65 persons were found to have suspected lesions and six persons were diagnosed with melanoma (all with a thickness less than 1 mm) (73). Rampen et al. (74) reported on screenings of 2564 persons in 1989 and 1990, finding 10 cases of melanoma and 43 cases of nonmelanoma skin cancer with no appreciable increase in workload.

D. Italy

In Italy, Cristofolini et al. (76,77) initiated an early diagnosis of melanoma campaign in the province of Trentino. Dermatologists were trained, general practitioners were informed, and public education was disseminated. They compared standardized mortality rates between Trentino and neighboring areas over time and made preliminary cost-effective ratio estimates. During this period, (1972 [6] versus 1982 [5]), Trentino experienced melanoma mortality rate increases of 44% (men) and 19% (women), whereas neighboring Lombardia had rate increases of 116% and 76%, respectively (76,77).

E. Sweden

In screening efforts in Sweden, Brandberg et al. (78,79) found increased attention to nevi in 190 persons after attendance at a skin cancer screening clinic. They found no increases in psychosomatic problems, anxiety, depressive symptoms, or signs of false security among 190 public screening participants surveyed before and after screenings (78,79).

F. Scotland

An educational campaign in Scotland to encourage skin cancer screening by primary providers found a trend toward reduction in thick tumors and mortality rate in women (80).

G. Germany

In 1991, 1467 persons from Bochum attended skin cancer screening sessions led by 36 dermatologists. Fourteen people (median age 53 years) were diagnosed with melanoma. Preparations for this campaign included 6 weeks of meetings with the local and national media as well as distribution of 30,000 flyers and 3000 posters (81).

IX. KEY CHALLENGES IN MELANOMA SCREENING

Some key issues in the challenges in melanoma screening are the following: who can screen, who should be screened, and where should screenings take place?

A. Who Can Screen?

The screening tool for melanoma remains the visual examination by the qualified observer (3,10,12,13). The quality of the visual examination (its sensitivity, specificity, and predictive value) differs according to type of screener and type of population screened (13). In addition to screening by dermatologists, screening could possibly be performed by a general physician, nurse, or the person himself or herself (self-screening) (62,63).

1. *Dermatologist in Clinical Settings*

As a starting point, some studies have documented the accuracy of dermatologists in diagnosing melanoma in the clinical setting. These studies describe the proficiency of dermatologists, not general physicians, in a diagnostic, not screening, setting. For example, Kopf et al. (82) noted an overall ''accuracy'' of 64%, whereas Grin et al. (83) documented improved clinical diagnosis of melanoma (from 1955 through 1982) with a sensitivity of 84.5% and specificity of 72.4%. Rampen and Rumke (84) observed variations in clinical accuracy by tumor thickness (58% for lesions less than 2 mm and 51% for those greater than 2 mm). Perednia et al. (85) noted significant differences among diagnostic sensitivity, reproducibility, and reliability of dermatologists in evaluating skin images.

2. *Dermatologists in Screening Setting*

Little data are available to evaluate the visual examination in the screening setting. Preliminary evaluation of AAD programs found that the predictive value positive of a visual examination by a dermatologist for melanoma was 27%–31% in Massachusetts screenings (1986–1989) and 17% in national AAD screenings (1992–1994) (54,55). Little data are relevant to specificity or sensitivity. However, in the Netherlands, investigators linked information on 1551 persons with a negative screening result (1990) with population-based registries (during 42 months of follow-up) and found no missed cases (86).

3. *Screening by Nondermatologist Physicians*

Professional education of medical providers could optimize accurate visual screening or case finding, that is, the testing of patients who have sought health care for disorders that may be unrelated to their chief complaints (19). ''The encounter is initiated by the patient and the purpose here is a comprehensive assessment of health'' (19). Integration of a skin examination into a routine physical examination by primary care providers may be the most practical and comprehensive strategy for reducing skin cancer morbidity and mortality rate (87,88). Many opportunities for such screening exist, in that patients with melanoma have contacts with physicians in the year before diagnosis (89). In one study of 216 patients, 87% had regular physicians and 63% had seen those physicians in the year before diagnosis, but only 24% of patients had examined their own skin before diagnosis and 20 % reported physician skin examinations (although data on patient recollection of physician skin examinations may be underestimated) (87). Studies of melanoma case-finding note that an estimated 14%–25% of melanoma is discovered by the physician (at a time when the patient had not noticed anything wrong) (40,89,90).

Few medical professionals have specific education in early detection of melanoma. A recent study evaluated 71 primary care residents (compared with 15 dermatologists and resident dermatologists) to determine their readiness to detect and triage suspicious lesions (91). Dermatologist scores were nearly double those of primary care residents, who failed 50% of the time to correctly diagnose nonmelanoma skin cancer and melanoma. In one-third of instances, primary care residents failed to recommend biopsies for cancerous lesions. The authors concluded that future studies should evaluate whether primary care training allows sufficient time for primary care physicians to learn the necessary skills (91). Earlier, Cassileth et al. (92) showed that only 12% of nondermatologists could correctly identify at least five of six melanomas compared with 69% of dermatologists. Girgis et al. (93) evaluated the effectiveness of an Australian postgraduate skin cancer training program in improving family doctors' (n=41) level of knowledge and clinical practice of melanoma. Even though physicians in the intervention group improved slide recognition of melanoma as well as their confidence levels to deliver advice, few changes were noted in clinical practice. In another study, 89% of second year students were capable of identifying a clinical photograph of a melanoma but only one fourth year student (in 285) detected a melanoma on a standardized patient (94).

Even though there are unresolved issues regarding skin cancer screening, a recent U.S. agenda-setting conference called for a new curriculum for the teaching of skin cancer prevention and detection for medical and nursing students, nurses, and allied health professionals (62,63). In addition, the National Institutes of Health Consensus Conference on the Diagnosis and Treatment of Early Melanoma (1992) suggested that physicians and nurses involved in primary care specialties (pediatrics, internal medicine, family practice, and obstetrics/gynecology) should communicate skin cancer information to patients and their families and identify members of high-risk groups (95).

The Australian Cancer Society has recommended that medical students learn about diagnosis and management of skin cancers and their precursors. Their education programs emphasize the role of the physicians to detect individuals at high risk within day to day medical practice, the role of ''opportunistic screening,'' and the ability to develop selective screening for high-risk individuals (96). Using national guidelines, prior physician screening had not occurred for 66% of Australian men and 70% of women (97).

4. *Self-Screening*

Recent studies have attempted to evaluate patients' ability to perform skin cancer examinations. One study evaluated patients' ability to apply a seven-point checklist (sensory change, diameter $\geqslant$ 1 cm, growth of the lesion, irregular edge, irregular pigmentation, inflammation, and crusting, oozing, or bleeding) to a skin lesion to prompt appropriate referrals to a dermatologist (98). Healsmith et al. (99) compared a seven-point checklist and the ''ABCDE system'' (E = elevation) and found all melanomas (n=65) to have at least one of the three major criteria of the seven-point list. Five melanomas were not detected by the ABCDE system (99). Keefe et al. (100) found that a seven-point checklist had a sensitivity of 71%, specificity of 99%, and positive predictive value of 7% for melanoma, using the dermatologist's clinical diagnosis as the benchmark.

Asking patients to count trunk moles could be one feasible way of identifying patients at high risk of melanoma (101). Lawson et al. (102) concluded that self-counts of large nevi are comparable to physician's counts and may be useful for melanoma screening. However, another study recently assessed the interobserver reliability between trained interviewers and dermatologists of nevi counts on the arms and found poor correlation (kappa=0.19) (103).

Gruber et al. (104) studied the ability of dermatology patients without melanoma to accurately record factors associated with an increased risk for melanoma (the number of freckles on the right forearm, the number of palpable arm nevi, and nevi greater than 5 mm in diameter on the entire body). These persons were first asked to examine their own skin and then they were subsequently examined by a physician. Using the physician examination as a standard, they found that specificity ranged from 83% to more than 95% for these three cutaneous markers (104). Sensitivity was 63% for detecting one or more palpable arm nevi and 68% for detecting large nevi.

B. Who Should Be Screened?

The most efficient cancer screening programs target individuals at high risk for the disease or death from the disease. For melanoma, high-risk persons include light skinned persons who suntan poorly or burn easily (RR 2–3), family history of melanoma (RR 2–8), number of large pigmented moles (RR 1–64), dysplastic nevi/atypical moles (RR 7–70), and excessive sun exposure (RR 3–5) (105). Research must better define risk markers associated with high attributable risk percent that identify a large proportion of individuals with cutaneous melanoma.

Some investigations have proposed the value of targeting future screening to those at risk for advanced disease, including older white men and possibly those of lower socioeconomic status (57,106–111).

1. *Middle-Aged and Older Men*

The National Cancer Institute's SEER Registry data indicate that the death rate for melanoma among men (1973–1988) increased faster than for any other cancer (57). Between 1973 and 1988, age-adjusted mortality rates increased 50% for white men (2.2 to 3.3 per 100,000) compared with 21% for women (1.4 to 1.7 per 100,000). Almost half (46%) of all melanoma deaths in the United States occur in men aged 50 years of age and older (57).

Similar gender differences in mortality trends occur in other countries. In New Zealand, almost half of all melanoma deaths occur among men 45 years of age and older (106). In Australia, the mortality rate for men aged 60 years and older has also increased, with a plateau in the rest of the population (8). In west Scotland, there may be a decline in melanoma mortality rate in women (80), whereas death rates have increased among men age 65 years and older (107).

Higher mortality rates in U.S. men may be attributable to their greater likelihood of presenting with regional and distant disease (14% of men versus 10% of women) ($P < 0.0001$) (6). A major survival analysis of more than 6500 melanoma patients found men had thicker lesions (median 1.4 mm versus 1.00 mm) as well as more ulcerated and fewer extremity lesions than women (11). In particular, men with back lesions had worse survival rates compared with women ($P=0.0163$). (11). Another Australian study found that men were more likely than women to be diagnosed with thick lesions (>3.00 mm) (108). An analysis of the South Australian Cancer Registry showed similar gender differences, particularly in elderly persons (109).

It remains unclear whether biological reasons, behavioral factors, or some combination explain why men appear to have thicker, later-stage lesions. Some researchers suggest hormonal differences can explain gender differences in melanoma survival. Others hypothesize that behavioral differences play a role, because one study found that men discover their own melanomas less frequently than women (40).

2. *Socioeconomic Status*

Socioeconomic factors may help target future melanoma screening programs. A recent study found that Massachusetts cases with the lowest socioeconomic status (SES) (census tracts with lowest proportion of high school graduates and households with low median income) were more likely to be diagnosed with advanced stage tumors and die of their melanoma (110). MacKie and Hole (111) also found higher incidence of melanoma among more affluent disease-free people, but poorer prognosis for persons of lowest SES (73% 5-year disease-free survival, versus 81%). Even after adjustment for known prognostic factors, the more affluent the group, the better the survival rate.

C. Where Should Screening Take Place?

Skin cancer screening can take place in a variety of settings: during physician examination (case finding), at health fairs, in occupational settings, or in mass efforts in which persons select themselves to attend (13,112).

X. CURRENT SKIN CANCER SCREENING RECOMMENDATIONS

In the absence of formal randomized trials, recommendations about skin cancer screening are diverse (Table 2). Some groups, such as the International Union Against Cancer (UICC), recomend no skin cancer screening (16,20,22). Others including the American Cancer Society, the American Academy of Dermatology, and the 1992 National Institutes of Health Consensus Conference on Early Melanoma do recommend some level of screening (16,22). The American Cancer Society recommends that all adults receive at least a baseline total body skin cancer screening examination from a physician, with subsequent skin exami-

Table 2 Skin Cancer Screening Guidelines

Organization (reference)	Recommends general population screening	Recommends screening for high-risk persons only	Does not recommend general population screening
American Cancer Society[a] 1980 (3)	x		
Canadian Task Force on the Periodic Health Examination, 1984 (3)		x	
National Cancer Institute, 1987 (3)	x		
Australian Cancer Society, 1991 (98)			x
International Union Against Cancer, 1992 (20)			x
American Academy of Dermatology, 1992 (3)	x		
National Institutes of Health (NIH) Consensus Development Panel on Early Melanoma, 1992 (97)	x		
United States Preventive Services Task Force, 1989 (3)		x	
United States Preventive Services Task Force, 1996 (2)			x

[a]The 1997 guidelines recommend all adults have a baseline complete skin examination by a physician, with future screenings to be determined by the physician.
Source: Balch C, Houghton A, Sober A, et al. Cutaneous Melanoma. 3d ed. St. Louis: Quality Medical Publishing, 1998.

nations at the physician's discretion as determined by risk status. Healthy People 2000 recommends an increase (to at least 40%) in the proportion of people aged 50 years and older visiting a primary care provider in the preceding year who have received skin examinations during one such visit (113).

The 1989 report of the U.S. Preventive Services Task Force (USPSTF) recommended screening for high-risk populations (family or personal history of skin cancer, clinical evidence of precursor lesions, and those with increased occupational or recreational exposure to sunlight) (3). Recently, however, the 1996 USPSTF found insufficient evidence to recommend for or against either routine screening for skin cancer by primary care providers or counseling patients to perform periodic skin self-examinations (2). They added that a recommendation to consider referring patients at substantially increased risk of malignant melanoma to skin cancer specialists for evaluation and surveillance may be made on other grounds (2). These recommendations will change and evolve over time as more data become available (22,58).

XI. CONCLUSION

Screening and early detection programs could potentially save lives otherwise lost to melanoma. Despite growing international interest in melanoma screening during the past decade, no randomized screening trial has been launched to demonstrate sustained reductions in melanoma mortality rates in a defined population. In lieu of such studies, evaluations must address a range of other design strategies to assess the potential impact of screening on mortality and late-stage melanoma.

REFERENCES

1. Koh HK. Cutaneous melanoma. N Engl J Med 1991; 325:171–182.
2. U.S. Preventive Services Task Force. Guide to Clinical Preventive Services. 2d ed. Baltimore: Williams Wilkins, 1996.
3. Koh HK, Geller AC, Miller DR, Lew RA. The early detection of and screening for melanoma. Cancer 1995; 75:674–683.
4. Koh HK, Miller DR, Geller AC, Lew RA. Screening for melanoma/skin cancer in the United States. In: Gallagher RP, ed. Epidemiology of Malignant Melanoma. Berlin: Springer-Verlag, 1994).
5. Parker SL, Tong T, Bolden S, Wingo P. Cancer statistics, 1997. CA Cancer J Clin 1997; 47:5–27.
6. Miller BA, Ries LAG, Hankey BF, Kosary CL, Edwards BK, eds. Cancer statistics review 1973–1989 [NIH Publication No. 92-2789]. National Cancer Institute, 1992.
7. Roush GC, McKay L, Holford TR. A reversal of the long-term increase in deaths attributable to malignant melanoma. Cancer 1992; 69:1714–1720.
8. Scotto J, Pitcher H, Lee JAH. Indications of future decreasing trends in skin-melanoma mortality among whites in the United States. Int J Cancer 1991; 49: 490–497.
9. Giles GG, Armstrong BK, Burton RC, et al. Has mortality from melanoma stopped rising in Australia? Analysis of trends between 1931 and 1994. BMJ 1996; 312: 1121–1125.
10. Roush GC, Berwick M, Koh HK, MacKie RM. Screening for melanoma. In: Balch CM, Houghton AN, Milton GW, Sober AJ, Soong SJ, eds. Cutaneous Melanoma. 2d ed. Philadelphia: JB Lippincott, 1992.
11. Balch CM, Soong SJ, Shaw HM, et al. An analysis of prognostic factors in 8500 patients with cutaneous melanoma. In: Balch CM, Houghton AN, Milton GW, Sober AJ, Soong SJ, eds. Cutaneous Melanoma. 2d ed. Philadelphia: JB Lippincott, 1992.
12. Friedman RJ, Rigel DS, Silverman MK, Kopf AW, Vossaert KA. Malignant melanoma in the 1990s: the continued importance of early detection and the role of physician examination and self-examination of the skin. CA Cancer J Clin 1991; 41:201–226.

13. Koh HK, Lew RA, Prout MN. Screening for melanoma/skin cancer: theoretic and practical considerations. J Am Acad Dermatol 1989; 20:159–175.
14. Andersen WK, Silvers DN. 'Melanoma? It can't be melanoma:' A subset of melanomas that defies clinical recognition. JAMA 1991; 266:3463–3465.
15. Bagley FH, Cady B, Lee A, Legg MA. Changes in clinical presentation and management of malignant melanoma. Cancer 1981; 47:2126–2134.
16. Elwood JM. Screening for melanoma and options for its evaluation. Journal of Medical Screening 1994; 1:22–38.
17. Burton RC, Armstrong BK. Recent incidence trends imply a non-metastasizing form of invasive melanoma. Melanoma Research 1994; 4:107–113.
18. Miller AB, ed. Screening for Cancer. Orlando, FL: Academic Press, 1985.
19. Sackett DL, Holland WW. Controversy in the detection of disease. Lancet 1975; 2:357–359.
20. Miller AB, Chamberlain J, Day NE, Hakama M, Prorok PC. Report on a workshop of the UICC project on evaluation of screening for cancer. Int J Cancer 1990; 46: 761–769.
21. Morrison AS. Screening in Chronic Disease. New York: Oxford University Press, 1985.
22. Koh HK, Geller AC, Miller DR, Lew RA. Early detection of melanoma: an ounce of prevention may be a ton of work. J Am Acad Dermatol 1993; 28:645–647.
23. Sackett DL, Haynes RB, Tugwell P. Clinical Epidemiology. Boston: Little, Brown, 1985:139–155.
24. Hakama M, Miller AB, Day NE. Screening for cancer of the uterine cervix. IARC Scientific Publications No. 76, Lyon, International Association for Research on Cancer, 1986.
25. Byar DP, Freedman LS. The importance and nature of cancer prevention trials. Sem Oncol 1990; 17:413–424.
26. Eddy DM. Screening for breast cancer. Ann Intern Med 1989; 111:389–399.
27. Hughes MD, Thompson SG, Pocock SJ. Optimal sequential screening guidelines for quantitative risk factors measured with error. J Am Stat Assoc 1995; 90:19–26.
28. Aron JL, Prorok PC. An analysis of the mortality effect in a breast cancer screening study. Int J Epidemiol 1986; 15:36–43.
29. Etzioni R, Self GS. On the catch-up time method for analyzing cancer screening trials. Biometrics 1995; 51:44–50.
30. Self GS, Etzioni R. A likelihood ratio test for cancer screening trials. Biometrics 1995; 51:44–50.
31. Berwick M, Begg CG, Fine JA, Roush GC, Barnhill RL. Screening for cutaneous melanoma by skin self-examination. J Natl Cancer Inst 1996; 88:17–23.
32. Begg CB, Huang Y, Berwick M. Separate estimation of primary and secondary cancer preventive impact: analysis of a case control study of skin self-examination and melanoma. J Am Stat Assoc 1996; 91:1381–1387.
33. Elwood JM. Skin self-examination and melanoma. J Natl Cancer Inst 1996; 88: 3–4.

34. Cole P, Morrison AS. Basic issues in population screening for skin cancer. J Natl Cancer Inst 1980; 64:1263–1272.
35. Sasco A, Day NE, Walter SD. Case-control studies for the evaluation of screening. J Chron Dis 1986; 39:399–405.
36. Albert VA, Koh HK, Geller AC, et al. Years of potential life lost: another indicator of the impact of cutaneous malignant melanoma on society. J Am Acad Dermatol 1990; 23:308–310.
37. Temoshok L, Heller BW, Sagebiel RW, et al. The relationship of psychosocial factors to prognostic indicators in cutaneous malignant melanoma. J Psych Res 1986; 29:139–153.
38. Davis N. Modern concepts of melanoma and its management. Ann Plast Surg 1978; 1:628–629.
39. Rigel DS, Friedman RJ, Kopf AW, Weltman R, Prioleau PG, Safai B, et al. Importance of complete cutaneous examination for the detection of malignant melanoma. J Am Acad Dermatol 1986; 14:857–860.
40. Koh HK, Miller DR, Geller AC, Clapp RW, Mercer MB, Lew RA. Who discovers melanoma? Patterns from a population-based survey. J Am Acad Dermat 1992; 26:914–919.
41. Kenet RO, Kang S, Kenet BJ, Fitzpatrick TB, Sober AJ, Barnhill RL. Clinical diagnosis of pigmented lesions using digital epiluminescence microscopy. Arch Dermatol 1993; 129:157–175.
42. Vasen HFA, Bergman W, Van Haeringen A, Scheffer E, Van Slooten EA. The familial dysplastic nevus syndrome: natural history and the impact of screening on prognosis. A study of nine families in the Netherlands. Eur J Cancer Clin Oncol 1989; 25:337–341.
43. Masri GD, Clark WH Jr, Guerry D IV, Halpern A, Thompson J, Elder DE. Screening and surveillance of patients at high risk for malignant melanoma results in detection of earlier disease. J Am Acad Dermatol 1990; 22:1042–1048.
44. MacKie RM, McHenry P, Hole D. Accelerated detection with prospective surveillance for cutaneous malignant melanoma in high-risk groups. Lancet 1993; 341:1618–1620.
45. Rigel DS, Rivers JK, Kopf AW, Friedman RJ, Vinokur AF, Heilman ER, et al. Dysplastic nevi: markers for increased risk for melanoma. Cancer 1989; 63:386–389.
46. Tucker MA, Crutcher WA, Hartge P, Sagebiel RW. Familial and cutaneous features of dysplastic nevi: a case-control study. J Am Acad Dermatol 1993; 28:558–564.
47. Kang W, Barnhill RL, Mihm MC, et al. Melanoma risk in individuals with clinically atypical nevi. Arch Dermatol 1994; 130:999–1001.
48. Marghoob AA, Kopf AW, Rigel DS, et al. Risk of cutaneous malignant melanoma in patients with 'classic' atypical-mole syndrome. Arch Dermatol 1994; 130:993–998.
49. Girgis A, Campbell EM, Redman S, Sanson-Fisher RW. Screening for melanoma: a community survey of prevalence and predictors. Med J Aust 1991; 154: 338–343.

50. Hennrikus D, Girgis A, Redman S, Sanson-Fisher RW. A community study of delay in presenting with signs of melanoma to medical practitioners. Arch Dermatol 1991; 127:356–361.
51. Miller DR, Geller AC, Wyatt SW, Halpern A, Howell JB, Cockerell CC, Reilley BA, Bewerse BA, Rigel D, Rosenthal L, Amonette R, Sun T, Grossbart T, Lew RA, Koh HK. Melanoma awareness and self-examination practices: results from a national survey of Americans. J Am Acad Dermatol 1996; 34:962–970.
52. Friedman LC, Bruce S, Webb JA, Weinberg AD, Cooper HP. Skin self-examination in a population at increased risk of skin cancer. Am J Prev Med 1993; 9:359–364.
53. Lee G, Massa MC, Welykyj S, Choo J, Greaney V. Yield from total skin examination and effectiveness of skin cancer awareness program. Findings in 874 new dermatology patients. Cancer 1991; 67:202–205.
54. Koh HK, Norton LA, Geller AC, Sun T, Rigel DS, Miller DR, Sikes RG, Vigeland K, Bachenberg EU, Menon PA, Billon SF, Goldberg G, Scarborough DA, Ramsdell WM, Muscarella VA, Lew RA. Evaluation of the American Academy of Dermatology's national skin cancer early detection and screening program. J Am Acad Dermatol 1996; 34:971–978.
55. Koh HK, Geller AC, Miller DR, Grossbart TA, Lew RA. Prevention and early detection strategies for melanoma and skin cancer. Arch Dermatol 1996; 132:436–443.
56. Bolognia JL, Berwick M, Fine JA. Complete follow-up and evaluation of a skin cancer screening in Connecticut. J Am Acad Dermatol 1990; 23:1098–1106.
57. Geller AC, Koh HK, Miller DR, Mercer MB, Lew RA. Death rates of malignant melanoma among white men—United States, 1973–1988. MMWR 1992; 41:20–21, 27.
58. Koh HK, Geller AC, Miller DR, Lew RA. The current status of melanoma early detection and screening. In: Weinstock M, ed. Dermatoepidemiology Dermatologic Clinics of North America. Philadelphia: WB Saunders, 1995: 623–634.
59. Koh HK, Geller AC, Miller DR, Caruso A, Gage I, Lew RA. Who is being screened for melanoma/skin cancer? Characteristics of persons screened in Massachusetts. J Am Acad Dermatol 1991; 24:271–277.
60. Weinstock MA. Prevalence of the early warning signs of melanoma among participants in the 1989 Rhode Island skin cancer screening. J Am Acad Dermatol 1990; 23:516–518.
61. Miller DR, Koh HK, Geller AC, Lew RA. National survey of skin cancer awareness (abstr). San Francisco: American Public Health Association, 1993.
62. Goldsmith L, Koh HK, Bewerse B, Reilley B, Wyatt SW, Bergfeld W, Geller AC, Walters PF. Full proceedings from the National Conference to Develop a National Skin Cancer Agenda. American Academy of Dermatology and Centers for Disease Control and Prevention, Washington, DC, April 8–10, 1995. J Am Acad Dermatol 1996; 35:748–756.
63. Goldsmith L, Koh HK, Bewerse B, Reilley B, Wyatt SW, Bergfeld W, Geller AC, Walters PF. Proceedings from the conference to develop a national skin cancer agenda (executive summary). American Academy of Dermatology and Centers for

Disease Control and Prevention. April 8–10, 1995. J Am Acad Dermatol 1996; 34:822–823.
64. Leffell DJ, Berwick M, Bolognia J. The effect of pre-education on patient compliance with full-body examination in a public skin cancer screening. J Dermatol Surg Oncol 1993; 19:660–663.
65. Koh HK, Geller AC, Miller DR, Rosenthal L, Lew RA. Health care access issues in national skin cancer screening programs. Abstract, presented at American Public Health Association meeting, Washington, DC, November, 1994.
66. Koh HK, Adame N, Geller AC, Clapp RW, Miller DR, Lew RA. Cancer registry data on melanomas. N Engl J Med 1990; 323:921–922.
67. Balch CM, Milton GW, eds. Cutaneous melanoma: clinical management and treatment results worldwide. Philadelphia: JB Lippincott, 1985.
68. Elwood JM. Screening and early diagnosis for melanoma in Australia and New Zealand. In: Miller AB, Chamberlain J, Day NE, Hakama M, Prorok PC, eds. Cancer Screening. Cambridge, MA: Cambridge University Press, 1991:243–255.
69. Tonks T, Duncan G, Crichton S. Public education and depth of primary malignant melanoma, Hutt Hospital, 1985–92. N Z Med J 1995; 108:344–346.
70. McGee R, Elwood M, Williams S, et al. Who comes to skin checks? NZ Med J 1994; 107:58–60.
71. Katris P, Crock JG, Gray BN. Research note: the Lions Cancer Institute and the Western Australian Society of Plastic Surgeons skin cancer screening programme. Aust N Z J Surg 1996; 66:101–104.
72. Kricker A, English DR, Randell PL, et al. Skin cancer in Geraldton, Western Australia: a survey of incidence and prevalence. Med J Aust 1990; 152:399–407.
73. Krol S, Keijser MT, Van Der Rhee HJ, Welvaart K. Screening for skin cancer in the Netherlands. Acta Derm Venereol (Stockh) 1991; 71;317–321.
74. Rampen FHJ, van Huystee BE, Kiemeney LA. Melanoma/skin cancer screening clinics; experiences in The Netherlands. J Am Acad Dermatol 1991; 25:776–777.
75. deRooij MJM, Rampen FHJ, Schouten LJ, Newumann HAM. Skin cancer screening focusing on melanoma yields more selective attendance. Arch Dermatol 1995; 131:422–425.
76. Cristofolini M, Bianchi R, Boi S, Decarli A, Hanau C, Micciolo R, et al. Analysis of the cost-effectiveness ratio of the health campaign for the early diagnosis of cutaneous melanoma in Trentino, Italy. Cancer 1993; 71;370–374.
77. Cristofolini M, Bianchi R, Boi S, Decarli A, Micciolo R, Cristofolini P, et al. Effectiveness of the health campaign for the early diagnosis of cutaneous melanoma in Trentino, Italy. J Dermatol Surg Oncol 1993; 19:117–120.
78. Brandberg Y, Bolund C, Michelson H, Mansson-Brahme E, Ringborg U, Sjoden PO. Psychological reactions in public melanoma screening. Eur J Cancer 1993; 29A:860–863.
79. Brandberg Y, Bolund C, Michelson H, Mansson-Brahme E, Ringborg U, Sjoden P. Psychological effects of participation in a prevention programme for individuals with increased risk for malignant melanoma. Eur J Cancer 1992; 28A:1334–1338.

80. MacKie RM, Hole D. Audit of public education campaign to encourage earlier detection of malignant melanoma. BMJ 1992; 304:1012–1015.
81. Hoffman K, Dirschka T, Schatz H, et al. A local education campaign on early diagnosis of malignant melanoma. Eur J Epidemiol 1993; 9:591–598.
82. Kopf AW, Mintzis M, Bart RS. Diagnostic accuracy in malignant melanoma. Arch Dermatol 1975; 111:1291–1292.
83. Grin CM, Kopf AW, Welkovich B, Bart RS, Levenstein MJ. Accuracy in the clinical diagnosis of malignant melanoma. Arch Dermatol 1990; 126:763–766.
84. Rampen FHJ, Rumke P. Referral pattern and accuracy of clinical diagnosis of cutaneous melanoma. Acta Derm Venereol (Stockh) 1988; 68:61–64.
85. Perednia DA, Gaines J, Rossum AC. Variability in physician assessment of lesions in cutaneous images and its implications for skin screening and computer-assisted diagnosis. Arch Dermatol 1992; 128:357–364.
86. Rampen FHJ, Casparie-van Velsen I, van Huystee B, Kiemeney L, Schouten LJ. False-negative findings in skin cancer and melanoma screening. J Am Acad Dermatol 1995; 33:59–63.
87. Geller AC, Koh HK, Miller DR, Clapp RW, Mercer MB, Lew RA. Use of health services before the diagnosis of melanoma. Implications for early detection and screening. J Gen Int Med 1992; 3:154–158.
88. Cockerell CJ, Howell JB, Balch CM. Think melanoma. South Med J 1993; 86: 1325–1333.
89. Elwood JM, Gallagher RP. The first signs and symptoms of melanoma: a population-based study. In: Elwood JM, ed. Naevi and Melanoma: Incidence, Interrelationships, and Implications. Pigment Cell No. 9, Karger: Basel 1988:140–152.
90. Green A. Incidence and reporting of cutaneous melanoma in Queensland. Australas J Dermatol 1982; 23:105–109.
91. Gerbert B, Maurer T, Berger T, et al. Primary care physicians as gatekeepers in managed care. Arch Dermatol 1996; 132:1030–1038.
92. Cassileth BR, Clark WH, Heiberger RM, et al. Relationship between patients' early recognition of melanoma and depth of invasion. Cancer 1982; 49:198–200.
93. Girgis A, Clarke P, Burton RC, Sanson-Fisher RW. Screening for melanoma by primary health care physicians: a cost-effectiveness analysis. Journal of Medical Screening 1996; 3:47–53.
94. Robinson JK, McGaghie WC. Skin cancer detection in clinical practice examination with standardized patients. J Am Acad Dermatol 1996; 34:709–711.
95. NIH Consensus Conference. Diagnosis and treatment of early melanoma. JAMA 1992; 268:1314–1319.
96. Marks R, Hill D. The Public Health Approach to Melanoma Control: Prevention and Early Detection. Geneva, Switzerland: International Union Against Cancer, 1992.
97. Heywood A, Sanson-Fisher R, Ring I, Mudge P. Risk prevalence and screening for cancer by general practitioners. Prev Med 1994; 23:152–159.
98. Higgins EM, Hall P, Todd P, et al. The application of the seven-point check-list in the assessment of benign pigmented lesions. Clinical and Experimental Dermatology 1992; 17:313–315.

99. Healsmith MF, Graham-Brown RAC, Osborne JE, et al. Further experience of public education for the early diagnosis of malignant melanoma in Leicestershire. Clinical and Experimental Dermatology 1993; 18:396–400.
100. Keefe M, Dick DC, Wakeel RA. A study of the value of the seven-point checklist in distinguishing benign pigmented lesions from melanoma. Clinical and Experimental Dermatology 1990; 15:167–171.
101. Little P, Keefe M, White J. Self screening for risk of melanoma: validity of self mole counting by patients in a single referral practice. BMJ 1995; 310:912–916.
102. Lawson DD, Moore DH, Schneider JS, et al. Nevus counting as a risk factor for melanoma: comparison of self-count with count by physician. J Am Acad Dermatol 1994; 31:438–444.
103. Byles JE, Hennrikus D, Sanson-Fisher R, Hersey P. Reliability of naevus counts in identifying individuals at high risk of malignant melanoma. Br J Dermatol 1994; 130:51–56.
104. Gruber SB, Roush GC, Barnhill RL. Sensitivity and specificity of self-examination for cutaneous malignant melanoma and risk factors. Am J Prev Med 1993; 9:50–54.
105. Rhodes AR, Weinstock MA, Fitzpatrick TB, Mihm MC Jr, Sober AJ. Risk factors for cutaneous melanoma. A practical method of recognizing predisposed individuals. JAMA 1987; 258:3146–3154.
106. Cooke KR, McNoe BM. Targeting early detection of malignant melanoma of the skin. N Z Med J 1991; 103:551–553.
107. McHenry PM, Hole DJ, MacKie RM. Melanoma in people aged 65 and over in Scotland, 1979–89. BMJ 1992; 304:746–749.
108. Hersey P, Sillar RW, Howe CG, Burton RC, Darbar SV, Foster HM, et al. Factors related to the presentation of patients with thick primary melanomas. Med J Aust 1991; 154:583–587.
109. Bonett A, Roder D, Esterman A. Epidemiological features of melanoma in South Australia: implications for cancer control. Med J Aust 1989; 151:502–509.
110. Geller AC, Miller DR, Lew RA, Wenneker M, Clapp RW, Koh HK. Cutaneous melanoma: another cancer that disproportionately affects the socioeconomically disadvantaged. Am J Public Health 1996; 86:538–543.
111. MacKie RM, Hole DJ. Incidence and thickness of primary tumors and survival of patients with cutaneous malignant melanoma in relation to socioeconomic status. BMJ 1996; 312:1125–1128.
112. Weary PE. A two-year experience with a series of rural skin and oral cancer detection clinics. JAMA 1971; 217:1862–1863.
113. Healthy People 2000. National Health Promotion and Disease Prevention Objectives. U.S. Department of Health and Human Services Public Health Service, Washington, DC, 1991.

15

Neuroblastoma

Isra G. Levy
Canadian Medical Association and University of Ottawa, Ottawa, Ontario, Canada

Geoffrey E. Dougherty
Montreal Children's Hospital and McGill University, Montreal, Quebec, Canada

Gerry B. Hill
University of Ottawa, Ottawa, Ontario, Canada

William G. Woods
South Carolina Cancer Center, Palmetto Richland Hospital, and University of South Carolina, Columbia, South Carolina

Mark L. Bernstein
Montreal Children's Hospital and McGill University, Montreal, Quebec, Canada

Neuroblastoma is unusual in that it is a pediatric malignancy for which a screening intervention has been proposed. The fact that this cancer is known to have a better prognosis when diagnosed clinically at an early age and stage, and that biochemical urine tests are available for early detection renders the condition a theoretical candidate for cancer screening (1–3). The disease compares with other serious conditions for which western society has embraced neonatal screening (e.g., phenylketonuria, hypothyroidism, and galactosemia) in terms of incidence (1,4), although the benefit of intervention in screen-detected cases remains in doubt. Although mass screening is national policy in Japan (5), the majority of opinion suggests that screening cannot currently be recommended and calls for additional research or anticipates results of current studies (2,6–

11). Evidence of screening effect from the major studies is summarized in section VI.

I. EPIDEMIOLOGY

Neuroblastoma is a malignant neoplasm of the ectodermal cells of the neural crest, the tissue from which the sympathetic nervous system is derived. Neuroblastoma occurs in any part of the paravertebral chain of sympathetic ganglia, but more than half of tumors begin in the abdomen, and, of these, the majority arise in the adrenal medulla (12). Only about 7% of all malignancies in childhood are neuroblastomas, but, in the 1st year of life, neuroblastoma is the most frequent histological type of cancer, accounting for about one-fourth of all infant malignancies (13).

The incidence rate of neuroblastoma is highest, about 35 per million, in infants and subsequently declines exponentially to about 1 per million at ages 10 to 14 years (14). In an international survey of 59 cancer registries (15), the cumulative incidence of neuroblastoma up to age 15 years ranged from 13 per million in the Philippines to 184 per million in New Zealand Maoris, much of the variation being concentrated in the 1st year of life. Over the whole survey, the median cumulative incidence to age 15 years was 102 per million, 111 in males and 91 in females. The median male-to-female ratio was 1.25.

The median cumulative incidence in both sexes was 133 in North America, 82 in Central and South America, 74 in Asia, 112 in Europe, and 137 in Oceania. The numbers of cases registered in three African registries were too small to permit rates to be calculated. However, in the United States Registries, the rates for blacks were similar to those for whites, but the rates for Hispanics were much lower.

Mortality rates for neuroblastoma are not published routinely, because cancer deaths are coded by site rather than histology. Table 1 summarizes the results of special studies in Denmark (16), Great Britain (17), and Canada (Gerry Hill, unpublished data). Over the past 5 decades, the incidence of neuroblastoma increased steadily, but mortality rates changed little. This may be due to a true increase in incidence, with a compensating decrease in case-fatality rate, but is perhaps more likely the result of increased ascertainment of cases with low case-fatality rate.

Apart from age, sex, and possibly ethnicity, no risk factors for neuroblastoma have been established definitively. Familial clusters do occur, and there is evidence to suggest that, like retinoblastoma, a two-mutational model applies (18). In the Danish series (16), there was some evidence of a U-shaped relationship between the incidence of neuroblastoma and maternal age. Also, in a case-control study in the Greater Delaware Valley (19), the odds ratio increased

Table 1 Neuroblastoma: Trends in Incidence and Mortality from Three Countries

Cumulative Rates of Neuroblastoma per Million at Risk to Age 15 (Denmark, Britain, and Canada)

	Incidence			Mortality		
Years	Denmark	Britain	Canada	Denmark	Britain	Canada
1946–1950	68			66		
1951–1955	55			44		
1956–1960	60			63		
1961–1965	98			72		
1966–1970	106			81		
1971–1975	114	105		72	85	
1976–1980	157	98		89	98	
1981–1985		109	154[a]		61	62[a]
1986–1990		129	154		73	60

[a] 1982–1985

Source: Denmark: Ref. 16; Britain: Ref. 17; Canada: Statistics Canada, special tabulations

with birth order. However, in neither of these studies were the trends statistically significant.

Some environmental exposures have been incriminated in case-control studies, including use of neurally active drugs, hormones, and hair dyes by the mother (19), and fathers' employment in occupations with exposure to electromagnetic fields (20). These findings from essentially hypothesis-generating studies need to be replicated.

II. NATURAL HISTORY OF NEUROBLASTOMA

Neuroblastoma appears to conform well with Cohnheim's theory that cancer in childhood ''is a congenital creation due to aberrations of embryonal cells'' (21). Adrenal crest cells histologically indistinguishable from neuroblastoma are common in fetal adrenal glands (22) and in autopsies in the newborn (23). Neuroblastoma is one of the few solid tumors that can regress spontaneously, either by differentiation to a benign ganglioneuroma or by disappearance of the neuroblastoma cells (24). Even in cases that become clinically apparent, it is clear that there are at least two categories of the cancer, one with an excellent prognosis (more common in the very young) and one with a very poor prognosis (25).

Neuroblastomas metastasize to regional lymph nodes, skin, liver, bone marrow, and bone. The staging schemes proposed by Evans (26) and others have been consolidated into the International Neuroblastoma Staging System (27).

This has the usual four major stages based on the degree of extension of the primary tumor and the presence of metastases, as in most other types of cancer. However, in the case of neuroblastoma, there is a special category IVS, with localized primary tumor and distant metastases limited to marrow, skin, or liver, and specifically excluding skeletal bone. This category carries a prognosis intermediate between that of stage I and stage II. Within each stage, survival varies inversely with age at first diagnosis (28).

In addition to age, stage, and histopathological grade (29), a variety of cytogenetic characteristics of neuroblastoma have been found to influence prognosis. Deletion of the short arm of chromosome 1, amplification of the oncogene N-myc, and DNA content (ploidy) are important examples, but others have been identified (30). These characteristics are correlated among themselves, as well as with age and stage. Brodeur and colleagues have grouped these prognostic factors into three categories based on outcome: good, intermediate, and bad,* and they have shown that cytogenetic characteristics are stable for individual cancers (31,32).

An important biological feature of neuroblastoma, in the context of screening, is the excretion by the tumor of the metabolic derivatives of norepinephrine and epinephrine (vanillylmandelic acid, or VMA) and of dopamine (homovanillic acid, or HVA) in the majority of cases (33). Tests for these substances assist in the differential diagnosis of patients with clinicall manifestations, but can also be used for routine screening of asymptomatic infants.

III. SCREENING TESTS

The VMA and HVA are excreted in urine (34). These substances are excreted in abnormally high quantities in 75%–90% of cases of clinically detected neuroblastoma (8,35,36) and can be measured in liquid urine or in urine collected on filter papers (36). Although the ability of urinary screening to detect small, early stage tumors has been questioned on the theoretical grounds that small tumors may not elaborate sufficient catecholamine metabolites (35), several studies have demonstrated the ability of urinary screening to detect neuroblastoma (37–42). As would be predicted, screening increases the reported incidence of the disease in the screened population (42,43).

Since the 1950s, there have been improvements in the methodologies and techniques used to measure VMA and HVA in the urine. The tests were first developed primarily to aid in the diagnosis and follow-up of clinical neuroblas-

---

*Good outcome disease is characterized by hyperdiploidy or near triploidy, normal N-myc copy number, and no deletion of the short arm of chromosome 1p. Bad outcome disease is characterized by near diploidy or tetraploidy, N-myc amplification, and loss of heterozygosity of chromosome 1p (31).

toma, and they originally involved 24-hour urine collections. Paper chromatography was used to measure the substances, and the tests were quite nonspecific. False-positive results were caused by several common foodstuffs, including bananas and vanilla (1). The first attempts to conduct mass screening occurred in Japan in the early 1970s; a qualitative ''spot test'' for VMA was used (5). This test, which was based on the color reaction of phenolic acids with diazotized aniline compounds, is positive only for VMA and resulted in a 3.8% false-positive rate when used in Japan between 1973 and 1983 (38,44). The development and use of chromatographic separations of organic acids by gas chromatography (GC) and high-performance liquid chromatography (HPLC) increased the sensitivity of the tests, and markedly reduced interference by exogenous (dietary) substances (1). Meanwhile, sensitivity had already been increased by the introduction of a thin layer chromatography (TLC) test for HVA (2). In the 1980s, a substantial contribution was made by Tuchman et al. (36), who demonstrated a 99% correlation between the results of quantitative analysis of dried filter samples for HVA and VMA and that of analysis of liquid samples. The same work supported a two-step screening procedure using TLC, followed by GC in TLC-positive specimens. Thus, the lower cost TLC is used to qualitatively detect abnormal or borderline concentrations of HVA or VMA. Gas chromatography is then used to quantitatively exclude or confirm elevations of the substances after normalizing the results to the urinary creatinine concentration to prevent possible inaccuracies resulting from variations in urine concentration. In the Quebec Screening Study (43), a second specimen was obtained from children for repeat GC analysis if the first was abnormally elevated.

There are no standard cut-off levels between positive and negative VMA and HVA tests. One recommendation is to use a VMA cut-off of 25 mu/mg creatinine and an HVA cut-off of 32 mu/mg creatinine. More usually, individual laboratories use a level of 2 standard deviations above the laboratory-specific, age-specific mean to identify specimens for reanalysis. On reanalysis, a cut-off of 3 standard deviations above the mean is used as the decision point to initiate a diagnostic evaluation (8).

The sensitivity of the screening procedure used in Japanese studies has been calculated (8) or reported (45) as ranging from 74% to 80%. Chamberlain (46) has pointed out that, to determine the true sensitivity of the test, it is necessary to have accurate data on ''interval'' cancers. True sensitivity is then the number of screen-detected cases as a proportion of the total number of cases found during that cohort's childhood. If screening detects cases that would have regressed before becoming clinically manifest, then the usual estimates of test sensitivity are, in fact, overestimates of the true sensitivity. Furthermore, other studies have shown sensitivities in the order of 40%–45%, even using newer testing procedures and relatively short periods of follow-up during which a symptomatic cancer could have arisen after a negative screen (43,46). On the other hand, the rate of false-positive tests using modern multistep screening strategies is less than 0.1% (i.e., specificity approximates 99.9%) (45,47). The proportion of tests requiring a repeat sample is higher than this, at approximately 2%, but repeat sampling can usually be obtained from the original filter paper (8). The predictive value of a positive test has been calculated as low as 0.13% from Japanese reports (47). However, with multistep testing strategies as used, for example, in the Quebec study, the positive predictive value exceeds 50% (43).

Historically, neuroblastoma has been considered a congenital tumor that is present and detectable at birth. This construct supports the theory that effective screening and detection of tumors in the sojourn phase can be achieved by a single screening test. A once-in-a-lifetime screening strategy, as used in screening for noncancerous conditions for which screening is performed in childhood, is logistically and financially far more practical than a strategy that involves more than one test at different intervals and ages. The optimal age at which to screen, and the need to screen a population more than once, remain uncertain. Screening has usually been performed once only, at 6 months of age, a standard set by the early Japanese interventions. Earlier screens should, theoretically, detect cancers that are present at birth, perhaps in time to affect the natural history of cases that would have become clinically apparent before 6 months of age. However, screening in the immediate neonatal period (which would be ideal from a logistical point of view because samples could be obtained at a time of almost universal contact with the health care system in Western society) is not advisable because there are large variations seen in urinary HVA and VMA levels immediately after birth because infants urinate minimal creatinine. In the Quebec study, where screening occurred at 3 weeks and at 6 months of age, only 17 of 43 cancers detected by screening in a cohort of 476,000 children were found at 3 weeks; 26 were detected at 6 months (43).

The standard 6-month screening age has, however, been challenged on the basis of the emerging knowledge of the natural history of the cancer (48). As discussed, the cytogenetic profile of screen-detected cases has led to conclusions that cancers do *not* all begin with favorable prognostic genotypes and phenotypes and then evolve into more aggressive tumors over time (49), rather, screening tends to detect tumors with a favorable prognosis (48–50). This observation about screen-detected cases and other suggestions that early screening results in diagnosis of cases that would have regressed spontaneously lead to the question of whether screening effectiveness could be increased by screening at an older age. Treuner and Schilling (48) have argued that screening at 12 months would be acceptable, because the cases occurring in the first 12 months of life have a good prognosis independent of stage.

A different argument for later screening, or repeat screening, has been advanced by other authors (2,7,51–56). This corollary of the first argument is that the false-negative (at 6-month screens) cases that have been detected clinically after screening tend to have poorer prognostic features. It is these aggressive tumors that could benefit most from early detection, but to identify such cases would require retargeting of screening to, or repeat screening of, children aged 12–18 months.

Repeat screening would greatly increase the intensity of effort and economic costs of a screening program without reducing the overdiagnosis burden. Retargeting of screening to a later age than 6 months may result in a reduced potential to effect a change in the natural history of the majority of cases. Only about 55% of normally presenting cases can be affected by a screening program even at 6 months of age (9); this proportion decreases with higher hypothetical screening ages. There is a dearth of information about the benefit of screening older children, so evidence-based conclusions are not permitted. Although some authors (9,57) have called for studies to address this question, others have argued that such studies, with the large numbers (1.5 to 2 million) of study subjects required to answer these questions, along with the substantial marginal cost of such studies at the expense of, for example, studies of treatment of advanced stage disease, are not warranted (58).

IV. PRACTICAL ASPECTS OF NEUROBLASTOMA SCREENING

Although there are logistical hurdles to overcome and there are expensive laboratory infrastructure and quality control program requirements, the feasibility of mass screening of urine samples obtained from infants has been well demonstrated in several countries (5,44,45,47,59,60).

The proper collection of liquid urine specimens from infants requires the education and careful cooperation of parents or the use of a health nurse visit to collect the urine. Quality control of specimen collection is difficult to maintain. In jurisdictions where routine home visits are performed by trained health visitors, the establishment of such a program is much easier. Even in such cases, however, approximately 11% of children require multiple repeat visits before the necessary specimens can be obtained, and specimens are never obtained from another 10% of children (44). Once liquid specimens are obtained, it is necessary to store them in a temporary refrigerated holding facility or to ship them to a central laboratory. Postage of such specimens requires special packing materials and is associated with higher costs.

The knowledge that there is a very high correlation between liquid urine concentrations of VMA and HVA and concentrations determined by analysis of dried filter samples (35) allows for a reliable screening initiative using the simpler method of ''blotting'' filter paper on a urine soaked diaper to collect specimens. This can easily be done by parents, who then are spared the need to collect urine and transfer it to a container. Mailing costs are reduced, because the filter paper can simply be dried and inserted in a plastic lined envelope, obviating the dangers of leakage or contamination. Furthermore, the central laboratory can easily store filter paper specimens in case of a future need for a retest.

Several studies have shown a high compliance rate in screening projects that use the filter paper blot technique for urinary sampling (5,41,43,47). This technique is used successfully when health visitors visit parents and either collect specimens or instruct parents in the collection of the specimens. In jurisdictions where the rate of hospital-based births is high (e.g., 99.9% in Quebec, Canada), mothers can be given a kit containing filter paper, return envelopes, and instructions; similarly, routine interactions with the health care system at later ages, such as the 3-month physical examination in Japan, afford opportunities to provide kits to parents. The use of health visitors to collect specimens has resulted in compliance rates as high as 97% (61). Relying on parental volunteerism following discharge from a hospital with kits and instructions also results in acceptable compliance rates, at least in some (62) societies, and it is substantially less resource intensive. In Quebec, the proportion of parents who complied with instructions for a 3-week screen was 91% using this approach, although it dropped to 74% at 6 months, despite there being several reminders (43). Compliance cannot, however, be universally assumed. Tuchman et al. (62) have shown that, in the United States, relying on parents to collect urine at home may not be very successful. Their experience showed compliance to be less than 15% for screens to be conducted between 3 weeks and 6 months after birth parents had been discharged with kits and collection instructions.

Compliance with requests for repeat samples is high. It is difficult, however, to predict the program requirements for obtaining repeat samples, because there is a high variability of the proportion of a screened population needing repeat specimens in different studies. Repeat sampling is required to obtain an adequate sample more frequently than to follow up a positive result. The increasing use of super absorbent diapers may result in higher rates of inadequate urine recovery using the ''blot'' technique in a mass screening context. This may greatly increase the logistical difficulties in obtaining at least one adequate specimen from large population birth cohorts.

Clearly, a screening program requires additional specialized central infrastructure, beyond the personnel and transport system required to obtain screening specimens. Specialist diagnostic follow-up facilities must be available and accessible to those children in whom screening identifies likely cases. Psychological support and counseling for families whose child tests positive is imperative.

In order to run large numbers of tests, a well-staffed and well-equipped biochemical laboratory, with a formal quality control process, is needed. Costs vary depending on the laboratory methods used. Establishment of a laboratory de novo may represent a substantial up-front expenditure, although the costs per person screened are likely to be small if large numbers are screened (8). This is especially likely if newer enzyme-linked immunosorbent assay (ELISA) kits compare favorably with gas and liquid chromotography, because screening of 1000 specimens per kit can be achieved in 1 week using ELISA assays (61).

V. ETHICAL CONSIDERATIONS

A. Potential for Benefit

The major possible benefit from any successful screening program for cancer in a population is the prevention of a substantial number of potential years of life lost. In the case of neuroblastoma, this would be accentuated by the very young age of most individuals suffering from the disease (63). For individuals, the potential benefits include the avoidance of death and the need for less aggressive therapy consequent to earlier diagnosis. Even for the relatively few who survive with advanced stage disease, the intensity and attendant concurrent morbidity of treatment is high, with the potential for significant adverse long-term sequelae (32). The avoidance or reduction of any of these represents an important potential for benefit of any successful program of screening.

B. Potential for Harm

The possibility that current screening strategies for identifying neuroblastoma in the 1st year of life find children who do not benefit from screening (10,43) is a source of concern. Such cases include good prognosis and early stage cases (stage IVS, I, and II) detected in screens at either 3 weeks or 6 months. As discussed, neuroblastoma is one of the few solid tumors known to regress spontaneously or mature to a benign form, and a substantial subset of these may so regress or mature into benign ganglioneuromas. As a result, they might not come to clinical attention in the absence of screening. In existing screening programs and studies, almost all of the screen-detected cases have received basic therapy, usually involving primary excision of the tumor, that is, general anesthesia and surgery, and, in many cases, some radiotherapy or chemotherapy (64). Although, at least in the Quebec experience, there have been no deaths related to treatment, it is virtually certain that some children have undergone operative procedures from which they derived no benefit (64).

The generic potential of screening to cause unnecessary iatrogenic disease is troubling, as is the labeling of a seemingly well individual as seriously ill and precipitating a change forever in the person's life. This ethical dilemma can generally be obviated through counseling and obtaining informed consent for screening. In the case of neuroblastoma screening, the subject has not attained the age of legal majority and a benevolent surrogate is required to digest information about risks and benefits of intervention. The child, who is potentially most affected by the consequences, has no control over this decision. If a lesion is identified in an apparently healthy child, it is, again, a third party that must weigh decisions of consent to possibly aggressive treatments. Although biological markers of prognosis provide useful discriminatory information, there are no characteristic clinical features that allow clinicians to definitively identify those who receive no benefit from basic treatments (65).

The psychological morbidity that is likely entailed by the false-positive rate of any screening program also constitutes potential harm. Even in the Quebec Neuroblastoma Screening Study, in which the specificity of the test was extremely high, the positive predictive value was only 50% (43), that is, for every two children identified, only one turned out to have neuroblastoma. False-positive cases are generally followed up for prolonged periods with serial noninvasive testing before a definitive diagnosis excluding cancer can be offered to the parents (64). Even though this aspect was not formally examined in the Quebec study, the risk of psychosocial morbidity in this circumstance appears high.

Somewhat different issues are raised by proposals for screening older children after the 1st year of life. These are related in large part to the low success rate of treating stage IV disease in older children. Because there is very little indication that neuroblastoma progresses from early to late stage disease or from favorable cytogenetic type to unfavorable type, at least once it is manifest (25), there is a risk that earlier detection of stage IV disease after the 1st year of life may have as its primary impact the earlier awareness of disease for which earlier treatment will incur little or no benefit. The primary implication of earlier detection would have been to extend the medical, psychological, and economic morbidity associated with the disease simply through a lead time effect. In addition, in part because of the rarity of the tumor and of the low effectiveness of treatment for stage IV disease, later screening programs for this condition would incur substantial expense per case identified, treated, and potentially cured. The resources devoted to screening for neuroblastoma could be used, perhaps more productively, in other areas with a greater chance of benefit to childhood populations. It is also possible that the diversion of these resources to the study of biological characterization of the tumor and its treatment would produce larger gains than pursuing studies of screening activities at this time.

VI. EVIDENCE OF SCREENING EFFECT

Studies of the feasibility and efficacy of screening infants for neuroblastoma have been accruing for several years. These include a series of descriptive studies of local and national programs in Japan, uncontrolled pilot experience at a number of sites in Europe and the United States, and an ongoing cohort study in Quebec, Canada (40–43,62,66–68). At the end of 1996, the evidence for a screening effect was limited to the observation in Japan and Quebec of an increase in incidence in infants, an increase in incidence of early stage disease, shifts in distribution by stage, and apparent increases in survival among those detected by screening (38,40,43,64,69–73). Despite the shift in stage distribution of cases detected by screening compared to those that are routinely detected, there is no evidence of reduction in the incidence of advanced stage cancers in either the Japanese or Quebec experience (10,42,43). Whereas downward trends in cause-specific mortality rates have been claimed to coincide with widespread screening in Japan after 1985, decreases of a similar magnitude have been reported in both England and Finland where widespread screening has not occurred (74–76). There is, as yet, no evidence from controlled studies or randomized trials of decreases in mortality rate associated with screening. Results of mortality analyses are expected from the Quebec Neuroblastoma Screening Study before the year 2000.

A. Quebec and Japanese Experiences with Screening

The most dramatic effect of the Quebec screening study and of the Japanese screening programs has been on the incidence of neuroblastoma in young infants. For example, with 63 months of follow-up, 75 cases of neuroblastoma were observed in children younger than 12 months of age in Quebec, compared with the 29 cases expected on the basis of the historical experience in the Greater Delaware Valley Tumor Registry (a long-established regional tumor registry covering several eastern American states) and data from the Surveillance, Epidemiology, and End Results Program of the National Cancer Institute (64). The standardized incidence ratios for the two age groups—0–6 months and 6–12 months—were 2.84 and 2.33, respectively (95% confidence intervals [CIs] 2.07–3.73 and 1.53–3.29, respectively). At the same time, the number of cases detected in two of the control groups for the Quebec study, the state of Minnesota and province of Ontario, were almost exactly the number expected—99 cases observed versus 100 cases expected. With another year of follow-up, the standardized incidence ratio for early stage (I, II, IVS) disease in the Quebec cohort of children compared to Greater Delaware Valley was 2.98 (95% CIs 2.34–3.70). Ratios in the control population groups of Minnesota and Ontario were not significantly different from 1.0 (43).

The potential to which screening can impact incidence of neuroblastoma exceeds the obvious. The Quebec study has shown that the overall incidence of the disease was increased in the study jurisdiction even after excluding patients detected by screening (43). The investigators of that study have attributed this "halo effect" to increased awareness of the disease in the entire province by health care providers.

If the observed increase in incidence in the screened population was the result of lead-time reductions, early detection of tumor might have been expected to increase the incidence of disease in young infants while correspondingly decreasing it in children older than 12 months of age, especially for those in the 1- to 4-year-old age group.

However, as shown for the Quebec study in Table 2, no such decrease in incidence in older children has been observed (64). In fact, despite the observation that incidence of early stage disease in younger children more than doubled, the incidence of advanced stage disease (stages III and IV) in older children in Quebec showed a statistically nonsignificant *increase* over that which would have been expected (43). These observations make it unlikely that a reduction in mortality rate will ultimately be observed in the screened cohort in the Quebec study.

With a maximum of 63 months of follow-up of the screened cohort in Quebec, all stages of disease were included among the 40 children diagnosed as having neuroblastoma through screening and, in the entire group (102 children), more advanced stage disease was represented—40 observed stage III and IV for ages 0–4 years (27 expected), with 18 of them being between the ages of 1 and 4 years (14 expected) (64). These observations in the Quebec study

Table 2 Age-Specific Incidences and Standardized Incidence Ratios of Neuroblastoma in the Quebec Screening Experience, May 1989 to August 1994

Age at diagnosis (months)	Person-years to date	Annual incidence per 10^6	Observed/expected	SIR	95% CI
0–5	227,905	72.6	47/16.6	2.84[a]	2.07–3.73
6–11	206,829	58.1	28/12.0	2.33[a]	1.53–3.29
12–17	183,662	30.8	8/5.7	1.41	0.59–2.59
18–23	161,027	29.1	9/4.7	1.92	0.85–3.41
24–29	136,889	13.8	4/1.9	2.12	0.53–4.76
30–35	13,234	24.2	2/2.7	0.73	0.06–2.13
36–41	88,685	17.5	1/1.6	0.64	0.00–2.58
42–47	65,053	14.0	1/0.9	1.10	0.00–4.39
48–53	43,240	22.9	1/1.0	1.01	0.00–4.04
54–59	21,636	10.6	1/0.2	4.36	0.00–17.44
Summary					
0–59	1,248,160		102/47	2.16[a]	1.75–2.61

[a] $P<0.05$
Source: Ref. 64.

broadly mirror the published experience of the uncontrolled studies in Japan (38,69–73).

With regard to the biology of screen-detected cases, almost all of the tumors with unfavorable ploidy (near diploidy or tetraploidy) and all of those with an amplified N-myc oncogene (known negative prognostic features of tumor biology) fell in the clinically detected group in Quebec, having been missed by screening. Similarly, almost all of the tumors histologically classified as ''unfavorable'' in the Shimada classification (37,38) were found in the clinically diagnosed group. All of the 12 deaths from neuroblastoma that occurred in the first 63 months in the screened cohort of the Quebec study fell within the clinically detected group (64), and all but two had presented with stage IV disease.

Similar to the findings in Quebec, several reviews of the biology of neuroblastomas detected through the mass Japanese screening program have reported that almost all of the tumors had favorable features (unamplified N-myc oncogene, near triploidy, and favorable histology) when compared with tumors found clinically, with all deaths occurring in the clinically detected group (39,50,77).

B. Other Findings from Observational Studies of Screened Populations

Bessho et al. (42) have reviewed data provided to the Kanto-Ko-Shin-Etsu branch of the Japan Children's Cancer Registry over different periods of time. This registry is not population based and there is no systematic pathology review. Rather, responses provided on a questionnaire are used. Comparing several time periods before national screening was instituted (1985) to time periods afterward, the investigators concluded that the incidence of neuroblastoma seemingly increased, because before screening, neuroblastoma represented approximately 10% of all reported malignant disease, whereas afterward, that percentage increased to 19.3% Moreover, eliminating all of the cases diagnosed by screening gave a relative proportion of 13.8%, similar to the figure obtained before the onset of screening. In addition, eliminating cases diagnosed by screening increased the proportion of children diagnosed after the age of 1 year from 50% to 75%. More recently, a review of national cancer registry data, adjusted for underreporting, revealed that widespread screening in Japan, initiated in the mid-1980s, was not associated with a diminished incidence of neuroblastoma after the 1st year of life (10), corroborating the findings of Woods et al. (43).

C. Observations on Mortality Rate

Improvements in survival rates following the commencement of population screening have been reported from several settings in Japan. For example, in reports of the early experience in Kyoto, Japan, long-term survival rates increased from 17% before 1974 (when screening was not used) to more than 70% after the implementation of screening (69,78). Similar results were observed in Sapporo City, where survival at 48 months improved from 21% to 87%, and at 60 months from 23% to 67%. During the same time interval, 48-month survival changed from 21% to 28% in the region surrounding Sapporo City, where screening was not in effect (73,79). Such observations are subject, especially, to lead time bias, length bias and, the extreme example of length bias, the overdiagnosis of cases through screening that would otherwise have spontaneously regressed. In addition, because increases in survival rate over a similar period have been reported from some jurisdictions where screening was not occurring, it has been suggested that this may also, in part, reflect improvements in clinical diagnosis and treatment (7).

A reduction in cause-specific mortality in a screened population compared to an unscreened control population remains the only definitive measure of screening efficacy. As discussed, final results from the Quebec study are expected soon; meanwhile, the investigators have concluded that its is unlikely that a reduction in mortality rate will be observed in the screened population (43). The *overall* (at-risk population defined as all children) mortality rate in Japan, where nationwide screening was implemented in 1985, from neuroblastoma has been decreasing (72,74). However, it has been decreasing in other industrialized countries as well (75). Reported overall rates of population-based survival and mortality were very similar in England and Finland to the 55% survival rate with annual mortality rates of nine per million for children aged 4 years or younger and five per million for children younger than 14 years for Quebec before the institution of a screening program (53,76,80). The recent Japanese figures of approximately six per million for children between the ages of 1 and 4 years falls within this range (81). Most authors think that this improvement represents the result of improved management of children with biologically favorable disease, because children older than 12 months of age diagnosed with advanced stage disease remain largely refractory to therapy and very difficult to cure. If only 10% of such children are cured, and they represent approximately one-third of the population of unscreened children with neuroblastoma, then an overall 55% cure rate implies the necessity of curing 80% of all other children with neuroblastoma. This is an achievable goal with modern treatment (32,82).

VII. RECOMMENDATIONS OF ORGANIZATIONS

The Japanese Ministry of Health and Welfare has recommended and funded mass screening of infants nationwide since 1985. At a 3-month check-up, parents are advised to provide urine samples from their child when the infant reaches the age of 6 months. Screening is recommended once only, at age 6 months (5). Recently, there has been an explicit call for a halt to the nationwide screening program (11).

Evidence-based groups in North American have not assessed neuroblastoma screening. Neither the U.S. Preventive Services Task Force (83) nor the Canadian Task Force on the Periodic Health Examination (84) make recommendations with respect to screening for this cancer.

In 1997, the U.S. National Cancer Institute's Physician Data Query (PDQ) (85) system developed a statement regarding neuroblastoma screening which concludes that a population-based mortality benefit from screening is unlikely based on available data. The American Cancer Society, through the consensus statement of a panel of experts, have taken the position that ''widespread implementation of screening for neuroblastoma cannot be recommended'' (7). This position is shared by the International Union Against Cancer (UICC) (6) and the International Society of Pediatric Oncology (SIOP) (86).

VIII. CONCLUSION

Inappropriate evaluation of early Japanese screening experiences focused on survival as an outcome of screening and likely resulted in overly optimistic estimates of screening benefit. Japan instituted nationwide screening in 1985. In North America, there have been reports of local physicians and lay people organizing screening programs or pressuring local public health agencies to do so (2). However, the Japanese studies have been criticized for the potential of classic biases (selection, lead time, length, overdiagnosis) having been inadequately addressed in design and analysis. These studies have, therefore, led to several additional research projects.

Although population-based mortality outcome data are still awaited from more recent screening studies, Woods et al (43) have shown that incidence of neuroblastoma does not increase in older children despite a more than two-fold increase in incidence in younger children associated with screening. It seems unlikely that screening infants younger than 1 year of age will reduce the risk of death in screened children. The risks of overdiagnosis of cases destined to regress or the diagnosis of biologically favorable disease likely to respond to treatment whenever in the course of the disease treatment is initiated, as well as the consequent risks of ''over'' treatment, renders it possible that screening for neuroblastoma may result in net harm.

Physicians and their patients, and the public at large, should be aware that the effectiveness of mass neuroblastoma screening remains unproven, and a significant theoretical risk of harm exists. As benefit of screening remains unproven, there is little imperative for third party health system payers to consider cost-benefit analyses. Almost by definition, marginal costs are prohibitively high in a circumstance in which benefit must be thought of as neutral at best. Furthermore, the ethical obligation of *primum non nocere*, especially when the subjects of intervention are not patients but asymptomatic individuals, dictates that system funders not invest in widespread screening based on currently available evidence.

ACKNOWLEDGMENTS

We are very grateful to the editors for their helpful suggestions and to Ms. Anne Lange, Mr. Andrew Gentile, Ms. Rita Riccio, and Ms. Madeleine Ranger for their assistance with preparing the manuscript for this chapter.

REFERENCES

1. Woods WG, Tuchman M. Neuroblastoma: The case for screening infants in North America. Pediatrics 1987; 79:869–873.
2. Goodman SN. Neuroblastoma screening data: an epidemiologic analysis. Am J Dis Child 1991; 145:1415–1422.
3. Tuchman M, Lemieux B, Woods WG. Screening for neuroblastoma in infants: investigate or implement? Pediatrics 1990; 86:791–793.
4. Woods WG. Neuroblastoma as a model for cancer screening: is an ounce of prevention worth a pound of cure? American Society of Clinical Oncology, Education Symposium, Washington DC, 1990.
5. Sawada T. Past and future of neuroblastoma screening in Japan. Am J Pediatr Hematol Oncol 1992; 14:320–326.
6. Miller AB, Chamberlain J, Day NE, Hakama M, Prorok PC. Report on a workshop of the UICC project on evaluation of screening for cancer. Int J Cancer 1990; 46: 761–769.
7. Murphy SB, Cohn SL, Craft AW, Woods WG, Sawada T, Castleberry RP, Levy HL, Prorok PC, Hammond GD. Do children benefit from mass screening for neuroblastoma? Lancet 1991; 337:344–345.
8. Chamberlain J. Screening for neuroblastoma: a review of the evidence. J Med Screening 1994; 1:169–175.
9. Parker L, Craft AW. Neuroblastoma screening: more questions than answers? Eur J Cancer 1991; 27:682–683.
10. Bessho F. Effects of mass screening on age-specific incidence of neuroblastoma. Int J Cancer 1996; 67:520–522.
11. Bessho F. Where should neuroblastoma screening go? Lancet 1996; 348: 1671.
12. Evans AE. Staging and treatment of neuroblastoma. Cancer 1980; 45:1799–1802.
13. Young JL, Miller RW. Incidence of malignant disease in US children. J Pediatr 1975; 86:254–258.
14. Stiller CA, Parkin DM. International variations in the incidence of neuroblastoma. Int J Cancer 1992; 52:538–543.

15. Parkin DM, Stiller CA, Bieber A, Draper GJ, Terracini B, Young JL, eds. International Incidence of Childhood Cancer. Lyon: International Agency for Research on Cancer, 1988.
16. Carlsen NLT. Epidemiological investigations on neuroblastoma in Denmark 1943–1980. Br J Cancer 1986; 54:977–988.
17. Stiller CA. Trends in neuroblastoma in Great Britain: incidence and mortality, 1971–1990. Eur J Cancer 1993; 29A:1008–1012.
18. Knudsen AG Jr, Strong LC. Mutation and cancer: neuroblastoma and pheochromocytoma. Am J Hum Genet 1972; 24:514–532.
19. Kramer S, Ward E, Meadows AT, Malone KE. Medical and drug risk factors associated with neuroblastoma: a case-control study. J Natl Cancer Inst 1987; 78: 797–804.
20. Spitz MR, Johnson CC. Neuroblastoma and paternal occupation: a case-control analysis. Am J Epidemiol 1985; 121:924–929.
21. Peller S. Cancer in Man. New York: International Universities Press, 1952:140.
22. Turkel SB, Itabashi HH. The natural history of neuroblastoma cells in the fetal adrenal gland. Am J Pathol 1974; 76:225–243.
23. Miller RW, Fraumeni JF Jr, Hill JA. Neuroblastoma: epidemiologic approach to its origin. Am J Dis Child 1968; 115:253–261.
24. Everson TC, Cole WH. Spontaneous regression of cancer. Philadelphia: WB Saunders, 1966.
25. Woods WG, Lemieux B, Tuchman M. Neuroblastoma represents distinct clinical-biologic entities: a review and perspective from the Quebec Neuroblastoma Screening Project. Paediatrics 1992; 89:114–118.
26. Evans AE, D'Angio GJ, Randolph J. A proposed staging for children with neuroblastoma. Cancer 1971; 27:374–378.
27. Brodeur GM, Seeger RC, Barrett A, Berthold F, Castleberry RP, D'Angio G, De Bernardi B, Evans AE, Favrot M, Freeman AI. International criteria for diagnosis, staging, and response to treatment in patients with neuroblastoma. J Clin Oncol 1988; 6:1874–1881.
28. Breslow N, McCann B. Statistical estimation of prognosis for children with neuroblastoma. Cancer Res 1971; 31:2098–2103.
29. Shimada H, Catten J, Newton WA Jr, Sachs N, Hamoudi AB, Chiba T, Marsden HB, Misugi K. Histopathological prognostic factors on neuroblastic tumors: definition of subtypes of ganglioneuroblastoma and age-linked classification of neuroblastoma. J Natl Cancer Inst 1984; 73: 405–416.
30. Brossard J, Bernstein ML, Lemieux D. Neuroblastoma: an enigmatic disease. Br Med Bull 1996; 52:789–803.
31. Brodeur GM, Nakagawara A. Molecular basis of clinical heterogeneity in neuroblastoma. Am J Pediatr Hematol Oncol 1992; 14:111–116.
32. Brodeur GM, Castleberry RP. Neuroblastoma. In: Pizzo PA, Poplack DG, eds. Principles and practices of pediatric oncology. 2d ed. Philadelphia: JB Lippincott, 1993:739–768.
33. Williams CM, Greer M. Homovanillic acid and vanillylmandelic acid in diagnosis of neuroblastoma. JAMA 1963; 183:836–840.

34. Tuchman M, Crippin PJ, Krivit W. Capillary gas-chromatographic determination of urinary homovanillic acid and vanillylmandelic acid. Clin Chem 1983; 29:828–831.
35. Carlsen NLT. Neuroblastoma: epidemiology and pattern of regression. Am J Pediatr Hematol Oncol 1992; 14:103–110.
36. Tuchman M, Auray-Blais C, Ramnaraine MLR, Neglia J, Krivit W, Lemieux B. Determination of urinary homovanillic and vanillylmandelic acids from dried filter paper samples: assessment of potential methods for neuroblastoma screening. Clin Biochem 1987; 20:173–177.
37. Sawada T, Todo S, Fujita K, Iino S, Imashuku S, Kusunoki T. Mass screening of neuroblastoma in infancy. Am J Dis Child 1982; 136:710–712.
38. Sawada T, Nakata T, Takasugi N, Maeda K, Hanawa Y, Shimizu K, Hirayama M, Takeda T, Mori T, Koide R, Tsunoda A, Nagahara N, Yamamoto K. Mass screening for neuroblastoma in infants in Japan. Lancet 1984; ii:271–273.
39. Hachitanda Y, Ishimoto K, Hata J, Shimada H. One hundred neuroblastomas detected through a mass screening system in Japan. Cancer 1994; 74:3223–3226.
40. Woods WG, Tuchman M, Bernstein ML, Leclerc JM, Brisson L, Look T, Brodeur GM, Shimada H, Hann HL, Robison LL, Shuster JJ, Lemieux B. Screening for neuroblastoma in North America: 2-year results from the Quebec Project. Am J Pediatr Hematol Oncol 1992; 14:312–319.
41. Parker L, Craft AW, Dale G, Bell S, Cole M, McGill AC, Seviour JA, Smith J. Screening for neuroblastoma in the North of England. BMJ 1992; 305:1260–1263.
42. Bessho F, Hashizume K, Nakajo T, Kamoshita S. Mass screening in Japan increased the detection of infants with neuroblastoma without a decrease in cases in older children. Pediatr 1991; 119:237–241.
43. Woods WG, Tuchman M, Robison LL, Bernstein M, Leclerc JM, Brisson LC, Brossard J, Hill G, Shuster J, Luepker R, Byrne T, Weitzman S, Bunin G, Lemieux B. A population-based study of the usefulness of screening for neuroblastoma. Lancet 1996; 348:1682–1687.
44. Craft AW, Dale G, McGill A, Seviour J, Spence EM. Biochemical screening for neuroblastoma in infants: a feasibility study. Med Pediatr Oncol 1989; 17:373–378.
45. Nishi M, Miyake H, Takeda T, Takasugi N, Hanai J, Kawai T. Mass screening for neuroblastoma and estimation of costs. Acta Paediatr Scand 1991; 80:812–817.
46. Chamberlain J. Neuroblastoma. In:Chamberlain J, Moss S, eds. Evaluation of cancer screening. London: Springer, 1996:145–149.
47. Scriver CR, Gregory D, Bernstein M, Clow CL, Weisdorf T, Dougherty GE, Auray-Blais C, Giguere R, Lemieux B, Laberge C. Feasibility of chemical screening of urine for neuroblastoma case finding in infancy in Quebec. Can Med Assoc J 1987; 136:952–956.
48. Treuner J, Schilling FH. Neuroblastoma mass screening: the arguments for and against. Eur J Cancer 1995; 31A:565–568.
49. Brodeur GM. Neuroblastoma—clinical applications of molecular parameters. Brain Pathol 1990; 1:47–54.
50. Hayashi Y, Hanada R, Yamamoto K. Biology of neuroblastomas in Japan found by screening. Am J Pediatr Hematol Oncol 1992; 14:342–347.

51. Ishimoto K, Kiyokawa N, Fujita H, Yabuta K, Ohya T, Miyano T, Shinohara T, Sera Y. Problems of mass screening for neuroblastoma: Analysis of false-negative cases. J Pediatr Surg 1990; 25:398–401.
52. Esteve J, Parker L, Roy P, Herrmann F, Duffy S, Frappaz D, Lasset C, Hill C, Sancho-Garnier H, Michaelis J, Philip T. Is neuroblastoma screening evaluation needed and feasible? Br J Cancer 1995; 71:1125–1131.
53. Huddart SN, Muir KR, Parkes S, Mann JR, Stevens MCG, Raafat F. Neuroblastoma: a 32-year population-based study—implications for screening. Med Pediatr Oncol 1993; 21:96–102.
54. Kerbl R, Urban C, Starz I, Ambros IM, Strehl S, Kovar H, Gadner H, Ambros PF. Neuroblastoma with N-myc amplification detected by urine: mass screening in infants after the sixth month of life. Med Pediatr Oncol 1993; 21:625–626.
55. Mera SL. Screening for cancer and pre-cancer. Br J Biomed Sci 1995; 52:120–141.
56. Matthay KK. Neuroblastoma: a clinical challenge and biologic puzzle. Ca Cancer J Clin 1995; 45:179–192.
57. Hayashi Y, Ohi R, Yaoita S, Nakamura M, Kikuchi Y, Konno T, Tsuchiya S, Shiraishi H. Problems of neuroblastoma screening for 6 month olds and results of second screening for 18 month olds. J Pediatr Surg 1995; 30:467–470.
58. Duncan MG. Screening for neuroblastoma (letter). BMJ 1993; 306:143–144.
59. Lemieux B, Auray-Blais C, Giguere R, Shapcott D, Scriver CR. Newborn urine screening experience with over one million infants in the Quebec Network of Genetic Medicine. J Inherit Metab Dis 1988; 11:45–55.
60. Craft AW, Parker L, Dale G, McGill AC, Seviour JA, Bell S, Cole M, Smith J. A pilot study of screening for neuroblastoma in the North of England. Am J Pediatr Hematol Oncol 1992; 14:337–341.
61. Seviour JA, McGill AC, Craft AW, Parker L, Bell S, Cole M, Smith J, Hawkins E, Brown J, Dale GA. Screening for neuroblastoma in the northern region of England. Am J Pediatr Hematol Oncol 1992; 14:332–336.
62. Tuchman M, Fisher EJ, Heisel MA, Woods WG. Feasibility study of neonatal neuroblastoma screening in the United States. Med Pediatr Oncol 1989; 17:258–264.
63. Davis S, Rogers MAM, Pendergrass TW. The incidence and epidemiologic characteristics of neuroblastoma in the United States. Am J Epidemiol 1987; 126:1063–1074.
64. Bernstein ML, Woods WG. Screening for neuroblastoma. In: Miller AB, ed. Advances in Cancer Screening. Boston: Kluwer Academic Publishers, 1996:149–163.
65. Suita S, Zaizen Y, Yano H, Akiyama H, Sera Y, Takamatsu H, Ueda K, Tasaka H, Miyazaki S, Kawakami K, Tsuneyoshi M, Ikeda K. How to deal with advanced cases of neuroblastoma detected by mass screening. A report from the Pediatric Oncology Study Group of the Kyushu area of Japan. J Pediatr Surg 1994; 29:599–603.
66. Takeda T. History and current status of neuroblastoma screening in Japan. Med Pediatr Oncol 1989; 17:361–363.

67. Mathieu P, Favrot M, Frappaz D, Chauvin F, Greffe J, Montegue A, Lacroic C, David L, Brunat-Mentigny M, Philip T. Le neuroblastome de l'enfant: aspects cliniques et biologiques. Une experience en France. Ann Biol Clin 1993; 51:665–688.
68. Schilling FH, Ertmann R, Ambros PF, Strehl H, Christiansen H, Kovar H, Kabisch H, Treuner J. Neuroblastoma with unfavourable prognostic parameters detected by mass screening: report of the German pilot study. Proc Am Soc Clin Oncol 1994; 13:421.
69. Sawada T, Kidowaki T, Sakamoto I, Hashida T, Matsumura T, Nakagawa M, Kusinoki T. Neuroblastoma; mass screening for early detection and its prognosis. Cancer 1984; 53:2731–2735.
70. Sawada T, Kidowaki T, Sugimoto T, Kusonoki T. Incidence of neuroblastoma in infancy in Japan. Med Pediatr Oncol 1984; 12:101–103.
71. Yamamoto K, Hayashi Y, Hanada R, Kikuchi A, Ichikawa M, Tanimura M, Yoshioka S. Mass screening and age-specific incidence of neuroblastoma in Saitama prefecture, Japan. J Clin Oncol 1995; 13:2033–2038.
72. Asami T, Otabe N, Wakabayashi M, Kakihara T, Uchiyama M, Asami K. Screening for neuroblastoma: a 9-year birth cohort-based study in Niigata, Japan. Acta Paediatr 1995; 84:1173–1176.
73. Naito H, Sasaki M, Yamashiro K, Takeda T, Takasugi N, Uchino J, Hata Y. Improvement in prognosis of neuroblastoma through mass population screening. J Pediatr Surg 1990; 25:245–248.
74. Hanawa Y, Sawada T, Tsunoda A. Decrease in childhood neuroblastoma death in Japan. Med Pediatr Oncol 1990; 18:472–475.
75. Cole M, Parker L, Craft A. Letter to the Editor: ''Decrease in childhood neuroblastoma death in Japan,'' Hanawa T, Sawada T, Tsunoda A, (1990). Med Pediatr Oncol 1992; 20:84–85.
76. Sankila R, Hakama M. Survival trends for neuroblastoma patients in Finland: negative reflections on screening. Eur J Cancer 1992; 29:122–123.
77. Nakagawara A, Zaizen Y, Ikeda K, Suita S, Ohgami H, Nagahara N, Sera Y, Akiyama H, Kawakami K, Uchino J. Different genomic and metabolic patterns between mass screening-positive and mass screening-negative later-presenting neuroblastomas. Cancer 1991; 68:2037–2044.
78. Sawada T, Matsumura T, Kawakatsu H, Sugimoto T, Kuroda H, Hosoi H, Mishimata T, Saida T, Hino T. Long-term effects of mass screening for neuroblastoma in infancy. Am J Pediatr Hematol Oncol 1991; 13:3–7.
79. Nishi M, Miyake H, Takeda T, Shimada M, Takasugi N, Sato Y, Hanai J. Effects of the mass screening of neuroblastoma in Sapporo City. Cancer 1987; 60:433–436.
80. Bernstein M, Leclerc JM, Bunin G, Brisson L, Robison L, Shuster J, Byrne T, Gregory D, Dougherty G, Scriver C, Lemieux B, Tuchman M, Woods WG. A population-based study of neuroblastoma incidence, survival, and mortality in North America. J Clin Oncol 1992; 10:323–329.
81. Hanawa Y. The recent trend of childhood neuroblastoma death in Japan. Presented at the Third International Symposium on Neuroblastoma Screening, Kyoto, Japan, 1993.

82. Matthay KK. Neuroblastoma. In: Pochedly C, ed. Neoplastic Diseases of Childhood. Montreaux, Switzerland, Harwood Academic Publishers, 1994:735–778.
83. U.S. Preventive Services Task Force. Guide to clinical preventive services. 2d ed. Baltimore: Williams & Wilkins, 1966.
84. The Canadian Guide to Clinical Preventive Health Care. Ottawa, Health Canada, 1994.
85. PDQ, Information Associates Program, 9030 Old Georgetown Road, Bethesda, MD 20814–1519.
86. Ninane J. Consensus statement of SIOP on neuroblastoma screening. SIOP Newsletter, Spring 1991.

16

Ovarian Cancer

Vicki L. Seltzer
Long Island Jewish Medical Center, New Hyde Park, and Albert Einstein College of Medicine, Bronx, New York

I. INTRODUCTION

Ovarian cancer is the fifth leading cause of cancer death in American women and the most common cause of death from pelvic gynecological malignancy (1). In 1995, there were approximately 26,600 new cases of ovarian cancer and 14,500 deaths (1).

If ovarian cancer is diagnosed when the disease is still stage IA, then 5-year survival rate may be as high as 75% to 90%. When it is diagnosed as stage IIIC disease, 5-year survival rate is less than 20% (2,3).

Most ovarian cancers are diagnosed at an advanced stage, at a point at which cure is unlikely. The reason for the relative rarity of early diagnosis is that women do not usually become symptomatic until they have an advanced stage ovarian cancer. In addition, symptoms of ovarian cancer are nonspecific, making early diagnosis even more difficult. Symptoms of ovarian cancer may be such nonspecific complaints as pelvic pressure and vague gastrointestinal discomfort.

Clearly, if the capability did exist to routinely identify ovarian cancer at an early stage, improved cure rates would result. However, there is no conclusive demonstration that the currently available modalities can be used for widespread screening to reduce the mortality rate from ovarian cancer. Many studies are underway to evaluate this.

II. RISK FACTORS

Women in the United States have a 1 in 70 lifetime risk of developing ovarian cancer. Several known factors increase this risk. Women of North American or northern European descent are at increased risk. Advancing age is associated with an increased risk. Ovarian cancer risk is also increased in nulliparous women; in women who have a personal history of breast, colon, or endometrial cancer; and in women with a family history of ovarian cancer. There is some suggestion that the use of fertility drugs is associated with increased risk, although the evidence is inconsistent.

Although the lifetime risk of ovarian cancer for U.S. woman is 1 in 70, this risk increases to 1 in 20 if the woman has one first degree relative with ovarian cancer, and to 7% if she has two first degree relatives. Furthermore, there are families at high risk for both breast and ovarian cancer as well as families with Lynch syndrome II (which includes early onset nonpolyposis colorectal cancer, upper gastrointestinal system cancers, urothelial carcinomas, endometrial cancer, and ovarian cancer). It is important for physicians and for the public to understand that there is a difference between family history and a familial syndrome, and that most women with a family history are not part of a family with a familial syndrome.

Less than 0.05% of women are at significantly increased risk because of a familial cancer syndrome (4). However, Claus et al. (5) have indicated that 10% of women with ovarian cancer carry a breast/ovarian cancer gene mutation. These individuals can be identified, and it is hoped that, in the future, the number of deaths in this subgroup of patients will be reduced.

Factors have also been identified that decrease ovarian cancer risk. Circumstances that reduce incessant ovulation (e.g., having greater than one full-term pregnancy, breast-feeding, and using oral contraceptives) reduce risk. One study estimated that, if oral contraceptives are used for more than 5 years, there is a 37% reduction in ovarian cancer risk (6). Tubal ligation has also been suggested as a potential protective factor.

III. CURRENTLY AVAILABLE SCREENING MODALITIES

The earliest available screening modalities for ovarian cancer, and ones that should be carefully used, are a thorough history and physical examination. As with the more sophisticated and expensive screening tools that are discussed in this chapter, there is no proof that use of these tools reduces ovarian cancer mortality rate. In addition to a careful family history, the physician should be alert for vague gastrointestinal symptoms and symptoms related to pelvic pressure or discomfort. Physical examination should include a bimanual pelvic examination, including a rectovaginal examination. In women who have ovarian cancer, subtle thickening or irregularity in the cul de sac may be noted on the rectovaginal examination. These physical findings are usually indicative of relatively late stage disease.

CA 125 is a serum tumor marker. CA 125 levels are elevated in 80% of women with epithelial ovarian cancer. This marker has been studied since the mid-1980s as a potential tool for screening for early ovarian cancer.

For almost 2 decades, ultrasound has been used to assist in the evaluation of ovarian morphology. With the more recent availability of the endovaginal probe and color Doppler imaging, a much more sophisticated evaluation is possible.

It has not yet been demonstrated that these modalities alone or in combination can reduce the mortality rate from ovarian cancer (4). However, continued investigation is very important because ovarian cancer is the fifth leading cause of cancer death in U.S. women and diagnosis at an earlier stage is associated with much higher cure rates.

IV. CA 125

The CA 125 radioimmunoassay was described by Bast et al. (7) in the early 1980s as being useful in monitoring the course of epithelial ovarian cancer. Eighty percent of women with epithelial ovarian cancer have elevated levels; however, levels are less likely to be elevated when women have mucinous tumors. In addition, levels are elevated in only approximately half of women who have early epithelial ovarian cancers. Levels of CA 125 may also be elevated when women have endometriosis, pelvic inflammatory disease, benign ovarian neoplasms, fibroids, liver disease, congestive heart failure, and a variety of other benign problems. It has been reported that CA 125 levels were elevated in 65% of women with endometriomas and 20% of women with serous cystadenomas (8).

Serum CA 125 levels may become elevated before women have disease that is clinically detected. In evaluations of sera that had been stored, it was discovered that approximately half of women in whom ovarian cancer developed had CA 125 levels greater than 35 U/mL 18 months to 3 years before the diagnosis of ovarian cancer (9–11). This fact prompted Einhorn to do a study prospectively evaluating serum CA 125 levels in 5550 women aged 40 years and older in Stockholm (12). A total of 175 women had abnormal CA 125 values, and they were paired with 175 age-matched control subjects who had normal CA 125 levels. These women were followed up with additional CA 125 determinations every 3 months and with transabdominal ultrasound and pelvic examinations every 6 months. Management was based on clinical and sonographic findings because the clinicians were not provided with the CA 125 results.

Ovarian cancer was ultimately found in six of the 175 women who had elevated CA 125 levels; each of these six was older than 50 years. The stage distribution of the six women with ovarian cancer and elevated CA 125 levels was as follows: two stage IA, two stage IIB, and two stage IIIC. Of those with normal CA 125 levels, three had ovarian cancer. All three were younger than 50 years of age (12). Einhorn urged further evaluation of this screening modality in postmenopausal women.

V. ULTRASOUND

Transvaginal and transabdominal ultrasound have been studied as possible non-invasive screening tools for the identification of early ovarian cancer. Earlier studies used the transabdominal approach. However, once the transvaginal probe was available, more sophisticated studies could be performed. The transvaginal approach is preferred in most instances, because it is more convenient for the patient (transabdominal pelvic ultrasound requires a full bladder and transvaginal ultrasound does not) and provides a better evaluation of the ovaries.

A major study using transabdominal ultrasound evaluated 5479 women, only 4% of whom were 45 years of age or younger. There were 338 positive screens, resulting in 326 women undergoing exploratory laparotomy. As a result of the 326 major abdominal surgeries, five stage I ovarian cancers were found. However, three of the five were for low malignant potential (LMP) lesions, so diagnosis at a later time may not have affected survival for these three women because LMP ovarian tumors usually have an indolent course. In this study of transabdominal sonography screening, 163 women had an exploratory laparotomy for each case of invasive stage I ovarian cancer identified (13). If the low malignant potential lesions are included, the result is one case of ovarian cancer for every 65 laparotomies performed.

A later study used transvaginal sonography (TVS) to evaluate 1601 pre-monopausal and postmenopausal women who had a family history of ovarian cancer. Sixty-one operations were performed, and five stage I ovarian cancers were identified, three of which were low malignant potential lesions (14).

DePriest et al. (15) evaluated 3220 postmenopausal women with endovaginal sonography and identified ovarian tumors in 44 women (1.4%). Of the 44 women who had surgery, 21 had serous cystadenomas and three had ovarian cancers (two stage IA and one stage IIIB). These three women all had normal pelvic examinations and normal CA 125 levels (15).

In a 1993 article in *Cancer*, van Nagell et al. (16) concluded that a large multi-institutional study is indicated to determine whether annual TVS screening will cause a significant decrease in the site-specific mortality rate from ovarian cancer.

Color flow Doppler imaging is being used in some centers as an investigational adjunct to endovaginal sonograms to measure flow patterns within ovarian vessels. It has been reported that malignant ovarian tumors had a significantly lower pulsatility index than benign ovarian lesions (17). The ultimate clinical utility of color flow Doppler in assisting to distinguish benign from malignant lesions is uncertain.

Although only a small number of asymptomatic women with early ovarian cancer have been detected by screening with ultrasound or CA 125 levels, it is hoped that, by combining modalities, utilizing new screening modalities, and perhaps targeting particular high-risk subsets of women for screening, ovarian cancer can be diagnosed earlier and mortality rate from the disease can be reduced.

VI. ROLE OF SCREENING FOR A BRCA1 MUTATION

BRCA1 is the first major breast cancer gene to be isolated. Individuals in high-risk families who carry a mutation of the BRCA1 gene have approximately an 80% risk of developing breast cancer by age 70 years and approximately a 45% to 60% risk of developing ovarian cancer. Families with BRCA1 mutations also have a three-fold increased risk of prostate cancer and a four-fold increased risk of colon cancer. The risk for developing ovarian cancer may differ for different BRCA1 mutations, with ovarian cancer occurring less often in families having mutations at the 3' end of the gene (18,19).

BRCA1 is a large gene found on the long arm of chromosome 17. It is believed that this gene codes for a protein that has a tumor suppressor function. It is also believed that an individual who inherits one mutated copy of the BRCA1 gene is at much greater risk of developing breast or ovarian cancer because a cell with one mutated allele does not have any backup tumor suppressor function if the other copy becomes altered at some point.

BRCA2 is located on chromosome 13. Women who have a BRCA2 mutation appear to have a similar risk of developing breast cancer as those who have BRCA1 mutations. These individuals are also at increased risk for ovarian cancer, but the risk is much less than for those with BRCA1 mutations.

Virtually all of what is known about risk level for both BRCA1 and BRCA2 mutations is from data obtained from high-risk families. The relative risks, given a mutation, may be far different in carriers from the general population.

It is currently believed that more than 90% of familial ovarian cancer is due to inherited mutations in the BRCA1 gene (20). Berchuck et al. (20) have concluded that, because the estimated carrier frequency of the BRCA1 mutation in the general population is 1 in 800, population screening for this mutation is not practical. One in 100 Ashkenazi Jewish women in the United States may carry an identical mutation in codon 185 (deletion AG) of the gene (21). Berchuck suggests that it might be reasonable to offer testing for the codon 185 mutation to this population, but that it may be premature to advocate population screening for BRCA1 mutations when it remains unproved that intervention can decrease cancer incidence or mortality rate (21).

What is known about testing for the BRCA1 mutation is somewhat in its infancy. As Berchuck (20) so significantly points out, ''It is important for zealous health care providers to remember that it has not been proved that cancer mortality can be prevented in BRCA1 carriers.'' However, hereditary cancer is appropriately a major area of concern, and hereditary cancer clinics exist at several major centers throughout the country.

Berchuck has written about the hereditary cancer clinic at Duke University (20). Women with concerns about a family history of breast or ovarian cancer have been invited to attend the clinic. First, a comprehensive family history is obtained. Afterward, it is sometimes determined that the likelihood of that individual carrying a BRCA1 mutation is too low to justify genetic testing. If a woman older than 50 years is the only individual in her family with ovarian or breast cancer, the probability of finding a BRCA1 mutation in that woman is less than 3%. On the other hand, in families with two cases of ovarian cancer and two cases of breast cancer, the probability of having a BRCA1 mutation is about 90%. In the Duke University hereditary cancer clinic, when careful pedigree analysis suggests a significant probability of the presence of a BRCA1 mutation, women in the family who have had breast or ovarian cancer are offered DNA screening. If a BRCA1 mutation is suspected on the initial screen, DNA sequencing is then used to confirm the presence of a significant sequence change and determine its location (16). Once the sequence change and its location have been identified, it becomes relatively simple to screen family members for the specific mutation (20).

Mutations in BRCA1 are believed to be responsible for more than 90% of cases of hereditary ovarian cancer. However, some familial ovarian cancers are due to mutations in other genes.

One problem with BRCA1 testing is that, if a woman does not carry this mutation, the physician cannot tell her that ovarian cancer will not develop. Another major concern is in how to best help those who are identified as carrying the mutation.

In October 1996, the American College of Obstetricians and Gynecologists issued a Committee Statement that included the following comment:

> The Committee on Genetics of the American College of Obstetricians and Gynecologists believes that genetic testing for breast-ovarian cancer should be performed only with the individual's full informed consent. Currently, such testing is best performed working under research protocols approved by an institutional review board. Individuals should be counseled before and after testing to ensure their understanding of the scope, implications, and limitations of the testing process. Counseling should include discussions of the uncertainties of applying genetic test results

to the prevention and treatment of breast and ovarian cancer and the potential for genetic discrimination and loss of insurance coverage. (22)

Once a woman has been identified as carrying a BRCA1 mutation, an attempt must be made to reduce the likelihood that she will die of cancer. It again must be stressed that not only is the ideal way to achieve this unclear, but also it is not yet proven that once the individual is identified as a carrier, that death from cancer is preventable.

Since the focus of this chapter is ovarian cancer screening, a discussion of the options that exist for the BRCA1 mutation carrier will be limited to those related to ovarian cancer. For women who have completed childbearing, prophylactic bilateral oophorectomy is usually recommended. In one series, peritoneal carcinomatosis went on to develop in 2% of women who were at high risk for ovarian cancer and had prophylactic bilateral oophorectomy (23); peritoneal carcinomatosis has a similar histological appearance and natural history as ovarian cancer.

Younger women usually prefer to delay prophylactic surgery and want to be screened. Although it is unclear exactly how much this screening will contribute to a reduction in mortality, it is known that ovarian cancer diagnosed at an earlier stage is associated with an improved survival rate. Therefore, until more complete data become available, I recommend that these very high-risk women who have not had prophylactic bilateral oophorectomy have serial screening with rectovaginal pelvic examination, endovaginal ultrasound, and CA 125 measurement every 6 months.

VII. CONCLUSION AND DIRECTIONS FOR THE FUTURE

A 1994 National Institutes of Health Consensus Statement on Ovarian Cancer concluded that, at that time, there was no evidence available that the current screening modalities of CA 125 and transvaginal ultrasound could be effectively used for widespread population screening to reduce the mortality rate from ovarian cancer nor that their use would result in decreased rather than increased morbidity and mortality rates. The statement noted that routine screening had resulted in a substantial amount of surgery being performed for women who were ultimately determined not to have a malignancy. The statement emphasized the urgent need for large clinical trials to determine whether, in fact, screening could reduce the ovarian cancer mortality rate (4). Such trials are underway, however, and it is hoped that this question can be answered. The PLCO trial is a large trial, initiated in late 1993, which is prospectively evaluating the benefits of screening for prostate, lung, colon, and ovarian cancers (24).

If a woman has one first degree relative with ovarian cancer, her risk of developing the disease is 1 in 20. The consensus statement again noted the absence of sufficient data to prove that the mortality rate can be reduced for these women by the currently available screening modalities, and urged participation in clinical trials. However, the consensus statement noted that, if a woman has one first degree relative with ovarian cancer and no clinical trials are available to her, she may wish to be screened and that this alternative and opportunity should be available to the woman and her physician (4).

The consensus statement noted that women with two first degree relatives with ovarian cancer have approximately a 7% chance of developing the disease and a 3% likelihood of having an autosomal dominant syndrome. The consensus statement advised that, despite the absence of data to indicate that screening can reduce the mortality rate for these very high-risk women, they should have at least an annual physical examination with a bimanual rectovaginal examination, CA 125 determinations, and transvaginal sonography. The statement also suggested that, when childbearing was completed, or at least by age 35 years, prophylactic bilateral oophorectomy is appropriate (4).

The consensus conference took place before BRCA1 screening was available. For women who appear to have a substantial family history, the capability exists to determine whether they carry a BRCA1 mutation. Even though the use of BRCA1 mutation evaluation is in its infancy, it is anticipated that the availability of this study will modify the approach for high-risk women.

In conclusion, ovarian cancer is the fifth leading cause of cancer death in women in the United States. Early detection results in a substantial increase in the potential for cure. There is not sufficient data that current modalities of screening can be used for widespread population screening to reduce the mortality rate. However, several studies are ongoing both to identify new screening modalities and to determine whether current screening techniques can result in a reduction in ovarian cancer mortality rates.

Medical organizations such as the American College of Obstetricians and Gynecologists (25), the American College of Physicians (26), and the Canadian Task Force on the Periodic Health Examination (27) do not recommend routine testing of asymptomatic, low-risk women for the early detection of ovarian cancer with studies such as pelvic ultrasound and serum CA 125 evaluation, because there is no evidence that such testing will reduce morbidity and mortality rates. Furthermore, false-positive studies sometimes lead to surgery, which may actually increase morbidity and mortality rates. It must be emphasized that a properly performed pelvic examination, including Papanicolaou smear, is an important component of regular care for women. Even though these studies do not always identify early ovarian cancer, they are very important for the early detection of other serious gynecological problems.

REFERENCES

1. Wingo PA, Tong T, Bolden S. Cancer statistics 1995. CA Cancer J Clin 1995; 45: 5–30.
2. Pettersson F, ed. Annual report on the results of treatment of gynecologic cancer. Int J Gynecol Obstet 1991; 36(suppl):238–277.
3. Goswamy RK, Campbell S, Whitehead MI: Screening for ovarian cancer. Clin Obstet Gynecol 1983; 10:621–643.
4. NIH Consensus Conference. Ovarian cancer: screening, treatment, and follow-up. JAMA 1995; 273:471–477.
5. Claus EB, Schildkraut JM, Thompson WD, et al. The genetic attributable risk of breast and ovarian cancer. Cancer 1996; 77:2318–2324.
6. Whittemore AS, Harris R, Itnyre J. Characteristics relating to ovarian cancer risk: collaborative analysis of twelve US case-control studies, IV: the pathogenesis of epithelial ovarian cancer. Am J Epidemiol 1992; 136:1212–1220.
7. Bast RC, Klug TL, St. John E, et al. A radioimmunoassay using a monoclonal antibody to monitor the course of epithelial ovarian cancer. N Engl J Med 1983; 309:883–887.
8. Gadducci A, Ferdeghini M, Prontera C, et al. The concomitant determination of different tumor markers in patients with epithelial ovarian cancer and benign ovarian masses: relevance for differential diagnosis. Gynecol Oncol 1992; 44:147–154.
9. Bast RC, Siegal FP, Runowicz C, et al. Elevation of serum CA-125 prior to diagnosis of an epithelial ovarian carcinoma. Gynecol Oncol 1985; 22:115–120.
10. Zurawski VR, Orjaseter H, Anderson A, et al. Elevated serum CA-125 levels prior to diagnosis of ovarian neoplasia. Relevance for early detection of ovarian cancer. Int J Cancer 1988; 42:677–680.
11. Helzllsour KJ, Bush TL, Alberg AJ, et al. Prospective study of serum CA-125 levels as markers of ovarian cancer. JAMA 1993;269:1123–1126.
12. Einhorn N, Sjovall K, Knapp RC, et al. Prospective evaluation of serum CA-125 levels for early detection of ovarian cancer. Obstet Gynecol 1992; 80:14–18.
13. Campbell S, Bhan V, Royston P, et al. Transabdominal ultrasound screening for early ovarian cancer. BMJ 1989; 299:1363–1367.
14. Bourne TH, Campbell S, Reynolds KM, et al. Screening for early familial ovarian cancer with transvaginal ultrasonography and colour blood flow imaging. BMJ 1993; 306:1025–1029.
15. DePriest PD, van Nagell JR, Gallion HH, et al. Ovarian cancer screening in asymptomatic postmenopausal women. Gynecol Oncol 1993; 51:205–209.
16. van Nagell JR, DePriest PD, Gallion HH. Ovarian cancer screening. Cancer 1993; 71:1523–1528.
17. Weiner Z, Thaler I, Beck D. Differentiating malignant from benign ovarian tumors with transvaginal color flow imaging. Obstet Gynecol 1992; 79:159–162.
18. Shattuck-Eidens D, McClure M, Simard J, et al. A collaborative survey of 80 mutations in the BRCA1 breast and ovarian cancer susceptibility gene: implications for presymptomatic testing and screening. JAMA 1995; 273:535–541.
19. Gayther SA, Warren W, Mazoyer S, et al. Germline mutations of the BRCA1 gene

in breast and ovarian cancer families provide evidence for genotype-phenotype correlation. Nat Genet 1995; 11:428–433.
20. Berchuck A, Cirisano F, Lancaster JM, et al. Role of BRCA1 mutation screening in the management of familial ovarian cancer. Am J Obstet Gynecol 1996; 175: 738–746.
21. Struewing JP, Abeliovich D, Peretz, et al. The carrier frequency of the BRCA1 185 del AG mutation is approximately 1 percent in Ashkenazi Jewish individuals. Nat Genet 1995; 11:198–200.
22. ACOG Committee on Genetics Committee Opinion Number 176. Breast-ovarian cancer screening. Washington, DC: American College of Obstetricians and Gynecologists, October, 1996.
23. Piver MS, Jishi MF, Tsukada Y, et al. Primary peritoneal carcinoma after prophylactic oophorectomy in women with a family history of ovarian cancer: a report of the Gilda Radner Familial Ovarian Cancer Registry. Cancer 1993; 71:2751–2755.
24. Kramer BS, Gohagan J, Provok, et al. A National Cancer Institute sponsored screening trial for prostatic, lung, colorectal, and ovarian cancers. Cancer 1993; 71:589–593.
25. American College of Obstetricians and Gynecologists. Routine cancer screening. Committee Opinion No. 128. Washington, DC: American College of Obstetricians and Gynecologists, 1993.
26. American College of Physicians. Screening for ovarian cancer. Ann Intern Med 1994; 121:141–142.
27. Canadian Task Force on the Periodic Health Examination. Canadian Guide to Clinical Preventive Health Care. Ottowa, Canada: Canada Communication Group, 1994: 869–881.

17

Pancreatic Cancer

Ralph H. Hruban, Charles J. Yeo, and Scott E. Kern
The Johns Hopkins Medical Institutions, Baltimore, Maryland

I. INTRODUCTION

Carcinoma of the pancreas is the fifth leading cause of cancer death in the United States (1). It is almost always a fatal disease, and the mortality rate approximates that of the incidence. For example, it was estimated that approximately 27,600 new cases of pancreatic carcinoma would be diagnosed in the United States in 1997 and that 28,100 patients would die of it (1). The high mortality (low survival) rate for carcinoma of the pancreas is, in large part, due to the fact that most patients with pancreatic cancer are not seen until late in the course of their disease, after the carcinoma has already metastasized or spread locally and is no longer curable (2,3). However, Yeo et al. (4) have recently demonstrated that, although the 5-year survival rate for persons with carcinoma of the pancreas is only 3%, the 5-year survival rate for patients with localized tumors who undergo pancreaticoduodenectomy can approach 25%. Additionally, the 5-year survival rate for patients with particularly low stage disease, that is, patients with small tumors (less than 3 cm), negative lymph nodes, and negative surgical margins, may be as high as 40% (4,5). Clearly, the earlier cancer of the pancreas is detected, the better the patient's chance of survival.

Most patients with pancreatic carcinoma present clinically with abdominal pain, with jaundice, or with protean symptoms related to the distant effects of the carcinoma (2). Once these patients come to medical attention, their pancreatic carcinomas can be imaged using a variety of techniques including spiral and conventional computed tomography (CT), magnetic resonance imaging (MRI), angiography, sonography, endoscopic ultrasonography, and endoscopic retrograde cholangiopancreatography (ERCP) (6–10). For example, the entire pancreas can be imaged using dynamic thin-section contrast-enhanced abdominal CT scanning, and ERCP can be used to accurately define the level of biliary obstruction. More recently, octreotide scanning and ultrafast echo-planar MRI have also shown promise in imaging the pancreas (11). Even though these imaging techniques can accurately diagnose established pancreatic cancers, their usefulness in detecting small pancreatic cancers and in screening for early pancreatic cancers is limited. For example, in a series reported by DiMagno et al. (12), abdominal ultrasonography and ERCP were able to detect pancreatic cancer in 89% of patients, but 87% of these patients had unresectable disease (12). In addition to low sensitivity, the major limitation of these imaging techniques is that they do not influence the timing of when a patient is first seen. Because these techniques are relatively invasive, costly, or involve the use of ionizing radiation, they are usually only applied to symptomatic patients, and, as discussed earlier, symptoms do not develop in most patients with pancreatic cancer until late in the course of their disease. It is therefore not surprising that these imaging techniques have not significantly changed the death rate from pancreatic cancer (1). The prognosis for patients with pancreatic cancer will not improve significantly until new tests are developed to screen for the disease *before* the patients become symptomatic. Even then, improved survival rates resulting from a lead time bias (a simple advance of the time of diagnosis without change in the course of disease) would have to be distinguished from a true benefit that decreases overall mortality rate.

Serum markers could be used for more widespread screening of asymptomatic individuals. However, the sensitivity and specificity of the currently available serum markers for pancreatic cancer, such as carbohydrate antigen (CA) 19–9, carcinoembryonic antigen (CEA), CA 125, DU-PAN-2, and pancreatic oncofetal antigen are unacceptably low (13–15). None of these markers is cancer or organ specific, and they can all be elevated in patients with benign gastrointestinal diseases. For example, Homma and Tsuchiya (16) screened more than 10,000 asymptomatic persons for pancreatic cancer using either ultrasonography or CA 19–19 and elastase-1, and they found that only 0.5% of the patients with a positive screening test actually had a pancreatic cancer. Furthermore, only one of the four cancers discovered could be curably resected. Even though serum markers are useful in following up patients with known disease, they simply are not reliable as screening tools.

There is an enormous need for new noninvasive screening tests to detect early pancreatic neoplasms. We believe that an understanding of the biology of pancreatic neoplasia will provide a rational basis for the development of such tests.

II. CLINICALLY EARLY TUMORS ARE BIOLOGICALLY LATE

Most neoplasms do not cause symptoms and cannot be detected using standard imaging techniques until they are at least 1 cm in size. By this time, the original tumor clone has already effectively undergone 30 doublings and contains approximately 10^9 cells (17,18). After only 10 more doublings, the tumor reaches a size that is no longer compatible with life. Therefore, in this theoretical representation, by the time the average tumor is clinically detected, it has exhausted three-fourths of its life span (17,18).

Cancer of the pancreas is no exception to this rule. Carcinomas of the pancreas that are clinically early are biologically late. For example, Pour et al. (19) examined the pancreata in a series of 83 consecutive autopsies from military veterans aged 35–88 years and found that 10% of these pancreata had clinically unsuspected exocrine pancreatic neoplasms (19). These included two infiltrating pancreatic adenocarcinomas as well as eight carcinomas in situ (19). This remarkable finding by Pour and his colleagues established that there is a long latency period between the development of a neoplasm in the pancreas and the development of clinical symptoms.

In the case of pancreatic cancer, we believe that these earliest preclinical neoplasms take the form of distinct lesions in the pancreatic ducts called ''duct hyperplasias'' (20). These lesions often accompany infiltrating adenocarcinomas of the pancreas and are characterized by the presence of an epithelial proliferation with varying degrees of architectural and cytological atypia in the pancreatic ducts and ductules. For example, Kozuka et al. (21) histologically examined the pancreata of more than 1170 autopsied patients and found an association between ''atypical papillary duct hyperplasia'' and infiltrating adenocarcinoma (21). These results have been confirmed by Cubilla and Fitzgerald (22) and more recently by Furukawa et al. (23). Furthermore, Furukawa et al. applied digital imaging and three-dimensional reconstruction techniques to histological sections, and they were able to demonstrate a step by step progression from ordinary duct epithelium, to mild dysplasia, to severe dysplasia, to infiltrating carcinoma (23). These lesions in the pancreatic ducts are not simply morphological curiosities, but instead Brat et al. (24) have shown that duct lesions in the pancreas can progress to invasive cancers. Thus, it would appear that, just as there is a progression from adenoma to infiltrating carcinoma in the colorectum, so too is there progression from ''duct hyperplasias'' to infiltrating adenocarcinomas in the pancreas.

Although only a minority of these duct lesions progress to infiltrating cancer, the presence of such early microscopic neoplasms in the pancreata of patients without clinical symptoms presents a potential window of opportunity. If techniques can be developed to detect cancers based on properties other than their size, then smaller, and presumably more curable, tumors will be found

(25). Investigations are underway to determine whether there are sensitive markers of high-risk lesions that would have adequate specificity to avoid overdiagnosis.

III. GENETIC CHANGES

Despite the striking histological appearance of early neoplasms of the pancreas, histological changes cannot form the basis of a screening test for early pancreatic neoplasia. Histology requires direct sampling of the pancreas, and, because of the high costs and risks involved, direct sampling of the pancreas simply cannot be a practical component of any population-based screening test. Infiltrating carcinomas and even the earliest duct lesions in the pancreas have been shown to harbor a number of genetic alterations, and, unlike histological changes, these alterations can form the basis of a screening test (25,26). The remainder of this chapter first describes the genetic alterations that have been identified in infiltrating adenocarcinomas of the pancreas and in duct lesions in the pancreas, then it describes how these genetic alterations could form the basis of a screening test for early pancreatic neoplasia.

The genetic alterations in infiltrating adenocarcinomas of the pancreas are distinct from those seen in other neoplasms. The *K-ras* oncogene is activated in the vast majority of pancreas cancers, and the *p16, p53*, and *DPC4* tumor suppressor genes are frequently inactivated (27–51). Activating point mutations in *K-ras* have been reported in 80%–100% of pancreatic carcinomas (28–35). This is the highest reported prevalence of *K-ras* mutations in any tumor type, and the vast majority of these mutations are confined to codon 12 of the *K-ras* oncogene (28). The next most frequently altered gene in pancreatic carcinoma is the *p16* tumor suppressor gene (36–40). *p16* function is lost in 80% of pancreatic cancers: in 40% by homozygous deletions and in 40% by mutation of one allele and deletion of the second (36). The *p53* tumor suppressor gene is also frequently inactivated in infiltrating pancreatic carcinoma (41–47,52). For example, Redston et al. (42) identified *p53* mutations in 70% of 27 primary pancreatic adenocarcinomas and Rozenblum et al. (53) in 75% of 38 human cancers expanded by xenografting into mice. The *p16* and *p53* gene products both help control the cell cycle, and loss of their function in pancreatic neoplasia disrupts important cell cycle checkpoints (27). *DPC4* is a third tumor suppressor gene frequently inactivated in pancreatic neoplasms (49–51). *DPC4* is believed to play a role in a transforming growth factor-β (TGF-β)-like signaling pathway and it is inactivated in approximately 50% of pancreas cancers (49). In 30% of pancreatic carcinomas, this inactivation occurs by homozygous deletion, and in the remaining 20% by mutation of one allele and loss of the second (50). Although *DPC4* is a common target of inactivation of pancreatic carcinoma, it is only occasionally inactivated in other types of cancer (51). Thus, the molecular genetic profile of infiltrating adenocarcinomas of the pancreas is distinctive (53). Pancreatic carcinomas have almost uniform activation of the *K-ras* oncogene by mutations at codon 12, and frequent ($\geq$50%) inactivation of *p53, p16*, and *DPC4*.

If these molecular alterations of pancreatic carcinoma occur early in tumorogenesis, they could theoretically be used as markers for the early diagnosis of pancreatic carcinoma. Indeed, this appears to be the case. Many of the same genetic alterations that are present in infiltrating adenocarcinomas of the pancreas are also present in the pancreatic duct lesions (the ''duct hyperplasias'') (41,54–60). For example, DiGiuseppe et al. (56) microdisected seven foci of papillary and nonpapillary duct hyperplasia from a surgically resected pancreas that did not have an associated infiltrating carcinoma and found that five of the seven foci had activating point mutations in codon 12 of the *K-ras* gene. Four of the five mutations were a G to A transition and the fifth was a G to C transversion (56). Similarly, Yangagisawa et al. (60) reported that 10 (62.5%) of the 16 duct lesions they examined harbored *K-ras* mutations. More advanced duct lesions with cytological and architectural atypia may also harbor *p53* and *p16* mutations (41,55,58). Hameed et al. (58) studied a series of more than 180 pancreata with infiltrating carcinomas for the *p53* gene product by immunohistochemistry and found that, when the infiltrating carcinoma overexpressed *p53*, 67% of the *in situ* carcinomas also overexpressed *p53*. Although immunohistochemical staining for *p53* is an indirect, and therefore somewhat imperfect, marker for *p53* mutations, these data suggest that a significant percentage of *in situ* adenocarcinomas of the pancreas contain *p53* mutations. Furthermore, Moskaluk et al. (55) examined 10 pancreatic resection specimens for *K-ras* and *p16* mutations and they found three duct lesions with *p16* mutations. Remarkably, in Moskaluk's series, one pancreas was found with two separate duct lesions, each of which harbored a different *p16* mutation.

Thus, both the infiltrating carcinomas of the pancreas and the early duct lesions from which these carcinomas presumably arise can harbor mutations in *K-ras, p53, p16*, and possibly *DPC4*. These genetic alterations, because they are common in pancreatic neoplasms and because they occur early in tumor development, make ideal targets for the development of novel molecular tests to screen for early, and therefore potentially curable, pancreatic cancers. The sections that follow describe some potential applications of molecular screening tests for pancreatic cancer.

IV. MOLECULAR TESTS FOR MUTANT *K-ras*

There are no commercially available molecular tests being used to screen for pancreatic carcinomas, but recent advances in understanding of the molecular genetics of pancreatic neoplasia provide several avenues for developing such a test (25,27). Probably the best example of this is the *K-ras* oncogene. *K-ras* is an attractive gene for a molecular screening test for pancreatic cancer for three reasons (28). First, as described earlier, *K-ras* is the most frequently mutated gene in cancer of the pancreas (28–35). More than 85% of pancreatic carcinomas harbor mutations in *K-ras*, suggesting that *K-ras* alterations will be sensitive genetic markers (28–35). Second, these mutations occur early in the development of pancreatic neoplasia, suggesting that *K-ras* alterations could be used to detect early, and therefore potentially curable, pancreatic cancers (54,56,60–62). Third, the vast majority of mutations in *K-ras* in pancreatic cancers are restricted to a single codon (codon 12), and so a limited number of probes can be used to detect these mutations, greatly simplifying these analyses (27). Indeed, molecular-based techniques have been developed that can detect one mutant copy of *K-ras* admixed with as many as 10,000 wild-type copies of this same gene (54,61).

The utility of molecular-based tests for mutant *K-ras* in screening a variety of different types of samples has already been examined in a number of patients with pancreatic cancer (54,61–68). For example, Berthelemy et al. (62) screened more than 75 samples of pancreatic secretions collected during endoscopic retrograde pancreatography (ERCP), an imaging procedure, for mutant *K-ras*. They detected mutant *K-ras* in 17 (77%) of the 22 samples obtained from patients with pancreatic cancer, whereas none of the 53 samples obtained from patients without pancreatic cancer harbored mutant *K-ras* (62). Of note, they detected mutant *K-ras* in the pancreatic secretions of two patients without pancreatic masses. One of these two patients had undergone surgical exploration at the time the pancreatic secretion sample was obtained and no mass was identifiable, and the second patient was examined extensively using both CT and endosonograpy. Remarkably, pancreatic cancer went on to develop in both of these patients with *K-ras* mutations in their pancreatic secretions and no pancreatic mass. In one of the patients, a cancer developed 18 months after the mutant *K-ras* was detected and, in the other, after 40 months (62). This study and a similar one by Brentnall et al. (63) demonstrates the potential for molecular screening tests to detect early pancreatic neoplasms before they are clinically apparent. The ERCP is, however, an invasive procedure and it is not practical to obtain samples of pancreatic secretions on a wide scale basis. Caldas et al. (54), therefore, examined the utility of testing stool samples for *K-ras* mutations. Using a plaque hybridization assay for *K-ras* mutations, Caldas et al. (54) were able to detect mutant *K-ras* in six (55%) of 11 stool samples obtained from patients with pancreatic cancer. These same patients underwent pancreaticoduodectomy, giving Caldas and his colleagues the opportunity to examine the source of the mutations in the stool. In five of the six cases in which mutant *K-ras* was detected in the stool, the same mutation was demonstrated in the patient's primary infiltrating pancreatic cancer (54). In the sixth case, the mutant

K-ras in the stool matched the mutation found in duct lesions (papillary hyperplasias) in the patient's pancreas.

These studies demonstrate the scientific basis for a clinical test. The DNA shed from pancreas cancers and from pancreatic duct lesions can be detected in pancreatic secretions and in the stool. Furthermore, early and clinically silent cancers could be identified using such tests. Caution, however, must be exercised before *K-ras* based tests are applied clinically (69). Mutant *K-ras* sequences can arise from extrapancreatic sites such as adenomas in the colon (70). Furthermore, even if the source of *K-ras* mutations could be localized to the pancreas, these mutations could originate from a benign duct lesion. It is not clear how frequently and how rapidly duct lesions with *K-ras* mutations progress to infiltrating carcinomas (20). It would be inappropriate to surgically explore or otherwise invasively examine a patient's pancreas because of a genetic alteration that arose in the colon or because of a genetically altered duct lesion that, by itself, is relatively harmless. Thus, further work is needed to define the timing and significance of *K-ras* mutations in the development of pancreatic neoplasia.

While acknowledging the limitations of the *K-ras* alterations in screening for pancreatic neoplasia, several investigators have suggested that tests for mutant *K-ras* may still be useful as an adjunct to other tests such as cytology tests (71–73). For example, van Es et al. (71) assayed archival cytological brushings obtained at ERCP from 17 patients with established pancreatic carcinomas for *K-ras* mutations. In this study, the *K-ras* codon 12 mutation identified in cytology preparations corresponded to the mutations identified in the resected carcinomas in all cases, and, in two cases with an inconclusive cytology diagnosis, the molecular markers suggested the presence of a neoplasm (71). Thus, molecular markers such as *K-ras* can be a valuable adjunct to diagnostic cytopathological and histopathological study (72,73).

V. TUMOR SUPPRESSOR GENES

As described earlier, several tumor suppressor genes, including *p53, p16*, and *DPC4*, are also frequently altered in pancreatic carcinomas (35–47,49–53). These genetic alterations are potentially attractive targets for molecular tests for pancreatic cancer because some appear to be relatively organ specific. For example, as discussed earlier, mutations in the *DPC4* gene are relatively restricted to pancreatic cancer (51). The genetic changes in the *p53, p16*, and *DPC4* genes usually inactivate the function of the tumor suppressor gene and include homozygous deletions and loss of one allele accompanied by an interstitial mutation in the second allele (35–37,39–51,53). Unlike the *K-ras* gene, however, in which the mutations are confined to a single codon, the mutations in the *p53, p16*, and *DPC4* genes in pancreatic neoplasia are not restricted. Instead, these genetic alterations lie dispersed within a large number of sites throughout these genes. The genetic changes in the *p53, p16*, and *DPC4* genes are therefore much more difficult to detect than are mutations in the *K-ras* gene.

Despite the challenges presented in detecting mutations in these tumor suppressor genes, molecular tests to detect alterations in these genes have been used in screening tests and they may be particularly useful in cases in which the mutational status of the primary carcinoma is known (26). Indeed, *p53* mutations have been detected in the urine of patients with bladder cancer, in the stool of patients with colon cancer, and at surgical margins in patients with head and neck carcinomas (26,74–76). For example, Eguchi et al. (75), using a combination of polymerase chain reaction (PCR) amplification and single-strand confirmation polymorphism (SSCP), detected *p53* mutations in stool samples from seven (28%) of 25 patients with adenocarcinoma of the colon (75). The same patients were also screened for colon cancer using a fecal blood test and three (16%) of the five patients who had a negative fecal blood test had stools with a detectable *p53* mutation (75). Thus, molecular genetic testing of stool for *p53* mutations may be more sensitive than fecal blood tests in detecting colon cancer (25,75). The technical hurdles inherent in the detection of mutant tumor suppressor genes admixed with much larger numbers of wild-type tumor suppressor genes will eventually be overcome. It is therefore conceivable that molecular genetic tests for genetic alterations in tumor suppressor genes might one day also be used to screen for pancreatic cancers.

VI. TELOMERASE

The expression of other genes may also form the basis of screening tests for pancreatic cancer. For example, telomerase is an enzyme that is believed to be involved in the de novo synthesis of the tandemly repeated sequence GGTTAG onto telomeric chromosomal ends. The expression and stabilization of telomerase has been associated with the attainment of immortality in cancer cells (77,78). Hiyama and colleagues (77) determined the telomerase activity in a series of benign and malignant pancreatic specimens and found that telomerase activity was detectable in 41 (95%) of 43 pancreatic cancers, in one (33%) of three pancreatitis samples, and in none (0%) of 11 benign tumors of the pancreas. They then examined the telomerase activity in pancreatic duct brushings and were able to detect telomerase activity in eight (100%) of eight brushings from patients with pancreatic cancer, whereas none of the four brushings from patients with benign lesions had telomerase activity (77). This finding suggests that, as we learn more about gene expression in pancreatic neoplasia, we will be able to develop novel biological tests for pancreatic cancer. One particular technique that holds promise in this regard is serial analysis of gene expression (SAGE).

VII. SERIAL ANALYSIS OF GENE EXPRESSION

SAGE is a method for simultaneously characterizing the expression of multiple genes in a tissue or tumor (79). SAGE can be used to determine which genes are expressed in a tissue or tumor, and, because it controls for certain biases in PCR amplification, the levels of expression of these same genes can be determined (79). SAGE is a powerful technique for developing new screening tests for pancreatic neoplasia for two reasons. First, it can be used to identify novel genes that are expressed by the neoplasms, and second, because it can be used to determine the levels of gene expression, it can be used to identify known genes that are differentially expressed in pancreatic neoplasia. For example, Zhang et al. (80) used SAGE to identify and quantify 303,706 transcripts derived from normal tissue and from a series of pancreatic and colon cancers. These transcripts represented more than 48,000 different genes. Many were novel in that only 54% matched known entries in GenBank (80). Furthermore, they were able to identify more than 400 transcripts that were expressed at higher levels in the pancreatic cancers than they were in other normal tissues, and the majority of these transcripts appeared to be specific to pancreas cancer (80). If these differentially expressed transcripts are shed into the serum, then the SAGE technique could be used to develop new clinical markers for pancreatic cancer. Thus, just as there is prostate-specific antigen marker for prostate cancer, so too could SAGE be used to discover a ''pancreas-specific antigen'' (25).

VIII. RISK FACTORS FOR PANCREATIC CANCER

All of these advances in developing new screening tests for early pancreatic neoplasia raise the fundamental question of ''who should be screened?'' (Table 1). It is premature to offer an answer in the absence of an efficacious screening test, but there has been initial success in defining subpopulations at greater risk for development of this cancer. In the United States, the known risk factors are older age, male gender, African-American race, Jewish ancestry, low socioeconomic status, high-fat/low-fiber diet, occupational exposure to certain chemicals and, most importantly, tobacco use (81).

In addition to these demographic and environmental risk factors, several recent studies have suggested that heredity also plays a role in the development of pancreatic cancer (82,83). For example, Ghadirian et al. (84) reported that 7.8% of patients with pancreatic carcinoma have a family history of pancreas cancer whereas only 0.6% of patients without pancreatic cancer have such a history. Similarly, Fernandez et al. (85) conducted a study in Northern Italy on 362 patients with histologically confirmed pancreatic carcinoma and 1408 control subjects admitted to the hospital for nonneoplastic nondigestive tract disease.

Table 1 Clinical Risk Factors for Pancreatic Cancer

The elderly, African Americans, males, Jewish ancestry
History of chronic pancreatitis
History of smoking
Distinctive clinical syndromes
Peutz-Jeghers syndrome
Familial atypical multiple mole melanoma (FAMMM)
Ataxia telangiectasia
Familial pancreatitis
Hereditary nonpolyposis colorectal cancer (HNPCC)
Family history of pancreatic cancer
Family history of breast cancer

They found that significantly more of the patients with pancreatic cancer had a family history of pancreatic cancer than did the control subjects (relative risk = 3.0). From these studies and others it has been estimated that as many as 10% of pancreatic carcinomas are caused by inherited factors (82,83). These families, in which there is an aggregation of pancreatic cancer, are particularly attractive targets for molecular screening tests because they can be screened at multiple diagnostic levels. That is, in addition to screening using the molecular and non-molecular tests described thus far, individuals from families with an aggregation of pancreatic cancer can also be screened for germline mutations in cancer-causing genes, which could be detected in circulating nucleated blood cells. For example, accurate presymptomatic genetic testing for *APC* gene mutations is available for individuals in families known to be at risk for development of familial adenomatous polyposis coli (FAP) (86). Prophylactic surgery on high-risk *asymptomatic* carriers of *APC* mutations reduces the incidence of colorectal cancer in families, and unaffected low-risk family members are relieved of the trouble and the uncertainty of long-term surveillance and repeated endoscopic examinations (86). It is similarly reasonable to expect that the identification of the genes responsible for familial forms of pancreatic carcinoma will provide a molecular test that would be invaluable in counseling families with pancreas cancer.

Even though the genes responsible for the development of most cases of familial pancreatic carcinoma are not yet known, germline mutations in patients with pancreatic cancer using a ''candidate gene approach'' have been identified. This approach is based on Knudson's classical hypothesis that at least two mutations in a tumor-suppressor gene are required to transform a normal cell into a cancer cell (87). In sporadic cancers, two normal alleles of the tumor-suppressor gene are inherited, and two somatic mutations must therefore occur.

In cancer families, the affected family members inherit a germline mutation in one allele of gene, followed by an acquired somatic mutation in the second allele (87). Knudson's hypothesis implies both that the inherited first mutation in cancer gene carriers can be detected in somatic cells and that, at the molecular level, familial and sporadic forms of cancer often involve the same genes. In the candidate gene approach, tumor suppressor genes that are known to be inactivated in sporadic pancreatic cancers are selected as the "candidate genes," and then DNA from members of families with familial pancreatic cancer are screened for mutations of this candidate gene (27). Using this approach, germline mutations in *p16* and in *BRCA2* have been identified in patients with pancreatic carcinoma (88,89). In contrast, a number of families have been examined for germline mutations of *DPC4* and *p53*, and, to date, none have been found (90).

p16 was a very promising "candidate gene" to examine for familial pancreatic cancer; *p16* is frequently mutated in sporadic pancreas cancers (36) and germline mutations in *p16* have been reported to be associated with the familial atypical multiple mole melanoma syndrome (FAMMM). The FAMMM syndrome is, in turn, associated with an increased risk of pancreatic carcinoma (91,92). Moskaluk et al. (88) therefore analyzed 21 kindred with familial pancreatic cancer for germline *p16* mutations. Kindred with the FAMMM syndrome were excluded. A germline *p16* mutation was identified in one (<5%) of the 21 families, and in this family the mutation segregated with pancreas cancer (88). One of the patients with pancreas cancer in this family also had a history of melanoma. Thus, germline *p16* mutations are associated with the development of pancreatic carcinoma and melanoma, and families with the FAMMM syndrome or an aggregation of melanoma and pancreatic cancer may benefit from testing for germline *p16* mutations. The overall contribution of germline *p16* mutations to familial pancreatic carcinoma in which there is no history of a melanoma appears to be too low to warrant more widespread screening for these mutations.

BRCA2 was another logical gene to examine in families with an aggregation of pancreatic cancer. The *BRCA2* locus has been reported to be homozygously deleted in a pancreas cancer (93) and several reports have demonstrated that the risk of pancreatic carcinoma development is increased in carriers of *BRCA2* mutations (94–97). Goggins et al. (89) therefore screened a series of 41 pancreas cancers for *BRCA2* mutations. Four (10%) of the 41 patients had an inactivating intragenic mutation of one allele of *BRCA2* and loss of the second (89). Remarkably, when normal tissue from the same patients was examined, three of the four mutations were inherited germline mutations (89). Two of the three germline mutations involved the 6174delT change at codon 1982. A panel of normal tissues from 265 patients with pancreatic carcinoma was then screened for *BRCA2* mutations near the 6174 location. Two additional germline mutations in *BRCA2* were identified. None of the patients with germline *BRCA2* mutations came from kindreds with a significant family history of breast or pancreas carcinoma. Thus, germline mutations in *BRCA2* represent the most common inherited predisposition to pancreatic carcinoma identified to date and, based on the small numbers of affected family members in these kindreds, it appears that the disease penetrance of these mutations can be low (27,89).

IX. CONCLUSION

We believe that germline mutations in *p16 and BRCA2* as well as germline mutations in additional yet-undiscovered tumor suppressor genes will account for most cases of familial pancreatic carcinoma and also for many cases of pancreatic carcinoma that appear to be sporadic. Once these germline mutations are characterized, they can be used to screen at-risk individuals. Such screening in the future is likely to involve a combination of both molecular detection assays and sensitive imaging approaches to achieve early diagnosis or to offer prophylactic treatment for high-risk preinvasive lesions. Importantly, those individuals without germline mutations could be relieved of their concerns.

Studies of families with an aggregation of pancreatic cancer will therefore provide new insights into the genetics of pancreatic carcinoma and new avenues for screening patients with this disease. In order to identify additional families with an aggregation of pancreatic carcinoma for study, we have established the National Familial Pancreas Tumor Registry at Johns Hopkins.* The National Familial Pancreas Tumor Registry has already accrued more than 80 families with an aggregation of pancreas carcinoma and we are actively recruiting additional families. Patients and physicians aware of such families are encouraged to refer kindreds to the registry.

Although the challenges are great, so too will be the rewards if a new test can be developed to screen for early pancreatic carcinomas. Currently, no organizations have recommended routine screening for pancreatic cancer. We believe that, in the future, early pancreatic carcinomas will be detected using a combination of testing for germline mutations, screening of secondary sources such as blood and stool for acquired mutations, testing of serum for novel tumor-specific gene products, and advanced imaging using extremely sensitive techniques. To learn more about pancreas cancer, visit our Web site at http://www.path.jhu.edu/pancreas.

REFERENCES

1. Parker SL, Tong T, Bolden S, Wingo PA. Cancer statistics, 1997. CA Cancer J Clin 1997; 47:5–27.

*1 National Familial Pancreas Tumor Registry, c/o Ralph Hruban, M. D., Meyer 7–181, The Johns Hopkins Hospital, 600 N. Wolfe Street, Baltimore, Maryland 21287

2. Warshaw AL, Castillo CFD. Pancreatic carcinoma. N Engl J Med 1992; 326: 455–465.
3. Niederhuber JE, Brennan MF, Menck HR. The national cancer data base report on pancreatic cancer. Cancer 1995; 76: 1671–1677.
4. Yeo CJ, Cameron JL, Lillemoe KD, Sitzmann JV, Hruban RH, Goodman SN, et al. Pancreaticoduodenectomy for cancer of the head of the pancreas: 201 patients. Ann Surg 1995; 221:721–733.
5. Fortner JG, Klimstra DS, Senie RT, Maclean BJ. Tumor size is the primary prognosticator for pancreatic cancer after regional pancreatectomy. Ann Surg 1996;223: 147–153.
6. Freeny PC, Marks WM, Ball TJ. Impact of high-resolution computed tomography of the pancreas on utilization of endoscopic retrograde cholangiopancreatiography and angiography. Radiology 1982; 142:35–39.
7. Wyatt SH, Fishman EK. Spiral CT of the pancreas. Semin Ultrasound CT MRI 1994; 15:122–132.
8. Palazzo L, Roseau G, Gayet B, Vilgrain V, Belghiti J, Fékété F, et al. Endoscopic ultrasonography in the diagnosis and staging of pancreatic adenocarcinoma: results of a prospective study with comparison to ultrasonogrpahy and CT scan. Endoscopy 1993; 25:143–150.
9. Rösch T, Braig C, Gain T, Feuerbach S, Siewert JR, Schusdziarra V, et al. Staging of pancreatic and ampullary carcinoma by endoscopic ultrasonography Comparison with conventional sonography, computed tomography, and angiography. Gastroenterology 1992; 102:188–199.
10. Bluemke DA, Cameron JL, Hruban RH, Pitt HA, Siegelman SS, Soyer P, et al. Potentially resectable pancreatic adenocarcinoma: spiral CT assessment with surgical and pathologic correlation. Radiology 1995; 197:381–385.
11. van Eijck CHJ, Lamberts SWJ, Lemaire LCJM, Jeekel H, Bosman FT, Reubi JC, et al. The use of somatostatin receptor scintigraphy in the differential diagnosis of pancreatic duct cancers and islet cell tumors. Ann Surg 1996; 224:119–124.
12. DiMagno EP, Malagelada JR, Taylor WF, Go VLW. A prospective comparison of current diagnostic tests for pancreatic cancer. N Engl J Med 1977; 297:737–742.
13. Ritts R Jr. Jacobsen D, Ilstrup D, DiMagno E. A prospective evaluation of MoAb CA 19–9 to detect GI cancer in a high risk clinic population (abstr). Cancer Detect Prev 1984; 7:525A.
14. Steinberg W. The clinical utility of the CA 19–9 tumor associated antigen. Am J Gastroenterology 1990; 85:350–355.
15. Steinberg WM, Gelfand R, Anderson KK, Glenn J, Kurtzman SH, Sindelar WF, et al. Comparison of the sensitivity and specificity of the CA 19–9 and carcinoembryonic antigen assay in detecting cancer of the pancreas. Gastroenterology 1986; 90:343–349.
16. Homma T, Tsuchiya R. The study of the mass screening of persons without symptoms and of the screening of outpatients with gastrointestinal complaints or icterus for pancreatic cancer in Japan using CA 19–9 and elastase-1 or ultrasonography. Int J Pancreatol 1991; 9:119–124.
17. Tannock IF. Biology of tumor growth. Hosp Pract 1983; 18(4):81–93.

18. Tannock IF. The basic science of oncology. In: Tannock IF, Hill RP, eds. Cell Proliferation. 2d ed. New York: McGraw-Hill, 1992:154
19. Pour PM, Sayed S, Sayed G. Hyperplastic, preneoplastic and neoplastic lesions found in 83 human pancreases. Am J Clin Pathol 1982; 77:137–152.
20. DiGiuseppe JA, Yeo CJ, Hruban RH. Molecular biology and the diagnosis and treatment of adenocarcinoma of the pancreas. Adv Anat Pathol 1996; 3:139–155.
21. Kozuka S, Sassa R, Taki T, Masamoto K, Nagasawa S, Saga S, et al. Relation of pancreatic duct hyperplasia to carcinoma. Cancer 1979; 43:1418–1428.
22. Cubilla AL, Fitzgerald PJ. Morphological lesions associated with human primary invasive nonendocrine pancreas cancer. Cancer Res 1976; 36:2690–2698.
23. Furukawa T, Chiba R, Kobari M, Matsuno S, Nagura H, Takahashi T. Varying grades of epithelial atypia in the pancreatic ducts of humans. Classification based on morphometry and multilvariate analysis and correlated with positive reactions of carcinoembryonic antigen. Arch Pathol Lab Med 1994; 118:227–234.
24. Brat DJ, Lillimoe KD, Yeo CJ, Warfield PB, Hruban RH. Progression of pancreatic intraductal lesions (high grade PanIN) to infiltrating adenocarcinoma of the pancreas Am J Surg Pathol 1998; 22:163–169.
25. Wilentz RE, Slebos RJC, Hruban RH. Screening for pancreatic cancer using techniques to detect altered gene products. In: Reber HA, ed. Pancreatic Cancer: Pathogenesis, Diagnosis, and Treatment. New York: The Humana Press, 1998.
26. Hruban RH, van der Riet P, Erozan Y, Sidransky D. Brief report: molecular biology and the early detection of carcinoma of the bladder—the case of Hubert H. Humphrey. N Engl J Med 1994; 330:1276–1279.
27. Hruban RH, Yeo CJ, Kern SE. Pancreatic cancer. In: Scriver C, Beaudet A, Sly D, Valle D, eds. Molecular and Mendelian Basis of Inherited Disease, 7th ed., CD-ROM. New York: McGraw-Hill, 1997.
28. Hruban RH, van Mansfeld ADM, Offerhaus GJA, van Weering DHJ, Allison DC, Goodman SN, et al. K-ras oncogene activation in adenocarcinoma of the human pancreas. A study of 82 carcinomas using a combination of mutant-enriched polymerase chain reaction analysis and allele-specific oligonucleotide hybridization. Am J Pathol 1993; 143:545–554.
29. Smit VTHBM, Boot AJM, Smits AMM, Fleuren GJ, Cornelisse CJ, Bos JL. K-ras codon 12 mutations occur very frequently in pancreatic adenocarcinomas. Nucleic Acids Res 1988; 16:7773–7782.
30. Mariyama M, Kishi K, Nakamura K, Obata H, Nishimura S. Frequency and types of point mutation at the 12th codon of the c-Ki-ras gene found in pancreatic cancers from Japanese patients. Jpn J Cancer Res 1989; 80:622–626.
31. Grünewald K, Lyons J, Fršhlich A, Feichtinger H, Weger RA, Schwab G, et al. High frequency of Ki-ras codon 12 mutations in pancreatic adenocarcinomas. Int J Cancer 1989; 43:1037–1041.
32. Nagata Y, Abe M, Motoshima K, Nakayama E, Shiku H. Frequent glycine-to-aspartic acid mutations at codon 12 of c-Ki-ras gene in human pancreatic cancer in Japanese. Jpn J Cancer Res 1990; 81:135–140.
33. Tada M, Yokosuka O, Omata M, Ohto M, Isono K. Analysis of ras gene mutations

in biliary and pancreatic tumors by polymerase chain reaction and direct sequencing. Cancer 1990; 66:930–935.
34. Motojima K, Urano T, Nagata Y, Shiku H, Tsurifune T, Kanematsu T. Detection of point mutations in the Kirsten-*ras* oncogene provides evidence for the multicentricity of pancreatic carcinoma. Ann Surg 1993; 217:138–143.
35. Pellegata NS, Sessa F, Renault B, Bonato M, Leone BE, Solcia E, et al. K-*ras* and *p53* gene mutations in pancreatic cancer: ductal and nonductal tumors progress through different genetic lesions. Cancer Res 1994; 54: 1556–1560.
36. Caldas C, Hahn SA, da Costa LT, Redston MS, Schutte M, Seymour AB, et al. Frequent somatic mutations and homozygous deletions of the p16 (MTS1) gene in pancreatic adenocarcinoma. Nat Genet 1994; 8:27–32.
37. Bartsch D, Shevlin DW, Callery MP, Norton JA, Wells SA, Goodfellow PJ. Reduced survival in patients wiht ductal pancreatic adenocarcinoma associated with CDKN2 mutation. J Natl Cancer Inst 1996; 88:680–682.
38. Huang L, Goodrow T, Zhang S, Klein-Szanto A, Chang H, Ruggeri B. Deletion and mutation analyses of the P16/MTS-1 tumor suppressor gene in human ductal pancreatic cancer reveals a higher frequency of abnormalities in tumor-derived cell lines than in primary ductal adenocarcinomas. Cancer Res 1996; 56:1137–1141.
39. Naumann M, Savitskaia N, Eilert C, Schramm A, Kalthoff H, Schmiegel W. Frequent codeletion of p16/MTS1 and p15/MTS2 and genetic alterations in p16/MTS1 in pancreatic tumors. Gastroenterology 1996; 110:1215–1224.
40. Bartsch D, Shevlin DW, Tung WS, Kisker O, Wells SA, Goodfellow PJ. Frequent mutations of CDKN2 in primary pancreatic adenocarcinomas. Genes Chromosomes Cancer 1995; 14:189–195.
41. DiGiuseppe JA, Hruban RH, Goodman SN, Polak M, van den Berg FM, Allison DC, et al. Overexpression of p53 protein in adenocarcinoma of the pancreas. Am J Clin Pathol 1994; 101:684–688.
42. Redston MS, Caldas C, Seymour AB, Hruban RH, da Costa L, Yeo CJ, et al. p53 mutations in pancreatic carcinoma and evidence of common involvement of homocopolymer tracts in DNA microdeletions. Cancer Res 1994; 54:3025–3033.
43. Casey G, Yamanaka Y, Friess H, Kobrin MS, Lopez ME, Buchler M, et al. p53 mutations are common in pancreatic cancer and are absent in chronic pancreatitis. Cancer Lett 1993; 69:151–160.
44. Scarpa A, Capelli P, Mukai K, Zamboni G, Oda T, Iacono C, et al. Pancreatic adenocarcinomas frequently show p53 gene mutations. Am J Pathol 1993; 142: 1534–1543.
45. Weyrer K, Feichtinger H, Haun M, Weiss G, Ofner D, Weger AR, et al. p53, Ki-ras, and DNA ploidy in human pancreatic ductal adenocarcinomas. Lab Invest 1996; 74:279–289.
46. Sugano K, Nakashima Y, Yamaguchi K, Fukayama N, Maekawa M, Ohkura H, et al. Sensitive detection of loss of heterozygosity in the TP53 gene in pancreatic adenocarcinoma by fluorescence-based single strand conformation polymorphism analysis using blunt end DNA fragments. Genes Chromosomes and Cancer 1996; 15:157–164.
47. Lundin J, Nordling S, von Boguslawsky K, Roberts PJ, Haglund C. Prognostic

value of immunohistochemical expression of p53 in patients with pancreatic cancer. Oncology 1996; 53:104–111.
48. DiGiuseppe JA, Redston MS, Yeo CJ, Kern SE, Hruban RH. p53-independent expression of the cyclin dependent kinase inhibitor p21 in pancreatic carcinoma. Am J Pathol 1995; 147:884–888.
49. Hahn SA, Schutte M, Hoque ATMS, Moskaluk CA, da Costa LT, Rozenblum E, et al. DPC4, a candidate tumor suppressor gene at human chromosome 18q21.1. Science 1996; 271:350–353.
50. Hahn SA, Hoque ATMS, Moskaluk CA, da Costa LT, Schutte M, Rozenblum E, et al. Homozygous deletion map at 18q21.1 in pancreatic cancer. Cancer Res 1996; 56:490–494.
51. Schutte M, Hruban RH, Hedrick L, Cho KR, Nadasdy GM, Weinstein CL, et al. DPC4 gene in various tumor types. Cancer Res 1996; 56:2527–2530.
52. DiGiuseppe JA, Redston MS, Yeo CJ, Kern SE, Hruban RH. p53-independent expression of the cyclin-dependent kinase inhibitor p21 in pancreatic carcinoma. Am J Pathol 1995; 147:884–888.
53. Rozenblum E, Schutte M, Goggins M, Hahn SA, Lu J, Panzer S, et al. Tumor suppressive pathways in pancreatic cancer. Cancer Res 1997; 57:1731–1734.
54. Caldas C, Hahn SA, Hruban RH, Redston MS, Yeo CJ, Kern SE. Detection of K-ras mutations in the stool of patients with pancreatic adenocarcinoma and pancreatic ductal hyperplasia. Cancer Res 1994; 54:3568–3573.
55. Moskaluk CA, Hruban RH, Yeo CJ, Kern SE. +K-ras gene mutation in the intraductal precursors of human pancreatic adenocarcinoma. Submitted 1997; 7897998.
56. DiGiuseppe JA, Hruban RH, Offerhaus GJA, Clement MJ, van den Berg FM, Cameron JL, et al. Detection of K-ras mutations in mucinous pancreatic duct hyperplasia from a patient with a family history of pancreatic carcinoma. Am J Pathol 1994; 144:889–895.
57. Cerny WL, Mangold KA, Scarpelli DG. K-ras mutation is an early event in pancreatic duct carcinogenesis in the Syrian golden hamster. Cancer Res 1992; 52: 4507–4513.
58. Hameed M, Marrero AM, Conlon KC, Brennan MF, Klimstra DS. Expression of p53 nucleophosphoprotein in *in situ* pancreatic ductal adenocarcinoma: an immunohistochemical analysis of 100 cases (abstr) Lab Invest 1994; 70:132A.
59. Tada M, Ohashi M, Shiratori Y, Okudaira T, Komatsu Y, Kawabe T, et al. Analysis of K-ras gene mutation in hyperplastic duct cells of the pancreas without pancreatic disease. Gastroenterology 1996; 110:227–231.
60. Yanagisawa A, Ohtake K, Ohashi K, Hori M, Kitagawa T, Sugano H, et al. Frequent c-Ki-ras oncogene activation in mucous cell hyperplasias of pancreas suffering from chronic inflammation. Cancer Res 1993; 53:953–956.
61. Tada M, Omata M, Kawai S, Saisho H, Ohto M, Saiki RK, et al. Detection of ras gene mutations in pancreatic juice and peripheral blood of patients with pancreatic adenocarcinoma. Cancer Res 1993; 53:2472–2474.
62. Berthélemy P, Bouisson M, Escourrou J, Vaysse N, Rumeau JL, Pradayrol L. Identification of K-ras mutations in pancreatic juice in the early diagnosis of pancreatic cancer. Ann Intern Med 1995; 123:188–191.

63. Brentnall TA, Chen R, Kimmey MB, Hecker L, Kowdley KV, Bronner MB, et al. Ras mutations and microsatellite instability detected in ERCP-derived pancreatic juice from patients with pancreatic cancer (abstr). Gastroenterology 1995; 108: A452.
64. Iguchi H, Sugano K, Fukayama N, Ohkura H, Sadamoto K, Ohkoshi K, et al. Analysis of Ki-ras codon 12 mutations in the duodenal juice of patients with pancreatic cancer. Gastroenterology 1996; 110:221–226.
65. Kondo H, Sugano K, Fukayama N, Kyogoku A, Nose H, Shimada K, et al. Detection of point mutations in the K-ras oncogene at codon 12 in pure pancreatic juice for diagnosis of pancreatic carcinoma. Cancer 1994; 73:1589–1594.
66. Suzuki H, Yoshida S, Ichikawa Y, Yokota H, Mutoh H, Koyama A, et al. Ki-ras mutations in pancreatic secretions and aspirates from two patients without pancreatic cancer. J Natl Cancer Inst 1994; 86:1547–1549.
67. Uehara H, Nakaizumi A, Baba M, Iishi H, Tatsuta M, Kitamura T, et al. Diagnosis of pancreatic cancer by K-ras point mutation and cytology of pancreatic juice. Am J Gastroenterol 1996; 91:1616–1621.
68. Wilentz RE, Chung CH, Strum PDJ, Musler A, Sohn TA, Offerhaus GJA, Yeo CJ, Hruban RH, Slebos RJC. K-ras mutations in duodenal fluid from patients with pancreas cancer. Cancer 1998; 82:96–103.
69. Gutiérrez AA, Martinez F, Mas-Oliva J. Identification of K-ras mutations in pancreatic juice. Ann Intern Med 1996; 124:1014–1015.
70. Fearon ER, Vogelstein B. A genetic model for colorectal tumorigenesis. Cell 1990; 61:759–767.
71. van Es JM, Polak MM, van den Berg FM, Ramsockh TB, Craanen ME, Hruban RH, et al. Molecular markers for diagnostic cytology of neoplasms in the head region of the pancreas: mutation of K-ras and overexpression of the p53 protein product. J Clin Pathol 1995; 48:218–222.
72. Villaneuva A, Reyes G, Cuatrecasas M, Martinez A, Erill N, Lerma E, et al. Diagnostic utility of k-ras mutations in fine-needle aspirates of pancreatic masses. Gastroenterology 1996; 110:1587–1594.
73. Apple SK, Hecht JR, Novak JM, Nieberg RK, Rosenthal DL, Grody WW. Polymerase chain reaction-based K-ras mutation detection of pancreatic adenocarcinoma in routine cytology smears. Am J Clin Pathol 1996; 105:321–326.
74. Sidransky D, von Eschenbach A, Tsai YC, Jones P, Summerhayes I, Marshall F, et al. Identification of p53 gene mutations in bladder cancers and urine samples. Science 1991; 252:706–709.
75. Eguchi S, Kohara N, Komuta K, Kanematsu T. Mutation of the p53 gene in the stool of patients with resectable colorectal cancer. Cancer 1996; 77(suppl):1707–1710.
76. Brennan JA, Mao L, Hruban RH, Boyle JO, Eby YJ, Koch WM, et al. Molecular assessment of histopathological staging in squamous cell carcinoma of the head and neck. N Engl J Med 1995; 332:429–435.
77. Hiyama E, Kodama T, Shinbara K, Iwao T, Itoh M, Hiyama K, et al. Telomerase activity is detected in pancreatic cancer but not in benign tumors. Cancer Res 1997; 57:326–331.

78. Kim NW, Piatyszek MA, Prowse KR, Harley CB, West MD, Ho PLC, et al. Specific association of human telomerase activity with immortal cells and cancer. Science 1994; 266:2011–2015.
79. Velculescu VE, Zhang L, Vogelstein B, Kinzler KW. Serial analysis of gene expression. Science 1995; 270:484–487.
80. Zhang L, Zhou W, Velculescu VE, Kern SE, Hruban RH, Hamilton SR, et al. Gene expression profiles in normal and in cancer cells. Science 1997; 276:1268–1272.
81. Gold EB. Epidemiology of and risk factors for pancreatic cancer. Surg Clin North Am 1995; 75:819–843.
82. Lynch HT, Smyrk T, Kern SE, Hruban RH, Lightdale CJ, Lemon SJ, et al. Familial pancreatic cancer: a review. Semin Oncol 1996; 23:251–275.
83. Lumadue JA, Griffin CA, Osman M, Hruban RH. Familial pancreatic cancer and the genetics of pancreatic cancer. Surg Clin North Am 1995; 75:845–855.
84. Ghadirian P, Boyle P, Simard A, Baillargeon J, Maisonneuve P, Perret C. Reported family aggregation of pancreatic cancer within a population-based case-control study in the francophone community in Montreal, Canada. Int J Pancreatology 1991; 10:183–196.
85. Fernandez E, La Vecchia C, D'Avanzo B, Negri E, Franceschi S. Family history and the risk of liver, gallbladder, and pancreatic cancer. Cancer Epidemiology Biomarkers Prevention 1994; 3:209–212.
86. Powell SM, Petersen GM, Krush AJ, Booker S, Jen J, Giardiello FM, et al. Molecular diagnosis of familial adenomatous polyposis. N Engl J Med 1993; 329:1982–1987.
87. Knudson AG. Antioncogenes and human cancer. Proc Natl Acad Sci U S A 1993; 90:10914–10921.
88. Moskaluk CA, Hruban RH, Lietman A, Jackson C, Yeo CJ, Lynch HT, et al. Novel germline $p16^{INK4a}$ mutations in familial pancreatic carcinoma. Hum Mutat. In press.
89. Goggins M, Schutte M, Lu J, Moskaluk CA, Weinstein CL, Petersen GM, et al. Germline BRCA2 gene mutations in patients with apparently sporadic pancreatic carcinomas. Cancer Res 1996; 56:5360–5364.
90. Moskaluk CA, Hruban RH, Schutte M, Lietman A, Smyrk T, Fusaro L, et al. Genomic sequencing of DPC4 in the analysis of familial pancreatic carcinoma. Diagn Mol Pathol 1997; 6:85–90.
91. Whelan AJ, Bartsch D, Goodfellow PJ. Brief report: a familial syndrome of pancreatic cancer and melanoma with a mutation in the CDKN2 tumor-suppressor gene. N Engl J Med 1995; 333:975–977.
92. Goldstein AM, Fraser MC, Struewing JP, Hussussian CJ, Ranade K, Zametkin DP, et al. Increased risk of pancreatic cancer in melanoma-prone kindreds with p16 mutations. N Engl J Med 1995; 333:970–974.
93. Schutte M, da Costa LT, Hahn SA, Moskaluk C, Hoque ATMS, Rozenblum E, et al. Identification by representational difference analysis of a homozygous deletion in pancreatic carcinoma that lies within the BRCA2 region. Proc Natl Acad Sci U S A 1995; 92:5950–5954.
94. Thorlacius S, Olafsdottir G, Tryggvadottir L, Neuhausen S, Jonasson JG, Tavitigian

SV, et al. A single BRCA2 mutation in male and female breast cancer families from Iceland with varied cancer phenotypes. Nat Genet 1996; 13:117–119.
95. Phelan CM, Lancaster JM, Tonin P, Gumbs C, Cochran C, Carter R, et al. Mutation analysis of the BRCA2 gene in 49 site-specific breast cancer families. Nat Genet 1996; 13:120–122.
96. Berman DB, Costalas J, Schultz DC, Grana G, Daly M, Godwin AK. A common mutation in BRCA2 that predisposes to a variety of cancers is found in both Jewish Ashkenazi and non-Jewish individuals. Cancer Res 1996; 56:3409–3414.
97. Couch FJ, Farid LM, DeShano ML, Tavtigian SV, Calzone K, Campeau L, et al. BRCA2 germline mutations in male breast cancer cases and breast cancer families. Nat Genet 1996; 13:123–125.

18

Prostate Cancer

Fritz H. Schröder
Erasmus University and Academic Hospital, Rotterdam, The Netherlands

I. INTRODUCTION

Screening for prostate cancer is technically feasible. Recent advances in diagnostic strategies have led to a significant stage shift at the time of diagnosis in the direction of disease that is confined to the prostate clinically and histologically. Much of this diagnostic advancement is due to the use of prostate-specific antigen (PSA) as a diagnostic tool. The development was pioneered by Catalona and coworkers (1), who have shown for the first time that the application of PSA-related screening to populations of healthy men recruited by advertisement is useful in the early detection of prostate cancer. However, diagnostic progress has not been paralleled by an increase of knowledge with relation to the effectiveness of early detection measures in achieving the major goal of screening for any malignant disease: a decrease of cancer mortality rate. The discrepancy between the capability of detecting localized prostate cancer early and the uncertainty about the usefulness of early detection and early treatment with relation to improving the fate of men suffering from prostate cancer and especially with respect to decreasing prostate cancer mortality has led to considerable international controversy (2–4). In some geographical areas, screening for prostate cancer has virtually become a health care policy and large proportions of the male population older than age 50 years undergo screening procedures at regular intervals, yet heavy opposition against the introduction of population-based screening exists in other parts of the world.

The conflict cannot be resolved at this moment. The key lies in the results of randomized treatment studies comparing early treatment by radical prostatectomy or radiotherapy to delayed forms of treatment (watchful waiting) and in the results of randomized screening studies that compare early detection regimens to routine health care patterns. Data from final analysis of such studies are as yet unavailable. This chapter updates available information relevant to the issue of early detection and screening for prostate cancer.

II. PROSTATE CANCER: A SIGNIFICANT HEALTH PROBLEM

Recent epidemiological surveys are available from Europe and from the United States (5–8). In the United States, it was estimated that 334,500 men would be diagnosed to have prostate cancer in 1997, and more than 42,000 would die of the disease. The lifetime risk of having prostate cancer has increased from 4% to more than 10%, the ratio between incidence and mortality has increased to 7.7. In the United States, prostate cancer is the most frequently diagnosed nonskin malignancy and the second most frequent cause of cancer death. In the European community, prostate cancer is the second most common form of cancer in men after lung cancer. More than 35,000 men die of this disease every year and the ratio between incidence and mortality is roughly 2 to 1. A similar tendency of increasing incidence is observed as in the United States. In the region of Amsterdam, for example, between 1989 and 1994, the absolute number of cases diagnosed increased from 671 to 929, an increase of 38%. In most European countries, prostate cancer is either the second or third most common cause of death of males from malignant tumors. Incidence and mortality rate are strongly age dependent. Competing causes of death related to age have always been the reason for a discrepancy between incidence and mortality rate. The large number of men dying of this disease every year should be the main target of diagnostic and therapeutic efforts.

A. Epidemiological Aspects

A number of trends and recent developments in prostate cancer epidemiology are relevant for strategies of early detection and early treatment or are even the results of such efforts. The review of epidemiological trends is limited to such features.

1. *Time Dependence of Incidence and Mortality*

Prostate cancer incidence is increasing rapidly, and there is a slow increase in prostate cancer mortality rate in the United States and other countries of the world, mainly those with a Western type of lifestyle (Figure 1). Both incidence and mortality are strongly dependent on age, and part of the observed increase is based on the increasing longevity in many countries of the world. Boyle et

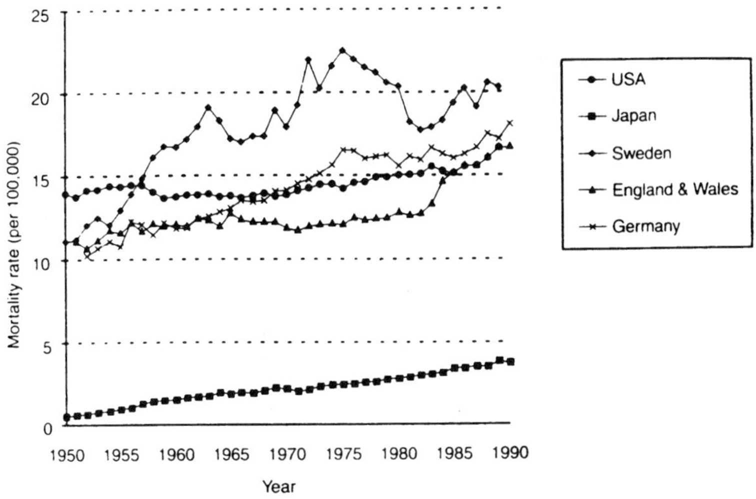

Figure 1 Mortality rates from prostate cancer, 1950–1990, in the United States and selected European countries (all ages, age-adjusted rates). Adapted from Ref. 9.

al. (9) predict an ''epidemic of prostate cancer'' up to the year of 2020 as a result of increasing male longevity. The data presented in Table 1 assume that prostate cancer incidence remains stable at 1980 levels. Even so, with this unrealistic assumption, for men 65 years or older, incidence in the European Community would increase by 67% to the year of 2020 on the basis of the expected increase in longevity. The presently observed increase of incidence, however, is not exclusively determined by the increasing age of the male population. Increasing awareness of prostatic cancer, an increasing application of early detec-

Table 1 Cancer of the Prostate in the European Community (EC) and in 6 Large EC Countries: Estimated Number of New Cases in Men >65 Years Old

Country	1990	2000	2100	2200
France	16,444	19,507	20,497	26,157
Germany	17,658	21,727	27,416	28,927
Italy	11,321	13,510	14,836	16,313
Netherlands	3,559	4,087	4,861	6,512
Spain	8,056	9,938	10,454	11,933
United Kingdom	14,125	14,421	15,352	18,099
Total EC	79,453	92,240	102,341	118,175

Source: Adapted from Ref. 9.

tion techniques to the general population, and a true increase in prostate cancer incidence are other important factors. Because current knowledge of risk factors related to the pathogenesis of prostate cancer is limited, no proper explanations for a true increase in incidence and mortality rate can be given. Considering the strong geographical variations in incidence and the observation that migrants from low-incidence to high-incidence areas experience an increase in prostate cancer incidence and mortality rate, it is likely that a "Western lifestyle" is an important risk factor (10).

2. *Geographical Variation*

Geographical variation in incidence is impressive and amounts to about 100-fold if the areas of highest and lowest incidence are compared. Table 2, adapted from the work of Parkin et al. (11), gives an indication of the four areas of the world with the highest and the lowest incidence rates of prostate cancer based on the year 1985. It is remarkable that the four highest incidence rates are all related to African American populations in the United States. It is unclear how much this phenomenon is determined by racial as opposed to lifestyle background. Also, it is remarkable that Japan does not belong anymore to those areas of the world with the lowest incidence rates. Lifestyle changes in Japan probably are responsible, together with adoption of intense screening in some areas. In this chapter, the reasons for these geographical variations are not discussed. Mass screening in Japan by means of transrectal ultrasonography (TRUS) was proposed as early as 1981, and initial data were reported by Watanabe et al. (12). Imai et al. (13) stated that PSA screening should be evaluated in the

Table 2 Four Highest and Lowest Incidence Rates of Prostate Cancer in 1985

		Cases	Rate
United States			
Atlanta	African American	832	102.0
	Whites	497	93.5
San Francisco Bay area	African American	944	95.6
Detroit	African American	2210	94.2
Seattle	Whites	7712	82.4
Utah	Whites	3019	77.9
China			
Shanghai		323	1.7
The Gambia		6	1.2
Tjanjin		84	1.2
Qidong		20	0.8

Source: Ref. 11.

Japanese male population. However, if screening for prostate cancer would ever be established as a population policy, it is clear that policies will have to be determined separately for those areas and populations with a high and a low incidence and mortality rate.

B. Impact of Early Detection on Incidence and Mortality

Through the application of early detection regimens, if they are effective, cancer will be diagnosed at an earlier stage in its biological development. The time added until a clinical diagnosis is made by routine clinical means is called ''lead time'' and is variable from tumor to tumor and from patient to patient. Earlier diagnosis in this sense may be reflected in the more frequent detection of organ-confined disease and of tumors that have not gone through all those steps of their biological development that are necessary to reach the capability to metastasize and kill the patient. In this sense, overdiagnosis with any type of screening policy is unavoidable. Over diagnosis can be defined in this context as the proportion of cancers detected by early diagnostic regimens that would not be diagnosed clinically during the life of a given man. For an early detection policy of prostate cancer to be effective, its application should result in a significant effect on the major endpoint, which is prostate cancer mortality. To reach this goal, it may be necessary to diagnose and treat a number of cancer cases that do not represent a threat to the patient, either because of their low biological potential, or because the life span of an individual is too short to experience the final state of the disease—the development of lethal metastatic prostate cancer—or for other reasons. A policy for screening for a malignant disease is likely to be introduced into the general health care systems only if a positive balance can be clearly shown between benefits and risks, and above all, if screening is shown to reduce prostate cancer mortality rate.

On this background, a change in incidence and in the proportion of incidence and mortality with changing health care policies or detection regimens is not unexpected. Documentation of this phenomenon is provided in a recent publication by Helgesen et al. (14). This group used the Swedish National Cancer Registry to identify 80,901 traceable cases of prostate cancer diagnosed between 1960 and 1988. Survival rates, relative survival rates, life expectancy for cohorts of men according to 5-year periods of diagnosis, as well as hazard ratios according to periods of diagnosis, age groups, incidence areas, and diagnostic methods were calculated. Relative survival is defined as the ratio of observed to expected survival. Expected survival is based on individuals in the general population who were matched to the study group in 5-year age groups and calender years of observation. The study shows that, besides a dramatic increase in incidence of prostate cancer over the study period, survival also improved dramatically. Evidence is presented that shows that improvement in relative survival and in the life expectancy of patients with prostate cancer is not due to the effect of improved treatment but is due to the effect of earlier diagnosis. It is shown that changing health care policies, the increase in the use of transurethral resection for the treatment of benign prostatic hyperplasia (BPH), and the subsequent increase of diagnosis of incidental prostate cancers as well as the more frequent use of rectal examination have led to this advancement of lead time in Sweden. However, considering the very small numbers of men that underwent treatment in Sweden during the observation period with curative intentions, an effect of treatment is not likely. Still, the impressive improvement in relative survival over time shown in Figure 2 clearly demonstrates that historical comparison of cohorts of patients treated or untreated will always be biased by the phenomenon described in this study. Even though this study shows that survival, relative survival, and the hazard of dying of prostate cancer depend on early diagnosis, even in an environment in which treatment with curative intent was uncommon at that time, the data do not contribute to the issue of whether early diagnosis and early treatment in a considerable number of these patients might have led to a significant reduction in prostate cancer mortality rate.

1. *PSA-Driven Screening in the United States*

In the areas of the world where screening policies are officially promoted, the proportion of the male population older than age 50 years who are actively screened has been steadily increasing. A recent study based on Medicare participants older than age 65 years documents the application of prostate-specific antigen (PSA)-driven screening from 1988 to 1991 in four different areas of the United States and shows a steep increase up to 1991 with some geographical variation (15). As Figure 3 shows, the steep increase in the application of the PSA test is paralleled by an increase of the use of prostatic needle biopsies. It is likely that the use of PSA-related screening has increased in a similar fashion after the year of 1991. In some areas of the United States where screening had been promoted by special campaigns, such as Prostate Cancer Awareness Weeks, the proportion of men using the PSA test has increased considerably. Exact figures on the proportion of men using PSA-based screening are not known.

Many reports connect the recent increase in prostate cancer incidence, especially in the United States but also in European countries, to the use of PSA-driven screening (15–18). Screening in the United States follows the American Cancer Society's policies, which are addressed later in this chapter and which recommend yearly screening by digital rectal examination (DRE) and PSA to men older than 50 years. The study by Jacobsen et al. (17) relates to the Olmsted County area in Minnesota, a Mayo Clinic study area. Besides clearly showing that the increase of prostate cancer incidence in that area is at least in part due

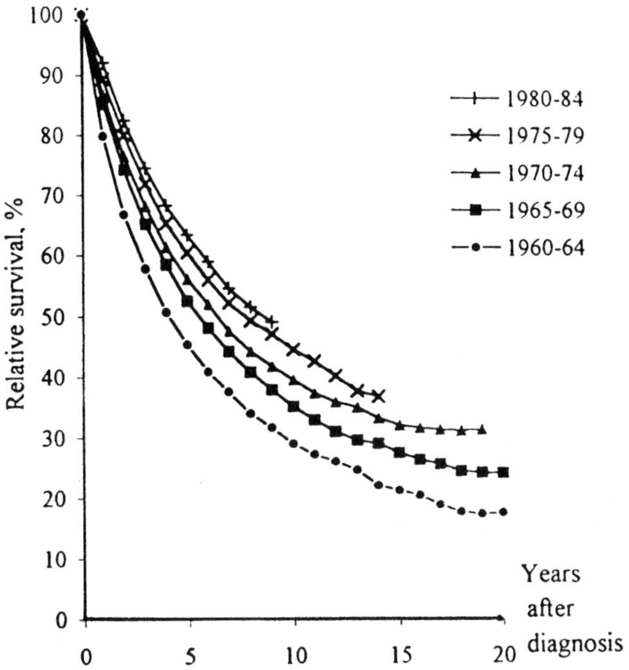

Period of diagnosis					
1960-64	7612	2350	924	374	151
1965-69	10249	3674	1521	667	233
1970-74	12841	4928	2164	742	
1975-79	14937	6206	2197		
1980-84	17749	5996			

Figure 2 Relative survival by period of diagnosis of 80,901 men identified with prostate cancer in the Swedish National Cancer Registry. Numbers of patients at risk per year after diagnosis are given below the diagram. Adapted from Ref. 14.

to the increased utilization of screening tests, the study also shows that this phenomenon is age dependent. Surprisingly, men older than age 65 years most frequently underwent testing. The utilization rate in the age groups of 70 and older was the highest and amounted to almost 50% of all men. Several studies show a decline of the incidence with annual screening. This decline in the Olmsted County series was first seen in men older than age 80 in whom a downward

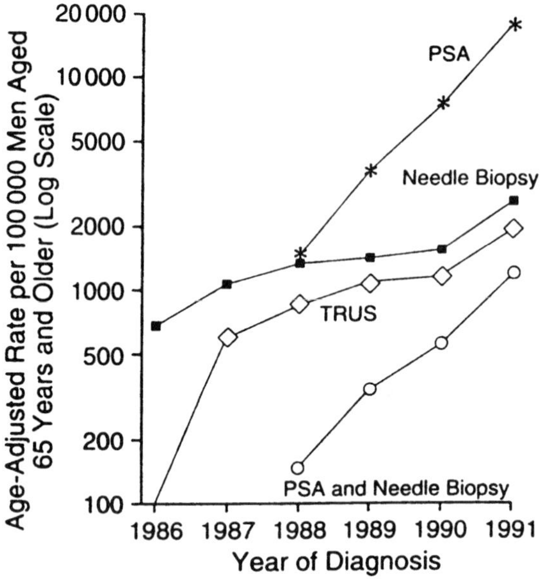

Figure 3 Trends in the use of procedures to detect prostate cancer in U.S. Medicare patients at least 65 years of age. Data are from four Surveillance, Epidemiology, and End Results (SEER) study areas of the National Cancer Institute. Adapted from Ref. 15.

trends occurred after 4 years. The same downward turn of incidence rate was, however, also seen in men between 60 and 80 years of age. Stephenson et al. (18) compared the Utah Cancer Registry, one of the centers of the Surveillance, Epidemiology, and End Result (SEER) study areas of the National Cancer Institute (NCI) of the Unites States to the incidence found in all SEER centers. In the area of the Utah Cancer Registry where screening has been prevalent and the increase of incidence has been witnessed since 1989, a downward turn is observed from 1992 on and is maintained (Figure 4). Screening leads to a favorable stage shift (19). In November 1996, the National Cancer Institute announced that the cancer death rate in the United States decreased by nearly 3% between 1991 and 1995 (20). A decrease of prostate cancer deaths contributed by a decrease of 6.3% during the same period of time, which was mainly attributed to men younger than age 75 years in whom mortality rate declined by 7.4%. Up to 1991, an increase of prostate cancer mortality rate had occurred. The decrease seen in 1995, however, represented a decrease in the mortality rate below the level of 1987. This phenomenon remains unexplained. The possibility that frequent early detection contributes to this favorable development, however, cannot be excluded. An epidemiological evaluation of these data is given in

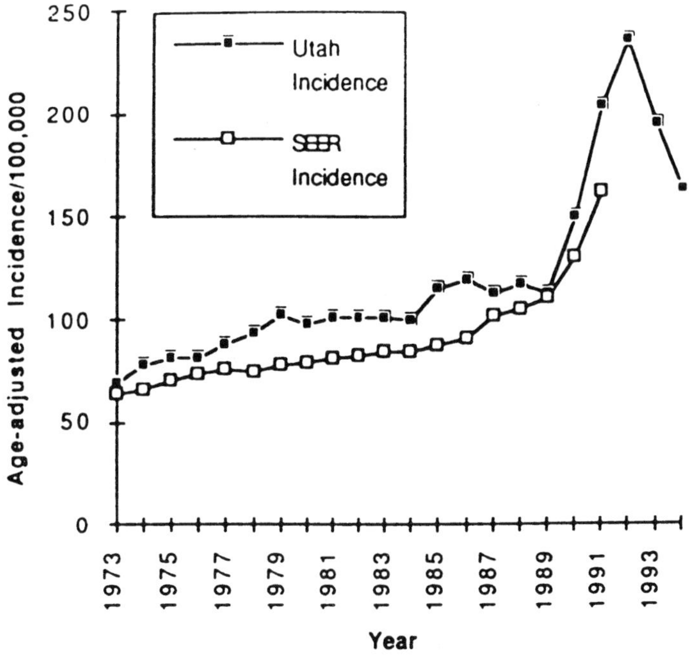

Figure 4 Age-adjusted incidence of prostate cancer per 10,000 for the Utah Cancer Registry (1973–1994) and the SEER program (1973–1991). Adapted from Ref. 18.

Chapter 9, which concludes that the changes seen are probably artificial and, because of their early occurrence, cannot be related to screening (21).

III. NATURAL HISTORY OF PROSTATE CANCER

Knowledge of the natural history of a given disease is a prerequisite for the application of screening as a public health policy (22). In recent years, a number of valuable contributions have been made to a better understanding of the natural history of prostate cancer. However, these publications refer to the time period before the early 1990s before PSA-related screening techniques were developed. This implies that, in evaluating the natural course, the older systems of classification were used and available data do not directly relate to the types of tumors detected with screening techniques. Progression rates, disease-specific mortality rates, and survival data have been produced from a number of retrospective studies, either based on selected groups of patients (23–25) or on defined populations (26). None of these studies takes account of the stage shift that occurs with the application of PSA-related screening, especially with the detection of

Table 3 T Categories of Prostate Cancer

TX	Primary tumor cannot be assessed	
T0	No evidence of primary tumor	
T1	Clinical unapparent tumor not palpable nor visible by imaging	
	T1a	Tumor incidental histological finding in 5% or less of tissue resected for a benign condition
	T1b	Tumor incidental histological finding in more than 5% of tissue resected for a benign condition
	T1c	Tumor identified by needle biopsy (e.g., because of elevated PSA)
T2	Tumor confined within prostate	
	T2a	Tumor involves half of a lobe or less
	T2b	Tumor involves more than half a lobe but not both lobes
	T2c	Tumor involves both lobes
T3	Tumor extends through the prostate capsule	
	T3a	Unilateral extracapsular extension
	T3b	Bilateral extracapsular extension
	T3c	Tumor invades seminal vesicles
T4	Tumor is fixed or invades adjacent structures other than seminal vesicles	
	T4a	Tumor invades bladder neck and/or external sphincter and/or rectum
	T4b	Tumor invades levator muscles and/or is fixed to pelvic wall

Source: Adapted from Ref. 27.

T1c prostate cancer (TNM 1992) (27). T categories are listed in Table 3. T1c lesions are nonvisible on transrectal ultrasound and nonpalpable and are diagnosed by an elevation of serum PSA only, usually above a cut-off value of 4.0 ng/ml, which is used as a biopsy indication. T1c lesions account for about 40% of screen-detected cancers. A literature-based meta-analysis of surveillance studies was carried out by Chodak and coworkers (28). The study is based on 828 case records from six surveillance studies carried out in different areas of the world. Disease-specific survival and risk ratios were calculated and compared to life expectancy. The main risk factor identified was grade of malignancy. Ten-year cause-specific survival for grade 1 and 2 lesions amounted to 87% and to 34% for patients with grade 3 tumors. Metastasis-free survival for grade 1, and grade 2, and grade 3 disease was 81%, 58%, and 26%, respectively. The study considered only patients with disease that was clinically confined to the prostate. The results of these surveillance studies and of the pooled analysis were criticized because of the high average age of 70 years, an overrepresentation of grade 1 lesions (59.4%), and an underrepresentation of grade 3 lesions (7.6%). Aus and coworkers (29) recently criticized the fact that, in the pooled analysis, age was not a risk factor with respect to the endpoints, and they clearly showed that cause-specific mortality is strictly age dependent and may be expected to be higher in populations of younger men.

A. T1c Disease

A number of autopsy studies show convincingly that prostate cancer is extremely prevalent in older men and that prevalence is dependent on age. In the study by Franks (30), the cumulative autopsy prevalence for men older than age 50 years was 34%. Most of the autopsy studies are older and have not applied exact determinations of prognostic factors that are relevant for the present clinical situation, such as tumor volume and grade of malignancy. Kabalin et al. (31) and Montie et al. (32) presented careful pathological studies of prostate cancer that was incidentally found in unsuspicious prostates removed with cystoprostatectomy specimens in the treatment of invasive bladder cancer. The volume determinations carried out in such studies and in another study by McNeal et al. (33) have been used by Barry et al. (34) to estimate the prevalence of prostate cancer at autopsy in age groups and by tumor volume. Autopsy prevalence rates estimated in this way amount to 12%, 15%, 22%, 39%, and 43% for men in the age groups 40–49, 50–59, 60–69, 70–79, and older than 80 years of age, respectively. The weighted average autopsy prevalence for men older than age 50 is 30% (547 of 1811). Sixty percent of all tumors included are estimated to have a volume less than 0.5 cm^3. The median volume of tumors recovered in prostatic specimens after cystoprostatectomy and its range amount to 0.04 mL (0.001–6.1) (35). These tumors have not led to prostate cancer related morbidity or mortality during the lifetime of their carriers. A worrisome aspect of the application of early detection regimens is the possibility of identifying substantial numbers of such lesions that cannot be differentiated from more extensive and more aggressive forms of prostate cancer in biopsy specimens.

It is unlikely that very small tumors with a volume less than 0.5 cm^3 are palpable on digital rectal examination or visible on transrectal ultrasonography. Even so, as data in Table 4 show, even the group of palpable and visible lesions in the report by Ohori et al. (35) contain some very small tumors ($<$0.5 mL). Their prevalence in radical prostatectomy series is estimated to be about 9% by Stamey and coworkers (36). However, significant differences exist in the volumes of those tumors that were palpable or visible as opposed to those who were diagnosed on the basis of an elevated PSA only (nonpalpable, nonvisible). Interestingly, the data also show that tumors with a PSA level less than 4.0 ng/mL are significantly smaller than those with higher PSA levels and also are less frequently poorly differentiated. Ohori et al. (35) show convincingly that T1c lesions, as far as their volume and grade is concerned, are intermediate between the T1a lesions found incidentally at the time of surgery for benign prostatic hyperplasia and those lesions that are palpable.

Table 4 Volume and Gleason Grade of 306 Prostate Cancers Detected by Different Diagnostic Tests and Treated by Radical Prostatectomy

	No.	Median tumor volume (mL) (range)	Gleason grade 4 or 5 present N (%)
DRE			
Palpable	246	2.37 (0.03–16.9)	151 (61)
Not palpable	60	1.67 (0.05–13.9)	33 (55)
TRUS			
Visible	260	2.33 (0.03–16.6)	161 (62)
Not visible	46	1.52 (0.05–16.9)	23 (50)
PSA			
≥4.0 ng/mL	238	2.54 (0.05–16.9)	157 (66)[a]
<4.0 ng/mL	68	1.0 (0.03–11.1)	27 (40)[a]

Source: Adapted from Ref. 35.
Abbreviations: DRE = digital rectal examination; TRUS = transrectal ultrasonography; PSA = prostate-specific antigen
All differences in volumes are significant within test category with *P*-values ranging 0.01–0.0001.
[a] $P = 0.0005$

Table 5 Distribution of Histologically Specimen-Confined Tumors, Grade, and Positive Margins in Men With T1c, >T1c PC, and Tumors Found with PSA Values <4.0 ng/mL (n = 96, Radical Prostatectomy)

	T1c (%) (n = 44)	>T1c (%) (n = 52)	PSA <4 ng/mL (%) (n = 37)
<pT3 (n = 80)[a]	27 (61)	24 (46)	29 (78)
≥pT3 (n = 53)[a]	17 (39)	28 (54)	8 (22)
G1 (n = 67)	17 (39)	20 (39)	38 (81)
G2 (n = 50)	20 (45)	23 (44)	7 (29)
G3 (n = 16)	7 (16)	9 (17)	0 (0)
Margin + (n = 38)	12 (27)	18 (35)	8 (22)

[a] pT = pathologically (surgically) determined T-status.
Source: Unpublished data, Rotterdam, European Randomized Study of Screening for Prostate Cancer, 1997.

Table 5 shows the pathological features of T1c cases in comparison with tumors that were larger than T1c and tumors identified in men with PSA levels > 4.0 ng/mL within the European Randomized Study of Screening for Prostate Cancer (ERSPC), section Rotterdam, and who have undergone a radical prostatectomy. A selection bias may be present because 9% of this population-based series of patients have been treated by watchful waiting. The remaining tumors may be more advanced and not entirely representative. It is noteworthy that tumors detected in the PSA range <4.0 ng/mL have more favorable characteristics. Still, it is evident that T1c lesions and palpable lesions identified in a population-based screening situation do not differ much. Although such data suggest that most of T1c lesions are clinically relevant with respect to current parameters and are seldom compatible with tumors found at autopsy, the natural history of this entity as well as the long-term effects of radical prostatectomy and radiotherapy are unknown. Epstein et al. (37) in a clinical series of 157 T1c cases considered, on the basis of tumor volume and grade, 16% as clinically insignificant and 10% as minimal lesions. Volumes were either below 0.2 or 0.5 mL, the tumors were locally confined and did not contain Gleason patterns 4 or 5 (see Section IV A).

B. Stage Distribution

With the application of modern PSA-based screening technology, a significant stage shift of prostate cancer at the time of diagnosis occurs. The proportion of clinically locally confined cases increases from 30%–40% to 70%–80% (38). Still, about 30% of all cases identified, even in a population-based setting (38), are too advanced for potential cure by radical prostatectomy. Inevitably, this stage shift in the direction of more locally confined disease will lead to a prolongation of survival by the creation of lead time and length time. The duration of lead time, defined as the time that elapses until clinical diagnosis would have been made without screening, is unknown and depends on the tumor doubling time of screen-detected lesions. In the screening arm of ERSPC Rotterdam, metastatic lesions are identified in only 2% of cases at diagnosis. In a previous survey of the stage distribution of prostate cancer in the same area, based on the years 1989 to 1993, metastatic prostatic cancer was found in 22%.

C. Biological Progression of Prostate Cancer

Well-differentiated focal disease, prostatic intraepithelial neoplasia (PIN), and atypical adenomatous hyperplasia (AAH), are considered precursors of cancers derived from the peripheral zone and of the transition zone (AAH) of the prostate. Well-differentiated focal prostate cancer has a similar prevalence in areas of a high and low incidence of clinical prostate cancer (39). This suggests that the pathogenesis of such lesions is part of the initiation process whereas promotion to larger lesions is dependent on different largely unknown factors. Most of the autopsy lesions must be considered precursor lesions in this sense. Prostatic intraepithelial neoplasia (PIN) grade 3 is seen in about 30% of cancer cases. If found in a biopsy specimen without evidence of prostate cancer, the presence of PIN prompts rebiopsy because of this relationship. The role of PIN as a precursor lesion is unclear (40). Atypical adenomatous hyperplasia is considered a precursor lesion of transition zone carcinoma (41). These tumors are usually well differentiated, can grow large, and lead to an increase of PSA, but they have in general a more favorable prognosis. They are rarely diagnosed by biopsy unless the transition zone is biopsied or during transurethral resection (TUR).

The most important factor determining prostate cancer morbidity and mortality in surveillance studies is the presence or absence of poorly differentiated prostate cancer. This category is usually defined as grade 3 cancer or, in the Gleason system, by the presence of the Gleason patterns 4 or 5. Grade 3 prostate cancer is found in locally confined disease detected in the screening situation in 13%–14% of cases [37,40]. This is compatible with the proportion of patients who have been found to have grade 3 disease in a pooled analysis of radical prostatectomy cases with locally confined disease (42). The overview of Chodak et al. (43) concerning surveillance studies shows that the presence of grade 3 disease has an important impact on cancer-related and metastasis-free survival. Well-differentiated and moderately differentiated prostate cancer may only have minor influence on disease-specific metastases-free survival. On this background, the results of management of poorly differentiated prostate cancer will be crucial for possible future screening and treatment strategies.

D. Poorly Differentiated Prostate Cancer

Poorly differentiated prostate cancer is more frequently encountered in advanced disease. The morphogenesis of Gleason pattern 4 and 5 (grade 3 disease) is incompletely understood but relevant for the application of screening tests. McNeal et al. (44) has suggested on the basis of careful anatomical studies that the progression to poorly differentiated disease results from genetic instability and a number of mutational steps that take place in previously well-differentiated disease. It is suggested that poorly differentiated cancer occurs more frequently in tumors that exceed 1 cm^3 in volume and that fully malignant behavior is not present at inception but is acquired with the passage of time. This hypothetical view is contradicted by Epstein (45) who presents evidence that grade 3 disease was present in 43 of 198 foci (21.7%) with a volume of <1 mL. These foci were identified in 153 radical prostatectomy specimens removed for T1c disease and within a total of 700 tumor foci found in these cases. Considering tumor volume doubling times in the order of magnitude of several years for well and moderately differentiated prostate cancer, the theory of McNeal et al. would offer a larger window of opportunity for early detection and treatment of potentially aggressive disease. However, if the policy is one of watchful waiting, repeated biopsies would be indicated to identify the development of poorly differentiated disease. To the contrary, the evidence presented by Epstein et al. indicates that poorly differentiated cancer may be present in very small volume foci, and its presence early on during the development of prostate cancer would give a very small window of opportunity. The more frequent presence of poorly differentiated disease in larger lesions in this model would be explained by early and rapid overgrowth of poorly differentiated clonal populations, which, in this mutifocal disease, might be present next to well-differentiated formations from the beginning on. The truth might be a mixture of both concepts.

Based on poor results obtained by means of radiotherapy (46) and radical prostatectomy (47), grade 3 prostate cancer has generally been considered almost incurable. Recent evidence, however, suggests the contrary. The experience obtained in three major U.S. institutions with large series of patients treated by means of radical prostatectomy shows that median-term disease-free survival can be obtained in about 50% of poorly differentiated cases as long as those tumors are confined to the prostate [35,48,49]. In addition, a comparative pooled analysis of surveillance studies with radical prostatectomy trials presents evidence that the effect of radical prostatectomy may be most pronounced in grade 3 disease. Clearly, if poorly differentiated prostate cancer could be detected in a screening situation in a substantial number of locally confined cases, selective screening for such lesions would be indicated. Unfortunately, due to insufficient sampling of current biopsy techniques, small grade 3 lesions are often missed at the time of biopsy (50,51). Undergrading occurs in 50%–60% of T2 and 20%–30% of T3 cases.

IV. DIAGNOSTIC TESTS AND TEST PERFORMANCE

Besides the paucity of knowledge about the effectiveness of screening in terms of lowering prostate cancer mortality, the application of detection techniques to population-based screening policies is complicated by several specific features of this disease:

1. Within the prostate of men of the age groups at risk, there is a reservoir of focal prostatic cancerous lesions present which are all likely to progress with time but may never threaten the health and life of their carrier.
2. A differentiation between clinically relevant and irrelevant tumors at the time of diagnosis is not possible in most of the cases. Some features of aggressive tumors, for example, poor differentiation or large tumor volume, are relevant if identified. However, sampling at the time of prostatic biopsy and

clinical staging are not sufficiently representative for the proper identification of such parameters.

3. The sensitivity and specificity of the available diagnostic tests are adjustable within a wide range. Invariably, an increase in sensitivity (more cancers diagnosed) leads to a decrease in specificity (more biopsies done), and vice versa. An increase in sensitivity may also lead to the more frequent diagnosis of small lesions with unknown clinical relevance.

4. On this background, if these diagnostic tests are used in clinical practice, the adjustment of the level of sensitivity and specificity becomes a philosophical and ethical question: how many cancers is one ready to miss and how many unnecessary biopsies are acceptable?

An answer to the pending questions can only be obtained by randomized prospective studies that allow the evaluation of the test procedures in comparison to appropriate control groups and with respect to relevant endpoints. Randomized screening studies that compare an early diagnostic and early treatment arm to a control arm that reflects regional health care patterns represent one appropriate approach.

A. How Many and Which Cancers Are to Be Detected?

Considering the frequent clinical use of early diagnostic procedures in hundreds of thousands of men at risk, an immediate answer to this question for clinical use is desirable. Within randomized screening studies, the possibility of reducing prostate cancer mortality rate could be missed by adjusting the sensitivity of the diagnostic procedure at too low a level. Even so, in the given situation, a heavy responsibility rests on those who design and conduct such studies to avoid excessive degrees of overdiagnosis with the result of overtreatment. A comparison of the prevalence and characteristics of tumors found at autopsy with those detected at screening and in clinical routine is warranted and contributes to the better understanding of the existing uncertainties. The TNM classification (see Table 3) serves as a matrix in this respect.

1. *Prevalence and Characteristics of Tumors Found at Autopsy*

Table 6 (30,52–54) gives a summary of four representative autopsy series indicating the prevalence of prostate cancer in age groups. The overall prevalence varies between 16.7% in the series of Holund (53) and 37.6% in the series of Franks (30). If, in the series of Franks, the age groups that are clinically at risk of development of prostate cancer and of undergoing treatment (age 50–79), are considered, then the prevalence amounts to 55 in 161 males, or 34%. In all three series, it was a prerequirement for entry into the study that prostate cancer was not diagnosed during the man's life time. It is for this reason that prostate cancer found unexpectedly at autopsy is called "latent."

Table 6 Latent Prostatic Carcinoma (PC) by Age Groups in Several Autopsy (aut) Series

Age	Franks (30)			Gaynor (52)			Holund (53)			Sakr (54)		
(years)	N (aut)	N (PC)	%	N (aut)	N (PC)	%	N (aut)	N (PC)	%	N (aut)	N (PC)	%
20–29	4	0	0	15	0	0	—	—	—	35	0	0
30–39	8	0	0	25	1	4	3	0	0	55	15	27
40–49	18	0	0	122	6	5	8	1	12.5	50	17	34
50–59	38	11	29	241	25	10	23	2	8.7	—	—	—
60–69	53	16	30	312	54	18	56	7	12.5	—	—	—
70–79	70	28	40	237	67	28	93	24	25.8	—	—	—
80–89	17	12	67	93	36	39	35	13	37.1	—	—	—
90+	2	2	100	5	2	40	5	3	60.0	—	—	—
Total	210	69	37.6	1050	191	18.4	173	2.9	16.7	152	32	21

In these autopsy series, except for the one of Sakr et al. (54) a careful characterization of the tumors identified according to modern standards has not been carried out. Their properties, however, may be reflected in the data obtained more recently with the careful study of prostate carcinomas found incidentally in prostates removed as part of a cystoprostatectomy specimen obtained in treating invasive bladder cancer (55). A review of the features of these lesions is given in Table 7. Tumor volumes ranged from 0.001 to 6.1 mL with median values of 0.11 and 0.04 mL. Between 84% and 100% of these tumors are confined to the prostate, and poorly differentiated tumors are seen in up to 27%. This is in confirmation of findings by Epstein et al. (44). The Gleason score and tumor volume (in this sequence) were the most important prognostic parameters in locally confined prostate cancer in two different studies. In terms of volumes, assuming that tumor foci are spherical, tumor diameters of 0.3, 0.5, 1.0, and 2.0 cm relate to volumes of 0.01, 0.1, 0.5, and 4 $cm.^3$ Tumors usually become palpable or visible on transrectal ultrasound with a diameter of 1 cm. Clearly, most of the tumors expected to be found in an autopsy are very small and usually well differentiated.

Epstein et al. (37) provides information that correlates these parameters with the currently used clinical T categories: T1a, T1c, and T2. T1a tumors are ''incidental'' cancers found unexpectedly at the time of surgery for benign disease. The natural history of these lesions is well understood, with a median time to progression of about 13 years (56). Active treatment in these tumors is only pursued if they are diagnosed in men younger than 60 years. Epstein, in following up on earlier suggestions by Stamey et al. (57), uses Gleason score and

Table 7 Morphological Characteristics of Prostatic Cancer Found Incidentally in Cystoprostatectomy Specimens

	Kabalin (31)	Ohori (35)[a]	Montie (32)
Mean age (years)	66		62
No. of cancers	25	90	39
% cancers	38%		46%
Tumor volume (mL)			
Median	0.11	0.04	—
Range	0.01–1.10	0.001–6.1	—
Gleason score			
4 or 5	0%	11%	27%
Confined	100%	97%	84%
SV+, N+, or M+[b]	0%	0%	1%

[a]Report of data by Kabalin et al. (31) (n = 55) and Ohori et al. (35) (n = 35), together N = 90.
[b]Seminal vesicle involvement (SV+), nodal metastases (N+), metastases present (M+).

tumor volume to subclassify locally confined prostate cancer, which may be clinically diagnosed. His data are summarized in Table 8. The terms ''insignificant'' and ''minimal'' relate to hypothetical clinical relevance of these tumors. These definitions are arbitrary and have not been established as being clinically relevant.

2. *Comparison of Clinical and Screen-Detected Tumors with Autopsy Lesions*

Comparison with the features of T1c tumors diagnosed in early detection or screening studies (Table 9) shows that the application of screening technology leads to the diagnosis of lesions with a volume of less than 0.5 mL in 11%–26% of cases. Some of these series are preselected with respect to the inclusion of T1c lesions. The ERSPC data are obtained from the screening arm of a population-based screening study and are shown in Table 5. In comparing Table 8 and Table 9, it is evident that screen-detected cancers are in general intermediate between the incidentally found tumors and the palpable/visible T2 lesions. Tumors with a volume less than 0.5 mL that are well differentiated and do not contain Gleason patterns 4 and 5 are often indicated as being ''clinically insignificant.'' Such tumors are seen in clinical series of radical prostatectomy at frequencies of 9%–10% [35,57]. This proportion of potentially ''clinically unimportant'' tumors, however, is much too small to explain the increasing gap between prostate cancer incidence and mortality rate in areas where screening is prevalent. Very slow progression rates have been described for larger lesions and even for cases of untreated prostate cancer metastatic to regional lymph nodes (61). The only sensible conclusion is that it is unknown which tumors are potentially aggressive and which ones are not. If prognostic factors become available that allow the determination of aggressiveness before treatment, selective screening will become possible by focusing on the identification and treatment of lesions that are aggressive and life-threatening and that might still be amenable to potentially curative management. Neither clinically nor with respect to randomized screening and treatment studies can the dilemma be resolved at this time. In randomized screening studies, which utilize prostate cancer mortality rate as their major endpoint, a decrease of the sensitivity with omission of the diagnosis of small lesions could mean that those lesions that have to contribute in a major way to an improvement of mortality rate would be eliminated. The observation that a substantial number of small focal tumors are poorly differentiated gives support to this thought. Also, Table 9 makes clear that despite a considerable stage shift that can be achieved with the application of PSA-driven screening, 30%–40% of cancers detected in this way are too advanced for cure. This proportion would increase with any loss of sensitivity of the screening procedure and subsequently the window of opportunity with respect to control arms would be narrowed.

Table 8 Morphological Classification of T1a, T1c, and T2 Prostatic Cancer According to Grade Tumor Volume and Local Extension

T-category	#	No Gleason 4 or 5, volume <0.2 mL confined ("insignificant")	No Gleason 4 or 5, volume 0.2–<0.5 mL confined ("minimal")	Confined, volume ≥0.5 mL *or*: FCP, Gleason <7 *or* ECP, Gleason <7, margins negative ("moderate")	FCP or ECP Gleason ≥7 *or* EP, margins+ *or* SV+ *or* LN+ ("advanced")
T1a	64	21 (33%)	16 (25%)	26 (41%)	1 (2%)
T1c	157	26 (16%)	15 (10%)	58 (37%)	58 (37%)
T2	369	6 (2%)	11 (3%)	137 (37%)	215 (58%)

Abbreviations: FCP = focal capsular penetration; ECP = extensive capsular penetration.
Source: Adapted from Ref. 37.

Table 9 Features of T1c Prostate Cancer

	Osterling (58)	Ohori (35)[a]	Scaletscky (59)	Epstein (37)	Humphrey (60)	ERSPC, Rotterdam
No. of prostate cancers	208	29	142	157	53	44
Age (mean)	66	—		61	?	64 (56–72)
PSA (ng/mL) (median, range)	10	—	8.7	8.1 (0.5–114.0)	?	7.6 (4.2–31.8)
Mean tumor volume (cm^3)	6	1.5	—	1.9	2.3	0.9 cm^3
Tumor volume $\leq$0.5 mL (%)	11	17	35[b]	26	19	4 (23%) (n = 17)
Gleason grade 2–4 (%)	44	52	15	6	8	17 (38.6%) (G1)
Confined (%)	53	79	68	51	57	27 (61%)
Positive surgical margins (%)	34	7	26	17	43[c]	12 (27%)
Nodes+ (%)	3	4	?	4	?	1 (2.3%)

[a] Tumors are nonpalpable and nonvisible on transrectal ultrasound
[b] <1.0 mL
[c] Extracapsular extension and positive margins

B. Screening Tests

Three different procedures are commonly used in the early diagnosis of prostate cancer, digital rectal examination (DRE), transrectal ultrasonography (TRUS) and serum prostate-specific antigen (PSA). These tests and their use in combination and with certain modifications have been the subject of intense clinical research efforts in recent years. One of the difficulties is that the mechanism of recruitment of study populations has a strong impact on the outcome of such evaluations. For example, PSA values show a substantial overlap between cancer cases and BPH cases [37,62]. This overlap seems to be less pronounced in population-based screening studies.

Reviews of the value of the three tests are available (63). Table 10 gives a summary of sensitivity and positive predictive values (PPVs) for the individual tests. The table is the result of an extensive literature review; the PPVs are averages of 15 different reports related to DRE or TRUS as a diagnostic test. Positive predictive values in this context are the number of cancers found divided by the number of indications for biopsies or the number of biopsies that were in fact obtained. It is evident that the PPVs of each individual test except for PSA with a cut-off of 10 ng/mL are very low. The PPV is dependent on the prevalence of the disease in a given study population. Surprisingly, during the initial phase of the European Randomized Study of Screening for Prostate Cancer (ERSPC), Bangma et al. (64) found in a general population–derived sample with an overall detection rate of 3.9%, 67 cancers with 308 biopsies.

Table 10 Sensitivity and PPVs of PC Screening Tools Indicated as Averages from a Literature Review of Early Detection Studies

Method	Sensitivity	Positive predictive value
Digital rectal examination	45%–82%	28%
Transrectal ultrasonography	60%–91%	31%
PSA >4 ng/mL	77%	24%[a]
PSA >10 ng/mL	43%	60%

[a]No median value; PSA between 4 and 10 ng/mL
Source: Adapted from Ref. 63.

This amounts to a PPV of 22%, 4.6 biopsies were necessary to find one cancer. All three screening tests were applied. Clearly, the combination of the tests does not lead to an increase of the PPV, at least in a population-based study.

C. Rectal Examination

Digital rectal examination is the classical technique for the detection of prostate cancer. Its value has been carefully studied. Most reports, however, date from the pre-PSA era. That the combination of DRE with PSA-based screening yields different results with respect to the detection rate and a higher proportion of locally confined tumors was clearly shown by Catalona et al. (65). In this series, clinical routine, which was essentially based on rectal examination, revealed 43% pathologically confined tumors whereas screen-detected cancers were confined in 63% of cases.

In the previously mentioned review (63), 11 studies that report on the use of rectal examination as an individual screening test have been reviewed and summarized. The detection rate in these studies amounts to 0.85% with a PPV of 28%. Remarkably, an average of 73% of all cases was clinically confined to the prostate. More recently, in a population-based study that used DRE only, Pedersen et al. (66) found 13 cancers in 1163 screenees, a detection rate of 1.1%. Rectal examination information was shared between urologists and general practitioners. The PPV was 29%, 27% for the general practitioners and 35% for the urologists. To find one cancer, 3.5 biopsies were necessary. A clear disadvantage of rectal examination as a screening test is its dependence on the examiner's experience and quality of performance.

Catalona et al. (65) compared the value of rectal examination and of serum prostate-specific antigen in a multicenter screening study of 6630 men. All men with a PSA value above 4.0 ng/mL underwent biopsy. The detection rate was 4.0%. The PSA alone detected 82% and DRE alone 55% of all 264 cancers.

The cancer detection rate for DRE was 3.2%, for PSA was 4.6% and for both methods combined was 5.8%. Pathologically organ-confined disease was detected overall in 71% of cases. The PSA-detected cancers and rectal examination–detected cancers were locally confined in 75% and 56% of cases, respectively ($P=0.003$).

From this information it is evident that DRE remains an important test and substantially increases the detection rate that can be achieved with PSA alone.

D. Transrectal Ultrasonography

In evaluating the relevant literature (60), it is found that detection rates and positive predictive values are extremely variable. Again, this may be due to differences in the study populations. However, it is well known that TRUS is a strongly investigator-dependent technique. Prostate cancer may appear as a hypoechogenic lesion on transrectal ultrasonography. Isoechoic and hyperechoic areas, however, may also be found on biopsy to be positive for the presence of cancer. On average, in seven studies reviewed, the PPV of TRUS as a single screening test averaged 31%.

In a number of study protocols, TRUS was used in addition to DRE in all study subjects. These protocols offer an opportunity to evaluate the value of TRUS as an additional test to DRE or to the combination of DRE and PSA (67–69). In a study by Bangma et al. (69), TRUS detected only 2 of 67 cancers that would not have been detected by other tests. The relatively high cost and low additional value of TRUS has led to the elimination of this test in screening strategies recommended in several countries and used in some studies.

Transrectal ultrasonography allows a determination of the total prostatic volume and of the volume of the transition zone. Whether PSA density (PSAD), PSA divided by total prostatic volume, adds meaningfully to the value of the screening procedure is still the subject of discussion (see Sections IV E and V).

Undisputed is the value of transrectal ultrasonography for the visualization of the prostate and in guiding the biopsy needle in the case of systematic sextant biopsies or when a hypoechogenic lesion is to be evaluated.

E. PSA as a Screening Test

A large amount of literature on the use of PSA as a test for the early detection of prostate cancer has recently been developed. This chapter discusses some of the most important aspects and recent studies.

Prostate-specific antigen is a protein with a molecular weight of approximately 36,000 and was first described by Papsidero et al. (70,71). The PSA is part of the human kallikrein gene family. This relationship, together with the biomolecular and physiological properties and molecular forms of PSA, have recently been reviewed by McCormack et al. (72). As shown by Riegman et al.

(73), the promoter region of the PSA gene contains two androgen-responsive elements, making the transcription of PSA an androgen-dependent process. This is contrary to acid phosphatase, another important marker of prostate cancer. On this background, it is surprising that, in healthy young men, serum PSA levels can apparently not be stimulated by exogenous testosterone (74). It is unknown whether other members of the human kallikrein family may also be useful markers of prostate cancer. This possibility is being explored at several centers (75). Another potential use of PSA as a screening test or as a parameter for tumor aggressiveness may relate to the detection of PSA-RNA containing cells by reverse Polymerase chain reaction (PCR) technology. Recent data on this relatively new technology are given in a review by Olsson et al. (76).

Sensitivity, specificity, and positive predictive value of PSA as a screening test have recently been reviewed by Schröder et al. (77). The calculation of sensitivity can only be based on the ''gold standard'' of a positive biopsy because the true prevalence of prostate cancer in the test sample remains unknown. Despite rather low sensitivities and PPVs in the range of 30%–40%, PSA was shown to be more effective in detecting prostate cancer and more effective in detecting locally confined disease by Catalona et al. (78). In comparing a group of 244 men with prostate cancer initially detected through PSA-based screening with a group of 266 patients referred because of locally confined prostate cancer detected by DRE, the proportion of cases clinically or pathologically advanced disease was 37% and 57% respectively.

A number of conditions have been shown to influence serum prostate-specific antigen levels. Aus et al. (79) have studied the effect of prostatectomy for benign disease on PSA levels and concluded that, in the presence of benign disease, PSA should decrease to the normal range of <4 ng/mL. It has been shown that prostatic biopsy and ejaculation may lead to a transient increase of serum PSA.

Labrie et al. (67) were first to suggest the use of PSA as a prescreening test for prostate cancer. This group, on the basis of a receiver operating curve (ROC) analysis, found that, with a PSA value of 3.0 ng/mL, sensitivity and specificity were identical and both were in the range of 80%. The Hybritech Tandem-R assay was used. In this concept, men with low PSA values are excluded from further analysis. The rationale and application are further discussed later in this chapter.

1. *Interassay and Biological Variations*

A large number of different PSA assays are in clinical use. In the absence of an international standard for PSA measurement, most of these assays are calibrated against other assays, mainly the Hybritech Tandem assays, which are thought to determine PSA in an equimola fashion (72). Equimolar determination depends on an assay design that is capable of determining free and α-antichymotrypsin (ACT) bound PSA. Both fractions together constitute "total PSA." Christensson et al. (80) first described that noncomplexed (free) PSA constitutes a minor fraction of total PSA, which is statistically significantly smaller in patients with untreated prostate cancer than in those with BPH.

The availability of multiple different assays for serum PSA results in the impossibility of comparing results coming from different institutions unless the type of assay used is known. Within the same assay systems, especially with those assays that are well standardized, the interassay variability amounts to confidence intervals between 5% and 10% (80–82). Komatsu et al. (83) studied the intraindividual variation, the biological variation of PSA values over a 3-week study period. The mean coefficient of variation of the first and the second blood drawn was 10%, 12% and 20% for the total PSA ranges 0–4, 4–10, and >10 ng/mL, respectively, by the use of the Tandem-R assay (Hybritech). This means that, in longitudinal comparisons of PSA values in individuals or cohorts of men over time, any changes of less than 20% should be considered irrelevant. Also, in the screening situation, one has to take into account the possibility that a PSA value of 4.0 could just as well be compatible with values of 3.6 or 4.4 ng/mL.

2. *PSA in Retrospective Cohort Studies*

Serum PSA increases slowly with age and increasing prostatic volume as shown by Oesterling et al. (84) and Bosch et al. (85) in community-based studies of men in whom the presence of prostate cancer was excluded by biopsy. That men who harbor prostate cancer have a faster and more pronounced increase of PSA than control populations was first shown in a small sample of men in whom prostate cancer was developing within the follow-up scheme of the Baltimore longitudinal study of aging (BLSA) by Carter et al. (86). The increase of PSA over time was termed "PSA velocity." An annual increase of more than 0.75 ng/mL was found to be critical for the later development of prostate cancer in this study. This concept was subsequently evaluated retrospectively by several research groups (87–89). The three studies have in common the use of blood samples that were drawn a long time ago in large groups of men, allowing a comparison of PSA or the increase of PSA over time in those in whom prostate cancer later developed, diagnosed by clinical means, and in those in whom it did not. For calculating sensitivities of PSA, prediction of later development of prostate cancer, in this case the "gold standard" of a positive biopsy in a screening study, is replaced by the "gold standard" of clinical development of cancer, which is likely to result in Prostate cancer–related death. As would be expected, the sensitivity and specificity in these studies is considerably higher than sensitivities and specificities found empirically in case-finding studies and are in the range of 70%–100% for clinical cancer being diagnosed within 4–6 years. In one study (88), a significant prediction of tumor aggressiveness was possible on the basis of a single PSA determination 4 years before clinical diagnosis. All three studies conclude that PSA is a suitable test for early detection. Gann et al. (88) confirm that the operating characteristics of PSA were acceptable, but did point out that efficacy in decreasing mortality rate is as yet unproven. Stenmann et al. (87) estimate the PSA doubling time in cancer cases to be 3 years. The median survival after diagnosis was 3.6 years. Cancer-specific survival correlated significantly with PSA levels at the time of diagnosis. In their study, with a cut-off for PSA of 4.0 ng/mL, the clinical development of cancer within 5 years could be predicted with a sensitivity of 95% and a specificity of 92%. In this series, the median time to presentation of prostate cancer after sampling was 5.7 (range 0.3–10.3) years. The PSA level exceeded 4.0 ng/mL on average 9 years before presentation. These estimates of preclinical periods of development of prostate cancer (lead time) must be confirmed prospectively. Their relevance is limited by the small number of 44 cancer cases. Also, the short time from diagnosis to death of 3.6 years on average shows that cases were usually diagnosed in a very advanced stage. The estimated PSA doubling time of 3 years is in line with clinical observations by Schmid et al. (90).

Several authors have attempted to use the concept of PSA velocity in early detection (91,92). The results of the use of an absolute increment depend on the initial PSA value and, as a consequence, on age and prostatic volume. Even so, Smith and Catalona (91) revealed a strong prediction of the presence of prostate cancer by the use of PSA velocity with an increment of 0.75 ng/mL year. Brawer et al. (92) found a better prediction with the use of an increment of 20% in the PSA level.

3. *Modifications of PSA-Based Screening*

As discussed, it was shown in population-based studies after exclusion of prostate cancer by means of biopsy that total serum PSA levels correlate with age and prostatic volume. If PSA values are corrected for age, the correlation with prostatic volume does not persist. This has given rise to the design and use of age-specific reference ranges by Oesterling et al. (84). On the other hand, the volume correlation can also be used. Benson et al. (93) recommended in 1992 to use PSA density, also described as PSA index, the total PSA value divided by prostatic volume, to improve the prediction of the presence of cancer (93–97). A variation of PSA density becomes possible with the measurement of the transition zone of the prostate, which is representative of the volume of benign prostatic hyperplasia which may be present in the same men who also harbor prostate cancer. This concept was used by Bangma et al. (64,69) and was shown in a logistical regression analysis to be superior to total prostate volume in predicting a positive biopsy.

Another possibility of improving on the predictive power of PSA-based screening is offered by the application of the ratio of free and total PSA instead of total PSA (80,84). The basic observation is that most of the free PSA seems to be derived from benign parts of the prostate whereas PSA that is complexed with α-antichymotrypsin PSA is primarily derived from cancer cells. The physiology of this phenomenon is poorly understood. The ACT is available in the cytoplasm of prostate cancer cells and in the blood. Binding seems to occur predominantly in the cancer cell. Theoretically, therefore, men who have relatively high free PSA are more likely to have benign disease, those with high levels of completed PSA are more likely to have prostate cancer. Recommended cut-offs for the ratio between free and total PSA vary between 0.18 and 0.23.

All three aforementioned concepts have been evaluated in case-finding studies and in the European Randomized Study of Screening for Prostate Cancer (sections Rotterdam and Göteborg, Sweden). The use of age-specific reference ranges proposed by Catalona et al. (98) leads to an increase of 15% in cancer detection in men aged 50–59 years and at the same time to a 45% increase in the number of biopsies done. In the old age groups, 8%–47% of organ-confined cancers would be missed by avoiding 8%–44% of biopsies. The study concludes that a PSA cut-off of 4.0 ng/mL as a biopsy indication is preferable (94).

The PSA density was also subject to intense investigation. Littrup et al. (95), in an analysis of the American Cancer Society National Prostatic Cancer Detection Project (NPCDP), concluded that PSA density was superior to other tests and test combinations. This view is not shared by several other authors including Ohori et al. (96), who found PSAD useful only in a small subset of patients, namely in those who had a high total PSA level (>10 ng/mL) and a PSAD below 0.15. In this group of patients, prostate cancer was rarely detected and biopsies could be avoided. Similar findings were obtained by Brawer et al. (97).

The ratio of free to total PSA or the use of free PSA alone as a modification of total PSA in PSA-based screening strategies has been evaluated extensively. Most authors show that 30%–50% of biopsies can be saved at the price of missing variable proportions of cancer cases. The result seems to be strongly dependent on the underlying study populations. Clearly, the utilization of free to total ratios in comparing well-defined cases of BPH to prostate cancer reveals a very good discrimination as shown by the work of Catalona et al. (98), who defined this BPH control group as men having a prostatic volume of at least 40 mL and biopsy-proven absence of prostate cancer. Similarly, positive results were obtained by Van Cangh et al. (99), who in a clinical series of 266 patients with BPH and 154 cancer cases showed a significant improvement in discriminating between the two disease entities. The study recommends variable cut-off values for free to total ratio according to prostate volume. The paper, however, points out that the discrimination is dependent on the prevalence of prostate cancer in a given cohort of men and that a less important difference must be expected within a screening situation in which the prevalence is considerably lower. Prestigiacomo and Stamey (100) found in their study of 104 men with a positive biopsy compared to 110 men with a negative biopsy in a clinical sample that, with the use of a 23% cut-off value for free to total PSA ratio, 23% of biopsies were preventable and 9.6% of the cancers would be missed. Results of similar studies have been reported by Björk et al. (101), Vashi et al. (102), and Prestigiacomo et al. (103). Lerner et al. (104) attempted in a retrospective study to predict stage, grade, and DNA ploidy through initial free and total PSA with negative results.

Evaluation of free to total PSA ratio in a population-based study has been carried out within the European Randomized Study of Screening for Prostate Cancer. In the area of Göteborg, Sweden, 5813 men, aged 50–68 years, were screened utilizing a PSA cut-off of ⩾3.0 ng/mL as an indication for biopsy (105). Of 545 men biopsied, 117 were found to have prostate cancer. A PSA free to total (F/T) ratio cut-off point of 18% correctly identified 67 of 79 cancers in the total PSA range of 3–10 ng/mL, but 12 of 79 (15.2%) tumors were missed. Fifty-one percent of the biopsies would have been avoided. In a similar study, Bangma et al. (64) carried out an ROC analysis and found, over the whole range of PSA values, a significant improvement by the use of free PSA and DRE above PSA alone, but not above the combination of PSA and DRE. This study relates to 1726 screenees in whom 67 prostate cancers were found. In a more recent update of this analysis on a sample of 4800 men a significant improvement was found for free to total ratio in the total PSA range of 4–10 ng/mL (106). From these data, it must be concluded that the use of PSA F/T ratio in a population-based screening situation leads to a less powerful discrimination between benign and malignant cases than in the clinical situation in which BPH may need to be differentiated from cancer cases.

F. Is There a "Best" Screening Strategy?

The use of the three screening tests, total PSA, DRE, and TRUS, with the possibility of using different PSA cut-off levels as a biopsy indication, offers a large range of possible test combinations. Figure 5 reproduces two Venn diagrams from Bangma et al. (106). This type of presentation visualizes the relative value of the screening tests individually and in combination. In this sample of 4800 men identified in the population registry of Rotterdam and invited to participate in the European Randomized Study of Screening for Prostate Cancer, 976 biopsies detected 191 cancers, a detection rate of 4.0%; 5.1 biopsies were necessary to detect one cancer. A total of 20.3% of all participants underwent biopsy, resulting in 16.3% false-positive biopsy indications. Biopsies were carried out if serum PSA was ⩾ 4 ng/mL and/or or if DRE and/or TRUS results were suspicious. The DRE and TRUS alone detected about equal numbers of

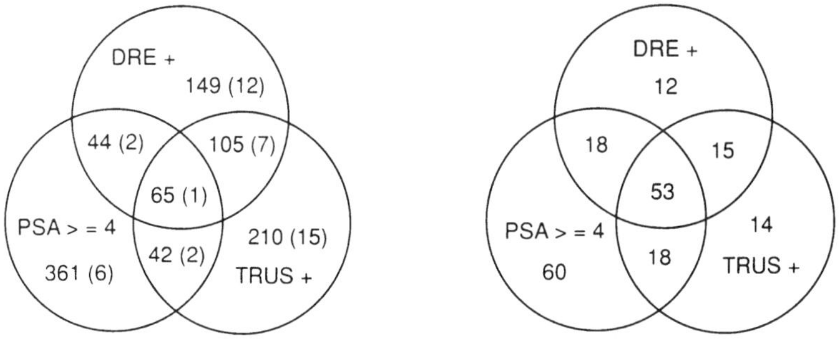

Biopsies (# biopsies per cancer) Cancers (#)

Figure 5 Venn diagram of 976 biopsies carried out in 4800 screened men from the general Rotterdam population; 191 cancers were detected. *Left*, Numbers of biopsy indications per test and test combination. Numbers in parentheses are the number of biopsies necessary to find one cancer. *Right*, number of cancers per screening test and combinations. Adapted from Ref. 106.

cancers; PSA alone detected more than two times as many. Clearly, any combination of tests, especially a situation in which all three tests are positive, improves the predictive value considerably. Positive predictive values can be derived from these Venn diagrams by dividing the numbers of cancers by the numbers of biopsies. With the biopsy indication used, it is possible to evaluate age-specific reference ranges. Free to total PSA ratios can only be evaluated for total PSA values >4. The results of a simulation experiment on a similar but smaller sample of men with 308 biopsies and 67 cancers is shown in Table 11 derived from Ref. 69. This table includes the regimen used in the present study in the top line. Subsequently, the gain in specificity (reduction in biopsies) and the change in sensitivity (missed prostate cancers) as well as the numbers of biopsies necessary to detect one cancer are indicated for different screening regimens. Clearly, this table could be expanded by numerous additional options. The PSA prescreen regimens have the advantage of avoiding rectal examination and transrectal ultrasonography in men with PSA values <2. Most of the options lead to an important reduction of biopsies ranging from 17% to 39%. However, this is offset by 6%–12% missed cancers that would otherwise have been detected. At the time this evaluation was done, it was thought that a PSA prescreen with a cut-off of 2, rectal examination and TRUS between 2 and 4, and a biopsy for men with a PSA $\geqslant$ 4.0 ng/mL or the same regimen with omission of TRUS were the best choices. Saving 30%–36% of all biopsies done, a considerable reduction of the numbers of biopsies needed to find one cancer and a loss of sensitivity of only 6% seemed attractive options.

Table 11 Results of Simulated Case Selection for Biopsy, Based on the Results by Screening with PSA, DRE, and TRUS in 1726 Men, of Whom 67 Men with Prostate Cancer Had 308 Biopsies Done

Prescreen[a]	Biopsy indicator	Reduction of biopsies (%) (SE)	PC missed (%) (SE)	Overall Bx/PC
—	PSA > 4, DRE, TRUS	—	—	4.6
—	PSA > 4, DRE	17 (2.1)	3 (2.1)	3.9
PSA > 2	PSA > 4, DRE, TRUS	30 (2.6)	6 (2.9)	3.4
PSA > 2	PSA > 4, DRE	36 (2.8)	6 (2.9)	3.1
—	PSA-age, DRE	37 (2.8)	12 (4.0)	3.3
—	PSAD, DRE	28 (2.4)	11[b] (3.8)	3.6[b]
—	PSA > 4 and PSAD, DRE	39 (2.8)	11 (3.8)	3.2
—	PSA > 4 and F/T, DRE	37 (2.8)	11 (3.8)	3.3
PSA > 3	PSA ≥ 3.0 ng/mL	21	10	

[a] Men with PSA values below this level receive the message ''all clear'' and are not examined further

[b] Real incidence of cancers unknown

Abbreviations: Bx = biopsy; DRE = abnormal result by digital examination; F/T = free/total PSA ratio 0.20 or less PC = prostate cancer; PSA > 4 = total serum PSA of 4.0 ng/mL or more; PSA-age = total serum PSA larger than age-specific reference range; PSAD = PSA density 0.12 ng/mL/cc or more; TRUS = lesion detected by transrectal ultrasound.

Source: Adapted from Ref. 69.

Conceptually, the screening test combination, which leads to the detection of most life-threatening and still curable prostate cancers, would have the potential of reducing prostate cancer mortality rate in a maximal fashion and would represent the best strategy. However a number of determinants remain unknown, and a ''best screening strategy'' can therefore not be determined on rational grounds. The type of tumors detected and especially the type of tumors omitted in the desire to increase the specificity of the screening tests are considered to be of great importance. Important parameters evaluable in radical prostatectomy specimens, such as perforation of the fibrous capsule, positive margins of resection, and tumor volume evaluated by morphometrical techniques show a strong positive correlation with increasing PSA values. In an evaluation of 80 such cases in ERSPC trials with a PSA value <3 on average were smaller than 0.5 cm^3, the largest lesions being in the range of 1.0 cm^3.

No poorly differentiated tumors were found at PSA values <3. About 30% of all cancers were detected by DRE and TRUS at PSA values <4.0 ng/mL, and 25% had PSA levels of 3 or lower. The mechanism of detection is not clear in that it is unlikely that most of the smaller lesions (minimum 0.04 mL) would be palpable or visible by DRE or TRUS.

Another important consideration in attempting to optimize screening procedures is the rescreening interval. With yearly rescreening as recommended by the American Cancer Society and current knowledge of PSA-related tumor doubling times, less sensitive screening tests may be acceptable. Tumors initially missed may be found during the next screening round after 1 year, still in a potentially curable state. Within the European Randomized Study of Screening for Prostate Cancer, a multicenter effort randomizing between screening and no screening and attempting to recruit more than 200,000 men in the age group 55–70 years, agreement has been reached to use a rescreening interval of 4 years. With a PSA doubling time that may reflect tumor doubling times of 2–4 years in mind, a lesion with a volume of 1 cm^3 might represent 4 years later as a lesion with a volume of 4 cm^3, a tumor that is likely to produce lymph node metastases.

Further evaluation of almost 10,000 screened men in the Rotterdam area has led to a recommendation for adaptation of the screening regimen within ERSPC, which amounts to a drastic change. The PSA prescreening with a PSA cut-off of 3.0 will be used. Men with a PSA <3.0 will not undergo any more testing and receive the message of ''all clear'' for 4 years. Every participant with a PSA $\geqslant 3.0$ ng/mL will be biopsied. It is anticipated that the detection rate will be 4.75%. Nineteen percent of all participants in the screening arm will be biopsied and simultaneously undergo DRE and TRUS, and 81% will have a PSA test as the only screening measure. The rate of false-positive biopsy indications will decrease to 14.2%, and 4.2 biopsies will be needed to find one cancer. With respect to the use of all three screening tests in a population projected on 10,000 screened cases and 475 cancers, there will be a loss of 32 cancers (6.7%). This regimen was chosen on the basis of increasing knowledge of the types of tumors that will be missed, with the wish to decrease false-positive biopsy indications and considering the great advantage of not having to expose 81% of participants to DRE and TRUS. The possibility that those cancers that are missed by this strategy may have a potential to contribute to the decrease in mortality rate cannot be excluded. The change in procedure allows continuous evaluation of tests and test combinations.

G. Prostatic Biopsy

Prostatic biopsy is an invasive but well-accepted procedure. The risk of ultrasound-guided transrectal biopsies is reviewed in Section VII. In the context of the screening procedure, the accuracy with which prostate cancer can be detected in the prostate gland by biopsy is of relevance. As elaborated previously, men older than 50 years have a chance of more than 30% of harboring prostate cancer. These tumors usually are very small with a volume of much less than 0.5 cm^3. The natural history of these very small lesions is unknown. From autopsy data, however, it must be assumed that most of these tumors will never pose a threat to the patient. How many and which ones will progress and kill is unknown, however. One of the dilemmas of early detection of prostate cancer is that, with knowledge of this large reservoir of cancers, the sensitivity of the detection procedure is a matter of choice but a sound basis for making choices is not available.

The concept of systematic sextant biopsies as introduced by Terris et al. (107) provides standardization and has been shown to detect less than 10% of small cancers with volumes of 0.5 mL or less. With this technique, three biopsy specimens are obtained from each side of the prostate. In addition, one specimen from a suspicious lesion and one from the transition zone are obtained if necessary or desired. Daneshgari et al. (108) studied the concept of sextant biopsies in a computer simulation experiment. They found that the size of the prostate, the size of the lesions and the distribution of the lesions within the gland determine the chance of diagnosis by biopsy. Aggregate tumor volumes between 0.034 and 5.1 mL were simulated. The needle angle played an important role, a 30-degree angle produced significantly better results than a 45 or 60 degree biopsy angle. Sextant biopsies were shown to be better than random biopsies. Within the volume range indicated, the chance of missing tumor was 27%. Only 20% of the simulated situations had a tumor distribution for which systematic biopsies had a 95% probability of tumor detection.

One of the most obvious difficulties relates to the great variation of prostatic volume. A small tumor can more easily be detected in a small prostate than in a large one. Some groups have therefore increased the numbers of specimens obtained according to total prostatic volume determined by ultrasound. The concept of five zonal biopsy specimens was proposed by Eskew in 1997 (109). With this concept, 13–18 samples are obtained; in a comparison to sextant biopsies, the five zonal biopsy technique increases the detection rate by 35%. These authors have carried out a comparison of tumor characteristics and do not find any differences in volume or Gleason score between those tumors detected by sextant and those detected by biopsies of five prostatic zones. This information comes from one center and requires confirmation. Also, the potential complications of obtaining such a large number of specimens need to be studied in a large series.

Several attempts have been made to use the outcome of biopsy to predict tumor stage, grade, and other parameters of aggressiveness. Ravery et al. (110) have carried out sextant biopsies in 158 men with prostate cancer and found that the fact that only one of six biopsy results is positive does not predict a low tumor volume. The degree to which biopsies are representative with respect to the determination of tumor grade has already been discussed. With locally confined disease, 50%–60% of cancers are undergraded at biopsy with respect to radical prostatectomy specimens. Dugan et al. (111) have developed a computer model for the prediction of clinically insignificant prostate cancer, which includes age, life expectancy, Gleason score, and assumed cancer volume, taking into account tumor doubling times of 2,3,4, and 6 years. This modeling cannot exclude the uncertainties around the pretreatment evaluation of important prognostic factors, which are part of the model. To come to a proper pretreatment prediction of tumor aggressiveness and outcome, methodology providing better sampling of available prognostic parameters and the development of better prognostic parameters are necessary.

H. Rescreening

Rescreening should be differentiated from reevaluation of patients or participants in an early detection program who have a negative biopsy and in whom a strong suspicion remains based on the presence of other indicators of prostate cancer. Such reevaluations have been subject to a study by Keetch et al. (112). A total of 427 of 1136 men had an initial negative biopsy result but persistently elevated PSA levels >4.0 ng/mL and abnormal rectal or ultrasound examination findings. On the basis of this suspicion, reevaluation and rebiopsy was carried out if abnormalities persisted. At the second biopsy done after 6 months, 82 men (19%) were found to have cancer. At a third and fourth biopsy, an additional 8% and 7% of cancers were detected. Ninety-six percent of the cancers found were detected through the first or second biopsy done.

Several studies that have used yearly rescreening have reported on rescreening detection rates. This information is important for the future determination of an appropriate rescreening interval. Mettlin et al. (113), in reporting on the National Prostatic Cancer Detection Project of the American Cancer Society, revealed detection rates of 2.8%, 2.0%, 1.1%, 2.1% and 1.9% (total = 9.9%) in the first to fifth yearly screening round. This demonstration project included 2990 men, 55–70 years old at entry, 9937 examinations led to 1215 biopsies and detected 203 cancers. In this program, at least initially, biopsies were not indicated with PSA levels >4. Labrie et al. (114), with a slightly different application of screening tests to a population-based sample, described three yearly rescreening detection rates of 0.6%, 0.7%, and 0.3% after an initial detection rate of 3.5% (cumulative rate = 5.1%).

Brewster et al. (115) in a United Kingdom based demonstration project used a 3-year rescreening interval. No interval cancers were detected in 472 eligible men. A total of 340 men were available for rescreening. On the basis of an elevated PSA, 17 men underwent biopsy and six cancers were found, of which five were localized and one was metastatic. The PSA had increased in this man from 2 to 120 ng/mL over the 3-year period. I am not aware of any other report on interval cancers. These are not seen apparently with yearly rescreening, a clear indication that a 1-year rescreening interval is too short. A future determination of a proper rescreening interval in prostate cancer can be based on the experience to be obtained within the European randomized study with a rescreening interval of 4 years, present knowledge of PSA doubling times, and tumor characteristics.

V. CASE-FINDING STUDIES, DEMONSTRATION PROJECTS, AND SCREENING STUDIES

A review of all case-finding studies and demonstration projects is not within the scope of this chapter. Several recent reviews of studies and concepts are available (77,114,115,117–121).

A. Case Finding Studies

Important lessons have been learned from large case-finding studies and demonstration projects. Cooner et al. (122) applied systematic screening to a clinic population of symptomatic urological patients of 1807 men and described a detection rate of 14.6%. This study better than any other demonstrates that detection rates depend on population characteristics. The American Cancer Society National Prostate Cancer Detection Project (NPCDP) has been the first large-scale multicenter demonstration project. The cooperative group eventually recruited 2999 men aged 55–70 years at entry. At least initially, biopsies were only indicated on the basis of abnormal rectal examinations and/or TRUS. The results were subsequently correlated with serum PSA levels. The overall detection rate was 2.4% as described by Mettlin et al. in 1991 (123). The latest update of this project is given in reference 113. Another case-finding study was carried out by Brawer et al. (97). The initial report on this project, which included 1249 men, was given in 1992 (124). Thirty-two carcinomas were found, accounting for a detection rate of 2.6%. The study contributes to the understanding of the screening procedure and to the characterization of screen-detected cancers. Another case-finding study was carried out in Sweden by Gustafsson et al. (125) and reported in 1992. The protocol compared rectal examination and ultrasound as screening tests. The TRUS increased the detection rate of rectal examination by 50%. The overall detection rate was 3.6%.

The largest project carried out so far is related to the prostate cancer awareness weeks conducted on a multicommunity basis in the United States (126). Data on the years of 1993 and 1994 relating to 13,953 eligible subjects are reported. Levels of PSA and rectal examination were used. Biopsies were carried out when there was suspicious rectal examination and/or an elevated serum PSA value >4.0. However, not all men with a PSA <4 underwent biopsy. A detection rate of 3.6% resulted. The PPV for elevated PSA levels was 31.6% and for an abnormal DRE was 25.5%. These findings are in line with data obtained in similar studies.

Case-finding studies and demonstration projects cannot contribute to the key question of whether early detection and population-based screening are useful in lowering prostate cancer mortality. This information can only be obtained in randomized screening studies and possibly in studies carried out in secluded geographical areas where historical comparison may give important hints. Such a study was described by Reissigl et al. in 1995 (127). In the area of Innsbruck, Austria, 21,078 men aged 45–75 years underwent screening by PSA only as an initial test. Nine percent had a PSA elevation above 4.0 ng/mL. Biopsies were carried out in 778 of the 1618 men (48%) who had an elevated PSA and a biopsy indication. The overall cancer detection rate was 1.2%. It is unclear why so many men who had a biopsy indication did not undergo biopsy. Selection factors might have contributed to the low detection rate.

Results of randomized screening studies with respect to the main endpoint, prostate cancer mortality rate are not available. Several authors have attempted to approach the problem by decision analysis (128–130). The most recent decision analysis on screening for prostate cancer includes quality adjusted life expectancy (QALE) and cost utility ratios next to life expectancy. Screening prolonged adjusted life expectancy in men aged 50–70 years but the benefit diminished drastically after correction for QALE. The paper has created significant public discussion. Like other similar studies, the authors use natural history data, which may not be compatible with the risk of screened populations of men. Decision analytic studies are bound to be flawed because of the uncertainty of input data. They cannot replace prospective randomized screening studies.

B. Randomized Screening Studies

The first randomized screening study was initiated in the region of Quebec City, Canada. More than 8000 men have been screened. After the initial use of PSA, DRE, and TRUS without the application of PSA-based biopsies, the screening protocol was changed. The PSA prescreening with a cut-off value of 3.0 ng/mL for exclusion of men from further screening is used. The participants are identified from the electoral role, randomized, and invited for screening. The control group is merely identified administratively. Follow-up is based on cancer registry. Several aspects of the Quebec randomized screening study have already been discussed in this chapter (19,67,114).

In 1991, the Division of Cancer Prevention and Control of the National Cancer institute of the United States initiated the Prostate, Lung, Colorectal, and Ovarian (PLCO) cancer screening trial, a controlled, randomized screening study (131). The study aims at including 74,000 men and 74,000 women, age 55–74 years old. The study has a power of 90% to determine a 20% reduction in prostate cancer mortality rate from a baseline and three subsequent annual screens. Both PSA and DRE are used. The major endpoint is prostate cancer mortality, but other aspects such as quality of life, pathological features of tumors, and the set-up of a biorepository are included. Biopsy samples are obtained in men with a suspicious rectal examination or an elevated serum PSA value. The study is described by Gohagan et al. (131). This report includes sample size considerations and a sample size calculation. In the United States, where screening is prevalent, recruitment may be difficult because of opportunistic screening in the screening and control arm of the study. The study group has taken action to at least exclude men who have already been screened within the last 3 years at the time of entry and before randomization. Preliminary data are not available.

The European Randomized Study of Screening for Prostate Cancer is conducted by a European consortium of six centers, which are listed together with an update of participation as of February 1, 1997 in Table 12. A Spanish center is scheduled to start participation. In this study, men between the age of 55 and 70 years are randomized to screening or no screening. Prostate cancer mortality is the main endpoint; quality of life studies, studies of prognostic factors, and a detailed evaluation of the screening tests are also part of the protocol. To show a difference in prostate cancer mortality rate of 20% with a power of 90%, approximately 65,000 men need to be recruited to each arm. Recent evaluation showed a contamination rate by the use of screening tests in the population of participants of about 10%. This increases the sample size to ± 190,000 in total. The study group has a capability of recruiting more than 200,000 men. A number of aspects of this study have already been discussed in Section IV when

Table 12 European Randomized Study of Screening for Prostate Cancer (ERSPC) Recruitment per May 31, 1997

Country/center	Start date	No. of men randomized	No. of men screened	Expected contribution
Antwerp, Belgium	1/04/95	9,264	4,619	20,000
Finland	1/05/96	14,989	4,989	75,000[a]
Florence, Italy	1/10/96	2,739	831	40,000
Göteborg, Sweden	1/01/95	32,298	10,000	32,000[b]
Portugal	1994	2,059	901	20,000
Rotterdam, Netherlands	1/06/94	19,393	9,804	40,000[c]
Total		54,561	31,144	227,000

[a] Randomization control/screening = 2:1
[b] Age 50–65 years
[c] Includes age 71–74 years

dealing with the screening procedure. The group has established a minimal data set and a prospective evaluation plan, which also includes the PLCO study and the randomized study in Quebec (132,133). Tables 13 and 14 indicate the algorithm for the prevalence screen and for rescreening after 4 years as it is used in the Rotterdam section of ERSPC. The ERSPC started in 1994. The first recruitment period is 4 years. A second round of screening after 4 years will follow. A first evaluation with respect to the main endpoint is expected to be published after 10 years of follow-up in the year 2008. I act as national and international coordinator of ERSPC. The study is supported by national funding in the individual countries, by Hybritech Incorporated, and the by European Union's program ''Europe Against Cancer.''

VI. RISK OF SCREENING

The risk of the screening procedure must be subdivided into the psychological impact of screening and the actual physical damage that can be caused by screening or subsequent treatment. The physical impact of the screening test depends on the procedures that are chosen for any given detection regimen. These may include venous puncture for the PSA determination, a rectal examination, and the transrectal ultrasonography. In patients who are identified as being at risk, the biopsy then follows. The risk of treatment comes down to the complications of radical prostatectomy and radiotherapy.

A. Complications of Screening and Treatment

Serious complications of rectal examination and transrectal ultrasonography have not been described. Venous puncture is frequently practiced and can be considered an innocent procedure. Nevertheless, DRE and TRUS are potentially annoying and pain is sometimes mentioned. Within ERPSC, a complete process evaluation was carried out, which included a study of the reasons for refusal (134) and the perception of the individuals of the screening procedure together with the willingness to be rescreened (135). The resulting data are favorable. Ninety-six percent of all participants are willing to be rescreened; the average score for the bother experienced through the entire screening process was 8.4 $\pm$ 1.3 (best possible: 10.0).

Prostatic biopsies, however, do have significant complications. Reports from large prospectively collected series have recently published. Rietbergen et al. (136) described that moderate pain during and after the procedure, usually not requiring medication, as well as hematospermia and hematuria are common. Serious bleeding, however, is a rare complication. A transient increase in body temperature is common, even after various forms of antibiotic prophylaxis. However, serious septic complications are relatively rare and occur in 0.3% to

Table 13 Algorithm Screening Procedure (Continuation First Round, Prevalence Screen)

Study based		*Health care system based*
Letter of invitation ↓ Informed consent (written) ↓ Randomization ↘ ↓		
Screening / No screening	→→→	If desired
↓ PSA test, blood drawn on location ↘ ↓		
PSA ≥ 3.0 ng/mL / PSA < 3.0 ng/mL "all clear" re-screen 4 years	→→→	Letter to general practitioner (GP), consultation possible
↓ ↓ Biopsy (Combination DRE + TRUS) ↘ ↓		
Cancer + / Cancer −	→→→	Letter to GP *Cancer* −: advice on complications, meaning of negative biopsy may be asked. *Cancer* +: referral to a regional urology service through family physician for treatment. ↓ ↓
↓ ↓ Concluding conversation with staff urologist—referral		
If radical prostatectomy pathological evaluation of the specimen in cooperation *within* the study *Follow-up* of all participants (screening and control group)		Posttreatment care and follow-up

Table 14 Algorithm for Rescreening After 4 Years

Informed consent was obtained prior to the first round participants.
Participants are those randomized to the screening.

Study based			*Health care system based*
Letter of invitation to PSA test			
↓			
PSA test (at location of participant)			
↓	↘		
PSA ≥ 3.0 ng/mL	PSA < 3.0 ng/mL "all clear" no more studies	→→→	Letter to general practitioner (GP)
↓			
↓			
Biopsy (Combination DRE and TRUS)			
↓	↘		
Cancer +	Cancer −	→→→	Letter to GP *Cancer* −: advice on complications, discussion result *Cancer* +: referral to regional urologist through GP, discussion treatment, treatment
↓	↓		
Result discussed with staff urologist			↓
If radical prostatectomy Pathological evaluation of the specimen in cooperation *within* the study			
Follow-up of all participants (screening + control group)			Posttreatment care and follow-up

3% of cases. Aus et al. (137) collected data on complications in 500 patients who underwent transrectal core biopsies and determined risk factors. Two different regimens of antibiotic pretreatment (1 week duration versus 2 tablets directly prior to the procedure) were compared. A recent use of an indwelling catheter, a history of urinary tract infection, diabetes mellitus, and prostatitis were risk factors. The frequency of the occurrence of febrile infections after biopsy was significantly influenced by the duration of antibiotic treatment.

The complications of treatment are well documented. However, the numbers of complications vary markedly. Radical prostatectomy is associated with a mortality rate between 0.1% and 1%; if older age groups and high-risk patients are included, this figure may be 2%. Urinary incontinence, impairment of sexual function, and strictures on the anastomosis between urethra and bladder neck are the most common long-term complications. After radiotherapy, proctitis, dysuria, loss of potency, and rarely incontinence and stricture are similar long-term complications. An independent assessment of the perception of the results after radical prostatectomy by 1072 Medicare patients who underwent radical prostatectomy was carried out by Fowler et al. (138). Despite significant complications, 81% of the interviewed patients appreciated the treatment as positive; 89% would choose surgery again.

B. Quality of Life

Quality of life assessment in prostate cancer and especially in the screening situation is still in the process of being established. Litwin et al. (139) studied 214 men after radical prostatectomy and 273 age-matched control patients without prostate cancer. Although there were no significant differences in general health-related quality of life parameters, differences between the treatment and control group were statistically significantly different as far as urinary incontinence and sexual function was concerned. Otnes et al. (140) studied the burden of prostate cancer from diagnosis until death. They concluded that symptomatic, conservatively treated prostate cancer imposes a considerable burden on patients and health care resources.

VII. COST OF SCREENING

A number of estimates of the cost of screening for prostate cancer are available. These estimates are all preliminary and may serve methodological purposes more than the actual appreciation of the expenses of screening. The uncertainties result from the fact that essential information is not yet available. Even with the assumption of different scenarios, the problem cannot be resolved. The cost of screening depends on the type of screening procedure used, the rescreening interval, the age of the population screened, the geographical variation and incidence of prostate cancer, and many other factors. Cost of screening is often expressed in terms of the expenses incurred to find one cancer. The method of calculation is further complicated by the fact that detection rates are variable with the choice of screening procedure and with the choice of populations of men. Cost of screening has been studied in Sweden and in The Netherlands (141,142). Several publications are based on ongoing screening programs in the United States.

Benoit et al. (143) estimated the cost of detection of one cancer with the use of rectal examination and PSA as screening tests in a population of men aged 50–59 years. Littrup et al. (144) presented calculations of potential expenses depending on different assumed screening scenarios. These included the use of three different PSA cut-off values alone or in combination with DRE. An attempt is made to include the cost of treatment and to compensate for avoided expenses related to progressive, metastatic disease. Lubke et al. (145) attempted calculations of the cost of nationwide screening programs for prostate cancer in the United States. Again, studying different scenarios, the annual cost of such a program is estimated to run between $8.5 and $25.7 billion respectively. Woolf (146) assessed the public health perspective of prostate cancer screening. He concluded that the billions of dollars required to screen 28 million men older than 50 years in the United States once a year would divert precious resources away from health care services of proven benefit. He stressed the necessity of conducting prospective randomized treatment and screening studies and concluded that a cost-benefit analysis is not possible until information resulting from such trials becomes available. Handley et al. (147) studied the effect of PSA testing on cost in the population of a large health maintenance organization. Details on the cost of PSA screening are given in this publication. Also, the complications of treatment are put into a cost perspective. The authors have seen a significant decrease of the use of PSA-based screening after full information about the evidence of effectiveness of screening for prostate cancer is provided to physicians.

VIII. RECOMMENDATION OF VARIOUS GROUPS

Considering the present controversy on screening for prostate cancer, it is not surprising that national and group policies vary greatly around the world. Both the American Cancer Society and the American Urological Association recommended an annual screen for prostate cancer by means of rectal examination and PSA screening as of age 50 years in 1992 (148–150). Both organizations agree that in men in high-risk groups such as African Americans or with a strong family history, screening may start at an earlier age (40 years). The ACS policy statement concludes: ''Generally, men with a life expectancy of at least 10 years after detection, may benefit from examination.'' There is, however, by no means unanimity on the North American continent. The American College of Physicians has established guidelines for the screening of prostate cancer based on a model approach (151,152). The model is based on relevant studies identified from a Medline search (1966–1995) as well as from other sources. In modeling cost-effectiveness for a one-time digital rectal examination and PSA measurement, the group concludes that, with the most positive assumptions, average life expectancy may be increased by approximately 2 weeks in men between 50 and 69 years of age. The paper (152) concludes doubtfully: ''Put simply, should we encourage wide-spread screening until we can prove that it does not work, or should we avoid it until we can prove that it does work.'' In its final recommendations, the American College of Physicians (152) does not recommend routine screening because ''the balance of benefit and harm for early treatment is unknown.'' In the meantime, it is recognized that prostate cancer is an important health problem, that there may be benefits of one time or repeated screening, which have not been proven, that early detection may save lives, and that early detection and treatment may avert future cancer-related illness. Furthermore, the American College of physicians encourages all physicians to help to enroll eligible men in ongoing randomized studies, especially the PLCO trial of the National Cancer Institute. Men who request screening are to be fully informed about potential risks and benefits; the use of the PSA test without notification and prior discussion is discouraged. The Canadian Task Force on Periodic Health Examination (CTF) pleads against screening of the general population by means of PSA and TRUS. However, this group thinks that there is not enough evidence to suggest to physicians and urologists to discontinue the use of rectal examination as a routine in clinical practice (153). The U.S. Preventive Services Task Force determined its viewpoints in two publications by Steven H. Woolf (154,155). The final conclusion is that routine screening for prostate cancer cannot be recommended at this time. If screening is performed in a well-informed patient, the best approach is the use of PSA and DRE in men who have a life expectancy of more than 10 years. A rescreening interval cannot yet be determined. Also in Europe, contradictory policies exist and are practiced. In Germany screening for prostate cancer had already been introduced as a health care provision in the 1970s, whereas other European countries, especially the Scandinavian countries and Great Britain, developed contrary views. Recently, the Health Technology Assessment Panel of the National Health Service Executive in the United Kingdom established a working group to advise on diagnosis, management, and treatment as well as costs of prostate cancer in England and Wales. The resulting report (156) recommends that ''Opportunistic screening should be discouraged and commissioning agencies should not purchase screening services for prostate cancer, other than in the context of a randomized controlled trial.'' Similar views have been taken in the Netherlands where a committee of the Dutch Health Council recommended to conduct a randomized screening study with prostate cancer mortality as the primary endpoint (157).

IX. WHERE DO WE STAND?

Prostate cancer represents a significant health problem. Cure of the disease is impossible in advanced stages, but possible with locally confined disease. Unfortunately, results of ongoing randomized studies comparing radical prostatectomy or radiotherapy to watchful waiting are not available and the effectiveness of treatment in terms of lowering prostate cancer mortality is therefore not established. Early detection is technically feasible and leads to the diagnosis of about 60% of cancers in a histologically organ-confined state, an improvement above clinical routines of at least 20%. The optimal use of the available detection procedures, which are all based on PSA, is still up for discussion. Optimal screening strategies are in the process of being determined. Application of screening to the general population and to clinic populations seem to yield different results. It is unknown whether screening for prostate cancer will lower prostate cancer mortality. Certainly will only be obtained from prospective randomized studies or from large-scale application of screening to defined geographical areas. Randomized screening studies have been initiated in 1993 and 1994 in Europe and in the United States. Mortality results, however, will not be available before the year 2008. In the meantime, the available early detection tests should be applied with caution. It is uncertain whether early detection and early treatment is of benefit or if, in fact, damage to health and well-being prevails. Men who wish to be screened for whatever reason should be carefully informed about this situation before the use of PSA-based early detection measures. The outcome of the presently ongoing randomized screening studies may be expected to make a major contribution to the resolution of the issue of application of early detection tests and early treatment to the general population. Suitable data upon which health care policies can be based in the future are urgently needed.

ACKNOWLEDGMENTS

The author is grateful to all collaborators within the ''European Randomized Study of Screening for Prostate Cancer (ERSPC),'' especially to Dr. C. Bangma, Drs. R. Hoedemaker, Ing. R. Kranse, and J. Rietbergen, who have provided data used in this chapter. Mrs. H. J. van Alphen has provided expert secretarial help.

REFERENCES

1. Catalona WJ, Smith DS, Ratliff TL, Dodds KM, Coplen DE, Yuan JJJ, Petros J A, Andriole GL. Measurement of prostate-specific antigen in serum as a screening test for prostate cancer. N Engl J Med 1991; 324,1156–1161.
2. Walsh PC. Prostate cancer kills: strategy to reduce deaths. Urology 1994; 44(4): 463–466.
3. Adami HO, Baron JA, Rothman KJ. Ethics of a prostate cancer screening trial. Lancet 1994; 343:958–960.
4. Schröder FH. Prostate cancer: to screen or not to screen? BMJ 1993; 306:407–408.
5. Moller Jensen O, Estève J, Moller H, Renard H. Cancer in the European community and its members states. Eur J Cancer 1990; 26(11/12): 1167–1256.
6. Parker SL, Tong T, Bolden S, Wingo PhA. Cancer statistics. CA Cancer J Clin 1997; 47:5–27.
7. American Cancer Society. Cancer Facts and Figures—1996. Atlanta: American Cancer Society, 1996.
8. Visser O, Horenblas S. Incidentie en behandeling van prostaatcarcinoom in de regio van het Integraal Kankercentrum Amsterdam, 1989–1994. Ned Tijdschr Geneeskd 1996; 40(52):2627–2631.
9. Boyle P, Maisonneuve P, Napalkov P. Geographical and temporal patterns of incidence and mortality from prostate cancer. Urology 1995; 46(suppl 3A): 47–55.
10. Adlercreutz H. Western diet and Western diseases: some hormonal and biochemical mechanisms and associations. Scand J Clin Lab Invest 1990; 50(suppl 201): 3–23.
11. Parkin DM, Muir CS, Whelan S, Gao YT, Ferlay J, Powell J, eds. Cancer Incidence in Five Continents. IARC Scientific Publication 120, Vol VI. Lyon, IARC, 1992.
12. Watanabe H, Mishina T, Ohe H. Mass screening of prostatic diseases. In: Hafez ESE, Spring-Mills E, eds. Prostatic Carcinoma Biology and Diagnosis. Boston, Martinus Nijhoff Publishers, 1981:71–78.
13. Imai K, Ichinose Y, Kubota Y, Yamanaka H, Sato J. Diagnostic significance of prostate specific antigen and the development of a mass screening system for prostate cancer. J Urol 1995; 154:1085–1089.
14. Helgesen F, Holmberg L, Johansson JE, Bergström R, Adami HO. Trends in prostate cancer survival in Sweden, 1960 through 1988: evidence of increasing diagnosis of nonlethal tumors. J Natl Cancer Inst 1996; 88 (17):1216–1221.
15. Potosky AL, Miller BA, Albertsen PC, Kramer BS. The role of increasing detection in the rising incidence of prostate cancer. JAMA 1995; 273 (7):548–552.
16. Lu-Yao GL, Greenberg ER. Changes in prostate cancer incidence and treatment in USA. Lancet 1994; 343:251–254.
17. Jacobsen SJ, Katusic SK, Bergstrahl EJ, Oesterling JE, Ohrt D, Klee GG, Chute CG, Lieber MM. Incidence of prostate cancer diagnosis in the eras before and after serum prostate-specific antigen testing. JAMA 1995; 274 (18):1445–1449.
18. Stephenson RA, Smart ChR, Mineau GP, James BC, Janerich DT, Dibble R. The

fall in incidence of prostate carcinoma. On the down side of a prostate specific antigen induced peak in incidence—data from the Utah Cancer Registry. Cancer 1995; 77:1342–1348.
19. Labrie F, Candas B, Cusan L, Gomez JL, Diamond P, Suburu R, Lemay M. Diagnosis of advanced or noncurable prostate cancer can be practically eliminated by prostate-specific antigen. Urology 1996; 47(2):212–217.
20. National Cancer Institute. Cancer death rate declined for the first time ever in the 1990s [press release]. National Cancer Institute, November, 1996.
21. Miller T. Interpreting recent changes in prostate cancer incidence and mortality trends. In: Kramer BS, Gohagan JK, Prorok PC, eds. Cancer Screening: Theory and Practice. New York: Marcel Dekker, 1999.
22. Wilson JMG, Jungner G. Principles and practice of screening for disease [Public Health Paper No. 34]. Geneva: World Health Organization 1968.
23. George NJR. Natural history of localized prostatic cancer managed by conservative therapy alone. Lancet 1988; March 5:494–497.
24. Whitmore WF Jr, Warner JA, Thompson IM Jr. Expectant management of localized prostatic cancer. Cancer 1991; 67(4):1091–1096.
25. Adolfsson J, Carstensen J. Natural Course of clinically localized prostate adenocarcinoma in men less than 70 years old. J Urol 1991; 146:96–98.
26. Johansson JE, Adami HO, Andersson SO, Bergström R, Holmberg L, Krusemo UB. High 10-year survival rate in patients with early, untreated prostatic cancer. JAMA 1992; 267(16):2191–2196.
27. Schröder FH, Hermanek P, Denis L, Fair WR, Gospodarowicz MK, Pavone-Macaluso M. The TNM classification of prostate cancer. The Prostate 1992; 4 (suppl):129–138.
28. Chodak GW, Thisted RA, Gerber GS, Johansson JE, Adolfsson J, Jones GW, Chisholm GD, Moskovitz B, Livne PM, Warner J. Results of conservative management of clinically localized prostate cancer. N Engl J Med 1994; 330:242–248.
29. Aus G, Pileblad E, Hugosson J. Impact of competing mortality on the cancer-related mortality in localized prostate cancer. Urology 1995; 46(5):672–675.
30. Franks LM. Latent carcinoma of the prostate. J Pathol Bact 1954; 68:603–616.
31. Kabalin JN, McNeal JE, Price HM, Freiha FS, Stamey TA. Unsuspected adenocarcinoma of the prostate in patients ondergoing cystoprostatectomy for other causes: incidence, histology and morphometric observations. J Urol 1989; 141: 1091–1094.
32. Montie JE, Wood DP, Pontes JE, Boyett JM, Levin HS. Adenocarcinoma of the prostate in cystoprostatecomy specimens removed for bladder cancer. Cancer 1989; 63:381–385.
33. McNeal JE, Bostwick DDG, Kindrachuk RA, Redwine EA, Freiha FS, Stamey T A. Patterns of progression in prostate cancer. Lancet 1086; 1:60–63.
34. Barry MJ, Fleming C, Coley CC, Wasson JH, Fahs MC, Oesterling JE. Should medicare provide reimbursement for prostate-specific antigen testing for early detection of prostate cancer? Part I: framing the debate. Urology 1995; 46 (1):2–13; Part II: early detection strategies. Urology 1995; 46 (2):125–141; Part III: management strategies and outcomes. Urology 1995; 46(3):277–289; Part IV:

estimating the risks and benefits of an early detection program. Urology 1995; 46(4):445–461.
35. Ohori M, Wheeler TM, Dunn JK, Stamey TA, Scardino PT. The pathological features and prognosis of prostate cancers detectable with current diagnostic tests. J Urol 1994; 152:1714–1720.
36. Stamey TA, Freiha FS, McNeal JE, Redwine EA, Whittemore AS, Schmid HP. Localized prostate cancer, relationship of tumor volume to clinical significance for treatment of prostate cancer. Cancer 1993; 71(suppl):933–938.
37. Epstein JI, Walsh PC, Carmichael M, Brendler CB. Pathologic and clinical findings to predict tumor extent of nonpalpable (stage T1c) prostate cancer. JAMA 1994; 271(5):368–374.
38. Schröder FH, Damhuis RAM, Kirkels WJ, Koning HJ de, Kranse R, Nijs GHT, Blijenberg BG. European randomized study of screening for prostate cancer—the Rotterdam pilot studies. Int J Cancer 1996; 65:145–151.
39. Breslow N, Chan CW, Dhom G, Drury RAB, Franks LM, Gellei B, Lee YS, Lundberg S, Sparke B, Sternby NH, Tulinius H. Latent carcinoma of prostate at autopsy in seven areas. Int J Cancer 1977; 20:680–688.
40. Bostwick DG. Prostatic intraepithelial neoplasia (PIN). Urology 1989; 34 (6 suppl):16–22.
41. Kastendieck H, Helpap. Prostatic ''dysplasia/atypical hyperplasia''. Urology 1989; 34(6 suppl):28–42.
42. Adolfsson J. Deferred treatment of low grade stage T3 prostate cancer without distant metastases. J Urol 1993; 149:326–329.
43. Chodak GW, Ronald A. Comparison of cancer specific survival (CSS) following radical prostatectomy (RP) or watchful waiting (WW) based on two multi-institutional pooled analyses (abstr 995). J Urol 1996; 155 (suppl):559A.
44. McNeal JE. Cancer volume and site of origin of adenocarcinoma in the prostate: relationship to local and distant spread. Hum Pathol 1992; 23:258.
45. Epstein JI, Carmichael MJ, Partin AW, Walsh PC. Small high grade adenocarcinoma of the prostate in radical prostatectomy specimens performed for nonpalpable disease: pathogenetic and clinical implications. J Urol 1994; 151:1587–1592.
46. Bagshaw MA, Cox RS, Ramback JE. Radiation therapy for localized prostate cancer. Urol Clin North Am 1990; 17(4):787–802.
47. Jewett HJ, Bridge RW, Gray GF, Shelley WM. The palpable nodule of prostatic cancer. Results 15 years after radical excision. JAMA 1968; 203(6):403–406.
48. Partin AW, Lee BR, CarMichael M, Walsh PC, Epstein JI. Radical prostatectomy for high grade disease: reevaluation 1993. J Urol 1994; 151:1583–1586.
49. Zincke H, Bergstralh EJ, Blute ML, Myers RP, Barrett DM, Lieber MM, Martin SK, Oesterling JE. Radical prostatectomy for clinically localized prostate cancer: Long-term results of 1,143 patients from a single institution. J Clin Oncol 1994; 12(11):2254–2263, 1994.
50. Ackermann R, Müller HA. Der Wert der perinealen Stanzbiopsie zur Beurteilung des histologischen Differenzierungsgrades des Prostata-Carcinoms. Verhandl Deutsch Ges Urol 1981; 32:79–81.

51. Ouden D vd, Davidson PJT, Hop W, Schröder FH. Radical prostatectomy as a monotherapy for locally advanced (stage T3) prostate cancer. J Urol 1994; 151: 646–651.
52. Gaynor EP. Zur Frage des Prostatakrebs. Virchows Arch Pathol Anat 1939; 301: 602-652.
53. Holund B. Latent prostatic cancer in a consecutive autopsy series. Scand J Urol 1980; 14:29–35.
54. Sakr WA, Haas GP, Cassin BF, Pontes JE, Crissman JD. The frequency of carcinoma and intraepithelial neoplasia of the prostate in young male patients. J Urol 1993; 50:379–385.
55. Kabalin JN, McNeal JE, Price HM, Freiha FS, Stamey TA. Unsuspected adenocarcinoma of the prostate in patients ondergoing cystoprostatectomy for other causes: incidence, histology and morphometric observations. J Urol 1989; 141: 1091–1094.
56. Lowe BA, Liström MB. Incidental carcinoma of the prostate: an analysis of the predictors of progression. J Urol 1988; 140:1340–1344.
57. Stamey TA, Freiha FS, McNeal JE, Redwine EA, Whittemore AS, Schmid HP. Localized prostate cancer, relationship of tumor volume to clinical significance for treatment of prostate cancer. Cancer 1993; 71(Suppl):933–938.
58. Oesterling JE, Suman VJ, Zincke H, Bostwick DG. PSA-detected (clinical stage T1c or B0) prostate cancer. Urol Clin North Am 1993; 20:687–693.
59. Scaletscky R, Koch MO, Eckstein CW, Bicknell SL, Gray GF, Smith JA. Tumor volume and stage in carcinoma of the prostate detected by elevations in prostate specific antigen. J Urol 1994; 152:129–131.
60. Humphrey PA, Keetch DW, Smith DS, Shepard DL, Catalona WJ. Prospective characterization of pathological features of prostatic carcinomas detcted via serum prostate specific antigen based on screening. J Urol 1996; 155:816–820.
61. Davidson PJT, Hop W, Kurth KH, Fossa SD, Waehre H, Schröder FH, and the European Organization for Research and Treatment of Cancer Genitourinary Group. Progression in untreated carcinoma of the prostate metastatic to regional lymph nodes (stage T0 to 4, N1 to 3, Mo, D1). J Urol 1995; 154:2118–2122.
62. Bogdanowicz JFAT, Bentvelsen FM, Oosterom R, Schröder FH. Evaluation of prostate-specific antigen and prostatic acid phosphatase in untreated prostatic carcinoma and benign prostatic hyperplasia. Scand J Urol Nephrol Suppl 1991; 138: 97–103.
63. Bentvelsen FM, Schröder FH. Modalities available for screening for prostate cancer. Eur J Cancer 1993; 29A(6):804–811.
64. Bangma ChrH, Kranse R, Blijenberg B, Schröder FH. The value of screening tests in the detection of prostate cancer. Part I: Results of a retrospective evaluation of 1726 men. Urology 1995; 46(6):773–778.
65. Catalona WJ, Richie JP, Ahmann FR, Hudson MA, Scardino PT, Flanigan RC, DeKernion JB, Ratliff TL, Kavoussi LR, Dalkin BL, Waters WB, MacFarlane MT, Southwick PC. Comparison of digital rectal examination and serum prostate specific antigen in the early detection of prostate cancer: results of a multicenter clinical trial of 6,630 men. J Urol 1994; 151:1283–1290.

66. Pedersen KV, Carlsson P, Varenhorst E, Löfman O, Beglund K. Screening for carcinoma of the prostate by digital rectal examination in a randomly selected population. BMJ 1990; 300:1041–1043.
67. Labrie F, Dupont A, Suburu R, Cusan L, Tremblay M, Gomez J-L, Emond J. Serum prostate specific antigen as pre-screening test for prostate cancer. J Urol 1992; 147:846–852.
68. Mettlin C, Lee F, Drago J, Murphy GP. The American Cancer Society National Prostate Cancer Detection Project. Cancer 1991; 67:2949–2958.
69. Bangma ChrH, Kranse R, Blijenberg B, Schröder FH. The value of screening tests in the detection of prostatc cancer. Part II: Retrospective analysis of free/total prostate-specific analysis ratio, age-specific reference ranges, and PSA density. Urology 1995; 46(6):779–784.
70. Papsidero LD, Wang MC, Merrin C, Valenzuela LA, Murphy GP, Chu TM. A prostate antigen in sera of prostatic cancer patients. Proc Am Assoc Cancer Res 1979; 20:209.
71. Papsidero LD, Wang MC, Valenzuela LA, Murphy GP, Ming Chu T. A prostate antigen in sera of prostatic cancer patients. Cancer Res 1980; 40:2428–2432.
72. McCormack RT, Rittenhouse HG, Finlay JA, Sokoloff RL, Wang TJ, Wolfert RL, Lilja H, Oesterling JE. Molecular forms of prostate-specific antigen and the human kallikrein gene family: a new area. Urology 1995 45(5):729–744.
73. Riegman PH, Vlietstra RJ, Korput JA van der, Romijn JC, Trapman J. Characterization of the prostate-specific antigen gene: a novel human kallikrein-like gene. Biochem Biophys Res Comm 1989; 159:95.
74. Cooper CS, Maclndoe JH, Perry PJ, Yates WR, Williams RD. The effect of exogenous testosterone on total and free prostate specific antigen levels in healthy young men. J Urol 1996; 156:438–442.
75. Young CYF, Seay T, Hogen K, Charlesworth MC, Roche PC, Klee GG, Tindall DJ. Prostate-specific human kallikrein (hK2) as a novel marker for prostate cancer. The Prostate 1996; 7 (suppl):17–24.
76. Olsson CA, Vries GM de, Benson MC, Raffo A, Buttyan R, Cama C. The use of RT-PCR for prostate-specific antigen assay to predict potential surgical failures before radical prostatectomy: molecular staging of prostate cancer. Br J Urol 1996; 77:411–417.
77. Schröder FH, Albrecht W, Auvinen A, Bartsch G, Bowsher WG, Haillot O, Imai K, Mettlin C. Screening and early detection of prostate cancer. Committee 5, First International Consultation on Prostate Cancer. Murphy G, Denis L, Chatelain C, Griffiths K, Khoury S, Cockett AT, eds. United Kingdon: Scientific Communication International, 1997, pp. 179–188.
78. Catalona WJ, Smith DS, Ratliff TL, Basler JW. Detection of organ-confined prostate cancer is increased through prostate-specific antigen-based screening. JAMA 1993; 270(8):948–954.
79. Aus G, Bergdahl S, Frösing R, Lodding P, Pileblad E, Hugossson J. Reference range of prostate-specific antigen after transurethral resection of the prostate. Urology 1996; 47(4):529–531.
80. Christensson A, Björk T, Nilsson O, Dahlén U, Matikainen MT, Cockett ATK,

Abrahamsson PE, Lilja H. Serum prostate specific antigen complexed to α1-antichymotrypsin as an indicator of prostate cancer. J Urol 1993; 150:100–105.
81. Vessella RL, Noteboom J, Lange PH. Evaluation of the Abbott IMx[R] automated immunoassay of prostate-specific antigen. Clin Chem 1992; 38(10):2044–2054.
82. Prestigiacomo AF, Stamey TA. Physiological variation of serum prostate-specific antigen in the 4.0 to 10.0 ng/ml range in male volunteers. J Urol 1996; 155:1977–1980.
83. Komatsu K, Wehner N, Prestigiacomo AF, Chen Z, Stamey TA. Physiologic (intraindividual) variation of serum prostate-specific antigen in 814 men from a screening population. Urology 1996; 47:343–346.
84. Oesterling JE, Martin SK, Bergstrahl EJ, Lowe FC. The use of prostate-specific antigen in staging patients with newly diagnosed prostate cancer. JAMA 1993; 269(1):57–60.
85. Bosch JLHR, Hop WCJ, Bangma CH, Kirkels WJ, Schröder FH. Prostate specific antigen in a community-based sample of men without prostate cancer: correlations with prostate volume, age, body mass index, and symptoms of prostatism. The Prostate 1995; 27:241–249.
86. Carter HB, Pearson JD, Metter EJ, Brant LJ, Chan DW, Andres R, Fozard JL, Walsh PC. Longitudinal Evaluation of Prostate-Specific Antigen Levels in Men with and Without Prostate Disease. JAMA 1992; 267:2215–2220.
87. Stenman UH, Leinonen J, Alfthan H, Rannikko S, Tuhkanen K, Alfthan O. A complex between prostate-specific antigen and α1-antichymotrypsin is the major form of prostate-specific antigen in serum of patients with prostatic cancer: assay of the complex improves clinical sensitivity for cancer. Cancer Res 1991; 51:222–226.
88. Gann PH, Hennekens CH, Stampfer MJ. A prospective evaluation of plasma prostate-specific antigen for detection of prostate cancer. JAMA 1995; 273(4):289–294.
89. Parkes CJ, Wald NJ, Murphy P, George L, Watt HC, Kirby R, Knekt P, Helzlsouer KJ, Tuomilehto J. Prospective observational study to assess value of prostate specific antigen as screening test for prostate cancer. BMJ 1995; 311:1340–1343.
90. Schmid HP, McNeal JE, Stamey TA. Observations on the doubling time of prostate cancer: the use of serial prostate-specific antigen in patients with untreated disease as a measure of increasing cancer volume. Cancer 1993; 71:2031–2040.
91. Smith DS, Catalona WJ. Rate of change in serum prostate specific antigen levels as a method for prostate cancer detection. J Urol 1994; 152:1163–1167.
92. Brawer MK, Beatie J, Wener MH, Vessella RL, Preston SD, Lange PH. Screening for prostatic carcinoma with prostate specific antigen: results of the second year. J Urol 1993; 150:106–109.
93. Benson MC, Whang IS, Pantuck A, Ring K, Kaplan SA, Olsson CA, Cooner WH. Prostate specific antigen density: a means of distinguishing benign prostatic hypertrophy and prostate cancer. J Urol 1992; 147:815–816.
94. Catalona WJ, Hudson MA, Sardino PT, Richie JP, Ahmann FR, Flanigan RC, DeKernion JB, Ratliff TL, Kavoussi LR, Dalkin BL, Waters WB, MacFarlaine MT, Southwick PC. Selection of optimal prostate specific antigen cutoffs for early

detection of prostate cancer: receiver operating characteristic curves. J Urol 1994; 152:2037–2042.
95. Littrup PJ, Kane RA, Mettlin CJ, Murphy GP, Lee F, Toi A, Badalament R, Babaian R, the investigators of the American Cancer Society National Prostate Cancer Detection Project. Cost-effective prostate cancer detection, reduction of low-yield biopsies. Cancer 1994; 74:3146–3158.
96. Ohori M, Dunn JK, Scardino PT. Is prostate-specific antigen density more useful than prostate-specific antigen levels in the diagnosis of prostate cancer? Urology 1995; 46:666–671.
97. Brawer MK, Araburu EΛG, Chen GL, Preston SD, Ellis WJ. The inability of prostate specific antigen index to enhance the predictive value of prostate specific antigen in the diagnosis of prostatic carcinoma. J Urol 1993; 150:369–373.
98. Catalona WJ, Smith DS, Wolfert RL, Wang TJ, Rittenhouse HG, Ratliff TL, Nadler RB. Evaluation of percentage of free serum prostate-specific antigen to improve specificity of prostate cancer screening. JAMA 1995; 274(15):1214–1220.
99. Van Cangh PJ, De Nayer P, Sauvage P, Tombal B, Elsen M, Lorge F, Opsomer R, Wese FX. Free to total prostate-specific antigen (PSA) ratio is superior to total-PSA in differentiating benign prostate hypertrophy from prostate cancer. The Prostate 1996; 7(suppl):30–34.
100. Prestigiacomo AF, Stamey TA. Can free and total prostate specific antigen and prostatic volume distinguish between men with negative and positive systematic ultrasound guided prostate biopsies? J Urol 1997; 157:189–194.
101. Björk T, Piironen T, Pettersson T, Lövgren T, Stenman UH, Oesterling JE, Abrahamsson PA, Lilja H. Comparison of analysis of the different prostate-specific antigen forms in serum for detection of clinically localized prostate cancer. Urology 1996; 48:882–888.
102. Vashi AR, Wojno KJ, Henricks W, England BA, Vessella RL, Lange PH, Wright GL, Schellhammer PF, Weigand RA, Olson RM, Dowell BL, Borden KK, Oesterling JE. Determination of the ''reflex range'' and appropriate cutpoints for percent free prostate-specific antigen in 413 men referred for prostatic evaluation using the AxSYM system. Urology 1997; 49:17–27.
103. Prestigiacomo AF, Lilja H, Pettersson K, Wolfert RL, Stamey TA. A comparison of the free fraction of serum prostate specific antigen in men with benign and cancerous prostates: the best scenario. J Urol 1996; 156:350–354.
104. Lerner SE, Jacobsen SJ, Lilja H, Bergstrahl EJ, Ransom J, Klee GG, Piironen T, Blute ML, Lieber MM, Zincke H, Pettersson K, Peterson D, Oesterling JE. Free, complexed, and total serum prostate-specific antigen concentrations and their proportions in predicting stage, grade and deoxyribonucleic acid ploidy in patients with adenocarcinoma of the prostate. Urology 1996; 48:240–248.
105. Hugosson J. The value of the ratio of free/total PSA in screening for prostate cancer. Personal communication, 1996.
106. Bangma CH, Kranse R, Blijenberg BG, Schröder FH. Retrospective analysis of free/total PSA ratio, PSAD, and age specific reference ranges for detecting prostate cancer in a screening population (abstr 244). J Urol 1996; 155; 5(suppl):317A.
107. Terris MK, McNeal JE, Stamey TA. Detection of clinically significant prostate

cancer by transrectal ultrasound-guided systematic biopsies. J Urol 1992; 148:829–832.
108. Daneshgari F, Taylor GD, Miller GJ, Crawford ED. Computer simulation of the probability of detecting low volume carcinoma of the prostate with six random systematic core biopsies. Urology 1995; 45(4):604–609.
109. Eskew LA, Bare RL, McCullough DL. Systematic 5 region prostate biopsy is superior to sextant method for diagnosing carcinoma of the prostate. J Urol 1997; 157:199–203.
110. Ravery V, Szabo J, Toublanc M, Boccon-Gibod LA, Billebaud T, Hermieu JF, Delmas V, Boccon-Gibod L. A single positive prostate biopsy in six does not predict a low-volume prostate tumour. B J Urol 1996; 77:724–728.
111. Dugan JA, Bostwick DG, Myers RP, Qian J, Bergstrahl EJ, Oesterling JE. The definition and preoperative prediction of clinically insignificant prostate cancer. JAMA 1996; 275(4):288–294.
112. Keetch DW, Catalona WJ, Smith DS. Serial prostatic biopsies in men with persistently elevated serum prostate specific antigen values. J Urol 151(6): 1994; 1571–1574.
113. Mettlin C, Murphy GP, Babaian RJ, Chesley A, Kane RA, Littrup PJ, Mostofi FK, Ray PS Shanberg AM, Toi A. The results of a five-year early prostate cancer detection intervention. Cancer 1996; 77:150–159.
114. Labrie F, et al. Optimized strategy for detection of early stage, curable prostate cancer: role of prescreening with prostate specific antigen. Clin Invest Med 1994; 16(6):426–441.
115. Brewster SF, Kemple T, Maclver AG, Astley JP, Gingell JC. The Bristol prostate cancer pilot screening study—a 3 year follow-up. Br J Urol 1994; 74:556–558.
116. Haas, GP, Montie JE, Pontes JE. The state of prostate cancer screening in the United States. Eur Urol 1993; 23:337–347.
117. Denis LJ, Murphy GP, Schröder FH. Report of the Consensus Workshop on Screening and Global Strategy for Prostate Cancer. Cancer 1995; 75(5):1187–1207.
118. Schröder FH. Screening, early detection, and treatment of prostate cancer: a European view. Urology 1995; 46:62–70.
119. Hudson MA. Screening for prostate cancer. AUA Update Series 1996; 15(lesson 25):198–203.
120. Menon M, Parulkar BG, Bakker S. Should we treat localized prostate cancer? An opinion. Urology 1995; 46(5):607–616.
121. Scardino PT, Weaver R, Hudson MA. Early detection of prostate cancer. Hum Pathol 1992; 23(3):211–222.
122. Cooner WH, Mosley BR, Rutherford CJ Jr, Beard JH, Pond HS, Terry WJ, Igel TC, Kidd DD. Prostate cancer detection in a clinical urological practice by ultrasonography, digital rectal examination and prostate specific antigen. J Urol 1990; 143:1146–1154.
123. Mettlin C, Lee F, Drago J, Murphy GP. The American Cancer Society National Prostate Cancer Detection Project. Cancer 1991; 67:2949–2958.
124. Brawer MK, Chetner MP, Beatie J, Buchner DM, Vessella RL, Lange PH. Screen-

ing for prostatic carcinoma with prostate specific antigen. J Urol 1992; 147:841–845.
125. Gustafsson O, Norming U, Almgard LE, Fredrikssson A, Gustavsson G, Harvig B, Nyman CR. Diagnostic methods in the detection of prostate cancer: a study of a randomly selected population of 2,400 men. J Urol 1992; 148:1827–1831.
126. Crawford ED, DeAntoni EP, Etzioni R, Schaefer VC, Olson RM, Ross CA, the Prostate Cancer Education Council. Serum prostate-specific antigen and digital rectal examination for early detection of prostate cancer in a national community-based program. Urology 1996; 47:863–869.
127. Reissigl A, Pointner J, Horninger W, Ennemoser O, Strasser H, Klocker H, Bartsch G. Comparison of different prostate-specific antigen cutpoints for early detection of prostate cancer: results of a large screening study. Urology 1995; 46(5):662–665.
128. Fleming C, Wasson JH, Albertsen PC, Barry MJ, Wennberg JE, for the Prostate Patient Outcomes Research Team. A decision analysis of alternative treatment strategies for clinically localized prostate cancer. JAMA 1993; 269:2650–2658.
129. Eddy DM. Finding cancer in asymptomatic people: estimating the benefits, costs, and risks. Cancer 1983; 51:2440–2445.
130. Krahn MD, Mahoney JE, Eckman MH, Trachtenberg J, Pauker SG, Detsky AS. Screening for prostate cancer: a decision analytic view. JAMA 1994; 272(10): 779–790.
131. Gohagan JK, Prorok PC, BS Kramer, Cornet JE. Prostate cancer screening in the prostate, lung, colorectal and ovarian cancer screening trial of the National Cancer Institute. J Urol 1994; 152:1905–1909.
132. Auvinen A, Rietbergen JBW, Denis LJ, Schröder FH, Prorok PhC. for the International Prostate Cancer Screening Trial Evaluation Group. Prospective evaluation plan for randomised trials of prostate cancer screening. J Med Screening 1996; 3: 97–104.
133. Auvinen A, Calais da Silva F, Denis LJ, Hugosson J, Schröder FH, on behalf of the Scientific Committee ERSPC and all other associates of the study. The European Randomized Study of Screening for Prostate Cancer (ERSPC): international co-operation and preliminary data. In: Schröder FH, ed. Recent Advances in Prostate Cancer and BPH, Proceedings of the IV Congress on Progress and Controversies in Oncological Urology (PACIOU IV), held in Rotterdam, The Netherlands, April 11–13, 1996. New York: Parthenon Publishing, 1996:167–172.
134. Nijs HGT, Tordoir DMR, Schuurman JH, Kirkels WJ, Schröder FH. Randomised trial of prostate cancer screening in the Netherlands: assessment of acceptance and motives for attendance. J Med Screening 1997; 4:102–106.
135. Nijs HGT, De Koning HJ, Essink-Bot ML, Kirkels WJ, Schröder FH. Acceptability of two community programmes for prostate cancer screening. In press.
136. Rietbergen JBW, Boeken Kruger AE, Kranse R, Schröder FH. Complications of ultrasound-guided transrectal core biopsies of the prostate. Submitted to Urology, 1997.
137. Aus G, Ahlgren G, Bergdahl S, Hugossson J. Infection after transrectal core bi-

opsies of the prostate—risk factors and antibiotic prophylaxis. Br J Urol 1996; 77:851–855.
138. Fowler FJ, Barry MJ, Lu-Yao G, Wasson J, Roman A, Wennberg J. Effect of radical prostatectomy for prostate cancer on patient quality of life: results from a medicare survey. Urology 1995; 45(6):1007–1015.
139. Litwin MS, Hays RD, Fink A, Ganz PA, Leake B, Leach GE, Brook RH. Quality-of-life outcomes in men treated for localized prostate cancer. JAMA 1995; 273(2): 129–135.
140. Otnes B, Harvei S, Fossa SD. The burden of prostate cancer from diagnosis until death. Br J Urol 1995; 76:587–594.
141. Pedersen KV, Carlsson P, Varenhorst E, Löfman O, Beglund K. Screening for carcinoma of the prostate by digital rectal examination in a randomly selected population. BMJ 1990; 300:1041–1043.
142. Newling DWW. Cost-effective diagnosis and management in patients with prostate cancer. Reviews on Endocrine-Related Cancer 1992; 41:23–30.
143. Benoit RM, Naslund MJ. An economic rationale for prostate cancer screening. Urology 1994; 44(6):795–803.
144. Littrup PJ, Goodman AC, Mettlin CJ, Murphy GP, the investigators of the American Cancer Society National Prostate Cancer Detection Project. Cost analyses of prostate cancer screening: frameworks for discussion. J Urol 152:1873–1877.
145. Lubke WL, Optenberg SA, Thompson IM. Analysis of the first-year cost of a prostate cancer screening and treatment program in the United States. J Natl Cancer Inst 1994; 86(23):1790–1793.
146. Woolf SH. Public health perspective: the health policy implications of screening for prostate cancer. J Urol 1994; 152:1685–1688.
147. Handley MR, Stuart ME. The use of prostate specific antigen for prostate cancer screening: a managed care perspective. J Urol 152:1689–1692.
148. American Cancer Society. Guidelines for the cancer-related checkup: an update. Atlanta: American Cancer Society, 1993.
149. Mettlin C, Jones G, Averette H, Gusberg SB, Murphy GP. Defining and updating the American Cancer Society guidelines for the cancer-related checkup: prostate and endometrial cancers. Ca Cancer J Clin 1993; 43:42–46.
150. American Urological Association. Executive Committee report. Baltimore: American Urological Association, January 1992.
151. Coley CM, Barry MJ, Fleming CC, Fahs MC, Mulley AG (American College of Physicians). Clinical guidelines: early detection of prostate cancer, estimating the risks, benefits, and costs (Part II). Ann Intern Med 1997: 126:468–479.
152. Coley CM, Barry MJ, Fleming CC, Fahs MC, Mulley AG (American College of Physicians). Clinical guidelines: screening for prostate cancer (Part III). Ann Intern Med 1997; 126:480–484.
153. Canadian Task Force on the Periodic Health Examination. Canadian guide to clinical preventive health care. Ottawa: Canada Communication Group, 1994:812–823.
154. Woolf SH. U.S. Preventive Services Task Force. Screening for prostate cancer: commentary on the recommendations of the Canadian Task Force on the Periodic Health Examination. Am J Prev Med 1994; 10:187–193.

155. Woolf SH. Screening for prostate cancer. In: Screening. Guide to Clinical Preventive Services. 2d ed. Report of the U.S. Preventive Services Task Force. Washington, DC: U.S. Department of Health and Human Services, 1995:119–134.
156. Chamberlain J, Melia J, Moss S, Brown J. The diagnosis, management, treatment and costs of prostate cancer in England and Wales. Report prepared for the Health Technology Assessment Panel of the NHS Executive. Sutton, UK, October 1996.
157. Proeve Bevolkingsonderzoek Prostaatcarcinoom. Gezondheidsraad Nederland, 1997. (Dutch Government Publication.)

19

Gastric Cancer

D. Maxwell Parkin and Paola Pisani
International Agency for Research on Cancer, Lyon, France

I. EPIDEMIOLOGY AND BURDEN OF DISEASE

Stomach cancer was the second most common cancer in the world in 1985, with an estimated 755,000 new cases and 620,000 deaths, about two-thirds of which occurred in the developing countries. The areas of highest incidence and mortality rate are in eastern Asia (e.g., Japan, Korea, China) where it is the most common cancer of men, as well as in parts of South America (particularly in the Andean regions) and eastern Europe (1,2). Incidence in men is about twice that of women in both high-and low-risk countries, although close inspection of age-specific incidence data shows that rates in women exceed those of men in the youngest age groups (younger than ages 35–44 years).

Survival from stomach cancer is generally poor. In the United States, the overall 5-year relative survival is about 20% (17% in men, 25% in women) (3), and figures are similar in European countries (19%), ranging from around 11% in the United Kingdom to 27% in Saarland, Germany (4). Survival in developing countries is lower, around 7%–17% (5), probably because of late presentation of disease rather than poor results of therapy. In Japan, survival is considerably higher, even in population-based series such as that from Osaka where it was 39% (6). In part, this is the result of much earlier diagnosis (or lead time bias) in Japan, because stage-specific survival rates are not markedly superior to those in the United States.

In general, incidence and mortality rates from stomach cancer are declining, with very few exceptions (7,8). Analysis of Japanese data suggested that the decline in incidence was principally due to declines in the intestinal type of gastric carcinoma, itself the component most responsible for geographical variations in risk (9). Similar, although less clear, changes were seen in U.S. data (8), but there were no clear differences in rate of change by histological subtype in Sweden (10).

II. NATURAL HISTORY OF DISEASE

The pathological changes that are assumed to precede cancer development are chronic atrophic gastritis, metaplasia, and dysplasia. Their natural history was studied in a follow-up study of 1400 asymptomatic subjects in an area of high gastric cancer risk in Colombia (11). Endoscopy and biopsy were performed twice, at an average interval of 5 years, and showed an overall progression from less to more advanced lesions (although some regression also occurred). Atrophic gastritis preceded gastric cancer in a Japanese prospective study (12), which reported a relative risk of 2.3 of development of cancer for those affected by severe atrophy. The less severe types of lesions, nonatrophic as well as mild atrophic gastritis, have also been shown to be more common in cases of gastric cancer than in control subjects (13).

The risk factors for gastric cancer are likely to act at different stages of this process, by modifying progression or regression of precancerous lesions. Infection with *Helicobacter pylori*, as well as a high intake of irritants such as salt and nitrates, causes chronic inflammation and superficial gastritis, which can develop into atrophic gastritis (14). A salty diet induces atrophy in experimental animals (15). That dietary salt also influences the more advanced pathological changes of the gastric mucosa was demonstrated in the prospective study conducted in Japan where the risk of progression of atrophic gastritis to cancer was reduced in subjects who diminished their intake of salted food (12). Reduced acid secretion in the presence of atrophy increases gastric pH, thus favoring the growth of anaerobic bacteria, which reduce nitrate to nitrite, and can eventually form N-nitroso mutagens (14).

A high intake of fruits and vegetables rich in vitamin C and beta-carotene has consistently been associated with a reduced risk of gastric cancer in both high-and low-risk populations (16). These antioxidants prevent the formation of mutagens, carcinogens, and reactive oxygen metabolites in the stomach. Ascorbic acid is normally secreted into the stomach, but this is significantly decreased in the presence of chronic gastritis (17). The exact role of these vitamins in the etiology of gastric cancer is not known and it may be that low ascorbic acid in gastric juice is simply a marker of the extension and severity of the damage that precedes cancer.

The role of *H. pylori* in causing superficial and chronic gastritis is well documented, but its role in the more advanced precancerous lesions, that is, after the development of severe atrophy, is the subject of controversy. The development of atrophy seems to be a fundamental step in the carcinogenic process, and once it has occurred, eradication of the bacterial infection does not induce regression (18). It appears that the infection acts by producing chronic inflammation and an enhanced rate of transition to atrophy. Strains that have the gene encoding the cytotoxin Cag-C protein cause the most damage to the gastric mucosa in mice (19). In humans, subjects infected with Cag-A–positive strains had a higher prevalence of atrophic gastritis (and, probably, a higher risk of development of atrophic gastritis) and higher levels of IgG anti-HP antibodies than those with Cag-A–negative infections (20).

The question as to whether all the early cancers detected by screening might not have manifested themselves if they had not been so detected (overdiagnosis) is important in evaluating trends in incidence and mortality rate after the introduction of screening. Tsukuma et al. (21) identified a series of 56 early gastric cancer cases (22 detected by screening) who received no treatment for at least 6 months. Forty-three cases had some follow-up (diagnostic tests, late operation) and advanced gastric cancer occurred in 27 (63%); the median duration as early gastric cancer after diagnosis was 37 months. This suggests that most such cancers eventually progress, although with the relatively long time period by which screening advances the diagnosis of cancer, there is likely to be something of an excess incidence in a screened population, because of competing causes of death in individuals with undiagnosed cancers in the unscreened group.

III. EARLY DETECTION TESTS

A. Tumor Markers

Fetal sulfoglycoprotein antigen (FSA) is produced by a proportion of gastric cancers and its detection in gastric juice has been proposed as a screening test for early cancer (22). In a large-scale screening program in 46 municipalities of southern Finland, involving almost 40,000 examinations, some 9% of individuals were found to be positive (specificity $\approx$ 91%). After 1 year of follow-up, sensitivity was 73%, whereas after 3 years, it had decreased to 47%. The relative risk for cancer within 3 years, for individuals who were FSA positive versus FSA negative, was around 9 (23).

In a case-control study nested in a cohort of subjects, 94 stomach cancer cases were matched with 188 controls (24). Sera drawn up to 12 years previously were tested for various tumor markers (CEA, CA 19-9, CA50, TATI). The only difference observed was for CA 19-9 in specimens drawn up to 1 year before diagnosis, but sensitivity (74%) and specificity (73%) were modest.

B. Photofluorography

Photofluorography involves the use of double-contrast radiology to outline small lesions of the gastric mucosa, with detection by radiography, either directly or under an image intensifier (generally involving six or more views). Follow-up of suspicious lesions is by gastroscopy, with biopsy or cytology being done when abnormal areas are visualized (25). This technique was introduced in Japan in the early 1960s and has been extensively used since then, the target of the program being to examine 30% of the population aged 40 years and older each year (26). From the Japanese experience (25,27), it seems that some 13%–23% of subjects will require follow-up gastroscopies (specificity 77%–87%). In one study in Japan, 14 of 17 cancers diagnosed within 1 year of screening radiography were detected in subjects screening "positive" (although for three, the lesion was not thought to be cancer), corresponding to a maximum sensitivity of 82% and positive predictive value of 1.5% (27).

Murakami et al. (28) followed up subjects participating in the screening program conducted by the Centre for Adult Diseases in Osaka, by linkage with the Osaka Cancer Registry. False-negative tests were defined as subjects registered with cancer within 1 year of a negative test. With direct radiography, sensitivity was 90.8% and specificity 91.4% (PPV 2.3%), whereas with photofluorography by an image intensifier, they were 88.5% and 92%, respectively (PPV 1.4%).

C. Antibody to *Helicobacter pylori* and Serum Pepsinogens

The progression of changes observed in the gastric mucosa, from chronic superficial gastritis to atrophic gastritis, intestinal metaplasia and cancer are associated with changes in serum levels of pepsinogens, proenzymes of pepsin (29–31). Pepsinogen I (PGI) is produced in the gastric fundus (chief cells of gastric glands and mucous neck cells), and PGII in the antrum. Hence, atrophic gastritis of the fundic mucosa is associated with decreased serum levels of PGI, whereas PGII is reduced much less, or not at all. Thus, chronic atrophic gastritis is associated with a decrease in PGI and a reduction in the ratio of PGI to PGII (32–34). Prospective studies have demonstrated that low PGI levels are associated with an increased risk of gastric carcinoma (35–37).

Helicobacter pylori is recognized by IARC as being carcinogenic to humans, because of its association with carcinoma of the stomach (16). Several prospective studies have reported an association between the presence of antibody to *H. pylori* in serum, and the risk of a subsequent gastric cancer (37–40), although two studies in Chinese populations have been negative (41,42).

These findings have raised the possibility that serological tests might be valuable in screening, either as an alternative to radiological screening or to identify high-risk groups in whom other tests (e.g., radiography) could be concentrated.

Serum pepsinogens seem to be the more promising, because the changes in levels are most likely to be associated with severe atrophic gastritis and the presence of an early gastric cancer. In prospective studies involving follow-up of subjects with stored serum specimens, nested case-control studies suggest that changes in PG levels are most marked close to the time of diagnosis of stomach cancer. Thus, Aromaa et al. (37) in a follow-up of 39,268 individuals aged 15 years or older in Finland, found that a PGI level below 50 μg/I was associated with diagnosis of gastric cancer in the next 8–14 years (odds ratio [OR] 2.7, 95% CI = 1.3–5.3), and the association was strongest for cases diagnosed within 2 years (OR 6.3, 95% CI = 1.3–29.9). Stemmerman et al. (43), in a follow-up of 7498 Japanese-Americans in Hawaii, found that the PGI:PGII ratio had better sensitivity for predicting later gastric cancer than PGI levels alone. With a ratio <2, sensitivity was 48% and specificity 88%, but the association was strongest for cancers diagnosed within 4 years of testing, and for those diagnosed at more advanced stages (III and IV).

The performance of pepsinogen as a screening test has been evaluated in Japan. Miki et al. (44) screened 4647 company employees (mean age 49 years) by measuring serum pepsinogen levels, and submitting those with PGI levels <50 μg/I together with a PGI:PGII ratio of <3.0 to follow-up endoscopy. Overall, 18.8% of subjects were positive (29% aged > 50 years) and four gastric cancers were detected (PPV = 0.5%). Although there was no proper evaluation of sensitivity, the authors imply that it could have been equal to that of X-ray screening, because the detection rate (0.86 per 100) was somewhat better than mass screening in the same company, using X-ray study (0.66 per 1000). However, the comparability of the screened groups, even in terms of age, was unclear, and the PG test represented a prevalent screen, rather than a mixture of new and repeat tests, as for X-ray tests. In a later study, using the same methods and criteria of positivity, 18.7% of the 4876 subjects who were screened were positive, and a total of 12 cases of gastric cancer were found in this group, representing a predictive value of 1.3% (45). Six cases were found amongst those screening negative within 1 year, although follow-up was confined to scrutiny of medical records and insurance claims. With this limited definition, sensitivity was estimated at 66.7% and specificity as 81.5%.

As far as antibody to *H. pylori* is concerned, the relative risk for subsequent gastric cancer is not high, and in prospective studies using serum banks, its risk is related to the length of time before cancer diagnosis, being lowest in the preceding 5 years (46). This probably results from the lower levels of infection with *H. pylori* in the face of progressive gastric atrophy.

Anti-*H. pylori* antibody would not, therefore, provide a useful screening test, nor would it be helpful in identifying high-risk groups for surveillance by other screening modalities. A substantial percentage of the normal population (in whom gastric cancer will never develop) are antibody positive, so that reasonable sensitivity for subsequent cancer can only be obtained at the cost of low specificity and predictive values. Parsonnet et al. (39), testing for anti-HP

IgG in a cohort of 129,000 subjects obtained a sensitivity of 84% and specificity of 40% for adenocarcinomas. Nomura et al. (40), testing for IgG antibody, obtained a sensitivity of 94.5% and specificity of 24% with any positive test, and sensitivity of 39% and specificity of 74% with titers greater than 2.2. In the study of Aromaa et al. (37), using IgA antibody, sensitivity at the 75 percentile of values in normal subjects (75% specificity) was 23% and was 57% at the median value (specificity = 50%)

Parsonnet et al. (36) observed that the relative risk of distal gastric cancer was highest in persons both positive for anti-*H. pylori* and with PGI levels below 50 ng/I; however, few cases would be identified by this combination (only 38%), although 89% of normal individuals would be negative to one or both tests.

Based on the assumption of a causal association between *H. pylori* infection and gastric cancer, the possibility of treatment of infected individuals with established anti-*H. pylori* antibiotic regimens as a preventive measure for gastric cancer has been suggested. In this scenario, ''screening'' refers to the preliminary identification of *H. pylori* positive subjects who would receive therapy, under the assumption that some 50%–60% of gastric cancer cases would occur in this group (47). We are a long way from knowing the efficacy of such an intervention in the prevention of cancer, and it is likely that more specific strategies will emerge in the meantime.

IV. SCREENING STUDIES OF THE VARIOUS TESTS

A. Fetal Sulphoglycoprotein Antigen

Hakama and Pukkala (23) describe a nonrandomized study, in which individuals aged 40–70 years in 46 rural municipalities in southwestern Finland were offered screening. Almost 40,000 individuals (75%) accepted and received at least two tests between 1975 and 1983. The effectiveness of the screening was evaluated by examining incidence and mortality rates from stomach cancer in the screening municipalities, in comparison with those in other (unscreened) rural municipalities in southwestern Finland as well as the whole country. There was no indication of any effect of screening on mortality rate, although study power to detect small differences was considered to be low.

B. Photofluorography

1. *Randomized Controlled Trials*

Only one trial has been described (48). In 1985, 39 municipalities in Miyagi prefecture, Japan, were randomized to two groups, A and B. In group A, individuals aged 50–59 years received an invitation to attend gastric cancer screening, which had been specially written by the head of local government; in group

B, the invitation letter was sent to individuals aged 60–69 years. The objective was to follow up mortality rates in the intervention and control populations. However, results published to date show that compliance with screening was higher in intervention than control areas only in the first year, especially in the 50–59 year age group (about 30% versus 15%), and had largely disappeared in subsequent years. It is inconceivable that, with such a small difference in screening experience between the intervention and control groups, any difference in mortality rate will be demonstrable.

2. *Nonrandomized Interventions*

a. Cohort Design. In a study in Osaka prefecture, Oshima et al. (49) used the cancer registry to follow up 33,000 individuals screened at least once during 1967–1970 for an average of 6.1 years. The number of gastric cancer cases diagnosed in the cohort was 1.37 times higher than expected, based on the general population rates, whereas there was a modest reduction in the number of deaths (Observed:Expected = 0.91 for all age groups). Some of the excess incidence may be due to earlier diagnosis of cases, although with a follow-up extending for a minimum of 5 years after screening, this is unlikely to account for much of the excess. The explanation is more probably overdiagnosis of gastric cancer (detection of cases in the screened group that would not have otherwise been diagnosed during the life span of the individual). It is less likely to be the result of cancer-prone individuals choosing to be screened, because the number of other cancers diagnosed in the cohort was only 0.64 that expected (a ''healthy screenee'' effect). A survey in Miyagi prefecture by Fukao et al. (50) observed that, compared to nonparticipants, those taking part in the gastric cancer screening program had a lower prevalence of smoking and higher intakes of dietary items such as milk and fruit, which are usually associated with a reduced risk of gastric cancer.

b. Case-Control Design. Three case-control studies have been reported from Japan (Table 1). The first study (51) was conducted in Nose Town (Japan), an area covered by the population-based cancer registry of Osaka, which allowed the identification of all residents who died of stomach cancer in the period 1969–1981. A total of 87 deaths from gastric cancer (cases) and three times as many control subjects were included in the analysis. Control subjects were live individuals, matched to the cases by sex, age, and residence. The odds ratio for screened versus nonscreened subjects was 0.59 (95% CI = 0.34–1.04) in men and 0.38 (95% CI = 0.18–0.78) in women. The protective effect was greater for those who had more than one examination (the linear trend in the ORs was significant) and seemed somewhat greater also for those who had a test within 2 years of the date of diagnosis of the case.

Table 1 Retrospective Studies Evaluating Efficacy of Gastric-Cancer Screening Using Photofluorography

Reference and country	Cases	Controls	OR (95% confidence interval)	Exposure definition
Oshima et al., 1986, Japan (51)	87 deaths from gastric cancer	261 live controls. Matched by sex, age, and residence	0.59 (0.34–1.04)*—men 0.38 (0.16–0.78)*—women	Any test
Fukao et al., 1987, Japan (52)	367 advanced gastric cancers	367 live controls. Matched by sex, age, and residence	0.34 (0.25–0.48)	Any prediagnosis test
Fukao et al., 1995, Japan (53)	198 deaths from gastric cancer	577 live controls. Matched by sex, age, and residence	0.41 (0.28–0.61)—at least 1 0.22 (0.09–0.51)—4–5 tests	Tests within preceding 5 years, and >6 months prediagnosis
Pisani et al., 1994, Venezuela (54)	241 deaths from gastric cancer	2410 live controls. Matched by sex, age, and residence	1.26 (0.83–1.91)	Any test >6 months prediagnosis
	85 ever-screened only	375 ever-screeened only	0.25 (0.12–0.51)	Any test >6 months prediagnosis

*90% C.I.

The same sampling design was adopted in a second study, which compared screening histories of 367 cases of advanced stomach cancer with the same number of controls matched for sex, age, and residence (52). A significant protective effect was detected up to 3 years since the last negative test: OR=0.34 95% CI 0.25–0.48).

In a third study, Fukao et al. (53) compared 198 dead cases and 577 live control subjects matched by sex, age, and residence (district) with respect to their screening histories in the 5 years preceding the diagnosis of the case. The authors considered that this was the relevant exposure period, according to the results of their earlier case-control study. Recent examinations were associated with an odds ratio of 0.41 (95% CI = 0.28–0.61). No significant difference was observed by year since last examination within the 5 years considered, but the OR declined to 0.12 (95% CI 0.07–0.59) for those who had four or five tests compared to none.

There has been little systematic screening for gastric cancer outside Japan. However, a program using the same methodology was introduced in Tachira state, Venezuela, in 1980 (54), during which two mobile units were used, and screening was offered to volunteers aged 35 and older. The 241 persons who died of stomach cancer in the period 1985–1989 were compared with 2410 control subjects drawn from the electoral rolls of the same district of residence as the cases (matched by sex and age). An excess risk of dying of stomach cancer was observed for those who had tests occurring within the 6 months preceding the diagnosis of the case, and no benefit from screening was detected when such recent examinations were excluded from the screening histories of both cases and control subjects. This result suggested that the screening program was being used primarily by symptomatic individuals at high risk of death from gastric cancer (selection bias). In order to control for this, an analysis confined to subjects (cases and controls) who had attended for screening at least once was performed, by ignoring tests taking place after 1 month before diagnosis of the case (and the same date for the matching control subjects of that case). The odds ratio associated with tests before this date was 0.47 (C.I. 0.24–0.98), consistent with a protective effect. The strong negative selection bias is probably a consequence of the low population coverage—only 16% of the control subjects had ever been screened. In contrast, the programs in Japan reach a much larger proportion of the population, for example, 69% ever-screened in Osaka in 1983 (51) and 78% screened in the preceding 5 years in Miyagi in the mid-1980s (50).

c. Descriptive Studies. Mass screening was introduced in Japan after 1960 and was followed by a progressively greater proportion of cases of gastric cancers being diagnosed at an early stage of disease (48). Mortality rates from stomach cancer in Japan have been in decline in most age groups for many years (55). When examined for different areas of the country, the rate of decline appeared to be associated with the intensity of screening (56).

Hisamichi et al. (48) have analyzed time trends of incidence and mortality are for stomach cancer during the period 1960–1985 in Miyagi prefecture, for which population morbidity and mortality statistics are available. Both measures follow the general trend toward a decline of the disease, but, after 1970, the rate of decline was greater for mortality, particularly in the age groups (50–79 years) that are the principal target of the screening program. The greater decline in mortality rate than in incidence must be related to improving survival from gastric cancer since 1970, and it has been suggested that this was a consequence of the introduction of mass screening. It is possible, however, that incidence has been somewhat inflated by detection of cancers (by screening), which would not otherwise have been diagnosed within the life span of some individuals and that this might, in part, account for the differing trends between mortality and incidence.

V. RISK ASSOCIATED WITH SCREENING

The principal screening test used to date—radiographic examination following ingestion of barium—has the potential hazard of exposure to radiation, although modern equipment and image intensification have lowered exposure doses dramatically. Fukao et al. (57) conducted a population-based case-control study of leukemia in Miyagi prefecture. Exposure to screening X-rays (evaluated from records of the screening program) was similar in leukemia cases and control subjects, suggesting no radiation hazard (95% CI for those screened at least once within 10 years was 0.61–1.96).

With the specificity of X-ray screening (80%–90%) or screening with pepsinogens (75%–88%), a substantial number of persons without gastric cancer will be subjected to diagnostic investigation; usually by gastroscopy. This procedure is associated with complications such as perforation (usually of the cervical esophagus) in 0.02%–0.11% of cases and hemorrhage, usually following biopsy, in 0.01%–0.03%, with an overall mortality rate of 0.006%–0.01% (58,59). Presumably, these complications would occur with the lower frequencies in relatively healthy subjects who are being followed up for abnormal X-ray findings.

VI. COST/COST EFFECTIVENESS

Most analyses of cost effectiveness of X-ray screening have appeared in the Japanese literature. Hisamichi *et al.* (48) summarized the results of earlier cost-effectiveness analyses in Miyagi and presents marginal costs (screening program versus none) per "lives saved" for the Japanese program, assuming 85% sensitivity and 90% specificity of the examination, and 70% survival of screen-detected cancer versus 25% survival in those diagnosed with symptoms. The figure derived ($55,000 per life saved) is not compared with any other health care strategy. However, the same group (60) suggested that the cost-effectiveness of gastric cancer screening is superior to screening for colorectal cancer by fecal occult blood testing with or without total colonoscopy (cost per life saved 5 times more for men, 2–3 times more for women). Most of this difference can be accounted for by the incidence of gastric cancer being considerably higher than that of colon cancer in Japan. In any case, the evidence for the effectiveness of X-ray screening in preventing gastric cancer is based on observational studies only, whereas, for colorectal cancer, the issue is more certain, in consequence of the results of randomized trials (61,62).

VII. CONCLUSION

In general, the consensus view is that screening for gastric cancer cannot be recommended as public health policy in countries other than Japan (63). This conclusion stems in part from the absence of any randomized controlled trials identifying the efficacy of screening, although, as described previously, there is a considerable body of evidence from observational studies suggesting that regular screening of high-risk populations by X-ray study may be capable of reducing mortality rate from the disease. Nevertheless, it seems highly unlikely that, in countries other than Japan, which have the necessary resources to devote to X-ray screening, gastric cancer would be considered a priority. In most if these countries, incidence and mortality rates are already low and declining, and there is a general belief that concentration on primary prevention methods is more appropriate, even though there may be debate about whether the most appropriate approach is through dietary change or eradication of or vaccination against *H. pylori.*

REFERENCES

1. Parkin DM, Pisani P, Ferlay J. Estimates of the worldwide incidence of eighteen major cancers in 1985. Int J Cancer 1993; 54:594–606.
2. Pisani P, Parkin DM, Ferlay J. Estimates of the worldwide mortality from eighteen major cancers in 1985. Implications for prevention, and projections of future burden. Int J Cancer 1993; 55:891–903.
3. Kosary CL, Gloeckler Ries LA, Miller BA, Hankey BF, Harras A, Edwards BK, eds. SEER Cancer Statistics Review, 1973–92 Tables and Graphs. Bethesda, MD, National Institutes of Health, 1995.

4. Berrino F, Sant M, Verdecchia R, Capocaccia T, Hakulinen T, Estève J. Survival of Cancer Patients in Europe: The EUROCARE Study (IARC Publication No. 132). Lyon, International Agency for Research on Cancer, 1995.
5. Sankaranarayanan R, Swaminathan R, Black RJ. Global variations in Cancer survival. Cancer 1996; 78:2461–2464.
6. Hanai A, Tsukuma H, Hiyama T, Fujimoto I. Cancer survival in Osaka. In: Tominga S, *et al*, eds. Cancer Mortality and Morbidity Statistics. Japan and the World 1994. GANN Monograph on Cancer Research 1994; 41:159–165.
7. Coleman MP, Esteve J, Damiecki P, Arslan A, Renard H. Trends in Cancer Incidence and Mortality. Lyon, International Agency for Research on Cancer, 1993.
8. Correa P, Chen VW. Gastric cancer. Cancer Surv 1994; 19–20:55–76.
9. Hanai A, Fujimoto I, Taniguchi H. Trends of stomach cancer incidence and histological types in Osaka. In: Magnus K, ed. Trends in Cancer Incidence. New York: McGraw Hill 1982:143–154.
10. Lundergardh G, Lindgren A, Rohul A, Nyren O, Hensson LE, Bergstrom R, Adami HO. Intestinal and diffuse types of gastric cancer: secular trends in Sweden since 1951. Br J Cancer 1991; 64:1182–1186.
11. Correa P, Haenszel W, Cuello C, Zavala D, Fontham E, Zarama G, Tannenbaum S, Collazos T, Ruiz B. Gastric precancerous process in a high risk population: cohort follow-up. Can Res 1990; 50:4737–4740.
12. Inoue M, Tajima K, Kobayashi S, Suzuki T, Matsuura A, Nakamura T, Shirai M, Nakamura S, Inuzuka K, Tominaga S. Protective factor against progression from atrophic gastritis to gastric cancer—data from a cohort study in Japan. Int J Cancer 1996; 66:309–314.
13. Sipponen P, Riihela M, Hyvarinen H, Seppala K. Chronic nonatrophic ('superficial') gastritis increases the risk of gastric carcinoma. Scand J Gastroenterol 1994; 29:336–340.
14. Correa P. Human gastric carcinogenesis: a multistep and multifactorial process—First American Cancer Society Award lecture on cancer epidemiology and prevention. Cancer Res 1992; 52:6735–6740.
15. Kodama M, Kodama T, Suzuki H, Kondo K. Effect of rice and salty rice diet on the structure of mouse stomach. Nutr Cancer 1984; 6:135–147.
16. Schistosomes, liver flukes and *Helicobacter pylori*. IARC Monographs on the Evaluation of Carcinogenic Risks to Humans, Vol 61. Lyon, International Agency for Research on Cancer, 1994.
17. Sobala GM Schorah CJ, Sanderson M, et al. Ascorbic acid in the human stomach. Gastroenterology 1989; 97:357–363.
18. Sakaki N, Momma K, Egawa N, Yamada Y, Kan T, Ishiwata J. The influence of *Helicobacter pylori* infection on the progression of gastric mucosal atrophy and occurrence of gastric cancer. Eur J Gastroenterol Hepatol 1995; 7:S59–S62.
19. Marchetti M, Arico B, Burroni D, Figura N, Rappuoli R, Ghiara P. Development of a mouse model of *Helicobacter pylori* infection that mimics human disease. Science 1995; 267:1655–1658.
20. Kuipers EJ, Perez-Perez GI, Meuwissen SGM, Blaser MJ. *Helicobacter pylori* and atrophic gastritis: importance of the CagA status. J Natl Cancer Inst 1995; 87:1777–1780.

21. Tsukuma H, Mishima T, Oshima A. Prospective study of ‘‘early’’ gastric cancer. Int J Cancer 1983; 31:421–426.
22. Hakkinen IPT. An immunochemical method for detecting carcinomatous secretion from human gastric juice. Scand J Gastroenterol 1966; 1:28–32.
23. Hakama M, Pukkala E. Evaluation of an immunological screening for stomach cancer. In: Chamberlain J, Miller AB eds. Screening for Gastric Cancer. Toronto: Hans Huber Publishers, 1988:71–75.
24. Hakama M, Stenman UH, Knekt P, Jarvisalo J, Leion A, Hakulinen T, Maatela J, Aromaa A. Tumour markers and screening for gastrointestinal cancer: a follow up study in Finland. J Med Screen 1994; 1:60–64.
25. Hisamichi S, Sugawara N. Mass screening for gastric cancer by X-ray examination. Jpn J Clin Oncol 1984; 14:211–223.
26. Oshima A. Screening for stomach cancer: the Japanese program. In: Chamberlain J, Miller AB, eds. Screening for Gastrointestinal Cancer. Toronto: Hans Huber Publishers, 1988:65–70.
27. Yamagata S, Sugawara N, Hisamichi S. Mass screening for cancer in Japan—present and future. In: Yamagata S, Hirayama T, Hisamichi S, eds. Recent Advances in Cancer Control (Proceedings of 6th Asia Pacific Cancer Conference). Amsterdam: Excerpta Medica, 1983.
28. Murakami R, Tsukuma H, Ubukata T, Nakanishi K, Fujimoto I, Kawashima T, Yamazaki H, Oshima A. Estimation of validity of mass screening program for gastric cancer in Osaka, Japan. Cancer 1990; 65:1255–1260.
29. Samloff IM. Pepsinogens, pepsins and pepsin inhibitors. Gastroenterology 1971; 60:586–604.
30. Samloff IM. Pepsinogens I and II: purification from gastric mucosa and radioimmunoassay in serum. Gastroenterology 1982; 82:26–33.
31. Samloff IM, Secrist DM, Passaro E. A study of the relationship between serum group I pepsinogen levels and gastric acid secretion. Gastroenterology 1975; 69: 1196–1200.
32. Samloff IM, Varis K, Ibamaki T, Sicerala M, Rotter JI. Relationships among serum pepsinogen 1, serum pepsinogen 11 and gastric mucosal histology: a study in relatives of patients with pernicious anemia, Gastroenterology 1982; 83:204–209.
33. Westerveld BD, Pals G, Lamers CBHW, Defize J, Pronk JC, Frants RR, Ooms ECM, Kreuning J, Kostense PJ, Eriksson AW, Meuwissen SGM. Clinical significance of pepsinogen A isozymogens, serum pepsinogen A and C levels, and serum gastrin levels. Cancer 1987; 59:952–958.
34. Miki K, Ichinose M, Shimizu A, Huang SC, Oka H, Furihata C, Matsushima T, Takahashi K. Serum pepsinogens as a screening test of extensive chronic gastritis. Gastroenterol Jpn 1987; 22:133–141.
35. Nomura AMY, Stemmermann GN, Samloff IM. Serum pepsinogen I as a predictor of stomach cancer. Ann Intern Med 1980; 93:537–540.
36. Parsonnet J, Samloff IM, Nelson LM, Orentreich N, Vogelman JH, Friedman GD. *Helicobacter pylori*, pepsinogen, and risk for gastric adenocarcinoma. Cancer Epidemiol Biomarkers Prev 1993; 2:461–466.
37. Aromaa A, Kosunen TU, Knekt P, Maatela J, Teppo L, Heinonen OP, Harkonen M, Hakama MK. Circulating anti-*Helicobacter pylori* immunoglobulin A antibodies

and low serum pepsinogen I level are associated with increased risk of gastric cancer. Am J Epidemiol 1996; 144:142–149.
38. Forman D, Newell DG, Fullerton F, Yarnell JW, Stacey AR, Wald N, Sitas F. Association between infection with *Helicobacter pylori* and risk of gastric cancer: evidence from a prospective investigation. BMJ. 1991; 302:1302–1305.
39. Parsonnet J, Friedman GD, Vandersteen DP, Chang Y, Vogelman JH, Orentreich N, Sibley RK. *Helicobacter pylori* infection and the risk of gastric carcinoma. N Engl J Med 1991; 325:1127–1131.
40. Nomura A, Stemmermann GN, Chyou PH, Kato I, Pérez-Pérez GI, Blaser MJ. *Helicobacter pylori* infection and gastric carcinoma among Japanese Americans in Hawaii. N Engl J Med 1991; 325:1132–1136.
41. Lin JT, Wang LY, Wang JT, Wang TH, Yang CS, Chen CJ. Weak association between *Helicobacter pylori* infection and gastric cancer risk: epidemiologic evidence from Taiwan. Gastroenterology 1993; 107:A421.
42. Webb PM, Yu MC, Forman D, Henderson BE, Newell DG, Yuan JM, Gao YT, Ross RK. An apparent lack of association between *Helicobacter pylori* infection and risk of gastric cancer in China. Int J Cancer 1996; 67:603–607.
43. Stemmermann GN, Samloff IM, Normura AMY, Heilbrun LK. Serum pepsinogens I and II and stomach cancer. Clin Chim Acta 1987; 163:191–198.
44. Miki K, Ichinose M, Ishikawa KB, Yahagi N, Matsushima M, Kakei N, Tsukuda S, Kido M, Ishihama S, Shimizu Y, Suzuki T, Kurchkawa K. Clinical application of serum pepsinogen I and II levels for mass screening to detect gastric cancer. Jpn J Cancer Res 1993; 84:1086–1090.
45. Hattori Y, Tashiro H, Kawamoto T, Kodama Y. Sensitivity and specificity of mass screening for gastric cancer using the measurement of serum pepsinogens. Jpn J Cancer Res 1995; 86:1210–1215.
46. Forman D, Webb P, Parsonnet J. *Helicobacter* and gastric cancer. Lancet 1994; 343:243–244.
47. Parsonnet J, Harris RA, Hack HW, Owens DK. Modelling cost-effectiveness of *Helicobacter pylori* screening to prevent gastric cancer: a mandate for clinical trials. Lancet 1996; 348:150–154.
48. Hisamichi S, Fukao A, Sugawara N, Nishikouri M, Komatsu S, Tsuji I, Tsubono Y, Takano A. Evaluation of mass screening programme for stomach cancer in Japan. In: Miller AB, Chamberlain J, Day NE, Hakama M, Prorok PC, eds. Cancer Screening. Cambridge: Cambridge University Press, 1991:357–370.
49. Oshima A, Hanai A, Fujimoto I. Evaluation of a mass screening program for stomach cancer. NCI Monogr 1979; 53:181–186.
50. Fukao A, Hisamichi S, Komatsu S, Shimizu H, Satoh H, Nakatsuka H, Watanabe T, Fujisaku S, Ichinowatari Y, Kuroda S, Ida Y, Suda S, Kato K, Ikeda M. Comparison of characteristics between frequent participants and non-participants in screening program for stomach cancer. Tohoku J Exp Med 1992; 166:459–469.
51. Oshima A, Hirata N, Ubukata T, Umeda K, Fujimoto I. Evaluation of a mass screening program for stomach cancer with a case-control study design. Int J Cancer 1986; 38:829–833.
52. Fukao A, Hisamichi S, Sugawara N. A case-control study on evaluating the effect

of mass screening on decreasing advanced stomach cancer. J Gastroenterol Mass Survey, 1987; 75:112–116 (in Japanese).
53. Fukao A, Tsubono Y, Tsuji I, Hisamichi S, Sugahara N, Takano A. The evaluation of screening for gastric cancer in Miyagi prefecture, Japan: a population-based case-control study. Int J Cancer 1995; 60:45–48.
54. Pisani P, Oliver WE, Parkin DM, Alvarez N, Vivas J. Case-control study of gastric cancer screening in Venezuela. Br J Cancer 1994; 69:1102–1105.
55. Kuroishi T, Hirose K, Tajima K, Tominaga S. Cancer mortality in Japan (1950–1990). GANN Monogr Cancer Res 1994; 41:1–105.
56. Kuroishi T, Hirose K, Nakagawa N, Tominaga S. Comparison of time trends of stomach cancer death rates between the model area of the screening program and the control area. J Gastroenterol Mass Survey 1983; 58:45–52 (in Japanese).
57. Fukao A, Hisamichi S, Kamatsu S, Sugawara N, Takano A. Risk of leukaemia among participants of gastric mass screening survey in Japan: a population-based case control study. Cancer Detect Prevent 1992; 16:283–286.
58. Davis RE, Graham DY. Endoscopic complications. Gastrointest Endosc 1979; 25: 146–149.
59. Schuller KFR, Cotton PB, Salmon PR. The hazards of digestive fiberendoscopy: a study of British experience. Gut 1972; 13:1027–1032.
60. Tsuji I, Fukao A, Sugawara N, Shoji T, Kuwajima I, Hisamichi S. Cost-effectiveness analysis of screening for gastric cancer in Japan. Tohoku J Exp Med 1991; 164:279–284.
61. Kronborg O, Fenger C, Olsen J, Jørgensen OD, Søndergaard O. Randomised study of screening for colorectal cancer with faecal-occult-blood test. Lancet 1996; 348: 1467–1471.
62. Hardcastle JD, Chamberlain JO, Robinson MHE, Moss SM, Amar SS, Balfour TW, James PD, Mangham CM. Randomised controlled trial of faecal-occult-blood screening for colorectal cancer. Lancet 1996; 348:1472–1477.
63. Chamberlain J, Day NE, Hakama M, Miller AB, Prorok PC. UICC workshop of the Project on Evaluation of Screening Programmes for Gastrointestinal Cancer. Int J Cancer 1986; 37:329–334.

20

Endometrial Cancer

Joan L. Walker
University of Oklahoma Health Sciences Center, Oklahoma City, Oklahoma

Evelyn Rosas Nuñez
Tufts University School of Medicine and Tufts New England Medical Center, Boston, Massachusetts

I. EPIDEMIOLOGY AND BURDEN OF DISEASE

Endometrial adenocarcinoma is the most common gynecological malignancy and is the fourth most common malignancy in American women. The American Cancer Society predicted that 34,900 women would be diagnosed in 1998 and that 6000 women will die of this disease each year (1). If identified early, when the disease is confined to the endometrial cavity, uterine cancer has an excellent survival rate. Abnormal vaginal bleeding is an early symptom that usually prompts women to seek medical evaluation. If cancer is diagnosed at an advanced stage, survival rate is markedly decreased. The overall case fatality rate for endometrial cancer is 33%, but dramatically improved survival rate is seen for early stage disease.

Endometrial cancer is treated primarily with surgery, and it is surgically staged according to the International Federation of Gynecology and Obstetrics (FIGO) classification (2) (Table 1). The 5-year survival rate for patients who are surgically staged improves from 19% at stage IV to 85% when documented to be stage I. Prognosis is dependent on a number of factors including stage, depth of myometrial invasion, cytologic grade, and histological cell type. Staging defines prognostic factors and aids in treatment planning. Adjuvant therapy alternatives include chemotherapy, radiation therapy, and hormonal therapy. In contrast to adenocarcinomas, which arise from the uterine glandular cells, uterine sarcomas arise from the muscle, stroma, or supporting tissue. Sarcomas, which comprise 3%–5% of all uterine tumors (3), have a uniformly poor prognosis and are not considered further here.

Table 1 FIGO Classification of Endometrial Cancer

Stage I	I a	tumor limited to the endometrium
	I b	invasion of less than half of the myometrium
	I c	Invasion of more than half of the myometrium
II	II a	endocervical glandular involvement only
	II b	cervical stromal invasion
III	III a	tumor invades serosa and/or adnexae and/or positive peritoneal cytology
	III b	vaginal metastases
	III c	metastases to pelvic and/or paraaortic lymph nodes
IV	IV a	tumor invasion of bladder and/or bowel mucosa
	IV b	distant metastases including intraabdominal and/or inguinal lymph node

Source: Courtesy of Ref. 2.

A number of risk factors have been defined that identify women at increased risk of development of endometrial cancer. Most endometrial cancer cases are diagnosed in postmenopausal women. The median age is 63 years. The National Cancer Institute (NCI) Surveillance, Epidemiology, and End Results (SEER) program data documents that 11% of women with endometrial cancer are younger than 50 years old (4) and others have documented that 75% are postmenopausal at the time of diagnosis. Approximately 3% of endometrial cancers are found in women younger than 40 years old. Race and ethnicity have been reported to play a role as well. Age-adjusted incidence is 22.3 per 100,000 in white women and 14.4 per 100,000 in African American women (1). Survival rate among African Americans has been reported to be substantially lower (relative risk of death 2.7) compared with that of white women; this has not been explained by a delay in treatment or access to medical care. A greater proportion of high-grade disease, a poor prognostic factor for survival, has been described in African American women (5,6). This has recently been clarified with the help of the California Cancer Registry. Plaxe reports that the increased proportion of high-risk endometrial cancers in African Americans is secondary to a reduction in low-risk disease, not an increase in high-risk disease. The high-grade carcinoma rate is the same in African American and white women in California and is decreased in Hispanics and Asians. The low-grade endometrial cancer rate in whites is five times that of African Americans, which makes the prognosis appear better for whites overall (7).

It is widely accepted that patients with prolonged exposure to estrogen are at increased risk for hyperplastic endometrial disorders: polyps, endometrial hyperplasia, and, of most concern, endometrial adenocarcinoma. The association of estrogen with endometrial cancer can be demonstrated in the incidence rates from the SEER data. The rate held steady until the 1970s when an estrogen-associated 30% increase in incidence occurred. A decrease in incidence followed the Food and Drug Administration (FDA) announcement in the 1980s of this association. Mortality rate, however, decreased in association with the increase in estrogen use because the most favorable low-risk endometrial cancers occur in women on estrogen replacement therapy. In the National Cancer Institute's Black/White Cancer Survival Study (BWCSS) (6), African American women were much less likely to have received estrogen, and this may account for some of the difference between African Americans and whites with respect to tumor grade and low-risk disease.

Unopposed estrogen replacement therapy (i.e., without a progestational agent) in a woman with a uterus places her at excessive risk of endometrial cancer (relative risk of 2.3) (8) (Figure 1). Therefore, hormone replacement therapy in women with an intact uterus should include the concurrent use of a progestin. Women who take the combination of estrogen with progestin may still be at a higher risk of endometrial cancer than those who do not take hormone therapy at all (relative risk 1.6–1.8) (9,10). This is mainly true of thin women and smokers (10). Women who are at increased risk of endometrial

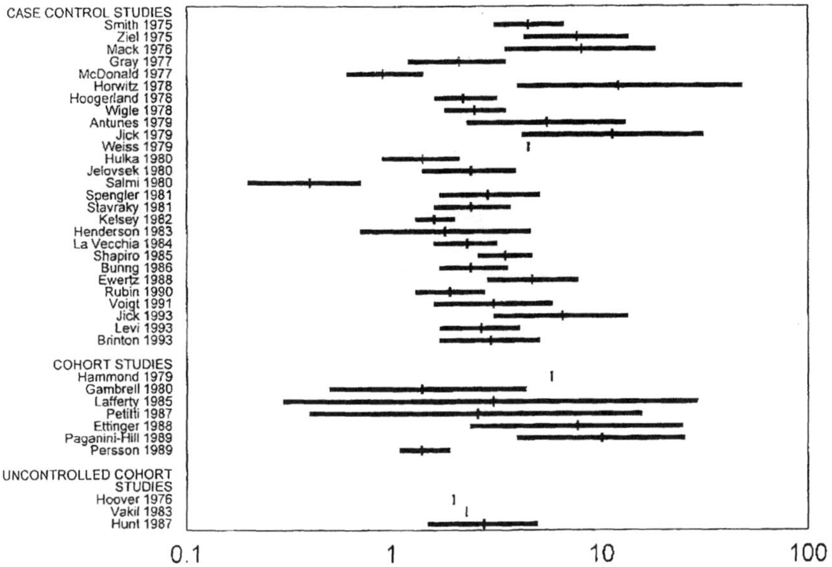

Figure 1 Estrogen replacement therapy and the relative risk or odds ratio of endometrial cancer. (Adapted from Grady D, Gebretsadik T, Kerlikowske K, Ernster V, Petitti D. Hormone replacement therapy and endometrial cancer risk: a meta-analysis. Obstet Gynecol 1995; 85:304–313.)

cancer secondary to obesity may actually achieve a beneficial effect of the progestin and decrease their risk of endometrial cancer by taking hormone replacement therapy. A subset of women do not take the progestin because of side effects such as depression, emotional lability, breast pain, and weight gain. Many physicians ask these patients to be screened annually with an endometrial biopsy. Some patients may not tell their physicians that they are not taking one of the prescribed hormones. The drug combination of a conjugated estrogen and medroxyprogesterone acetate (Prempro, Premphase) is available in a single tablet, thus eliminating this potential problem.

Gorsky evaluated the overall risks and benefits of estrogen replacement therapy (without progestin) using a hypothetical population based model and projected a 207% increase in endometrial cancer deaths, a 21% increase in breast cancer deaths, a 48% decrease in coronary heart disease deaths, and a 49% decrease in hip fracture deaths. No adjustments of the model were made for presence or absence of the uterus. Overall, a gain in quality-adjusted life-years was estimated (11). Col et. al., in another decision model, predicted that hormone replacement therapy (estrogen plus progestin in women with an intact uterus, and estrogen alone in women without a uterus) could increase life expectancy by 6 to 41 months, depending on the individual's risk factors for coronary heart disease and breast cancer (12).

Obese women have increased endogenous estrogen stimulation of the endometrium secondary to peripheral conversion of ovarian and adrenal androgens to estrone by aromatase in adipose cells. Relative risk is positively associated with the number of pounds over ideal body weight or their body mass index (BMI). The relative risk is 10 if 50 pounds overweight and 3 if 20–50 pounds overweight (13). Body mass index (also reported as Quetelet index) is calculated as weight in kg/(height in M^2). A review of reported studies of BMI by Ballard-Barbash documented both an increased risk of endometrial cancer and an increased risk of mortality secondary to endometrial cancer. The relative risks ranged from 2.5 to 3.5 for women with BMI $\geq$ 28 or $\geq$ 30 (14). Others have reported the increased incidence of endometrial cancer in obese women, but the overall prognosis is better for the obese endometrial cancer patient than the thin one, similar to other estrogen-induced cancers (15). A small association between risk and sustained physical inactivity, persisting even after adjustment for body mass index, has also been reported (16). Women with diabetes have an independently increased risk of 2 and higher if they are also obese (17). The mechanism is unclear, but it is possibly related to a growth factor–like effect of insulin on the endometrium. Hypertension has also been reported to increase endometrial cancer risk; however, if controlled for other variables (e.g., obesity), the increase is not confirmed (17).

Reproductive and menstrual factors alter the risk of endometrial cancer. Women who undergo late menopause have been reported to have a relative risk of 2.4, presumed to be due to increased estrogen exposure (13). Brinton did not find an increased risk with late menopause, but found early menarche and longer days of menstrual flow was associated with an elevated relative risk. The occurrence of one or more births is associated with a 60% decrease in risk compared with no prior births, decreasing somewhat further with additional births. Age at first birth does not appear to affect risk. Nulliparous women have a greater than two-fold increased risk of developing endometrial cancer (17). The increased risk may be secondary to infertility or anovulation, possibly resulting from a lack of exposure to high levels of progesterone during the luteal phase of the menstrual cycle and during pregnancy. Women who have a history of anovulation account for most of the premenopausal endometrial cancer patients. They may be seen by physicians during adolescence for control of heavy irregular bleeding or may seek assistance for associated infertility. If these conditions are untreated, the women are destined for long-term exposure to endogenous estrogen unopposed by progesterone. For a variety of reasons, many of these women fail to seek treatment, or the physician fails to counsel her about endometrial cancer risks. Presumably, most of these cancers are preventable with the use of cyclic progesterone or oral contraceptive pills. This therapy should prevent estrogen-induced premalignant endometrial lesions, hyperplasia, and cancer.

Perimenopausal women also become anovulatory for a number of years. In the perimenopause, the period immediately before and after the menopause, proliferative endometrium is likely to develop in response to estrogen, which fails to be cycled and sloughed in response to progesterone. Active medical management during this period of a woman's life can prevent the proliferative endometrium from progressing to hyperplasia or invasive adenocarcinoma. This active management has been advocated by Speroff (18). Physicians are encouraged to place women who are anovulatory and who do not require contraception on monthly cyclic progesterone therapy, usually in the form of medroxyprogesterone acetate, 10 mg orally per day, on days 1 through 10 of each month. This allows the perimenopausal woman to have regular monthly menses that are predictable and reduced in volume until they become unresponsive to the progesterone. The failure of the endometrium to bleed in response to the cyclic progesterone therapy documents a low estrogen production by the ovary and the woman is then offered estrogen replacement therapy as well. A woman who has symptoms of estrogen deficiency is offered combination hormone replacement therapy at that time without delay. Symptom relief can often be obtained and the effects of anovulation during the perimenopause can be prevented.

Smoking appears to confer a protective effect. Most studies evaluating the relationship between smoking and endometrial cancer have identified a decreased relative risk (RR) or odds ratio (OR) of approximately 0.5 in postmenopausal women who are current smokers. It is less clear whether the risk is decreased in premenopausal women or former smokers. The mechanisms are not definitively understood and are probably multifactorial, but smoking is associated with effects that reflect lower levels of estrogen exposure. Women who smoke have decreased urinary estrogen excretion, earlier menopause, and probably an increased risk of osteoporosis (19). Moderate alcohol intake has been reported to lead to sustained increases in circulating estradiol levels (20); however, studies of the effect on risk of endometrial cancer have conflicting results.

A number of studies have investigated the risk of endometrial cancer in women who have used oral contraceptive pills (OCPs). Sequential hormone preparations were available until 1976. They contained estrogen for 14–15 days followed by 5–6 days of estrogen and progestin. They were discontinued in the United States and Canada after a likely increased risk of adverse effects, including endometrial cancer, were noted and are of historical interest only (21). Oral contraceptives are currently formulated with daily estrogen and progesterone. Case-control studies of combination oral contraceptives consistently reveal a decreased endometrial cancer odds ratio (OR) or estimated relative risk (RR) of approximately 0.5 (Figure 2). No protection is associated with use of less than 1 year. Kaufman found that protection lasted at least 5 years following discontinuation. Several other studies found no change in relative risk over time since last use (latency period) (22–26) (Table 2). Jick in 1993 and the Cancer

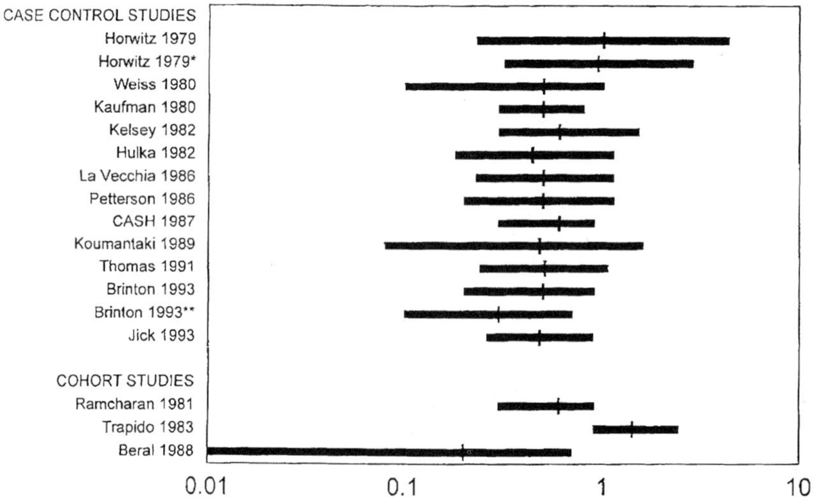

Figure 2 Oral contraceptive use and the relative risk or odds ratio of endometrial cancer. *Second study. **Use for more than five years. (Adapted from Grimes D A, Economy K E. Primary prevention of gynecologic cancers. Am J Obstet Gynecol 1995; 172:227–235.)

Table 2 Effect of Oral Contraceptive Pills on Endometrial Cancer Risk

Case Control	Cases	Controls	Relative risk	Odds ratio
Horowitz 1979	13/187	10/137		0.95
Kaufman 1980	16/152	105/516	0.50[a]	
			0.30[b]	
Kelsey 1982	6/167	8/903		0.60[d]
Hulka 1982	5/79	31/203	0.44	
Henderson 1983	43/110	88/110	0.28[c]	
LaVecchia 1986	7/170	178/1282	0.56	
CSHS 1987	183/433	2044/3191	0.50	
WHO 1988	12/130	180/835	0.55	
Thomas 1991	12/130	180/835	0.51	
WHO data				
Jick 1993	26/142	270/1042	0.48	

[a] 1–2 years use
[b] greater than 2 years
[c] greater than 6 years
[d] greater than 5 years

and Steroid Hormone Study (CSHS) evaluated factors known to increase the risk of endometrial cancer (including obesity, nulliparity, estrogen replacement therapy) and found that they did not confound the protective effect of OCPs (24,25). Women with these risk factors may derive benefit from OCPs. In the CSHS study, increased parity ($\geq$5) was associated with a loss of protection. It is thought that these women have an extremely low risk of endometrial cancer and OCPs could not further decrease this risk. In an analysis of hormone content, Rosenblat found that low progestin pills were associated with decreased risk after 2 years, but were not protective after 10 years. High progestin pills were associated with decreased risk in less than 2 years of use and the protective effect lasted greater than 10 years (27). There have also been three large prospective studies evaluating the effect of OCPs on endometrial cancer risk. Two of these, the Walnut Creek Contraception Drug Study from Northern California and The Royal College of General Practitioners' Oral Contraception Study in the United Kingdom (28,29) found a decreased risk in women who used oral contraceptive pills. The latter study reported a cohort of 47,000 women, half of whom used oral contraceptives, followed up from 1968 to 1987, a relative incidence of 0.2 was calculated. In contrast, Trapido (30) followed up a cohort of 97,300 women aged 25–57 years, 30% of whom used oral contraceptives. He reported an increased, although not statistically significant, rate ratio of 2.4 in nulliparous women and 1.4 in parous women. However, the cohort consists of a relatively young group of women who are at low risk of endometrial cancer because of their age. Overall, OCPs may be a strategy to decrease the risk of endometrial cancer in high-risk women (anovulation, obesity, diabetes, nulliparity) with no contraindication to hormone therapy. Low-dose oral contraceptive pills are being recommended more commonly in the 35 to 52-year-old nonsmoking population in order to impart protection against endometrial cancer and ovarian cancer.

Women on long-term tamoxifen have a 2–7.5 relative risk of endometrial cancer. The increased risk is thought to be due to local estrogenic effects of tamoxifen on the endometrium. The benefits of tamoxifen use far outweighs its risk in most breast cancer patients for 5 years in the adjuvant setting. The National Surgical Adjuvant Breast and Bowel Project (NSABP) B14 trial randomized women with primary, estrogen receptor–positive, axillary lymph node–negative breast cancers to either placebo or tamoxifen (20 mg/day). The annual hazard rate of endometrial cancer was 1.6 per 1000 compared to 0.2 per 1000 in the placebo control group. However, the 5-year cumulative hazard rate for disease-free survival in the randomized tamoxifen group was 38% less than in the placebo group (31). Women who take tamoxifen for more than 5 years may no longer have a favorable risk-to-benefit ratio; this is currently under debate. Furthermore, the risk-to-benefit ratio may vary by breast cancer stage and receptor status. Participants in the breast cancer prevention trial, comparing 5 years of tamoxifen to placebo in unaffected women at elevated risk of breast cancer secondary to strong family history, undergo an endometrial biopsy annually. It has been suggested that endometrial cancer in tamoxifen patients is more aggressive than those in the nonusers (32). Most evidence does not support this contention, but the issue is still under study (33). Multiple screening modalities in this population are being explored. Annual gynecological examinations, Papanicolaou (Pap) smears, endometrial biopsies, dilation and curettage, transvaginal ultrasound, sonohysterography, and hysteroscopy are being studied. Data are insufficient to give definitive guidelines (34). Current recommendations by the American College of Obstetrics and Gynecology for tamoxifen patients is to perform a biopsy only if the patient is symptomatic. Screening procedures or diagnostic tests should be performed at the discretion of the individual practitioner (35).

At least two distinct types of endometrial cancer have been recognized: an estrogen-dependent type, as described, which has the better prognosis, and an estrogen-independent type. These latter cancers are believed to be secondary to genetic predisposition, such as Lynch II syndrome, or to have a molecular basis such as having an associated K-*ras* or Her-2/*neu* oncogene or p53 tumor suppresser gene; these are thought to be more aggressive and confer a higher mortality rate. Although estrogen-independent tumors account for only 20% of endometrial tumors, they account for a disproportionate number of deaths (36).

Hereditary associations exist as well. The endometrial cancer patient often has a personal or family history of breast or colon cancer (37). The Lynch II syndrome, hereditary nonpolyposis colorectal cancer (HNPCC), includes a predisposition for endometrial cancer in which the lifetime risk of endometrial cancer is estimated to be 20%. The peak incidence is between 40 and 60 years of age, which is considerably younger than for the general population (38). Endometrial cancer is, however, an infrequent cause of death in HNPCC families. The 5-year survival rate has been reported to be equivalent to unselected sporadic cases (39). Ovarian cancer is also clustered in these HNPCC families. Some researchers have recommended that women found to be a member of such a family should have an endometrial biopsy and transvaginal ultrasound of the ovaries yearly starting at age 30 years. Hysterectomy along with bilateral salpingo-oophorectomy is often recommended after completion of childbearing because ovarian cancer risk is also increased, especially if prophylactic subtotal colectomy is planned (40). No data exist on the benefit of screening or prophylactic gynecological surgery, however. Birth control pills should be used until pregnancy is desired and they should be continued until menopause.

II. NATURAL HISTORY AND BIOLOGY OF THE DISEASE

Endometrial cancer is one of the most well described and best understood cancers because of the accessibility of the uterine cavity to evaluation. It has long been known that the endometrium proliferates in response to estrogen stimulation and in a large percentage of cases becomes hyperplastic or even malignant if this stimulation remains unopposed by progesterone for a long duration. This was documented in the late 1970s, with multiple case-control studies of postmenopausal estrogen therapy having up to a 15-fold increased risk of endometrial cancer (41–43). Previous observations of endometrial cancer occurring in other circumstances of unopposed endogenous estrogen stimulation, including anovulation, have also been documented. During the luteal phase of a normal menstrual cycle, the corpus luteum produces high levels of progesterone, thus inducing secretory phase endometrium. Ovulation is required for the ovary to create a corpus luteum, which induces the luteal phase. Young, anovulatory women who thereby fail to produce a corpus luteum and fail to have two weeks per cycle of high progesterone levels also illustrate this link between an endogenous unopposed estrogen source and endometrial cancer (44,45). Endometrial cancer is also associated with estrogen-producing ovarian neoplasms (46). Synchronous endometrial adenocarcinoma is found in 5%–25% of women when surgery is performed for the estrogen-producing ovarian granulosa cell tumor. Endometrial cancer precursor lesions are also seen in these patients.

Endometrial cancer has a known precursor lesion, hyperplasia, with a relatively long latency period. Hyperplastic lesions of the endometrium follow a continuum. The current histological classification system categorizes lesions as complex or simple based on histological complexity, with or without atypia based on nuclear abnormality. The pathophysiology of this disease demonstrates that continuous estrogen stimulation causes the endometrium to go from proliferative to hyperplastic, first simple hyperplasia followed by complex hyperplasia. The nuclei of the cells may then become atypical, with nuclear neoplastic features, and finally adenocarcinoma in a subset of cases. The risk of malignant transformation to adenocarcinoma from each of these states has been estimated by Kurman in a retrospective analysis of untreated women. He found adenocarcinoma developed in untreated women in 2% of patients with hyperplasia without atypia and in 23% of women having hyperplasia with atypia. There is also a higher likelihood of development of endometrial adenocarcinoma over time in women with complex hyperplasia with atypia (29%) compared with those with simple hyperplasia with atypia (8%). Progression from atypical endometrial hyperplasia to clinically evident carcinoma had a mean interval of 4 years, and from hyperplasia without atypia, 9.5 years (47). In other studies, 17%–23% of women with atypical endometrial hyperplasia in uterine curettings had a well-differentiated carcinoma in the uterus at the time of hysterectomy (48).

The Postmenopausal Estrogen/Progestin Intervention (PEPI) trial (49,50) recently prospectively documented the effect of estrogen alone on women with an intact uterus. Because of the presupposed improved beneficial cardiovascular effects of estrogen alone when compared with estrogen plus progesterone, the PEPI trial was designed to have a placebo arm, an estrogen alone arm, and multiple estrogen plus progestin arms. Because of the known endometrial cancer risk of estrogen therapy, annual endometrial biopsies were performed. The estrogen alone arm demonstrated a 34% endometrial hyperplasia rate in 3 or fewer years of use compared to 1% in the other arms. Also demonstrated was a 6% hysterectomy rate in the estrogen alone arm for hyperplasia and other reasons compared to 1% in the other arms. The conclusions were that, in women with a uterus, the use of cyclic medroxyprogesterone acetate gave the most favorable cardiovascular risk profile and the lowest endometrial cancer risk. The Comparative Effect on Bone Density, Endometrium, and Lipids of Continuous Hormones as Replacement Therapy (CHART) study also demonstrated a dose-response for ethinyl estradiol (EE), an estrogen commonly used in oral contraceptive pills, and endometrial hyperplasia. Endometrial biopsies were performed every 6 months in this study. At a 1-μg dose of unopposed ethinyl estradiol, there was one case of hyperplasia out of 64 patients, and at 10-μg, there were 10 cases of hyperplasia in 18 patients. The 2.5-μg and the 5-μg dose of EE had hyperplasia develop in 1 of 67 and 2 of 90 patients. A linear dose trend was identified at all time points in the unopposed estrogen group. The 10-μg arm of the study was closed secondary to this adverse outcome after an interim data analysis. A protective effect of combination therapy with norethindrone (a progestin) was identified at all the estrogen doses (51).

The treatment of hyperplasia can be surgical or medical. Hyperplastic lesions are reversible with progesterone therapy until neoplasia occurs. The dose, duration, and type of progesterone needed to reverse these lesions are insufficiently studied, as is the incidence of these precursor lesions. Most women with hyperplasia are asymptomatic and do not undergo endometrial biopsy and thus the incidence is unknown. The Gynecologic Oncology Group is conducting a prospective, randomized trial of complex, endometrial hyperplasia with atypia to study this disease and the treatment with progesterone. Currently, medical management consists of treatment with a progestational agent, usually for 3 months with rebiopsy at the time of completion. If persistent hyperplasia is found, the patient usually then undergoes hysterectomy. If the hyperplasia has resolved, there is no consensus on subsequent management; however, it is generally believed that she will require continued medical management with progesterone unless the cause is eliminated. Surgical treatment, hysterectomy, can be completed transvaginally (with or without laparoscopic assistance) in most patients. The procedure is well tolerated and has low morbidity, and most women can be discharged within 24 hours. There are ongoing molecular studies to determine whether a molecular marker for malignant transformation can be found (52,53). Ideally, those women whose histological abnormalities can be reversed with progesterone could be distinguished from those who are destined to require hysterectomy to treat the neoplastic process.

III. EARLY DETECTION TESTS

A. Papanicolaou Smear

Endometrial cells found on a Pap smear during the first half of the menstrual cycle in premenopausal women are thought to be due to normal shedding. However, data suggest that, if endometrial cells are identified on Pap smear either in the second half of the menstrual cycle (secretory phase) or at any time on the Pap smear of a postmenopausal woman, she is at risk for an endometrial abnormality (54,55). Ng first reported 836 women who had normal-appearing endometrial cells on cellular studies obtained by endocervical aspiration. Of the 696 patients with ''abnormal'' shedding, 26.5% had adenomatous polyp, 7.8% endometrial hyperplasia, and 3.2% adenocarcinoma. The percentage of adenocarcinomas increased with age. In women older than 49 years of age, 6.8% were diagnosed with adenocarcinoma, and 13.2% had endometrial hyperplasia. Gondos (55) subsequently reported on 238 cervicovaginal smears with associated histological material. In the group of women older than 40 years (n=147), 32 (22.8%) had an abnormal histology, and 1.4% were given the diagnosis of carcinoma. Abnormal bleeding was present in the majority of patients. Cherkis, in a retrospective review, found that 9 of 25 patients with endometrial cancer were asymptomatic and the presence of atypical endometrial cells triggered the evaluation leading to the diagnosis (56). Additionally, it has been reported that endometrial cancer patients with cervical cytology containing malignant endometrial cells have a higher risk of poor prognostic indicators when compared with patients with normal cervical cytology (57)

Several retrospective studies have evaluated the role of Pap smear in endometrial cancer screening (Table 3). All concluded that the sensitivity and positive predictive value (PPV) of the Pap smear is too low to effectively screen for endometrial cancer. A PPV ranging from 37% to 64% and sensitivity of 18%–75% has been reported (56,58–62). Two studies (56,63) have further evaluated the effectiveness of grading cytological atypia of the endometrial abnormalities to improve specificity. They also concluded that this increased the percentage of patients with documented endometrial pathology but once again was not sensitive enough for screening. In 1993, Mitchell (61) reported on 430,000 Pap smears from the Victorian Cervical Cytology Registry. Sixty-nine endometrial cancers were identified. In the same pool, 41 adenocarcinomas of the cervix and 214 squamous cell cancers of the cervix were found. Only 28% of women with endometrial cancer were detected by cytological study. Two subgroups of women were defined: those who were asymptomatic and had unsuspected endometrial cancer and those who were symptomatic and had Pap smears as part of their workup. There did appear to be a survival benefit to those women identified in the asymptomatic stage with an odds ratio of death equal to 0.78 ($P>0.05$). Even though the sensitivity is too low to rely on for screening asymptomatic women, it does suggest that early identification of endometrial carcinoma can lead to improved survival.

In summary, few women with endometrial cancer will benefit from Pap smear screening. However, if an abnormality is found on Pap smear, further evaluation is required. The endometrium should be biopsied and evaluated by histological study. Symptomatic women should have endometrial evaluation regardless of the Pap result.

B. Endometrial Biopsy

Dilation and curettage (D & C) has been the gold standard for endometrial tissue sampling. The false-negative rate for endometrial cancer has been reported to be 10% (13) presumed to be from sampling error. Stock reported 50 consecutive D & C specimens obtained immediately before hysterectomy and found that less than one-half of the endometrial cavity had been curetted, and in 84%, less than three-fourths of the cavity had been curetted (64). Office-based endometrial biopsies (EMB) have been shown to be equivalent to D & C in the ability to

Table 3 Pap Smears and the Detection of Endometrial Cancer

Study	N	Study population	Screening vs. diagnostic test	PPV	Sensitivity
Koss, 1962	205 patients	Symptomatic and asymptomatic	Mixed	42%	75%
Burk, 1974	171 pap smears in 154 patients with endometrial cancer	1. Alabama Cancer Registry 1972 2. Birmingham Metropolitan area endometrial cancer patients 1971	Diagnostic	N/A	18%
Hoffmeister, 1974	187 patients with endometrial cancer	Symptomatic Pre- and post-menopausal	Diagnostic	N/A	27%
Gondos, 1977	1. 238 pap smears with normal endometrial cells	UCSF Cytology lab	1. Mixed		1.4%
	2. 17 patients with endometrial cancer		2. Not specified	88%	
Vuoppala, 1977	718 pap smears 86 endometrial cancer patients	Symptomatic Pre- and post-menopausal		37%	57%
Bibbo, 1982	840 patients	Symptomatic and asymptomatic high risk patients	Diagnostic		
Cherkis 1987	300,000 screening pool 134 paps with atypical endometrial cells	All women undergoing pap smear		52%	
Kim 1991	1. 79 endometrial cancer patients	1. Baylor Tumor Registry	1. Diagnostic	62%	35%
	2. 130 with positive or suspicious paps	2. All paps done 1975–1986	2. Mixed		
Mitchell 1993	430,000 pap smears 69 endometrial cancer patients	Victorian Cervical Cytology Registry	not specified	64%	28%

identify cancer (65). The EMB has the advantage of being performed in an office setting, is relatively easy to perform, is less expensive, usually requires no anesthesia, and is well tolerated by patients. Many family physicians and any gynecologist can perform the test. A variety of commercial instruments exist, including the Pipelle instrument, Vabra aspirator, the Novak curette, and others. The instrument used currently is a disposable plastic tube of 3 to 4 mm in diameter and 10 cm long with a plunger that can be withdrawn to apply suction after the catheter is introduced into the endometrial cavity. Reported complication rates are low for both D & C and endometrial biopsy, although they are slightly higher for D & C. Reported complications include uterine hemorrhage (<4 per 1000), infection (0–5 per 1000), uterine perforation (0–13 per 1000), and unanticipated major operation (0–5 per 1000) (65).

Endometrial biopsy has been evaluated for endometrial cancer screening in a number of large clinical case series. Archer et al. (66) in 1991 reported on 801 asymptomatic women who were not using hormone replacement therapy and found an incidence of 0.72% of endometrial cancer and 0.63% of atypia. The authors suggest that the yield is so low that screening asymptomatic women is not warranted. Groonos (67) reported a series from Finland where 715 asymptomatic patients who were defined as high risk (either diabetic, hypertensive, or both) had an attempted endometrial biopsy. Of these women, samples were obtained in 597 patients. No endometrial cancers were diagnosed. Preinvasive lesions (adenocarcinoma in situ, atypical hyperplasia, adenomatous hyperplasia) were found in 3.2% of the 597 women in whom successful biopsies were performed. A higher proportion of diabetic women had abnormal results (6.3%) compared with the hypertensive women (1.7%). This study suggests that women at high risk may benefit from screening; however, a prospective randomized trial would be required to assess whether screening and treatment of preinvasive lesions will lead to a decrease in cancer incidence and mortality rate. One would predict that there would be a decreased risk of hysterectomy if endometrial hyperplasia could be found and reversed with progesterone. A subset of patients exist in whom the endometrial sample cannot be obtained because of cervical stenosis or distorted cervical canal. Additionally, a large percentage of patients have biopsies that are insufficient for diagnosis or yield only atrophic endometrium. Transvaginal ultrasound may have a role in confirming atrophy in this group of patients.

C. Endometrial Cytology

A variety of instruments are available to obtain endometrial samples for cytological evaluation. They consist of a metal or plastic brush or sponge that are used to obtain a small number of cells. They tend to be of a smaller diameter and are of a softer material than the endometrial biopsy instruments. It has been suggested that these instruments are better tolerated than endometrial biopsy and thus may be better for screening (68). Ferenczy and Gelfand (69) reported 100% of women returned for sampling after a brush was used and only 40% of those who underwent endometrial biopsy would consent to repeat screening.

Koss (70) prospectively evaluated 2586 asymptomatic women with endometrial sampling, using two commercially available devices (Mi-Mark, and Isaac's cannula) to obtain samples for cytological evaluation. Twenty-one percent of the women tested used estrogen replacement therapy. An endometrial sample was successfully obtained 92.7% of the time. Koss reported a prevalence of occult endometrial carcinoma of 6.96 per 1000 women and suggested that women older than age 50 years be screened at least once. Of interest, only 60% of study participants returned for a second screening. The women who did not return were more likely to have experienced discomfort at the time of the biopsy. By the third year, only 7% of the original population returned for screening.

Of more significance is the complexity in interpretation of cytological specimen. In general, the sample takes three to four times longer than a conventional cervical cytological smear to read and requires that an expert endometrial cytopathologist examine every specimen. In addition, there is decreased accuracy and decreased sensitivity in identifying endometrial hyperplasia. Ferenczy and Gelfand reported that nearly 20% of hyperplasias in their series were called normal endometrium. In addition, there was an inability to distinguish the different types of hyperplasia (69). With the newer endometrial biopsy instruments available (e.g., Pipelle) there does not appear to be an advantage to endometrial cytological study. The patient tolerance is comparable and the tissue quality is far superior.

D. Ultrasound

Ultrasound has been used extensively in the evaluation of gynecological diseases and symptoms. Pelvic ultrasound can be either transabdominal or transvaginal (TVS). Transvaginal ultrasound has been considered superior for the evaluation of the endometrium. In one study comparing transabdominal and TVS for endometrial cancer screening, TVS had a higher sensitivity and specificity of 100% and 54.1% compared with 83.3% and 58.8%, respectively (71). Transvaginal ultrasound has been considered as a screening test for endometrial disease (Table 4). Transvaginal ultrasound done in bleeding postmenopausal women has been shown to be effective in identifying nonmalignant conditions (72). Other studies have established the use of TVS to determine endometrial thickness. Transvaginal ultrasound has most extensively been studied as a substitute for endometrial biopsy in symptomatic postmenopausal women. A number of studies have suggested an endometrial thickness of <5 mm to be associated with the absence of endometrial disease (73–77). In these studies, TVS preceded endometrial

Table 4 Pelvic Ultrasound in Diagnosis of Endometrial Cancer

TVS EM≥5 mm	Goldstein 1990 (≥ 6 mm)	Varner 1991	Smith 1991 (≥4 mm)	Dørum 1993	Cacciatore 1994	Karlsson 1995
Sensitivity			100%	80%	95.7	96%
Specificity			61%	60%	45.5	68%
PPV	35%	40%	39%	26%	64.7	61%
NPV			100%	94%		97%
	PMB	sympt/asypt PM women	PMB		PMB	PMB

TVS EM<5mm	Goldstein 1990 (≤15mm)	Varner 1991	Granberg 1991
Sensitivity		96.7%	100%
Specificity		100%	96%
PPV		100%	87.3%
NPV	100%	90%	
	PMB	sympt/asypt PM women	

Abbreviations: PPV positive predictive value; NPV negative predictive value; PMB postmenopausal bleeding; EM endometrial stripe.

sampling and final histological study revealed atrophy. It has been suggested that, in this group of patients with endometrial thickness <5 mm, endometrial sampling can be avoided, thus preventing the associated costs, risks, and discomfort. Dørum et al. (78), however, identified three cases of malignancy in a group of 54 women with postmenopausal bleeding with TVS finding of endometrial thickness <5 mm. Clear guidelines for follow-up if bleeding persists have not been established, but rebiopsy or possibly hysteroscopy and fractional D & C should be considered if bleeding occurs. Evaluation of the cervix is required for women with unexplained abnormal bleeding.

E. Progesterone Challenge Test

The progesterone challenge test (PCT) has been described as a physiological test of uterine function and estrogen secretion in the evaluation of amenorrhea. In amenorrhea secondary to anovulation there is adequate endogenous estrogen that would cause proliferative endometrium. This test involves the oral administration of progestin tablets (medroxyprogesterone acetate, 10 mg) for 5 days as originally described by Speroff (79). A positive test, defined as any vaginal bleeding or spotting, indicates an estrogen stimulated endometrium. Theoretically, postmenopausal women at high risk of endometrial cancer have an increased exposure to estrogen, either endogenous or exogenous. In estrogen-induced excess proliferation of the endometrium, the PCT should result in bleeding.

Gambrell (80) reported on the use of progestogens to reduce the risk of adenocarcinoma of the endometrium. He concluded that, because progesterone counters the effects of estrogen and causes a sloughing of the endometrium, a PCT is a reliable indicator of whether estrogen (endogenous or exogenous) is present in sufficient quantity to proliferate the endometrium. In 1983, Hanna (81) proposed using PCT to identify endometrial disease in asymptomatic postmenopausal women. Thirty women underwent progesterone challenge test. Of these, 25 did not bleed and five had bleeding. Three of the women with the positive tests had endometrial hyperplasia. A separate analysis was performed on 10 women known to have endometrial hyperplasia; nine of them had a positive test (10% false-negative result). This small series suggests that the PCT could identify those patients whose endometrium is exposed and responding to endogenous estrogen and therefore are at increased risk for endometrial hyperplasia and cancer. Pansini (82) subsequently combined the use of PCT and TVS, showing that PCT-negative patients have thin endometrial linings, thus distinguishing patients not at risk for disease. This would suggest PCT can be used to identify the asymptomatic population that needs further testing either by sonography or endometrial biopsy. No studies have applied this test in the setting of screening and no large numbers of women have been screened. A subgroup of women at high risk for endometrial cancer, such as obese, postmenopausal, diabetic, women, and women on tamoxifen, may be the population that would benefit from this type of screening. However further studies need to be undertaken before this could be recommended as a screening test.

IV. COST AND COST EFFECTIVENESS ANALYSIS

There have been no formal cost or cost-effectiveness analyses performed for endometrial cancer screening. The costs for tests are much more difficult to find than the charges. Costs of screening for a disease includes the cost of the tests and the cost of reevaluating false-positive results, as well as costs of treatment of cases and the treatment of complications. Other factors, such as time off from work and the emotional consequences of the evaluation, are difficult to quantify. The Medicaid and Medicare allowable charges are slowly approaching the costs of the tests, and these are the best figures available (Table 5).

The Medicare allowable charges for a physician to see a woman for a well-woman examination at which a screening test would be performed is $22, the

Table 5 Cost Comparison of Potential Tests for Endometrial Cancer Screening

Test	Customary charge	Medicare allowable charges
Well woman exam	$45	$22
Pap smear	$22	$7
Endometrial biopsy procedure fee	$150	$54.6
Endometrial biopsy pathology fee	$80	$57.98
Progesterone challenge test	N/A	N/A
Transvaginal ultrasound	$275	$79.74

customary charge is $45. The Medicare reimbursement for a Pap smear to be processed and read by a cytotechnologist or pathologist is $7.31, and the customary charge is $22. The performance of a transvaginal ultrasound can be reimbursed for $79.74, and the customary charge is $275. The customary charge for an endometrial biopsy is $150, and the Medicare procedure fee for an endometrial biopsy is $54.60. The customary charge for a pathologist to process and read the endometrial biopsy is $80, and Medicare reimbursement is $57.98.

V. RECOMMENDATIONS FROM PROFESSIONALORGANIZATIONS

A. America Cancer Society

Beginning in 1988, the American Cancer Society recommended endometrial tissue sampling at menopause in women at high risk of development of endometrial cancer in addition to the annual pelvic examination by a health care professional that was recommended for all women. High-risk factors mentioned included obesity, infertility, anovulation, estrogen therapy, and uterine bleeding (83). In 1997, the guidelines were revised to the following: women 40 and older should have an annual pelvic examination by a health care professional. Women at high risk of development of endometrial cancer should have counseling to reduce risk, annual pelvic examinations, and immediate intervention for postmenopausal or irregular bleeding (1).

B. American College of Obstetrics and Gynecology

> Screening techniques to detect endometrial cancer, including ultrasonography and endometrial sampling, have not been demonstrated to be sufficiently accurate, cost-

> effective or acceptable for general use. Cervical cytology alone is not reliable. The detection of abnormal endometrial cells in a nonmenstruating woman requires endometrial sampling. (84)
>
> At present, there are no effective screening methods to detect endometrial cancer. Specifically, routine endometrial sampling has not been shown to be cost effective in asymptomatic patients. (85)

Endometrial biopsy as a diagnostic test is recommended for women with unexpected breakthrough bleeding during the perimenopausal period or while on estrogen replacement therapy
(86).

C. American College of Physicians

Women on unopposed estrogen should undergo diagnostic endometrial evaluation if uterine bleeding occurs and no recent endometrial evaluation has been done. If bleeding does not occur, screening endometrial evaluation should be done, probably every year. The optimal interval for endometrial evaluation, however, has not been assessed. Women who take estrogen plus progestin do not require baseline or routine endometrial evaluation. Women using cyclic estrogen plus progestin therapy should undergo endometrial evaluation if bleeding occurs outside the expected time of withdrawal bleeding (days 5 to 15 of the month if the progestin is given on days 1 to 10 of the month). For women on continuous estrogen plus progestin therapy, if bleeding is heavy (heavier than normal menstrual period), prolonged (longer than 10 days), or frequent (more often than monthly), and if bleeding persists longer than 10 months after beginning therapy (87).

The issue of endometrial cancer screening in asymptomatic women not on hormone replacement therapy has not been considered.

D. U.S. Preventive Services Task Force

In 1996, revised and expanded guidelines of the U.S. Preventive Services Task Force were published. Whereas endometrial cancer screening had not been previously considered, the new guidelines addressed both the issue of screening and prevention of endometrial carcinoma. The recommendations include the following (88):

> There is insufficient evidence to recommend for or against routine counseling of female patients about measures to reduce the risk of cervical, ovarian and endometrial cancer. . . . Clinicians counseling women about contraceptive practices should include information about the potential benefits of specific methods with respect to gynecologic cancers. These benefits include the reduced risks of ovarian and endometrial cancer in women using OCs. . . . All women should be counseled about . . . avoiding obesity . . . and avoiding tobacco use. . . .

E. Canadian Task Force on the Periodic Health Examination

In 1988, the fourth report of the Canadian Task Force on the Periodic Health Examination was issued (89). The issue of endometrial cancer screening was considered. It recommended against the screening of asymptomatic healthy individuals because

> no controlled trials have been conducted to evaluate the effectiveness of prevention or treatment in changing the outcome of endometrial cancer. . . . Screening of high-risk people would detect only 50% of the cases of endometrial cancer. Endometrial cancer manifests early. . . . The overall 5 year survival rate for all stages is about 75%.

Consequently, it is premature to recommend the inclusion of endometrial sampling in the periodic health examination of postmenopausal women, even if they are at high risk (e.g., taking estrogen).

The following research priorities are recommended:

1. Controlled studies of cytological screening of asymptomatic women in detecting endometrial cancer and of the relation of endometrial cancer to clinical outcomes
2. Studying the effect of treatment with progestational agents on the incidence and mortality rates of endometrial cancer in high-risk groups
3. Developing controlled studies of the effectiveness of the different treatments available for poorly differentiated tumors and for advanced stages of the disease

CONCLUSION

Endometrial cancer continues to be an important cause of morbidity and mortality. There is no currently recommended test for the screening of endometrial cancer. Prevention with treatment of anovulation and reduction of high-risk factors continues to be the most appropriate approach at this time. Progesterone therapy in a regimen of postmenopausal hormone replacement therapy and oral contraceptives are both important strategies for reduction of risk. Research priorities should include the development of a low-cost screening test that could be incorporated into well-women examinations of at-risk patients.

REFERENCES

1. American Cancer Society. Cancer Facts and Figures. ACS publication 97–300M-No. 5008.97. New York: American Cancer Society, 1997.
2. FIGO. Rio de Janeiro, Oct 1988. Int J Gynecol Obstet 1989; 28:189–190.

3. Levenback C F, Tortolero-Luna G, Pandey D K, Malpica A, Baker V V, Whittaker L, Johnson E, Follen Mitchell M. Uterine sarcoma. Obstet & Gynec Clinics of North America 1996; 23:457–473.
4. Kosary C L. FIGO stage, histology, histologic grade, age and race as prognostic factors in determining survival for cancers of the female gynecologic system: an analysis of 1973–87 SEER cases of cancers of the endometrium, cervix, ovary, vulva, and vagina. Sem Surg Oncol 1994; 10:31–46.
5. Bain R P, Greenberg R S, Chung K C. Racial differences in survival of women with endometrial cancer. Am J Obstet Gynecol 1987; 157:914–923.
6. Hill H A, Coates R J, Austin H, Correa P, Robboy S J, Chen V, Click L A, Barrett R J, Boyce J G, Kotz H L, Harlan L C. Racial differences in tumor grade among women with endometrial cancer. Gynecol Oncol 1995; 56:154–163.
7. Plaxe, S C, Saltzstein S L. Impact of ethnicity on the incidence of high-risk endometrial carcinoma. Gynecol Oncol 1997; 65:8–12.
8. Grady D, Gebretsadik T, Kerlikowske K, Ernster V, Petitti, D. hormone replacement therapy and endometrial cancer risk: a meta-analysis. Obstet Gynecol 1995; 85: 304–313.
9. Voigt L F, Weiss N S, Chu J, Daling J R, McKnight B, Van Belle, G. Progestagen supplementation of exogenous oestrogens and risk of endometrial cancer. Lancet 1991; 338:274–277.
10. Brinton L A, Hoover R N, and the Endometrial Cancer Collaborative Group. Estrogen replacement and endometrial cancer risk: unresolved issues. Obstet Gynecol 1993; 81:265–271.
11. Gorsky R D, Koplan J P, Perterson H B, Thacker S B. Relative risks and benefits of long term estrogen replacement therapy: a decision analysis. Obstet Gynecol 1994; 83:161–165.
12. Col N F, Eckman M H, Karas R H, Pauker S G, Goldberg R J, Ross E M, Orr R K, Wong J B. Patient-specific decisions about hormone replacement therapy in postmenopausal women. JAMA 1997; 277:1140–1147.
13. DiSaia P J, Creasman W T. Clinical Gynecologic Oncology. 4th ed. St. Louis: Mosby Year Book, 1993:157.
14. Ballard-Barbash R, Swanson C A. Body weight: estimation of risk for breast and endometrial cancers. Am J Clin Nutr 1996; 63(suppl)437S–441S.
15. Anderson B, Connor J P, Andrews J I, Davis C S, Buller R E, Sorosky J I, Benda J A. Obesity and prognosis in endometrial cancer. Am J Obstet Gynecol 1996; 174: 1171–1199.
16. Sturgeon S R, Brinton L A, Berman M L, Mortel R, Twiggs L B, Barrett R J, Wilbanks G D. Past and present physical activity and endometrial cancer risk. Br J of Cancer 1993; 68:584–589.
17. Brinton L A, Berman M L, Mortel R, Twiggs L B, Barrett R J, Wilbanks G D, Lannom L, Hoover R N. Reproductive, menstrual, and medical risk factors for endometrial cancer: results from a case-control study. Am J Obstet Gynecol 1992; 167:1317–1325.
18. Speroff L, Glass R H, Kase N G. Clinical Gynecologic Endocrinology and Infertility. 5th ed. Baltimore: Williams & Wilkins, 1994:593.

19. Baron J A, La Vecchia C, Levi F. The antiestrogenic effect of cigarette smoking in women. Am J Obstet Gynecol 1990; 162:502–14.
20. Ginsburg E S, Mello, N K, Mendelson J H, Barbieri R L, Teoh S K, Rothman M, Gao X, Sholar J W. Effects of alcohol ingestion on estrogens in postmenopausal women. JAMA 1996; 276:1747–1751.
21. Dickey R P. Managing Contraceptive Pill Patients. 8th ed. Durant, OK: Essential Medical Information Systems, Inc., 1994:18–19.
22. Kaufman D W, Shapiro S, Slone D, Rosenberg L, Miettinen O S, Stolley P D, Knapp R C, Leavitt T, Watring W G, Rosenshein N B, Lewis J L, Schottenfeld D, Engle R L. Decreased risk of endometrial cancer among oral-contraceptive users. N Engl J Med 1980; 303(18):1045–1047.
23. Thomas D B. The WHO collaborative study of neoplasia and steroid contraceptives: the influence of combined oral contraceptives on the risk of neoplasms in developing and developed countries. Contraception 1991; 43(6):695–710.
24. Jick S S, Walker A M, Jick H. Oral contraceptives and endometrial cancer. Obstetrics & Gynecology 1993; 82(6): 931–935.
25. Cancer and Steroid Hormone Study of the Centers for Disease Control and the National Institute of Child Health and Human Development. Combination oral contraceptive use and the risk of endometrial cancer. JAMA 1987; 257(6):796–800.
26. Henderson B E, Casagrande J T, Pike M C, Mack T, Rosario I, Duke A. The epidemiology of endometrial cancer in young women. Br J Cancer 1983; 47:749–756.
27. Rosenblatt K A, Thomas D B. Hormonal content of combined oral contraceptives in relation to the reduced risk of endometrial carcinoma. Int J Ca 1991; 49:870–874.
28. Ramcharan S, Pellegrin F A, Ray R M, Hsu J P. The Walnut Creek contraceptive drug study. J Repro Med 1980; 25:345–72.
29. Beral V, Hannaford P, Kay C. Oral contraceptive use and malignancies of the genital tract. Lancet 1988; 10:1331–1334.
30. Trapido E J. A prospective cohort study of oral contraceptives and cancer of the endometrium. Int J Epidemiol 1983; 12(3): 297–300.
31. Fisher B, Costantino J P, Redmond C K, Fisher E R, Wickerham L, Cronin W M. Endometrial cancer in tamoxifen-treated breast cancer patients: findings from the National Surgical Adjuvant Breast and Bowel Project (NSABP) B-14. J Natl Cancer Instit 1994; 86(7): 527–538.
32. Magriples U, Naftolin F, Schwartz P E, Carcanuiu M L. High-grade endometrial carcinoma in tamoxifen-treated breast cancer patients. J Clin Oncol 1993; 11:485–490.
33. Fisher B. Commentary on endometrial cancer deaths in tamoxifen-treated breast cancer patients. J Clin Oncol 1996; 14(3):1027–1039.
34. Barakat R R. The effect of tamoxifen on the endometrium. Oncol 1995; 9: 129–134.
35. American College of Obstetrics and Gynecology. Tamoxifen and Endometrial Cancer. ACOG Committee Opinion Number 169. February 1996.
36. Sherman M E, Sturgeon S, Brinton L, Kurman R J. Endometrial cancer chemo-

prevention: implications of diverse pathways of carcinogenesis. J Cell Biochem 1995, 23 (suppl): 160–164.
37. Schoen R E, Weissfeld J L, Kuller L H. Are women with breast, endometrial or ovarian cancer at increased risk for colorectal cancer? Am J Gastroent 1994; 89: 835–842.
38. Watson P, Vasen H F A, Mecklin J P, Järvinen H, Lynch H T. The risk of endometrial cancer in hereditary nonpolyposis colorectal cancer. Am J of Med 1994; 96: 516–520.
39. Vasen H F A, Watson, P, Mecklin J P, Jass J R, Green J S, Nomizu T, Müller H, Lynch H T. The epidemiology of endometrial cancer in hereditary nonpolyposis colorectal cancer. Anticancer Research 1994; 14: 1675–1678.
40. Lynch H T, Smyrk T. Hereditary nonpolyposis colorectal cancer (Lynch Syndrome): an updated review. Cancer 1996; 78: 1149–1167.
41. Antunes C M F, Stolley P D, Rosenshein M B, et al. Endometrial cancer and estrogen use. N Engl J Med 1979; 300–309.
42. Mack T M, Pike M C, Henderson B E, et. al. Estrogens and endometrial cancer in a retirement community. N Engl J Med 1976; 294: 1262–1267.
43. Food and Drug Administration. Estrogens and endometrial cancer. FDA Drug Bull 1976; 6: 18–20.
44. Gallop D G, Stock R J. Adenocarcinoma of the endometrium in women 40 years of age or younger. Obstet Gynecol 1984; 64: 417–420.
45. Crissman J D, Azoury R S, Barnes A E, Schellhas H F. Endometrial carcinoma in women 40 years of age or younger. Obstet Gynecol 1981; 57: 699–704.
46. Kurman, Robert J. Blaustein's Pathology of the Female Genital Tract. 4th ed. New York; Springer Verlag, 1994; 784.
47. Kurman R J, Kaminski P F, Norris H J. The behavior of endometrial hyperplasia: A long term study of untreated hyperplasia in 170 patients. Cancer 1985; 56: 403–412.
48. W J Hoskins, Perez C A, Young R C. Principles and Practice of Gynecologic Oncology. 2d ed. Philadelphia; Lippincott-Raven, 1997: 863.
49. The Writing Group for the PEPI Trial. Effects of estrogen and estrogen/progestin regimens on heart disease risk factors in post menopausal women. JAMA 1995; 273: 199–208.
50. The Writing Group for the PEPI Trial. Effects of hormone replacement therapy on endometrial histology in postmenopausal women. JAMA 1996; 275: 370–375.
51. Speroff L, Rowan J, Symons J, Genant H, Wilborn W. The comparative effect on bone density, endometrium, and lipids of continuous hormones as replacement therapy (CHART). a randomized controlled trial. JAMA 1996; 276: 1397–1403.
52. Baker V V. Potential criteria for cohort selection in chemoprevention trials of uterine adenocarcinoma. J Cell Biochem, 1995; 23 (suppl): 184–188.
53. Tezuka F, Chiba R, Takahashi T. Quantitative cytopathology of endometrial lesions. J Cell Biochem 1995; 23 (suppl): 147–150.
54. Ng A B P, Reagon J W, Hawliczek S, Wntz B W. Significance of endometrial cells in the detection of endometrial carcinoma and its precursors. Acta Cytol 1974; 18: 356–361.

55. Gondos B, King E. Significance of endometrial cells in cervicovaginal smears. annals of clinical and laboratory science 1977; 7: 486–490.
56. Cherkis R C, Patten S F, Dickinson J C, Dekanich A S. Significance of atypical endometrial cells detected by cervical cytology. Obstet Gynecol 1987; 69: 786–789.
57. DuBeshter B, Warshal D P, Angel C, Dvoretsky P M, Lin J Y, Raubertas R F. Endometrial carcinoma: the relevance of cervical cytology. Obstet Gynecol 1991; 77: 458–462.
58. Koss L, Dufree G. Cytologic diagnosis of endometrial carcinoma: result of ten years of experience. Acta Cytol 1962; 6: 519–528.
59. Vuopala S. Diagnostic accuracy and clinical applicability of cytological and histological methods for investigating endometrial carcinomas. Acta Scand Obst Gynaecol 1977; 70 (suppl): 1–70.
60. Kim H S, Underwood D. Adenocarcinomas in the cervicovaginal papanicolaou smear: analysis of a 12-year experience. Diag Cytopathol 1991; 7: 119–124.
61. Mitchell H, Giles G, Medley G. Accuracy and survival benefit of cytological prediction of endometrial carcinoma on routine cervical smears. Int J Gynecol Pathology 1993; 12: 34–40.
62. Burk J R, Lehman H F, Wolf F S. Inadequacy of papanicolaou smears in the detection of endometrial cancer. N Engl J Med 1974; 291 (4): 191–192.
63. Zucker P, Kasdon E, Feldstein M. The validity of pap smear parameters as predictors of endometrial pathology in menopausal women. Cancer 1985; 56:2256–2263.
64. Stock R J, Kanbour A. Prehysterectomy curettage. Obstet Gynecol 1975; 45(5): 537–541.
65. Grimes D. Diagnostic dilation and curettage: a reappraisal. Clinical Opinion. Am J Obstet Gynecol 1982; 142:1–6.
66. Archer D F, McIntyre-Seltman K, Wilborn W W Jr, Dowling E A, Cone F, Creasy G W, Kafrissen M E. Endometrial morphology in asymptomatic postmenopausal women. Am J of Obstet Gynecol 1991; 165(2):317–320Ascussion 320–322.
67. Gronroos M, Salmi T A, Vuento M H, Jalava E A, Tyrkko J E, Maatela J I, Aromaa A R, Siegberg R, Savolainen E R, Kauraniemi T V. Mass screening for endometrial cancer directed in risk groups of patients with diabetes and patients with hypertension. Cancer 1993; 71:1279–1282.
68. Herbst A, et al. Comprehensive Gynecology. 2d ed. St. Louis: Mosby-Year Book, 1992: 227–232.
69. Ferenczy A, Gelfand M M. Outpatient endometrial sampling with endocyte: comparative study of its effectiveness with endometrial biopsy. Obstet Gynecol 1984; 63(3):295–302.
70. Koss L G, Schreiber K, Oberlander S G, Moukhtar M, Levine H S, Moussouris H F. Screening of asymptomatic women for endometrial cancer. Obstet Gynecol 1981; 57:681–691.
71. Tsuda H, Kawabata M, Kawabata K, Yamaoto K, Hidaka A, Umesaki N. Comparison between transabdominal and transvaginal ultrasonography for identifying endometrial malignancies. Gynecol & Obstet Investigation 1995; 40 (4):271–3.
72. Varner R E, Sparks J M, Cameron C D, Roberts L L, Soong S J. Transvaginal son-

ography of the endometrium in postmenopausal women. Obstet Gynecol 1991; 78: 195–199.
73. Osmers R, Volksen M, Schauer A. Vaginosonography for early detection of endometrial carcinoma? Lancet 1990; 335:1569–1571.
74. Karlsson B, Granberg S, Wikland M, Ylöstalo P, Torvid K, Marsal K, Valentin L. Transvaginal ultrasonography of the endometrium in women with postmenopausal bleeding—a Nordic multicenter study. Am J Obstet Gynecol 1995; 172(3): 1488–1494.
75. Nasri M N, Coast G H. Correlation of ultrasound findings and endometrial histopathology in postmenopausal women. Brit J of Obstet Gyn 1989; 95:1333–1338.
76. Botsis d, Kassanos D, Pyrgiotis E, Zourlas P. Vaginal sonography of the endometrium in postmenopausal women. Clin Exp Obst Gyn 1992; 21:189–192.
77. Granberg S, Wikland M, Karlsson B, Norström A, Friberg L. Endometrial thickness as measured by endovaginal ultrasonography for identifying endometrial abnormality. Am J Obstet Gynecol 1991; 164(1):47–52.
78. Dørum A, Kristensin G, Langebrekke, A, Sørnes T, Skaar O. Evaluation of endometrial thickness measured by endovaginal ultrasound in women with postmenopausal bleeding. Acta Obstet Gynecol Scan 1993; 72:116–119.
79. Speroff L, Glass R H, Kase N G. Clinical Gynecologic Endocrinology and Infertility. 5th ed. Baltimore: Williams & Wilkins, 1994: 405.
80. Gambrell R D, Massey F M, Castaneda T A, Ugenas A J, Ricci C A, Wright J M. Use of the progestogen challenge test to reduce the risk of endometrial cancer. Obstet Gynecol 1980; 5:732–737.
81. Hanna J H, Brady W K, Hill J M, Phillips G L. Detection of postmenopausal women at risk for endometrial cancer by a progesterone challenge test. Am J Obstet Gynecol 1983; 147:872–875.
82. Pansini F, De Paoli D, Serra M M, Campobasso C, Levato F, Giulini D. Combined use of progesterone challenge test and endometrium thickness evaluated by transvaginal ultrasonography in the preventive management of postmenopausal women. Gynecol Obstet Invest 1992: 34:237–239.
83. American Cancer Society. Cancer Facts and Figures, Summary of Current Guidelines for the Cancer-Related Check-up: Recommendations. Atlanta: American Cancer Society, 1988.
84. American College of Obstetricians and Gynecologists Guidelines for Women's Health Care. 1996, rev. ed. of Standards for Obstetric-Gynecologic Services, 7th ed. 1989.
85. American College of Obstetrics and Gynecology. Carcinoma of the Endometrium. ACOG Technical Bulletin Number 162. Washington, DC: American College of Obstetricians and Gynecologists, December, 1991.
86. American College of Obstetrics and Gynecology, Committee on Professional Standards. Report of the task force on routine screening. In: Standards for Obstetrics and Gynecologic Services. 7th ed. Washington DC: American College of Obstetricians and Gynecologists, 1989: 97–104.
87. American College of Physicians. Guidelines for counseling postmenopausal women about preventive hormone therapy. Ann Intern Med 1992; 117:1038–1041.

88. U.S. Preventive Services Task Force. Guide to Clinical Preventive Services. 2d ed. Baltimore: Williams & Wilkins, 1996: 760–761, 838–839.
89. Canadian Task Force on the Periodic Health Examination. The Periodic Health Examination: 2. 1987 Update. Can Med Assoc J 1988; 138:618–26.

IV

INTO THE FUTURE

21

New Screening Technologies

John K. Gohagan, Sudhir Srivastava, and Susan C. Rossi
Division of Cancer Prevention, National Cancer Institute, Bethesda, Maryland

William C. Black
Dartmouth-Hitchcock Medical Center, Lebanon, and Dartmouth Medical School, Hanover, New Hampshire

I. INTRODUCTION

Technological advances in physics, engineering, and molecular biology in the late 20th century have expanded the traditional window of opportunity for cancer screening from early malignancy to premalignant conditions to even genetic predisposition (Figure 1). It has become possible to identify in newborns gene mutations that may heighten their cancer risk in adulthood. As Edward Golub depicts in ''Heroic Myths and Probability: Confronting the Limits of Screening'' (Chapter 2), technologic advances have outpaced our ability to assess and comprehend the ramifications, good and bad, that the application of these technologies may have on society and individuals. Through marketing and media coverage, technologies can come into use with unmeasured public risk and presumed, but unproven, benefit. Identifying, developing, and assessing the risks and benefits of technologies for cancer screening through clinical trials, technology development, and assessment and computational analysis, and subsequently educating both the public and medical communities in this regard is the mission of early detection research.

The evolution of new cancer screening technologies and their assessment occurs in a cyclic process, which can be envisioned as a technology research cycle (Figure 2). This learning and evaluation process has been described in other fields as an experiential learning model (1). The learning phases are cat-

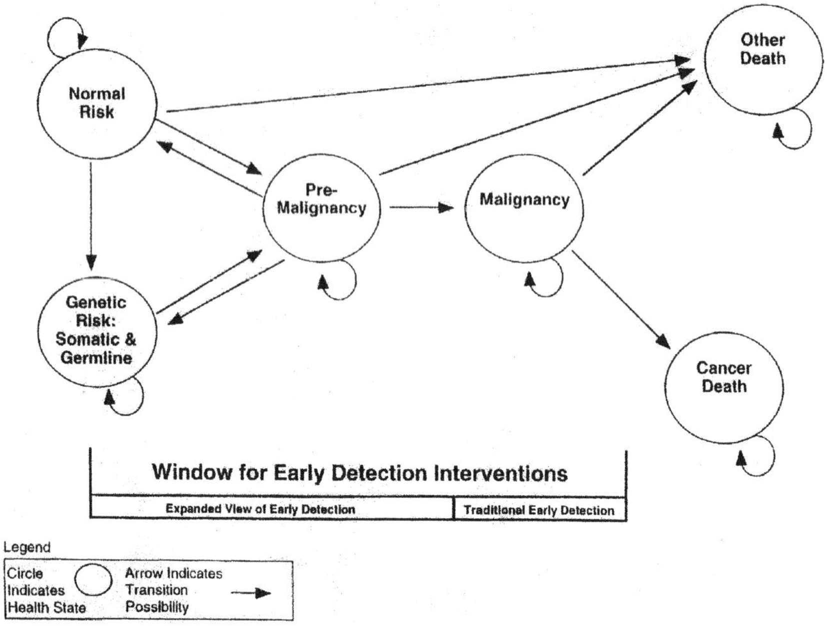

Figure 1 Window of early detection intervention opportunities.

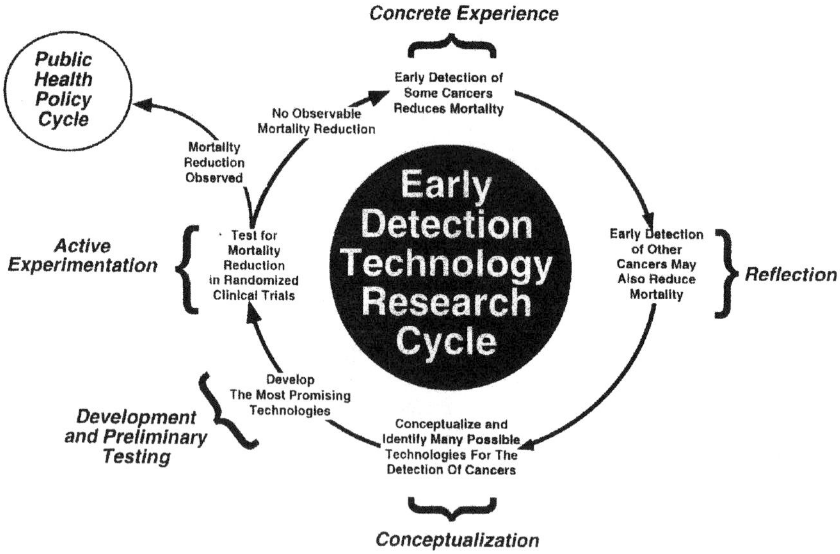

Figure 2 Early detection technology research cycle.

egorized as concrete experience, reflective observation, abstract conceptualization, and active experimentation. The process begins with a ''concrete experience,'' for example, the finding that early detection of breast cancer by mammography reduces breast cancer mortality rate. This concrete experience is generalized to the notion that early detection of other cancers, prostate cancer, for example, may also reduce mortality rate. This reflection results in an active search and subsequent identification or creation of new technologies that might detect cancer or even precancers at earlier points in disease progression. Many technologies are conceived of and undergo at least partial development and evaluation but relatively few produce sufficiently promising preliminary results to justify their evaluation for mortality rate reduction in randomized clinical trials. Two historical scenarios illustrate the process.

Prostatic acid phosphatase (PAP) was found wanting for early detection by clinical experience. Prostate-specific antigen (PSA), developed in the 1970s for monitoring disease burden in treated patients, because of its relative specificity for prostate tissue and its elevation in serum in association with prostate cell proliferation, was quickly reconsidered as a candidate early detection assay. Extensive laboratory and clinical evaluation established it as useful in preclinical disease detection. Its mortality rate-reducing potential is being evaluated in the National Cancer Institute (NCI) Prostate, Lung, Colorectal and Ovarian (PLCO) Cancer Screening Trial, a randomized controlled trial, (RCT) and in a European/United States collaboration called the International Prostate Screening Trials Evaluation Group (IPSTEG) (2).

Sputum cytology as an adjunct to chest radiography was evaluated in three NCI-sponsored randomized controlled trials in the 1970s and in Europe, with disappointing results. These trials, by their designs, could not measure the mortality rate-reducing effect of chest radiography alone, an issue being addressed in the PLCO Cancer Screening Trial.

Other technologies currently in the active experimentation clinical trials phase of the assessment cycle include CA, 125 and transvaginal ultrasound (TVU) for ovarian cancer detection in the United States and the United Kingdom, and flexible sigmoidoscopy and colonoscopy for polyps and colon cancer detection in the United States and the United Kingdom. Plane film mammography continues to be investigated in Europe, the United Kingdom, and Canada for breast cancer screening among women 40 to 49 years of age. And, a host of biomarkers, including BRCA1 and BRCA2, HNPCC, *ras*, p53, p105, IGF-I, and many more are under investigation, as discussed in other chapters in this book. Only mammography for women 50 years of age and older, Papanicolaou smears, and fecal occult blood test (FOBT) have been clearly proven to provide a mortality rate reduction benefit when applied as screening tests. The histories of a few contemporary technologies are listed in Table 1.

Table 1 Time Dimension to Technology Development and Application

1895	1912	1917	1930s	1940s
Roentgen discovers x-rays. First Medical films made within months.	Titanic sinks.	Math principle for cross-sectional imaging	Mammography films	
				SONAR for defense First medical applications MR spectroscopy discovered by Block and Purcell Fiber optics discovered
			Pap smear developed by Papanicolaou	

Abbreviations: HIP, Health Insurance Plan of New York; NIH, National Institutes of Health; CT, computed tomography; TVU, transvaginal ultrasound; TRUS, transrectal ultrasonography; MRI, magnetic resonance imaging; Pap, Papanicolaou smear; FOBT, fecal occult blood test; PSA, prostate-specific antigen; PET, positron emission tomography.

1950s	1960s	1970s	1980s	1990s
Egan improves mammography	HIP mammography tiral	Dedicated mammography Swedish and Canadian mammography trials		Digital mammography NIH Consensus Development Conference on mammography trials
	Laboratory CT	Lung trials Clinical CT		Spiral CT Virtual colonoscopy
			TVU TRUS	Color blood flow Dopplar
		MR image created by Lauterbur	Clinical MRI systems	
	Rigid sigmoidoscopy		Flexible sigmoidoscopy Pap effectiveness accepted/ NIH Consensus Development Conference	
		FOBT assay developed	NCI launches FOBT trial PSA assay developed by Wang CA 125 assay developed by Bass	FOBT proven effective NCI launches 16-year PLCO trial to evaluate CA 125, PSA, flexible sigmoidoscopy, chest X-ray, TVU
Cyclotrons produce radioactiproducts	First PET scanner by Pogosian	Clinical PET scanners		

The characteristics of a particular cancer have direct impact on its suitability for a screening and early detection approach to mortality reduction. Each cancer has a probabilistically characterized natural history of initiation, progression, and sometimes regression (see Figure 1), which affects the ultimate usefulness of a screening technology. The complexity of and counterintuitiveness of the interaction between the screening technology and the cancer are observed in the case of neuroblastoma. A cancer of infants and young children, neuroblastoma qualifies as a cancer that might benefit from an early detection intervention. It has a treatable preclinical phase, and there is a reliable and simple screening method of detecting catecholamine metabolites in urine. But, the several large studies undertaken to assess the risks and benefits of screening infants for neuroblastoma gave a counterintuitive result. The Quebec Neuroblastoma Screening Project set out to determine whether mass screening could reduce mortality rate in a cohort of infants (3). The screening program produced a sharp increase in neuroblastoma incidence from newly identified screen-detected cases. But mortality rate was not reduced in this cohort. This unexpected finding has led clinical researchers to hypothesize the existence of two forms of neuroblastoma, one being less aggressive, prone to spontaneous regression, and easily screen detected, and the other being more aggressive clinically and less responsive to therapy. Another study, this one conducted in Japan, examined the effect of the introduction of mass screening on the age-specific incidence of neuroblastoma (4). As in the Quebec study, a sharp increase in incident neuroblastoma cases was observed; the incidence of neuroblastoma increased from approximately nine cases per million to more than 19 cases per million after the introduction of mass screening and, in infants younger than 1 year of age, the incidence of neuroblastoma quadrupled. This study showed that overtreatment of disease that might spontaneously regress is a considerable hazard to children in such a program; in fact, six of the 10 deaths observed were therapy related (5). Finally, results from an observation program without surgery or chemotherapy were recently reported by Yamamoto and colleagues (6). After observing spontaneous regression almost uniformly for localized neuroblastomas detected by screening among infants by 1 year of age, the researchers recommended a wait-and-see policy. The experience with neuroblastoma illustrates how a reasoned presumption of benefit from early detection practices can be incorrect and lead to unintended consequences.

A common tendency is to assume that the ability to detect cancer, defined by the operating characteristics of a given technology, predetermines screening effectiveness and mortality reduction, but early detection does not preordain this benefit as both the lung (sputum) and neuroblastoma cases illustrate.

The primary operating characteristics of a technology are defined as its sensitivity to the presence of the condition or disease being screened for, its specificity for that condition, and its ability to predict the presence of the condition, the positive predictive value (see Chapter 3). Knowing these parameters is important to determining the reasonableness of conducting a randomized controlled trial (RCT) to assess the mortality impact of the detection technology.

Highly sensitive and specific technologies with substantial positive predictive values (PPYs) are good candidates for further study with respect to mortality rate reduction if it can be shown that the cancer is more curable when detected by screening. The predictive accuracy or the PPV of the technology is strongly influenced by the prevalence of the condition in the screened population, so the application of even a highly sensitive and specific screening technology may be inappropriate when the condition screened for is rare.

Table 2 summarizes incidence and mortality statistics for organ-specific cancers as well as median age at diagnosis, based on Surveillance, Epidemiology, and End Results (SEER) registries, as a way of depicting each cancer in one logical order of public health impact (7), and the technologies either in use or under development for screening for each cancer or preclinical condition are shown. This presentation helps to emphasize the ''many to one'' concept, that is, many technologies emerge as interesting but few have actually been shown to reduce mortality rate.

The remainder of this chapter describes, illustrates, and tabulates the status of the wide array of radiological, spectroscopic, and molecular-biological technologies in use and under development for early detection (by screening) of premalignant and malignant conditions.

II. IMAGING

In November 1895, the German physicist Roentgen discovered X-rays. Within 3 months, the first clinical X-ray study in America was made (8). During the past century, there have been numerous innovations in medical imaging and continual refinements in radiography and other imaging techniques. Following is a brief review of the imaging modalities that are being used or have strong potential for use as screening tests for cancer. This review considers imaging procedures that may be used as the initial screening test or to evaluate a positive or ambiguous result from the initial screening test. This review does not include all the imaging procedures used in cancer staging and surveillance, even though they may occasionally have a critical impact on the effectiveness or cost of a screening strategy.

A. Radiography

The essential elements of conventional radiography are an X-ray tube and radiographic film (9). The object of interest is placed between the X-ray tube and the film, and it absorbs and deflects some of the X-rays traveling toward the

Table 2 Screening Technologies by Disease Burden and Mortality Impact (1990–1994)

Tissue	U.S. incidence[a]	U.S. mortality[a]	Median age at diagnosis (year)	Radiography	Endoscopy	Serum/blood markers	Markers in bodily fluids, brushings, feces, secretions	Germ-line markers of at-risk people	Gene-environment markers of at-risk people	Technologies shown to reduce mortality
Prostate	160.7	26.5	71	TRUS		PSA; free PSA; PSAV	Ejaculate cells *ErbB*-2	HPC		
Breast	110.2	26.4	64	Mammography; digital mammography; MRI; ultrasound		CA15-3; CEA; HCG	Nipple aspirate cells; telomerase	BRCA1; BRCA2	ER	Mammography
Lung	58.2	49.9	68	Chest X-ray; virtual bronchoscopy	Bronchoscopy	CEA; ACTH; ADH; NSE	Sputum cells; bronchoscopy		CYP	
Colon	46.2	18.1	72	Barium enema; virtual colonoscopy	Sigmoidoscopy	CEA; CA72-4	FOBT; exfolia	APC; MLH1; MSH2		FOBT
Endometrium	21.1	1.8	66	TVU			Endometrial biopsy; endometrial brushing; PCT			
Bladder	17.0	3.3	71		Cytoscopy		BTA; NMP22; hyaluronidase			
Ovary	14.9	7.8	63	TVU		CA125; CA15-3; CA72-4; HMFG2	Telomerase	BRCA1; BRCA2		
Stomach	7.4	4.5	71		Gastroscopy	PGI	FSA			
Melanoma	12.2	2.2	55							
Pancreas	9.0	8.5	72			POA; CGI; RNAse; LAIP; CA19-9				
Kidney	9.1	3.5	66							
Cervix	8.3	2.9	47		Cervicography		PAP; telomerase			PAP
Liver	2.9	2.6	67	Ultrasound		AFP				

Abbreviations: TRUS, transrectal ultrasound; PSA, prostate specific antigen; PSAV, PSA velocity; HPC, human prostate cancer gene; MRI, magnetic resonance imaging; CA, cancer antigen; CEA, carcino-embryogenic antigen; HCG, human chorionic gonadotropine; BRCA1, breast cancer gene 1; BRCA2, breast cancer gene 2; ER, estrogen receptor; ACTH, adrenocorticotropic hormone; ADH, antidiuretic hormone; NSE, neuron-specific enolase; CYP, cytochrome P; FOBT, fecal occult blood test; APC, adenomatous polyposis coli; MLH1, human analogue of the yeast mismatch repair gene (homologue of bacterial MutL gene); hMSH2, human analogue of the *S. cervisia* MSH2 gene; TVU, transvaginal ultrasound; PCT, progesterone challenge test; BTA, bladder tumor antigen; NMP22, nuclear matrix protein 22; HMFG, human milk fat globule glycoproteins; PGI, pepsinogen I; FSA, fetal sulfoglycoprotein antigen; POA, pancreatic oncofetal antigen; RNAse, ribonuclease; LAIP, leukocyte adherence inhibition protein; PAP, papanicolaou smear; AFP, alpha-feto protein.

[a]Number per 100,000, age-adjusted to the 1970 U.S. Standard Population, SEER

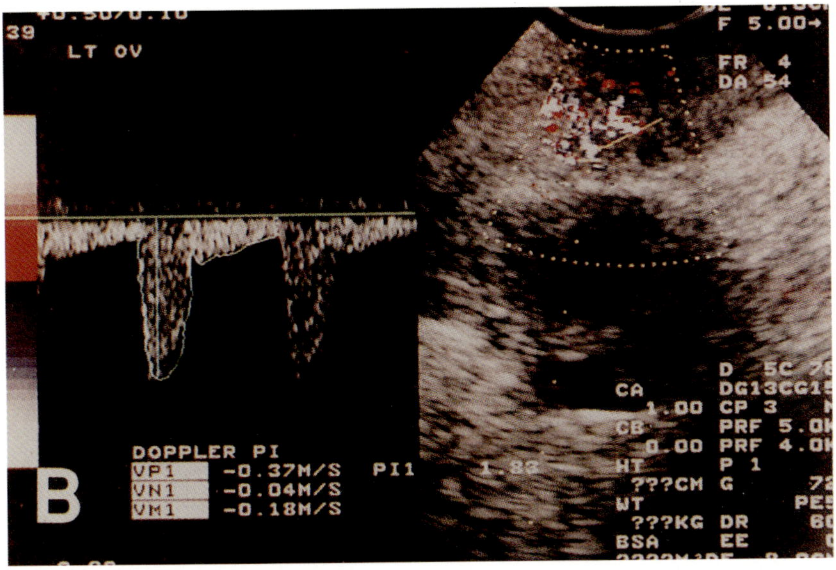

Figure 5B Transvaginal color Doppler ultrasonography depicts numerous vessels adjacent to cyst, with some displaying high velocity. (From Fleischer AC, Cullinan JA, Peery CV, Jones HW. Early detection of ovarian carcinoma with transvaginal color Doppler ultrasonography. Obstet Gynecol 1996; 174(1PT 1): 101–106.)

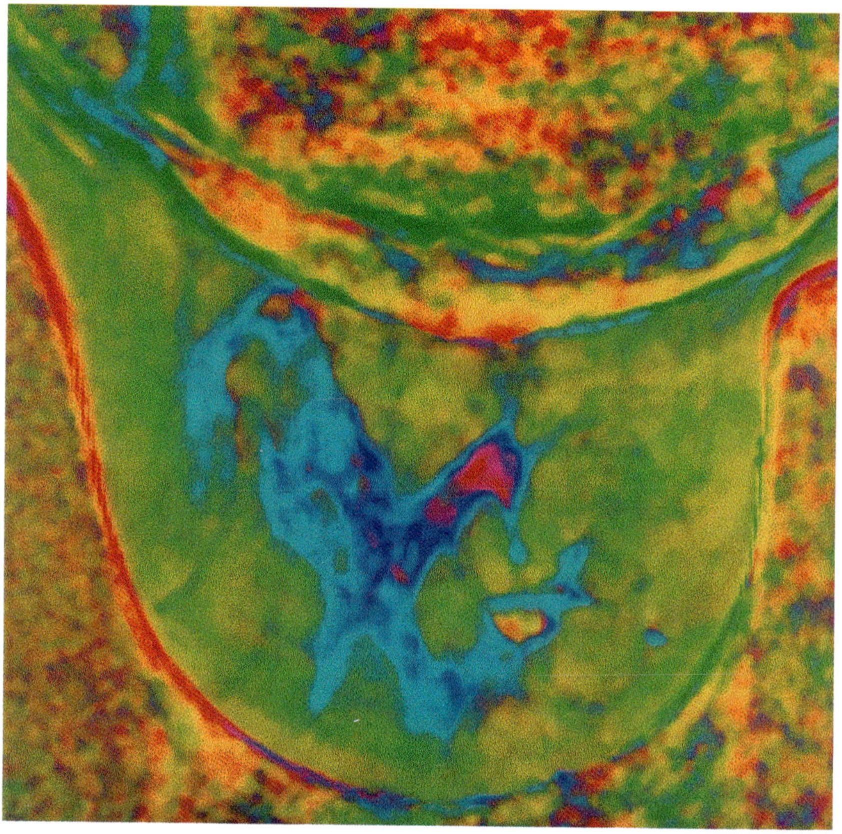

Figure 6 Magnetic resonance image of invasive ductal breast carcinoma. Breast is dependent, chest wall arching down from top of figure. False color based on composite image intensity (T1-weighted; T2 -weighted; DX-fat/water separation. Siemens 1.5 Tesla).

film. The resulting image depends on many technical factors as well as the size, shape, and X-ray attenuation characteristics of the object. Air has a very low attenuation coefficient and transmits almost all of the incident X-rays. The attenuation characteristics of body tissues depend mainly on their density. Thus, bone has a very high attenuation coefficient, whereas soft tissues have attenuation coefficients between those of air and bone. Radiography is particularly useful in the evaluation of the lungs, where there is a high degree of contrast between normal lung tissue, which is approximately air density, and lung pathology, most forms of which are approximately water density.

Chest radiography is usually performed with radiographic units designed for the chest. Chest radiography is safe, acceptable to persons being screened, widely available, and relatively inexpensive. The estimated effective dose equivalent is about 6 mrem according to the National Council on Radiation Protection and Measurements (NRCP) (10). The threshold size for visibility of a nodule is about 6 mm when it is projected over an intercostal space (11), although nodules measuring greater than 1 cm in diameter are often hidden by ribs or clavicle, hilar structures, mediastinum, or diaphragm (12). One recent development in chest radiography is an asymmetrical screen-film combination. One study comparing the new asymmetrical screen-film combination to a conventional system demonstrated that the former improved visualization of mediastinal, retrocardiac, and diaphragmatic regions while increasing contrast and resolution in lung parenchyma (13,14).

Chest radiography has been evaluated in the setting of lung cancer screening (see Chapter 13). In the context of randomized clinical trials, more lung cancer was detected in the screened groups than in the control groups and, where reported, the stage distribution of lung cancer cases was more favorable in the screened groups. However, none of these trials demonstrated a reduction in lung cancer mortality rate, and screening with chest radiography is no longer widely recommended in the United States (15). Recently, however, the design and interpretation of the Mayo Clinic Study, the largest RCT of chest radiographic screening, has been criticized (16). Whether the lack of mortality rate reduction was due to inadequate sample size, methodological problems, the limitations of chest radiography, or simply the lack of effectiveness of early detection in lung cancer remains controversial. Some of this controversy may be resolved by the ongoing PLCO trial, which includes annual chest radiographs for 4 consecutive years (an initial screen plus three annual screens). Some of the sites participating in the trial will use the new asymmetrical screen-film combination previously described.

B. Mammography

Although clinical breast radiography was first reported in 1930 (17), the images were of poor quality and not reproducible. In 1960, Egan described a high milliamperage–low kilovoltage technique that renewed interest in mammography (18). Ten years later, the first dedicated screen-film mammography system was marketed for screening. Further improvements in film quality, radiation dose, positioning, breast compression, and biopsy technique have occurred since.

Mammography is currently performed with dedicated mammography units that produce higher spatial resolution images (15 line pairs per millimeter versus 4 for chest radiography) and lower energy X-rays (20 keV versus 140 keV for chest radiography). These features are necessary for detecting small microcalcifications and discriminating the relatively subtle density differences between the soft tissue components of the breast (fat, glandular tissue, and cancer). Another important feature of mammography is breast compression, which holds the breast firmly against the film holder. Compression improves image quality (by decreasing motion, geometric unsharpness, and scatter radiation) and decreases the radiation dose to the breast. Grids in widespread use to reduce film blurring by X-ray scatter actually increase dose, although not beyond U.S. Food and Drug Administration standards.

Mammography is safe, widely available, and relatively inexpensive. The procedure is mild to moderately uncomfortable for most women, primarily because of the breast compression. When properly performed, mammography delivers a mean glandular dose of about 2.5 to 5.0 mGy (0.25 to 0.5 rad) per two-view examination (19). The accuracy of film-screen mammography is highly dependent on the composition of the breast, which varies with age, hormonal therapy, and other factors. Because fat is less dense than cancer, lesions as small as a few millimeters are detectable in breasts that are predominately composed of fat. However, because the density of glandular tissue is similar to that of cancer, lesions as large as a few centimeters may be undetectable in breasts that are predominately composed of glandular tissue.

Although there have been no major technical improvements in film-screen mammography during the past several years, the consistency of mammographic screening has been improved by quality control measures for the entire screening process, including the technical performance and the radiological interpretation. This has been achieved largely through a voluntary program established by the American College of Radiology and subsequent legislation establishing reimbursement standards through the Mammography Quality Standards Act (MQSA) enforced by the U.S. Food and Drug Administration. In addition, several recent studies suggest that computer-aided diagnosis (CAD) of radiologist-extracted features may improve the accuracy and consistency of mammographic interpretation in the future (20,21). One major advantage of CAD is that it can be programmed to operate at a predefined threshold that is optimal for the particular screening conditions.

One of the most promising areas of investigation is digital mammography (21,22). Digital mammography offers three major advantages over the conventional film-screen mammograph (22). First, digitized images can be processed to improve contrast and hence lesion detectability (Figure 3). Second, digitized images can be directly analyzed by a computer to improve the accuracy and consistency of mammographic interpretation. Third, digitized images can be transmitted rapidly over long distances for expert consultation. Although conventional mammograms and other radiographic films can be readily digitized with existing technology, this approach is limited by the quality of the radiographic image and the reproduction process. In the future, it is expected that film will be replaced by a receptor that directly transmits the information in the mammographic image to a computer. Experimental versions of digital imaging mammography exist, but further development of digital image receptors are needed.

Mammography has undergone more extensive evaluation than perhaps any other screening test for cancer (see Chapter 7). Several randomized clinical trials since the early 1960s have demonstrated a breast cancer mortality rate reduction

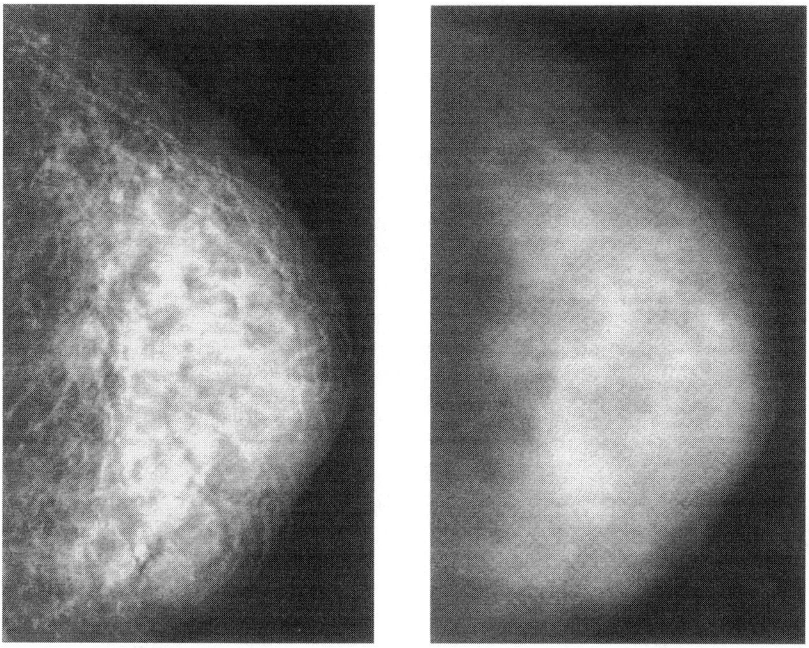

Figure 3 Film screen and digitized, enhanced reconstructed mammograms. Global enhancement via suboctave multiscale analysis. An enhanced image is reconstructed from analysis coefficients modified by adaptive nonlinear processing. *A*, Original digitized mammogram of a dense breast, 110 micron spot size. *B*, Processed digital mammogram, showing improved contrast of spicular mass, 1.5 cm diameter, and microcalcifications.

of 30% to 40% in women 50 to 69 years of age (23). Evidence for a mortality rate reduction in women 40 to 49 years of age is less conclusive; and, if present, the degree of mortality rate reduction appears to be lower than for older women (24). A randomized trial attempting to resolve this question is in its sixth year in the United Kingdom. Since all of the completed trials have been carried out, there have been major increases in the sensitivity of mammography, as evidenced by the decreasing size and stage distribution of screen-detected cases. The nearly four-fold increase in the reported incidence of ductal carcinoma in situ (DCIS) from 1983 to 1992 has been attributed to the increased use and sensitivity of screening mammography (25). However, how these improvements in mammographic sensitivity influence breast mortality rate and the overall quality of life for women who participate in screening is uncertain. How to evaluate continual refinements of mammographic technology and the impact of secondary tests for evaluating suspicious or equivocal mammographic findings, such as breast magnetic resonance imaging (MRI), has already become a major challenge.

C. Computed Tomography

Although the mathematical principles underlying computed tomography (CT) were proven in 1917 by J. Radon (9), it remained only a theoretical possibility for several decades. About 50 years later, Hounsfield, a British scientist, developed the first laboratory CT scanner, which had a scanning time of 9 hours (26). The first clinical prototype head scanner was installed in London in 1972 and it was an immediate success. In 1979, Hounsfield received the Nobel prize for his contributions to the development of CT.

A CT examination is performed differently than radiography. With CT, the subject is moved longitudinally through a gantry with a rotating X-ray tube and detector (9). The detector measures the amount of transmitted radiation throughout each rotation. This information is processed by a computer, which generates a cross-sectional image based on the density of the internal structures. The major advantages of CT over radiography are that it eliminates the superposition of anatomic structures and provides much greater contrast resolution. The CT scan can resolve density differences of 0.5% or less, whereas film-screen radiography requires density differences of about 10% (9). The major disadvantages of CT are the increased radiation dose and cost.

Since it was first used clinically in 1972, CT has undergone numerous refinements. A major recent development is spiral (helical) CT, in which slip ring coupling of the rotating and stationary components of the gantry enables subjects to be scanned continuously and rapidly (during a single breath-hold) (27). Spiral CT greatly reduces the problem of motion misregistration in conventional CT scanning and makes possible multiplanar and three-dimensional display of the anatomy. New computer graphics techniques can provide an intraluminal perspective with real-time ''fly-through'' capabilities for any hollow viscus. Virtual reality imaging is currently being applied to the bronchus, colon, stomach, bladder, larynx, and paranasal sinuses (28).

1. *Virtual Colonoscopy*

Various spiral CT techniques for the colon—often collectively referred to as *virtual colonoscopy*—are under investigation (28–32). Unlike sigmoidoscopy and colonoscopy, these CT techniques are minimally invasive, requiring only cleansing and air insufflation of the colon. Although the complex geometry of the colon makes it difficult to evaluate on axial CT images, investigators have developed static two-dimensional and three-dimensional and virtual reality fly-through computer processing techniques that greatly enhance the detection of polyps (Figure 4). In a recent study, Hara et al. (29) detected all 34 colonic

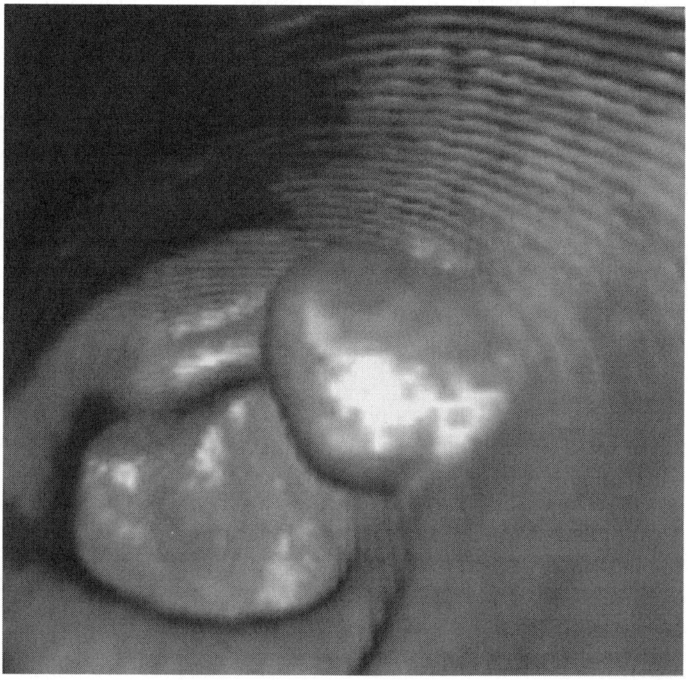

Figure 4 Spiral computed tomography (CT) colonography (virtual colonoscopy) is a novel technique capable of providing an endoscopic view of the colonic mucosa for detection of colorectal polyps. This is a CT image of a 2-cm sigmoid polyp using perspective volume rendering (Vitrea TM, Vital Images, Inc., Minneapolis, MN).

polyps greater than or equal to 5 mm and five of nine (56%) polyps less than 5 mm. The time required for a complete examination was less than 60 minutes: 15 minutes for patient data acquisition and axial CT reconstruction, 10 minutes for image processing, and 15 to 30 minutes for image interpretation. The estimated radiation dose was 187 mrad for men and 285 mrad for women, which is about the same as the radiation dose from the conventional barium enema and one-fourth the radiation dose from a standard CT of the abdomen and pelvis (29).

Much more research is needed before virtual colonoscopy can be tested in screening trials. The methods of colonic preparation, data acquisition, computer rendering, and image display (static two-or three-dimensional or virtual reality fly-through) and interpretation need to be improved and standardized (28,32). One major limitation of current techniques is a high rate of false-positive results. In a recent study involving a single 72-year-old male subject with computer-simulated polyps, one of the two observers had five false-positive findings in the rectosigmoid (31). In addition, current techniques are less sensitive than colonoscopy in the detection of small polyps (less than 5 mm) and flat lesions (32). However, the clinical significance of this lower sensitivity is unknown and probably cannot be determined without testing in a randomized clinical trial.

2. *Chest Computed Tomography*

Lung cancer screening with spiral CT is also under investigation (33). Spiral CT enables the lungs to be scanned during a single breath-hold, thereby eliminating the problem of respiratory misregistration in conventional CT scanning (which may cause a nodule to be missed). In a recent study (34) of 1,369 Japanese smokers who were screened with both chest radiography and low-dose spiral CT, only four peripheral lung cancers were detected by the former test, whereas 15 were detected by the latter (see Figure 7 in Chapter 13). The mean diameter of tumors detected by spiral CT alone was only 12 mm. Spiral CT also has potential for detecting endobronchial lesions. With perspective volume rendering, a new three-dimensional image processing technique, spiral CT provides an internal view of the tracheobronchial tree that simulates the fiberoptic bronchoscopic perspective and is sometimes referred to as *virtual bronchoscopy* (see Figure 8 in Chapter 13) (33,35).

Spiral CT of the chest is fast and probably the most sensitive noninvasive method of detecting lung cancer. However, the radiation dose from screening spiral CT of the chest is about 10 times (260 mrem) that of plain chest radiography (34). In addition, because spiral CT is more sensitive than chest radiography for benign as well as malignant lesions, false-positive results are a potentially serious problem. Furthermore, many of the small lung cancers that can be detected by spiral CT may be pseudodisease. As with colorectal cancer screening, much more research is needed before spiral CT can be tested in lung cancer screening trials.

D. Ultrasonography

In 1912, ultrasound (US) was used in a failed attempt to discover the sunken Titanic (9, pp 323–371). The first successful imaging application of ultrasound was sonar, which was developed during World War II. Shortly thereafter, during the late 1940s, the first successful applications in medicine were begun. Clinical ultrasound has since continued to progress dramatically.

In clinical US, the subject is scanned by a hand-held transducer (9). The transducer converts electrical energy into ultrasonic energy that is transmitted into body tissues and converts the ultrasonic energy reflected back from the tissues into an electrical signal, from which the image is derived. Thus, the resulting image is mainly determined by the acoustic impedance of the tissues being scanned. One major advantage of US over radiography and CT is that the former does not involve the use of ionizing radiation, which promotes carcinogenesis. Another advantage is that the images can be viewed as they are acquired (in real time) from multiple projections. One major disadvantage of ultrasound, however, is that it does not readily penetrate gas or bone so that the quality of the examination is dependent on the operator's skill and the subject's body habitus. Consequently, the US examination is relatively subjective and difficult to reproduce.

Real-time imaging can be coupled with color Doppler imaging, which is sensitive to the motion of moving red blood cells. The blood vessels in malignant tumors have relatively little smooth muscle in their walls and therefore less resistance to blood flow. In theory at least, color Doppler imaging could be helpful in distinguishing "malignant" from physiological and "benign" causes of increased blood flow. To date, however, reliable criteria for malignancy have not been developed (36,37).

1. *Transvaginal Ultrasonography*

Screening for ovarian cancer with transvaginal ultrasonography (TVU) is currently under investigation (36, 38, 39). Various morphological and Doppler criteria for malignancy have been developed and evaluated in different screening populations. Although it is well established that TVU can detect ovarian cancer earlier than pelvic examination and CA 125, there is much uncertainty about the natural history of ovarian cancer and the effectiveness of screening (Figure 5). In addition, there is a major problem with false-positive results and there may also be a problem with overdiagnosis. In a recent study of 1,601 women with a family history of ovarian cancer (40), 57% required follow-up of initial abnormalities on TVU, which included real-time and color Doppler imaging.

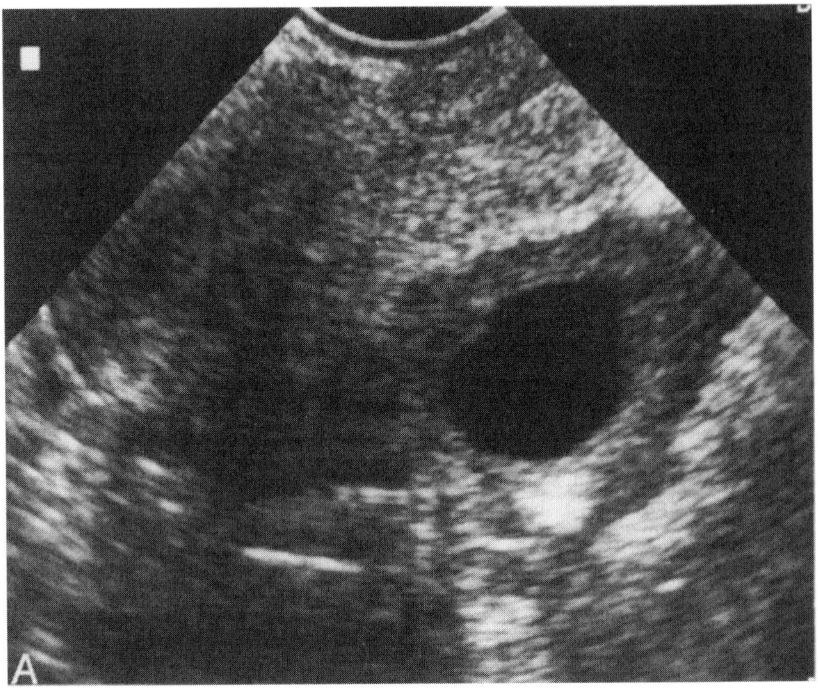

Figure 5 Stage I moderately differentiated ovarian serous cystadenocarcinoma in a 47-year-old woman. A) Transvaginal color Doppler ultrasonography shows 17-mm cystic area within left ovary. B) See color insert. (From Fleischer A C, Cullinan J A, Peery C V, Jones H W. Early detection of ovarian carcinoma with transvaginal color Doppler ultrasonography. Am J Obstet Gynecol 1996; 174(1 Pt 1): 101–106.)

After follow-up, 61 women (3.8%) were ultimately thought to have positive examinations and were referred for surgical exploration. Only six women had ovarian cancer, three of which were of low malignant potential. Benign ovarian neoplasms were found in 48 women and no abnormalities were found in the remaining seven.

In April 1994, a National Institutes of Health (NIH) Consensus Panel concluded that "there is no evidence available yet that the current screening modalities of CA-125 and transvaginal sonography can be effectively used to reduce the mortality from ovarian cancer nor that their use will result in decreased rather than increased morbidity and mortality" (41). TVU is currently being evaluated in the PLCO trial (38,42). However, the results of this trial will not be available for several years. Currently, no major medical organization recommends mass screening for ovarian cancer with pelvic examination, CA 125, or TVU. However, both the NIH Consensus Panel and American College of Physicians encourage women to participate in clinical trials and recommend that women with a family history of hereditary ovarian cancer syndrome consult a specialist to discuss their options for testing and prophylactic surgery (39,41).

A screening role for TVU may eventually be established if its specificity can be improved by advances in real-time or color Doppler imaging or new prediction criteria, or if other imaging techniques, such as MRI, can be used to better distinguish true-and false-positive results (36). If a screening role for TVU is established by RCT, it may be most valuable for women with a high risk for ovarian cancer on the basis of family history or genetic testing (36,39).

2. *Breast Ultrasonography*

Breast ultrasonography is used to evaluate abnormal findings on clinical and mammographic screening examinations (43). Breast ultrasonography is mainly used to diagnose simple cysts, which do not generally require a biopsy. Breast ultrasonography is also used to guide interventional diagnostic procedures, including preoperative needle localization, fine needle aspiration of cystic masses, and core needle biopsy of solid masses. A recent study suggests that breast ultrasonography may also be useful for excluding cancer in certain solid lesions (44), but this role is not currently accepted (45). Despite recent technological advances, breast ultrasonography lacks both the sensitivity and specificity of mammography and is not recommended as screening test by any group, even in women with mammographically dense breasts. Nevertheless, various ultrasound techniques, including color Doppler, are under intense investigation, and the role of ultrasonography in breast cancer screening could expand in the future.

3. *Transrectal Ultrasonography*

Transrectal ultrasound (TRUS) has been the most commonly used imaging modality in screening for prostate cancer. However, TRUS is not used as a primary screening test because it lacks sufficient specificity. Instead, TRUS is used to direct biopsy in men suspected to have prostate cancer on the basis of an elevated PSA level or an abnormal digital rectal examination (46).

E. Magnetic Resonance Imaging

The first experiments in nuclear magnetic resonance (MR) were performed in the 1940s, and the Nobel prize was awarded to physicists Purcell and Block in 1951 (47) for their discovery of the phenomenon. Subsequently, this technology became a valuable analytical tool in chemistry. The notion that MR might also have applications in medical imaging originated with Jackson, who generated the first MR signals from a live animal in 1967. Subsequently, Lauterbur produced the first two-dimensional MR images (of a water sample) in 1972 and the first images of a live animal in 1974. The first human images were reported in the late 1970s. Magnetic resonance imaging has since undergone numerous refinements that have dramatically increased image quality and reduced scanning time.

In an MRI examination, a subject is placed in a magnetic field (47). The subject is pulsed with radio frequency waves and the returning radio frequency signals are processed by a computer, which generates an image based mainly on the (hydrogen) proton density and T_1 and T_2 "relaxation" characteristics of the tissues (47). In comparison to CT, MRI does not use ionizing radiation, MRI is less degraded by bone artifact (and therefore produces the better images of the brain and spine), and MRI can better discriminate soft tissues that are of similar density. Intravenously injected gadolinium contrast agents accumulate within neoplasms because of their hypervascularity and expanded interstitial spaces (22). Although MRI generally provides less spatial resolution and takes longer to complete than CT, new three-dimensional imaging techniques eliminate the gaps between slices, provide thinner slices than two-dimensional imaging, and enable reformation of the image in any desired plane. In addition, imaging sequences are being made increasingly faster. However, MRI will probably not be used as a screening test in the foreseeable future because of its high cost and limited specificity. Nevertheless, MRI may play an increasing role in the evaluation of abnormalities found by screening examinations, particularly in breast cancer screening.

In this context, breast MRI is mainly being used to evaluate suspicious or equivocal mammographic findings (48,49). Several MRI sequences using high spatial resolution (1 mm in all directions), fat suppression and contrast enhancement, and rapid data acquisition (5 minutes) have been designed for these purposes. In studies of women with nonpalpable, suspicious, or equivocal findings on mammography, the sensitivity of MRI has been over 90% for malignant lesions. In a study of clustered microcalcifications, ranging in size from 3 to 50 mm (mean, 12 mm), MRI detected 95% of malignant lesions (invasive carcinoma and DCIS)(Figure 6) (50). In a recent study of DCIS, early contrast enhancement was demonstrated in 34 of 36 patients with DCIS and the enhancement was attributed to the presence of tumor angiogenesis in the stroma (51). In another recent study, MRI demonstrated enhancement in all cases of mammographically detected DCIS and helped to distinguish pure DCIS from that with an invasive component (52). Certain architectural features, such as smooth or lobulated borders, appear to be highly predictive of benign disease (53).

The major problem with breast MRI is that a variety of benign lesions—fibroadenomas, atypical hyperplasia, fibrosis—and even normal glandular tissue can enhance with gadolinium contrast (54). Consequently, the reported specificity has been low, around 50% (49). After further refinements and standardization of the technique, contrast-enhanced MRI may help eliminate unnecessary biopsies in the evaluation of women with suspicious mammographic findings (49).

F. Nuclear Medicine Imaging

In contrast to most imaging modalities, which provide primarily anatomic information, nuclear medicine imaging provides information about tissue metabolism and physiology. For this type of imaging, the subject is injected, usually intravenously, with a radioactive tracer attached to a compound that is active in tissue metabolism. The radioactive tracer emits high-energy photons (or positrons) that are imaged by an external camera. Because there are a variety of radioactive tracers and metabolically active compounds suitable for imaging, the potential oncological applications of nuclear medicine are vast. However, the potential screening applications of these procedures are limited because of their radiation dose, limited spatial resolution, cost, unavailability, or some combination of these factors. Consequently, no nuclear imaging procedure will be used as a first line screening test in the foreseeable future. Nevertheless, some of these tests may play an important role in the evaluation of positive or equivocal screening studies.

1. *Planar Imaging*

Conventional planar scintigraphy is similar to conventional radiography in that both provide two-dimensional images of a subject or a subject's body part. One form of planar imaging, scintimammography (SMM), has been investigated as a second line test in breast cancer screening (55). The radiopharmaceutical is Tc-99m sestamibi, a lipophilic molecule whose cellular uptake correlates with mitochondrial activity and density. Several studies of SMM have reported sensitivities and specificities around 90%. However, these studies included primarily palpable lesions. The sensitivity of SMM is much lower for lesions that are less than 10 mm in size or have little desmoplastic activity, such as DCIS. In a recent study of nonpalpable lesions detected by screening mammography, the sensitivity was only 56%, leading the authors to conclude that SSM was not useful for excluding biopsy for such lesions (55). The specificity of SMM is also limited by uptake of the agent in benign proliferative conditions (56).

2. *Single-Photon Emission Computed Tomography*

Single-photon emission computed tomography (SPECT) provides three-dimensional image information and is often used as an adjunct to planar imaging. In SPECT, standard radionuclides, including Tc-99m sestamibi (55), and one or more scintillation cameras are rotated around the subject. The projection data are reconstructed by a computer that provides cross-sectional images in different planes. The advantage of SPECT over routine planar imaging is that the former provides greater sensitivity and better anatomic localization. However, SPECT requires more costly instrumentation and longer imaging times than conventional planar scintigraphy. The SPECT technique is also particularly susceptible to imaging artifacts. To date, its role in oncological imaging has been restricted to staging.

3. *Positron Emission Tomography*

During the early 1950s, cyclotrons became available for the production of radioisotopes and the first planar positron scanner was built (57). The first cross-sectional imaging technique was described in the 1960s and positron emission tomography (PET) hardware has since undergone continual improvement.

Positron emission tomography provides higher resolution than SPECT and can be used with a greater varicty of basic metabolic substrates. The PET radiotracer most commonly used for tumor imaging is the glucose analog 2-[F-18]fluoro-2-deoxy-D-glucose (FDG), which is avidly taken up by malignant tumor cells. However, clinical applications of PET are limited by its relatively high cost and limited availability of cyclotron-produced positron emitting radionuclides.

With regard to cancer screening, PET has been most extensively investigated in the evaluation of the solitary pulmonary nodule (58). The reported sensitivity ranges from 80% to 100%. The reported specificity is slightly lower, 60% to 100%, because of FDG uptake by benign inflammatory lesions. If screening for lung cancer with chest radiography or spiral CT is eventually shown to be effective, PET may be useful in evaluating solitary pulmonary nodules with intermediate probabilities of being malignant. (PET has also been shown to be more accurate than CT in detecting metastasis to mediastinal lymph node and distant sites in patients with known lung cancer [58,59].) Positron emission tomography also appears to be highly sensitive in the detection of breast cancer, but most published studies have only included primary tumors greater than or equal to 1.0 cm (58). More research is needed to determine the accuracy of PET for smaller breast tumors. Combining PET metabolic images and anatomic images provided by other radiological modalities with greater spatial resolution is under investigation (21).

G. Optical Coherence Tomography

Light-based optical coherence tomography (OCT) is an exciting new technology that has great potential for medical imaging (60). The OCT is similar to ultrasonography except that infrared light waves are used instead of acoustic waves. An optical beam is focused into the tissue and the light reflected back from the internal microstructure is used to create a cross-sectional image, which has been described as an *optical biopsy*. The OCT provides a spatial resolution of 10 to 20μm, which is 10 times higher than that of any clinically available imaging modality. Its major limitation is that the depth of imaging is only 2 to 4 mm. Tearney, et al. (61) have recently obtained optical biopsies from the gastrointestinal and respiratory tracts of living rabbits using a catheter endoscope 1 mm in diameter (Figure 7). Because OCT is noninvasive, nontoxic, and relatively inexpensive, it has strong potential as a screening test. How accurate it is in detecting small cancers will undoubtedly be a subject of intense investigation.

III. BIOMARKERS

Whereas imaging technologies observe morphological features and sometimes metabolic information, biomarkers detect chemical signals of abnormalities in bodily fluids such as blood products, urine, and mucus. Cancer screening requires that excess risk not be incurred by otherwise healthy individuals. This requirement frequently limits the sampling to easily collected fluids, cells, and tissues. This is a major limitation to screening. The dramatic advances in the molecular biological characterization of tumorous tissue has not had a significant impact on early detection technologies because tissue cannot be easily tested without imparting real risk to the the person being screened.

A. Serum and Blood Markers

Tumor markers have long been recognized for their potential contribution to discriminating cancer cells from noncancerous cells. The use of antibodies to treat cancer with serum from animals that were immunized against human tumors was first reported in 1895 (62). Although the experiments did not succeed, the principle of raising xenographic antibodies to human tumors paved the way

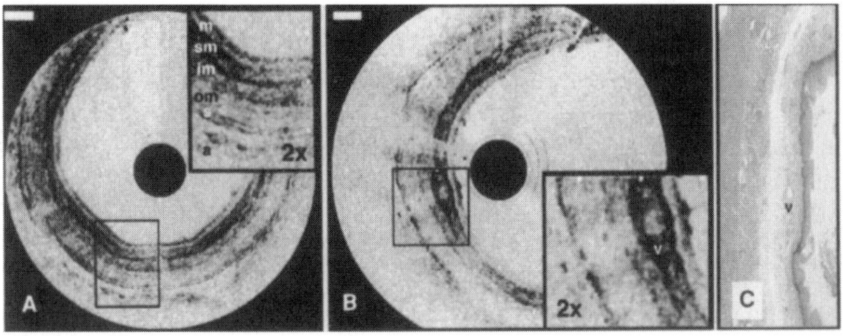

Figure 7 Optical coherence tomography (OCT) of the rabbit esophagus in vivo. A) Esophageal layer of the rabbit including the mucosa (m), the submucosa (sm), the inner muscular layer (im), the outer muscular layer (om), the serosa (s), and the adipose and vascular supportive tissues (a). B) Blood vessel (v) in the submucosa of the esophagus. C) Corresponding histology for B. (Reprinted with permission from Tearney GJ, Brezinski ME, Bouma BE, et al. In vivo endoscopic optical biopsy with optical coherence tomography. Science 1997; 276: 2037–9. Copyright 1997 American Association for the Advancement of Science.)

for studying and recognizing antigens expressed or secreted from the cell surface. In 1965, Gold and Freedman (63) identified carcinoembryonic antigen (CEA) by immunizing rabbits with extracts of colonic carcinoma and characterizing the antigen that reacted with the resulting antibodies. The rate of discovery of various types of CEA secreted by tumors has been phenomenal in the past 4 decades. Although the early hopes of finding tumor-specific biomarkers have been disappointing, there has developed an appreciation that tumor cells secrete a large number of biomolecules that are either absent in normal cells or present in very small amounts (64).

The CEAs belong to the mucin family, with some exceptions. Historically, mucins have been considered a mixture of mucoproteins and mucopolysaccharides. They have molecular weights of more than 200,000 dalton and are usually made up of more than 40% to 50% carbohydrate. Mucins generally contain oligosaccharide groups linked through O-glycoside with either serine or threonine of the peptide core. However, CEA (e.g., CA 19–9) also has N-linkage with oligosaccharide. CA 125 epitope contains less than 25% carbohydrate. The CEAs, glycoproteins of about 200,000 dalton, were first thought to be present exclusively in gastrointestinal carcinomas and the fetal digestive tract (63). However, with the advent of sensitive radioimmunoassays, CEAs were discovered in a variety of tumors, providing a rationale for the clinical applications of CEA in cancer diagnosis. The use of CEA as a serum marker in detection of primary tumors has been limited. Among several classes of CEA markers, CA 125 has been widely used in the monitoring of ovarian adenocarcinoma. The screening value of using CA 125 remains to be determined. The discovery of CEA markers set the stage for the current intense search for tumor-specific markers.

Serum and other blood products continue to be the single most used source of material for the development of biomarkers for cancer screening. The collection of blood from individuals carries a very low risk and is well accepted by persons being screened. Although tumors have a variety of characteristics that can be used to differentiate tumors and normal tissue, for screening, only those tumor-associated cellular products that are secreted or shed into the bloodstream can be easily sampled. This eliminates a large number of potential antigens, proteins and enzymes, associated with tumorigenesis, which are expressed and retained either within the cell or upon the cellular membrane. Serum markers that are developed for cancer screening may be used in a variety of configurations. Serum markers may be used as a singleton assay, sequentially with other screening tests or in parallel with other screening methods. The configuration of marker usage affects the sensitivity and specificity of a screening protocol.

Table 2 shows the variety of serum markers that have been identified and undergone some preliminary testing. Of these markers, none has been shown to reduce mortality rate. Even in the case of PSA, a widely used screening test for prostate cancer, there is not sufficient evidence of mortality rate reduction to allow a solid public health policy formulation. The PSA story is an illustrative example of the complexity of bringing a promising technology into practice. Prostate-specific antigen is a serum marker for prostate abnormalities; however, PSA is not specific for prostate cancer, as was originally believed, and a number of other benign conditions can cause elevation of PSA. Use of PSA in the general population has resulted in about 10% of men tested being identified as having an abnormal PSA. A positive PSA screen often leads to invasive follow-up (sextant biopsy, prostatectomy). There was a surge in prostate cancer incidence, which mirrored the introduction of PSA screening in the 1980s (7). The PSA screening detects a preclinical prostate cancer that is amenable to treatment. However, the natural history of prostate cancer is dominated by a large proportion of latent tumors found in older men. This high frequency of latent disease makes prostate cancer a difficult case in which to acheive a reduction in mortality rate by screening. Cancers that have later median ages of onset are likely to be eclipsed by competing morbidities, (e.g., heart disease), and, in general, the potential for reducing mortality rate by screening decreases with later onset cancers. In the case of PSA and prostate cancer, the PLCO screening trial will help determine whether PSA screening is associated with a mortality rate reduction (65).

A common theme that is emerging in the early detection field is the need for technologies or risk-factor criteria that separate high-risk and low-risk individuals for screening and follow-up. In the absence of an adequate triage procedure to separate these groups, the potential benefit of screening individuals in one risk group may be offset by the excess risks incurred by a group with different risk characteristics.

As portrayed in Figure 2, the development of new technologies is a nonlinear process, with new technologies emerging before previous techologies have been completely evaluated. For example, PSA has yet to be fully evaluated for mortality rate reduction, yet a serum test for free PSA has emerged and become available to the research and clinical communities (66).

Free PSA is an unbound form of PSA. Total PSA is made up of both the free and bound forms (ACT-PSA) of PSA. The measurement of the total and free PSA levels allows for the calculation of the ratio of free to total PSA (67). The use of a to-be-determined cut-point of the free-to-total PSA ratio is hypothesized to improve specificity over the total PSA value alone, thereby eliminating some men from undergoing unnecessary prostate biopsies (68). For free PSA, as is the case for total PSA, the evidence needed to assess mortality rate reduction is not available.

Insulin-like growth factors (IGFs) have long been recognized for their mitogenic and antiapoptotic effects on normal and transformed prostatic epithelial cells (69). In a case-control study on prospectively collected plasma, the level of IGF-1 was positively associated with an increased risk of prostate cancer and independently of PSA level (70). Further investigations of this issue are planned within the PLCO.

In addition to providing serum, blood samples also provide a source of circulating cancer cells. The existence of these circulating cancer cells are being actively pursued as a possible screening approach. Some cancers, as yet poorly defined, release tumor cells into the bloodstream at early points in their natural history (71,72). Detection of these cells may provide a useful early detection approach. Concentration of the circulating cancer cells by the use of aphretic enrichment using cell-volume characteristics may increase the limit of detection for relatively rare cells in the bloodstream. Aphresis is a relatively low-risk procedure but is time-consuming and costly. Once cancer cells are in an enriched fraction, the reverse transcriptase polymerase chain reaction (RT-PCR) may be used to measure tissue-specific mRNAs. The use of PCR after reverse transcription (RT) of mRNA into cDNA offers outstanding sensitivity, but it is difficult to use quantitatively. The key difficulty is that the net efficiency of the PCR process is inherently unstable, that is, small differences in the efficiency of DNA synthesis at each PCR cycle are amplified, resulting in variation in total yield with multiple PCR cycles. The magnitude and cause of these small differences vary depending on primer annealing efficiencies of different primers. To circumvent these problems, competitive PCR has been introduced. In this technique, a known amount of a synthetic DNA construct is added to the PCR reaction. The competitor is constructed such that it shares the primer sequence with the analyte DNA, but yields a distinct primer product, which allows cleaner quantitation.

B. Markers in Bodily Fluids, Brushings, Feces, and Secretions

In general, serological tumor markers as described in the previous section have not proven very effective in early detection because of their poor sensitivity and specificity. Their use has been limited mostly to patient surveillance and monitoring of recurrent tumors. In the early 1980s, the molecular and genetic studies of tumors led to the discovery of oncogenes (73–76) and tumor suppressor genes (77–80). Oncogenes capable of inducing the neoplastic transformation of cells, for example, *myc* and *ras*, and their gene products, such as the *ras* p21 protein, have been detected in human body fluids. Tumor suppressor genes such as RB1 in retinoblastoma, p53, and WT1, typically cell growth regulators, were discovered in studies of hereditary cancer. Gene products, typically nuclear regulatory proteins, interact with other proteins to alter transcription regulatory activity.

Genetic alterations can be detected in cells transported by bodily fluids and secretions. For example, urine contains cells shed from the bladder wall that may respond to appropriate molecular markers of genetic abnormality, although the degradation of these cells makes histology difficult (81). With the development of molecular markers that do not require histology, the usefulness of sputum for lung cancer screening would be enhanced (82). And, cells shed from the mucosal lining of the intestines and colon into feces (83), the consistent detection of which is confounded by time-variant changes in stool composition, may become a useful source of genetic material for improved technologies.

Over the past 2 decades, a variety of oncogenes and tumor suppressor genes have been implicated in the development of cancers. Applications have focused on early diagnosis, prognosis, and therapy. Among the many molecular markers (genes and gene products) identified, recent attention has been on genes known to be associated with tumor initiation, progression, and metastasis (Figure 8). The predominant genetic changes include deletions, rearrangements, and mutations of genetic loci, leading either to inactivation or activation of certain target genes. These changes result in unregulated cell proliferation and an increase in the amplification and expression of certain genes (e.g., *myc*, and HER2/neu) (84,85). More than 100 oncogenes have been identified and mapped within the

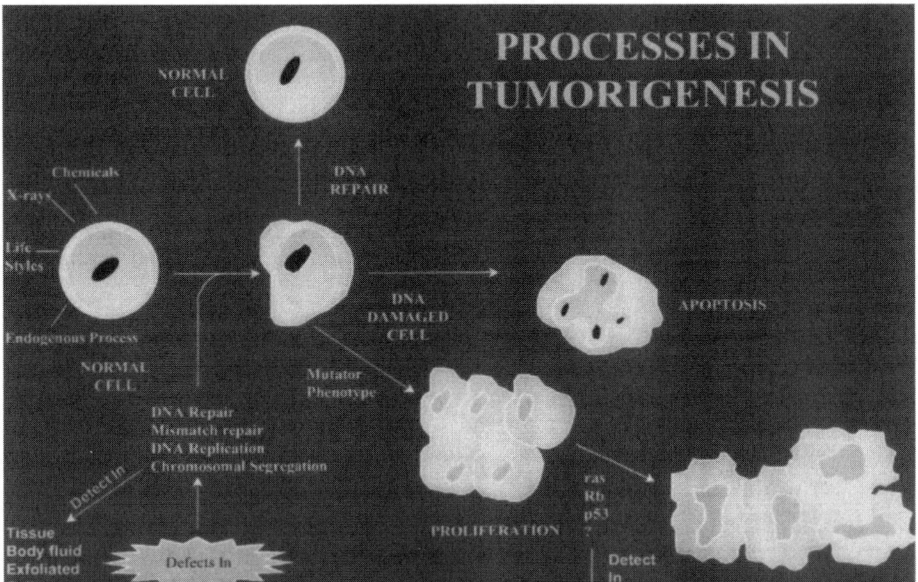

Figure 8 Processes in tumorogenesis. A simplified illustration of molecular and genetic changes during tumor initiation and progression. Genomic changes or gene products as indicated in the figure can be detected in tissues and body fluids.

Table 3 Potential Molecular Markers in Body Fluids

Cancer	Body fluid	Marker	Function
Head & neck	Saliva	Telomerase	Maintains telomeric length
		P53	Cell cycle check point
Lung	Sputum	ras	Growth factor signaling
		p53	Regulates cell cycle check point
		MSI[a]	Provides growth advantage
		FHIT[b]	Putative tumor suppressor gene
	Serum DNA	MSI	See above
Colon	Stool	ras	See above
		Telomerase	See above
Bladder	Urine	p53	See above
		MSI	See above
		Telomerase	See above
Pancreas	Stool/juice	ras	See above
Cervical	Cervical	Telomerase	See above

[a] Microsatellite instability
[b] Fragile histidine triad

human karyotype (86). Products of oncogenes include peptide growth factors, growth factor receptors, signal transduction factors, tyrosine kinases, and transcription regulatory proteins.

The ability to detect genetic alterations of preneoplastic tissue, in which the number of altered cells may be small, has been hampered by the lack of sensitive technologies. However, recent developments in nucleic acid-based methodologies have revolutionized detection capabilities. Techniques (e.g., PCR) are yielding new and potentially powerful nucleic acid-based markers in preneoplastic and neoplastic tissues. The PCR-based techniques can detect one altered cell among 10^6 normal cells in body fluids, including sputum, bladder wash, and serum (87–93). Table 3 lists a few markers of theoretical potential for early cancer detection based on these technological considerations; validation studies are lacking.

C. Germ-Line Markers to Identify At-Risk Individuals

Germ-line testing for cancer genes has been ushered into public consciousness by the availability of tests for the breast cancer genes BRCA1 and BRCA2. Germ-line testing for cancer genes is likely to become widely available as technologies make it simpler and less costly to screen for many different cancer as high-throughput procedures emerge.

Recent technological advances have made it possible to synthesize and assay biological molecules in a miniaturized format. Large arrays of biopolymers can be generated on microscopically small areas, for example, a 1 cm × 1 cm chip can contain 135,000 different oligonucleotides on the array. These arrays are synthesized robotically and analyzed with an automated system using computer programs. A chip with 20,000 different oligonucleotides can have data acquisition completed in about 15 minutes.

Nucleic acids (oligonucleotide array and cDNAs array) have been evaluated using the chip-microarray format. Recently, oligonucleotide arrays were used to analyze the genetic variability of human mitochondrial DNA (94). All possible polymorphisms were identified in the mitochondrial genome by synthesizing a number of different arrays and probing these arrays with fluorescently labeled reverse transcribed mitochondrial DNA samples (labeled RNA). This technology has potential to identify polymorphic loci contributing to an increased risk of cancer.

Tiled arrays present oligonucleotides that are degenerate in a central position and that change stepwise with increasing length with all possible sequences represented. The length of the oligonucleotides used in the array is determined by the empirically defined hybridization conditions for a given probe. Two-color analysis uses a reference probe that contains the known genome sequence labeled with a colored dye that is compared to the test sample improving fidelity as a result of probe competition. Most importantly, two-color analysis allows simultaneous comparison of loss of signal (binding) versus gain of signal data.

The power of oligonucleotide arrays for germ-line mutation detection was demonstrated by the use of high-density oligonucleotide arrays to detect mutations in the 3.45kB BRCA1 exon 11 region, which contains 60% of the BRCA1 coding sequence (95).

In summary, oligonucleotide arrays are most useful for the analysis of mutations in large regions of known sequence. This technique is critical for the future analysis of disease-causing mutations, particularly for genes such as BRCA1 for which mutations differ considerably between individual patients.

cDNA arrays, unlike oligonucleotide arrays, do not require the gene of interest to have a known sequence. cDNA arrays can be used to look at the differential expression of sequences of DNA with unknown function between different cells or conditions. This technique is useful for studying differential gene expression in a variety of tissues and conditions on the same chip. cDNAs are chemically coupled to glass slides and probed with fluorescently labeled total RNA, which has been isolated from the tissues of interest. Differentially expressed cDNAs can then be sequenced for identity. Hybridizations are done in hours, data collection in minutes, and data analysis is performed by computer.

In a recent paper, more than 1,000 unknown cDNAs were examined using a cDNA library constructed from human peripheral blood lymphocytes and PCR technology (96).

There are two broad categories of genes that might be amenable to cancer screening for cancer susceptibility. The first group is the ''cancer family'' germline mutations, which have been identified and confer a high level of risk to carriers. The second group are the ''gene-environment'' genes, which may modify cancer risk depending on the exposure history and lifestyle.

The cancer family genes are rare in the general population but tend to have high penetrance and an early age at onset. It might appear useful to identify these individuals at an early age to institute surveillance before cancer occurrence to optimize treatment. Screening for cancer family genes in the general population is not indicated because the frequency of these cancer genes is low. Individuals should only be considered for gene testing if they have a family history or they have a known gene carrier in their family. The advisability of screening will depend on the degree of disease penetrance for a particular cancer and the availability of an effective treatment for that early detected cancer. Lack of an effective treatment can eliminate the benefit of screening. Even a germline mutation with a highly penetrant and morbid disease can lose its usefulness in a screening setting without an effective treatment. For example, Huntington's disease, although not cancer, is a good example in that there are reliable gene tests for Huntington's disease but no useful treatments for gene carriers (97,98). As a result, many people who have a family history of the Huntington's gene have declined testing (99).

The gene-environment genes might prove useful in screening to help identify individuals susceptible to certain exposures (100). These genes include the cytochrome P-450 genes, glutathione-transferases, n-acetyl-transferases, and others (101). The way in which these genes might be used is to separate individuals by level of risk with respect to their personal genotype and phenotype. A screening program for particular cancers might be tailored to the individual's identified susceptibilities. In addition to identification of high-risk subsets for screening purposes, these individuals could be counseled about exposure avoidance for primary cancer prevention. An example of the potential for exposure avoidance by identification of a genetic metabolic disorder is the phenylketonuria (PKU) screening in newborns (102). Phenylketonuria is a metabolic disorder which can be well managed by avoidance of exposure to phenylalanine in the diet (103). Failure to identify the genetic susceptibility can result in phenylketonuria and mental retardation in the affected infants. Through newborn screening, the affected infants are identified at birth and appropriate precautions are taken to avoid the deleterious exposure. The gene-environment interactions associated with cancers are not likely to be as straightforward as the PKU screening situation, yet targeted screening of higher risk individuals is likely to increase benefit to those most likely to benefit and reduce risk to those unlikely to benefit from screening. Targeted screening may reduce the financial burden of generalized population screening while simultaneously avoiding overtreatment in individuals in whom disease is unlikely to develop.

D. Maturation of Molecular Technology

There is no systematic data on how long it takes for a laboratory-based test to mature (i.e., the time from the initial discovery to its transfer to clinical laboratories for routine practice). The point is illustrated by the following example. In 1913, the first description of cancer prone family "G" was reported (104). In 1960, mutations arising from "helix-with-loop" were reported (105), followed by the discovery of DNA polymerase slippage at repeated sequences and a first-ever report of "Cancer Family Syndrome" in 1966 (106), later named as hereditary nonpolyposis colorectal cancer (HNPCC) in 1984 (107). It was not until 1993 when a number of reports describing biological events responsible for this syndrome appeared. In 1993, microsatellite instability (MSI) was reported in the HNPCC syndrome (108) and the HNPCC locus was mapped to chromosome 2p (109). In 1993, a second locus was mapped to chromosome 3p and a human analog of yeast MSH2 (hMSH2) was discovered and implicated in the HNPCC syndrome (110). In 1994, three more HNPCC genes—hPMS1, hPMS2, and hMLH1—involved in DNA mismatch repair were reported (110–112). To date at least six HNPCC genes—hMSH1, hMSH2, hMLH1, hMLH2, hPMS1, and hMLH1—have been cloned and implicated in the HNPCC syndrome. In 1997, the genetic tests for these genes were made available. In summary, it took more than 60 years for the realization of laboratory tests for the HNPCC syndrome. However, this is not a typical timeline for the maturation of a laboratory test. The time involved in the discovery of associated genes in this example would have been reduced substantially had modern, high-throughput gene discovery technologies been available, but validation for early detection applications, not reflected in this time accounting, is yet years and maybe decades away.

IV. CONCLUSION

The past does not predict the future. Imaging technologies have traditionally been relatively free of government regulation, and medical markets have embraced new technologies, even very expensive ones, including CT and MRI, in the absence of proven value. These factors and the obvious need of diagnosticians and surgeons to visualize pathology have encouraged the development and marketing of imaging technologies, although that climate is changing toward a more conservative tone with the changes in the health care system in the United

States. Biomarkers for early detection, although conceived of decades ago, are only currently coming into serious play as the huge investments in biomolecular science and technology since the 1960s have begun to bear fruit. Small, entrepreneurial companies developing molecular and genetic diagnostic tests are increasingly common, but these companies typically lack the resources and expertise to validate their products for early detection. Developing collaborations among academic institutions, biomarker companies, and government may be the key to focusing biomolecular advances into clinically valid early cancer detection technologies.

REFERENCES

1. Kolb D A. On management and the learning process. Working paper 652–73. MIT Sloan School of Management, 1973.
2. Auvinen A, et al. Prospective evaluation plan for randomised trials of prostate cancer screening. The International Prostate Cancer Screening Trial Evaluation Group. J Med Screen 1996; 3:97–104.
3. Woods W G, Tuchman M, Robison L L, Bernstein M, Leclerc J M, Brisson L C, Brossard J, Hill G, Shuster J, Luepker R, Byrne T, Weitzman S, Bunin G, Lemieux B. A population-based study of the usefulness of screening for neuroblastoma. Lancet 1996; 348:1682–1687.
4. Bessho F. Effects of mass screening on age-specific incidence of neuroblastoma. Int J Cancer 1996; 67:520–522.
5. Bessho F. Where should neuroblastoma mass screening go? Lancet 1996; 348: 1672.
6. Yamamoto K, Hanada R, Kikuchi A, Masataka I, Toshinori A, Eiji O, Toshio M, Yoshihisa S, Masako T, Yasuhide H. Spontaneous regression of localized neuroblastoma detected by mass screening. J Clin Oncol 1998; 16:1265–1269.
7. Ries L A G, Kosary C L, Handkey B F, Miller B A, Harras A, Edwards B K, eds. SEER Cancer Statistics Review, 1973–1994, National Cancer Institute. [NIH Publication No. 97–2789] Bethesda, MD: National Institutes of Health, 1997.
8. Spiegel P K. The first clinical x-ray made in America—100 years. AJR Am J Roentgenol 1995; 164:241–243.
9. Curry T S, Dowdey J E, Murray R C. Christensen's Physics of Diagnostic Radiology. 4th ed. Philadelphia: Lea & Febiger, 1990.
10. National Council on Radiation Protection and Measurements (NRCP). Ionizing radiation exposures of the population of the United States. [Report No. 93] Washington, DC.: National Council on Radiation Protection and Measurements, 1987.
11. Fraser R S, Pare J A P, Fraser R G, Pare P D. Neoplastic diseases of the lungs. In: Synopsis of Diseases of the Chest. Philadelphia: W B Saunders, 1994: 445–538.
12. Woodring J H. Pitfalls in the radiologic diagnosis of lung cancer. AJR Am J Roentgenol 1990; 154:1165–1175.
13. Swensen S J, Gray J E, Brown L R, Aughenbaugh G L, Harms G F, Stears J. A new asymmetric screen-film combination for conventional chest radiography: evaluation in 50 patients. AJR Am J Roentgenol 1993; 160:483–486.

14. Gray J E, Stears J G, Swensen S J, Bunch P C. Evaluation of resolution and sensitometric characteristics of an asymmetric screen-film imaging system. Radiology 1993; 188:537–539.
15. United States Preventive Services Task Force. Screening for lung cancer. In: Guide to clinical preventive services. 2d ed. Baltimore: Williams & Wilkins, 1996:135–139.
16. Strauss G M, Gleason R E, Sugarbaker D J. Screening for lung cancer. Another look; a different view [see comments]. Chest 1997; 111:754–768.
17. Warren S L. Roentgenologic study of the breast. AJR Am J Roentgenol 1930; 24: 113–124.
18. Gold R H. The evolution of mammography. Radiol Clini North Am 1995; 30(1):1–19.
19. Mettler F A, Upton A C, Kelsey C A, Ashby R N, Rosenberg R D, Linver M N. Benefits versus risks from mammography: a critical reassessment [see comments]. Cancer 1996; 77:903–909.
20. Wu Y, Giger M L, Doi K, Vyborny C J, Schmidt R A, Metz C E. Artificial neural networks in mammography: application to decision making in the diagnosis of breast cancer. Radiology 1993; 187:81–87.
21. Adler D D, Wahl R L. New methods for imaging the breast: techniques, findings, and potential. AJR Am J Roentgenol 1995; 164:19–30.
22. Shtern F. Digital mammography and related technologies: a perspective from the National Cancer Institute. Radiology 1992; 183:629–330.
23. Fletcher S W, Black W C, Harris R, Rimer B K, Shapiro S. Report of the international workshop on screening for breast cancer. J Natl Cancer Inst 1993; 85: 1644–1656.
24. Anonymous. National Institutes of Health Consensus Development Conference Statement: Breast Cancer Screening for Women Ages 40–49, January 21–23, 1997. National Institutes of Health Consensus Development Panel. J Nat Cancer Inst 1997; 89(14):1015–1026.
25. Ernster V L, Barclay J, Kerlikowske K, Grady D, Henderson I C. Incidence of and treatment for ductal carcinoma in situ of the breast. JAMA 1996; 275:913–918.
26. Barnes G T, Lakshminarayanan A V. Computed tomography: physical principles and image quality cosiderations. In: Lee J K, Sagel S S, Stanley R J, eds. Computed Body Tomography. 2d ed. New York: Raven Press, 1989:1–22.
27. Heiken J P, Brink J A, Vannier M W. Spiral (helical) CT. Radiology 1993; 189: 647–656.
28. Fenlon H M, Ferrucci J T. Virtual colonoscopy: what wil the issues be? AJR Am J Roentgenol 1997; 169(2):453–458.
29. Hara A K, Johnson C D, Reed J E, Ehman R L, Ilstrup D. Colorectal polyp detection with CT colography: two-versus three-dimensional techniques. Work in progress. Radiology 1996; 200:49–54.
30. Ahlquist D A, Hara A K, Johnson C D. Computed tomographic colography and virtual colonoscopy. Gastrointest Endosc Clin North Am 1997; 7:439–452.
31. Hara A K, Johnson C D, Reed J E, Ahlquist D A, Nelson H, Ehman R L, Harmsen

W S. Reducing data size and radiation dose for CT colonography. AJR Am J Roentgenol 1997; 168:1181–1184.

32. Vining D J. Virtual endoscopy: is it reality? Radiology 1996; 200:30–31.
33. Vining D J, Liu K, Choplin R H, Haponik E F. Virtual bronchoscopy. Relationships of virtual reality endobronchial simulations to actual bronchoscopic findings. Chest 1996; 109:549–553.
34. Kaneko M, Eguchi K, Ohmatsu H, Katinuma R, Naruke T, Suemasu K, Moriyama N. Peripheral lung cancer: screening and detection with low-dose spiral CT versus radiography. Radiology 1996; 201:798–802.
35. Rubin G D, Beaulieu C F, Argiro V, Ringl H, Norbash A M, Feller J F, Drake M D, Jeffrey R B, Napel S. Perspective volume rendering of CT and MR images: applications for endoscopic imaging. Radiology 1996; 199:321–330.
36. Taylor K J, Schwartz P E. Screening for early ovarian cancer. Radiology 1994; 192:1–10.
37. Youssefzadeh S, Eibenberger K, Helbich T, Jakesz R, Wolf G. Use of resistance index for the diagnosis of breast tumours. Clin Radiol 1996; 51(6):418–420.
38. Westhoff C. Current status of screening for ovarian cancer. Gynecol Oncol 1994; 55:S34–S37.
39. Karlan B Y. The status of ultrasound and color Doppler imaging for the early detection of ovarian carcinoma. Cancer Inves 1997; 15:265–269.
40. Bourne T H, Campbell S, Reynolds K M, et al. Screening for early familial ovarian cancer with transvaginal ultrasonography and colour blood flow imaging [see comments]. BMJ 1993; 306:1025–1029.
41. Anonymous. NIH Consensus Conference. Ovarian cancer. Screening, treatment, and follow-up. NIH Consensus Development Panel on Ovarian Cancer. JAMA 1995; 273:491–497.
42. Kramer B S, Gohagan J, Prorok P C, Smart C. A National Cancer Institute sponsored screening trial for prostatic, lung, colorectal, and ovarian cancers. Cancer 1993; 71:589–593.
43. Jackson V P. The current role of ultrasonography in breast imaging. Radiol Clini North Am 1995; 33(6):1161–1170.
44. Stavros A T, Thickman D, Rapp C L, Dennis M A, Parker S H, Sisney G A. Solid breast nodules: use of sonography to distinguish between benign and malignant lesions [see comments]. Radiology 1995; 196(1):123–134.
45. Jackson V P. Management of solid breast nodules: what is the role of sonography? [editorial; comment]. Radiology 1995; 196(1):14–15.
46. Clements R. Has ultrasonography a role in screening for prostatic cancer? Eur Radiol 1997; 7:217–223.
47. Wehrli F W. Principles of magnetic resonance. In: Stark D D, Bradley W G, ed. Magnetic resonance imaging. 2d ed. St. Louis: Mosby–Year Book, 1992:3–21.
48. Harms S E. MRI in breast cancer diagnosis and treatment. [review]. Curr Probl Diagn Radiol 1996; 25:193–215.
49. Heywang-Kobrunner S H, Viehweg P, Heinig A, Kuchler C. Contrast-enhanced MRI of the breast: accuracy, value, controversies, solutions. Eur J Radiol 1997; 24:94–108.

50. Gilles R, Zafrani B, Guinebretiere J M, Meunier M, Lucidarme O, Rochard F, Vanel D, Neuenschwander S, Arriagada R. Ductal carcinoma in situ: MR imaging-histopathologic correlation. Radiology 1995; 196:415–419.
51. Gilles R, Meunier M, Lucidarme O. Clustered breast microcalcifications: evaluation by dynamic contrast-enhanced subtraction MRI. J Comput Assist Tomogr 1996; 20:9–14.
52. Soderstrom C E, Harms S E, Copit D S, Evans W P, Savino D A, Krakos P A, Farrell R S, Flamig D P. Three-dimensional RODEO breast MR imaging of lesions containing ductal carcinoma in situ. Radiology 1996; 201:427–432.
53. Nunes L W, Schnall M D, Siegelman E S, et al. Diagnostic performance characteristics of architectural features revealed by high spatial-resolution MR imaging of the breast. AJR Am J Roentgenol 1997; 169(2):409–415.
54. Gilles R, Guinebretiere J M, Lucidarme O, et al. Nonpalpable breast tumors: diagnosis with contrast-enhanced subtraction dynamic MR imaging. Radiology 1994; 191:625–631.
55. Waxman A D. The role of (99m)Tc methoxyisobutylisonitrile in imaging breast cancer. Semin Nucl Med 1997; 27:40–54.
56. Tolmos J, Cutrone J A, Wang B, Stunz M, Mishkin F S, Diggles L E, Venegas R J, Klein S R, Khalkholi, I. Scintimammographic analysis of nonpalpable breast lesions previously identified by conventional mammography. J Natl Cancer Inst 1998; 90:846–849.
57. Hoffman J M, Hanson M W, Coleman R E. Clinical positron emission tomography imaging. Radiol Clin North Am 1993; 31:935–959.
58. Hoh C K, Schiepers C, Seltzer M A, et al. PET in oncology: will it replace the other modalities? Semin Nucl Med 1997; 27:94–106.
59. Lewis P, Griffin S, Marsden P, Gee T, Nunan T, Malsey M, Dussek J. Whole-body 18F-fluorodeoxyglucose positron emission tomography in preoperative evaluation of lung cancer. Lancet 1994; 344(8932):1265–1266.
60. Benaron D A, Cheong W F, Stevenson D K. Tissue optics. Science 1997; 276: 2002–2003.
61. Tearney G J, Brezinski M E, Bouma B E, et al. In vivo endoscopic optical biopsy with optical coherence tomography. Science 1997; 276:2037–2039.
62. Hericourt J, Richet C. Comptes Rendus Hebdomadaires des Seances de, l'Academie des Sciences, 1895; 120:948–950.
63. Gold P, Freedman D S. Specific carcinoembryogenic antigens of the human digestive system. J Exp Med 1965; 122:468–481.
64. Begent R and Rustin G J S. Tumor Markers: from carcinoembryogenic antigen to products of hybridoma technology. Cancer Surveys 1989; 8:107–121.
65. Gohagan J K, Prorok P C, Kramer B S, Cornett J E. Prostate cancer screening in the prostate, lung, colorectal and ovarian cancer screening trial of the National Cancer Institute. J Urol 1994; 152:1905–1909.
66. Woodrum D L, French C M, Hill T M, Roman S J, Slatore H L, Shaffer J L, York L G, Eure K L, Loveland K G, Gasior G H, Southwick P C, Shamel L B. Analytical performance of the Tandem-R free PSA immunoassay measuring free prostate-specific antigen; Chem 1997; 43:1203–1208.

67. Partin A W, Catalona W J, Southwick P C, Subong E N, Gasior G H, Chan D W. Analysis of percent free prostate-specific antigen (PSA) for prostate cancer detection: influence of total PSA, prostate volume, and age, Urology 1996; 48:55–61.
68. Catalona W J, Smith D S, Ornstein D K, Prostate cancer detection in men with serum PSA concentrations of 2.6 to 4.0 ng/mL and benign prostate examination. Enhancement of specificity with free PSA measurements; JAMA 1997; 277:1452–1455.
69. Cohen S M, Ellwein L B. Cell proliferation in carcinogenesis. Science 1990; 249: 1007–1011.
70. Chan J M, Stampfer M J, Giiovannucci, Gann P H, Ma J, Wilkinson P, Hennekens C H, Pollak M. Plasma insulin-like growth factor-1 and prostate cancer risk: a prospective study. Science 1998; 279:563–566.
71. Seiden M, Sklar J L. PCR-and RT-PCR-based methods of tumor detection: potential applications and clinical implications. Important Adv Oncol 1996; 191–204.
72. Pelkey T J, Frierson H F Jr, Bruns D E. Molecular and immunological detection of circulating tumor cells and micrometastases from solid tumors. Clin Chem 1996; 42: 1369–1381.
73. Weiss W, Teich N, Varmus H, Coffin J, eds. RNA Tumor Viruses. Cold Spring Harbor, NY: Cold Spring Harbor Laboratory Press, 1982.
74. Bishop J M. Viral oncogenesis. Cell 1985; 42: 23–28.
75. Bishop J M. The molecular genetics of cancer. Science 1987; 235: 305.
76. Weinberg R A. Oncogenes, anti-oncogenes, and the molecular basis of multistep Carcinogenesis. Cancer Res. 1989; 49: 3713–3721.
77. Cavenee W K, Koufos A, Hansen M F. Recessive mutant genes predisposing to human cancer. Mutat Res 1986; 168: 3–14.
78. Knudson A G. Genetics of human cancer. Annu Rev Genet 1986; 20: 231–251.
79. Klein G. The approaching era of the tumor suppressor genes. Science 1987; 38: 1539–1545.
80. Friend S H, Bernard R, Rogelj S, Weinberg R A, Rapport J M, Albert D M, Dryja T P. A human DNA segment with properties of the gene that predisposes to retinoblastoma and osteosarcoma. Nature 1986; 323: 643–646.
81. Mao L. Genetic alterations as clonal markers for bladder cancer detection in urine. J Cell Biochem Suppl 1996; 25: 191–196.
82. Tockman M S. Clinical detection of lung cancer progression markers. J Cell Biochem Suppl 1996; 25: 177–184.
83. Davidson L A, Jiang Y H, Lupton J R, Chapkin R S. Noninvasive detection of putative biomarkers for colon cancer using fecal messenger RNA. Cancer Epidemiol Biomarkers Prev 1995; 4: 643–647.
84. Little, C, Nau M M, Carney D N, Gazdar A F, Minna J D. Amplification and expression of the c-*myc* oncogenes in human lung cancer cell lines. Nature 1983; 306: 194–196.
85. Slamon D J, Clark G M, Wong S G, Levin W G, Ullrich A, Mcguire W L. Human breast cancer: correlation of relapse and survival with amplification of the HER-2/neu oncogene. Science 1987; 235: 177–182.
86. Cooper G M. Oncogenes. Boston: Jones and Bartlett Publishers, 1990: 141–151.

87. Sidransky D. Molecular markers in cancer diagnosis. J Natl Cancer Inst Monogr 1995; 17: 27–29.
88. Sidransky D, Von Eschenbach A, Tsai Y C, Jones P, Summerhayes I, Marshal F, Green P, Hamilton S R, Frost P. Identification of p 53 gene mutations in bladder cancers and urine samples. Science 1991; 252: 706–709.
89. Mao L, Hruban R H, Boyle J O, Tockman M, Sidransky D. Detection of oncogene mutations in sputum precedes diagnosis of lung cancer. Cancer Res 1994; 54: 1634–1637.
90. Mao L, Lee D J, Tockman M S, Erozan Y S, Askin F, Sidransky D. Microsatellite alterations as clonal markers for the detection of human cancer. Proc Natl Acad Sci U S A 1994; 91: 9871–9875.
91. Steiner G, Schoenberg M P, Linn J F, Mao L, Sidransky D. Detection of bladder cancer recurrence by microsatellite analysis of urine. Nat Med 1997; 3: 621–624.
92. Chen X Q, Stroun M, Magnenat J L, Nicod L P, Kurt A M, Lyautey J, Lederrey C, Anker P. Microsatellite alterations in plasma DNA of small cell lung cancer patients. Nat Med 1996; 2: 1033–1035.
93. Nawroz H, Koch W, Anker P, Stroun M, Sidransky D. Microsatellite alterations in serum DNA of head and neck cancer patients. Nat Med 1996; 2: 1035–1037.
94. Chee M, Yang R, Hubbell E, Berno A, Huang X C, Stern D, Winkler J, Lockhart D J, Morris M S, Fodor S P. Accessing genetic information with high-density DNA arrays. Science 1996; 274: 610–614.
95. Hacia J G, Brody L C, Chee M S, Fodor S P, Collins F S. Detection of heterozygous mutations in BRCA1 using high density oligonucleotide arrays and two-color fluorescence analysis. Nat Genet 1996; 14: 441–447.
96. Schena M, Shalon D, Heller R, Chai A, Brown P O, Davis R W. Parallel human genome analysis: microarray-based expression monitoring of 1000 genes. Proc Natl Acad Sci U S A 1996; 93: 10614–10619.
97. Craufurd D. Huntington's disease. Prenat Diagn 1996; 16: 1237–1245.
98. Purdon S E, Mohr E, Ilivitsky V, Jones B D. Huntington's disease: pathogenesis, diagnosis and treatment. J Psychiatry Neurosci 1994; 19: 359–367.
99. Taylor S D. Demand for predictive genetic testing for Huntington's disease in Australia, 1987 to 1993. Med J Aust 1994; 161: 354–355.
100. Greenwald P. Cancer risk factors for selecting cohorts for large-scale chemoprevention trials. J Cell Biochem Suppl 1996; 25: 29–36.
101. Rothman N, Hayes R B. Using biomarkers of genetic susceptibility to enhance the study of cancer etiology. Environ Health Perspect 1995; 103 (suppl 8): 291–295.
102. Scriver C R. Science, medicine and phenylketonuria, Acta Paediatr Suppl 1994; 407: 11–18.
103. Smith I. Treatment of phenylalanine hydroxylase deficiency. Acta Paediatr Suppl 1994; 407: 60–65.
104. Warthin A S. Hereditary with reference to carcinoma. Arch Intern Med 1913; 12: 546–555.
105. Fresco J R, Alberts B M. The accommodation of noncomplimentary bases in helical polyribinucleotides and deoxyribonucleic acids. Proc Natl Acad Sci U S A 1960; 46: 311–321.
106. Lynch H T, Shaw M W, Magnusan C W, Larsen A L, Krush A J. Hereditary fac-

tors in cancer study of two large midwestern kindreds. Arch Intern Med 1966; 117: 206–212.

107. Boland C R, Troncale F J. Familial colonic cancer without antecedent polyposis. Ann Intern Med 1984; 100: 700–701.

108. Thibodeau S N, Bren G, Schaid D. Microsatellite instability in cancer of the prosimal colon. Science 1993; 260: 812–816.

109. Peltomaki P, Aaltonen L A, Sistonen P, Pyikkanen L, Mecklin J, Jarvinen H, Green J S, Jass J R, Weber J L, Leach F S, Petersen G M, Hamilton S R, de la Chapelle A, and Vogelstein B. Genetic mapping of a locus predisposing to human cancer. Science 1993; 260: 810–812.

110. Lindblom A, Tannergard P, Werelius B, Nordenskjold M. Genetic mapping of a second locus predisposing to hereditary nonpolyposis colon cancer. Nat Genet 1993; 5: 279–282.

111. Bronner C E, Baker S M, Morrison P T, Warren G, Smith L G, Lescoe M K, Kane M, Earbinoc C, Lipford J, Lindblom A, Tannergard P, Bollag R J, Godwin A R, Ward D C, Nordenskjold M, Fishel R, Kolodner R, Liskay R M. Mutation in the DNA mismatch repair gene homologue hMLH1 is associated with hereditary nonpolyposis cancer. Nature 1994; 368: 258–261.

112. Nicolaide N C, Papadopoulos N, Liu B, Wel Y, Carter K C, Ruben S M, Rosen C A, Haseltine W A, Fleischmann R D, Fraser C M, Adams M D, Venter J C, Dunlop M G, Hamilton S R, Petersen G M, de la Chapelle A, Vogelstein B, Kinzler K W. Mutations of two PMS homologues in hereditary nonpolyposis colon cancer. Nature 1994; 371: 75–80.

22

Genetic Inheritance of Cancer Risk: Applications to Cancer Screening

Emily L. Harris
Kaiser Permanente Center for Health Research, Portland, Oregon

I. INTRODUCTION

Current cancer screening recommendations are based in part on estimated cancer risk, because cancer risk in the screened population affects the performance characteristics of the screening test—the higher the risk, the higher the positive predictive value of a screening test. The most influential risk modifier is age—cancer-screening recommendations differ by age (1,2). Another influential risk modifier is family history of cancer—some organizations recommend that individuals with a family history of certain cancers (e.g., breast, prostate, colon) have earlier and/or more frequent screening (1,2)—such individuals are at increased cancer risk (relative risk of a specific cancer = 2 to 3 when one has a first-degree relative who has been diagnosed with that specific cancer [3–12]) and may be at higher risk of developing cancer at a younger-than-average age (4–7,10–12). As the biologic basis of a family history of cancer becomes understood and more cancer susceptibility genes are discovered, we will be able to make more specific risk predictions for individuals.

However, underlying cancer risk is not, in and of itself, proof that screening would be beneficial. Screening effectiveness hinges on proof of mortality, or at least morbidity, reduction when screening is coupled with effective therapy. There is little available information about the effectiveness of these enhanced cancer-screening strategies on reducing cancer morbidity or mortality.

II. BIOLOGIC BASIS OF HERITABLE RISK

Cancer results from a series of genetic or epigenetic changes that allows poorly controlled cell proliferation and invasion of normal tissue. Some of these genetic changes may be inherited; some may occur because an individual is genetically susceptible to environmental exposures that cause mutations; and some may occur by chance. Thus, the basis of the increased cancer risk associated with a family history of cancer may be genetic, environmental, or a combination of the two. Even individuals without a strong cancer family history may have an inherited sensitivity to certain exposures or a predisposition to certain cancers. Inherited genes with several types of functions affect cancer risk, including:

Tumor suppressor genes
Genes affecting repair of DNA damage
Carcinogen metabolism genes
Genes affecting immune function.

A. Tumor Suppressor Genes

Tumor suppressor genes are directly involved in cell growth and differentiation. As long as a cell has at least one normal-functioning copy of a tumor suppressor gene, abnormal cell proliferation is suppressed. Many documented or suspected tumor suppressor genes have been discovered, often through studies of tumor cell lines. These include the retinoblastoma (Rb) gene, inherited alterations associated with retinoblastoma; p53, inherited alterations associated with Li-Fraumeni syndrome; BRCA1 and BRCA2, inherited alterations associated with breast cancer and ovarian cancer; and VHL, inherited alterations associated with von Hippel–Lindau disease (13–20).

B. Genes Affecting Repair of DNA Damage

These genes may make an individual particularly sensitive to specific exposures; genes that are important in protecting against tumor initiation and progression (e.g., tumor suppressor genes) may be damaged through background or intentional exposures and continue through the cell cycle unrepaired, allowing a cell that can no longer function normally to replicate. For example, several genes have been identified that, when altered, greatly increase colon cancer risk (e.g., altered MLH1 and MSH2 genes and hereditary nonpolyposis colorectal cancer [HNPCC]) (21,22). These DNA mismatch repair genes produce proteins that are responsible for recognizing and correcting errors that arise during DNA replication. Alternatively, some individuals may inherit a sensitivity to certain kinds of exposures—i.e., they may not be able to repair exposure-induced DNA damage very well. For example, individuals with xeroderma pigmentosum are extremely sensitive to the sun and have a very high risk of developing skin cancer because they have inadequate repair of UV radiation-induced DNA damage (23). There is also epidemiologic evidence that DNA repair proficiency may influence breast cancer risk: individuals who carry one copy of an altered ataxia telangiectasia gene (ATM) may be at increased breast cancer risk (24,25); and the cells of women who have breast cancer or a family history of breast cancer more often show inadequate repair of ionizing radiation-induced DNA repair damage in vitro than those of other women (26).

C. Carcinogen Metabolism Genes

These genes affect how efficiently an individual activates potential carcinogens or gets them into his cells (e.g., genes for phase I enzymes) and how efficiently one detoxifies carcinogens or gets them out of his cells (e.g., genes for phase II enzymes). For example, glutathione S-transferase M1 deficiency (GSTM1 null status), an autosomal-recessive trait that nearly 50% of U.S. Caucasians have, is associated with increased lung cancer risk (27). Evidence is emerging that variation in certain carcinogen metabolism genes affects the function of other carcinogen metabolism genes—an example of gene-gene interaction (28).

D. Genes Affecting Immune Function

Immune function may affect cancer susceptibility in two ways: by influencing susceptibility to infection with viruses that activate oncogenes, and by influencing immune surveillance for abnormal cells. For example, immune response to HPV (human papilloma virus) antigens may be influenced by HLA (human leukocyte antigen) DR-DQ haplotype, which in turn affects the risk of developing cervical dysplasia and invasive cervical cancer (29,30).

III. PATTERNS OF GENETIC INHERITANCE

Genetically inherited traits may be transmitted through Mendelian inheritance (e.g., autosomal-dominant, -recessive, or -codominant trait; X-linked trait) or non-Mendelian inheritance (e.g., uniparental disomy; expansion of trinucleotide repeats). Genetically inherited traits may be transmitted through one genetic locus or by several loci (e.g., two-locus inheritance; polygenic inheritance). The models of inheritance most relevant to cancer susceptibility are autosomal-dominant (AD) inheritance and autosomal-recessive (AR) inheritance.

With AD inheritance, carrying only one copy of a cancer susceptibility gene alters one's cancer risk; for example, if an individual inherits an altered tumor suppressor gene from his mother or father, he is at increased cancer risk even though he also carries a normal-functioning copy of that gene (Figure 1). Well-known AD cancer susceptibility genes include: APC, MSH2, MLH1 (colon can-

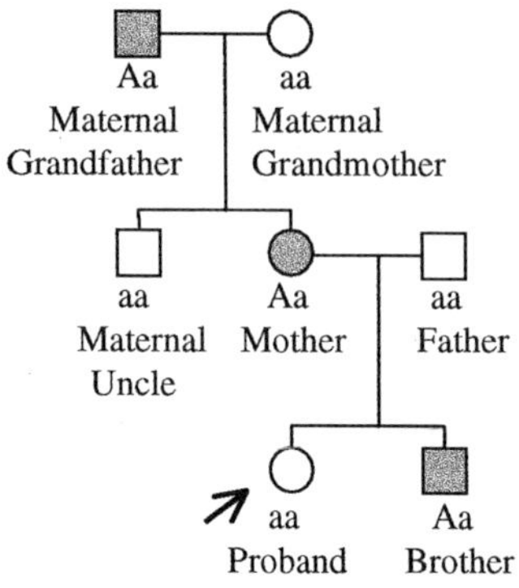

Figure 1 Autosomal-dominant inheritance of cancer susceptibility at genetic locus A. ‘‘A’’ is an altered gene or allele; ‘‘a’’ is a normal-functioning gene or allele. The shaded squares and circles indicate the persons who have an increased risk of cancer. The proband inherited normal-functioning alleles from her mother and her father, so she is not at increased cancer risk. The proband’s brother inherited an altered gene from his mother, so he has an increased risk of cancer. Note how cancer susceptibility is passed on from generation to generation.

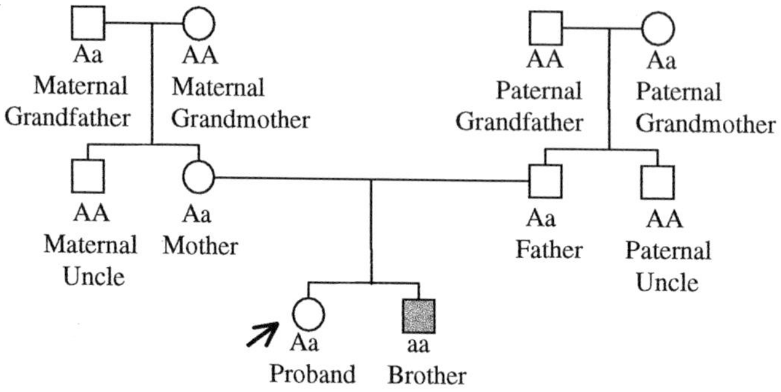

Figure 2 Autosomal-recessive inheritance of cancer susceptibility at genetic locus A. ‘‘A’’ is a normal-functioning gene or allele; ‘‘a’’ is an altered gene or allele. The shaded square indicates that that person has an increased risk of cancer. The proband inherited only one altered gene, so she is not at increased cancer risk. Her brother inherited an altered gene from his mother and his father, so he has an increased risk of cancer. Note how cancer susceptibility is only evident in one generation of the family.

cer) (21,22,31); BRCA1, BRCA2 (breast/ovarian cancer) (16,17); p53 (Li-Fraumeni syndrome) (15); VHL (von Hippel–Lindau disease) (18,19); and Rb (retinoblastoma) (13,14).

With AR inheritance, an individual must inherit a cancer susceptibility gene from each parent for his cancer risk to change (Figure 2). Well-recognized AR cancer susceptibility conditions include ataxia telangiectasia (AT [the ATM gene]), and xeroderma pigmentosum (XP) (32–34). These individuals are at extremely high cancer risk, and are sensitive to ionizing radiation and ultraviolet radiation, respectively.

Interestingly, there is evidence that individuals who carry one copy of the ATM gene (i.e., are heterozygous for the mutation in the AT gene that causes ataxia telangiectasia) may have a higher cancer risk; in particular, women who are heterozygous appear to be at increased breast cancer risk, although perhaps not for early-onset breast cancer (24,25). There is biologic plausibility; *in vitro*, the cells of individuals who presumably are ATM homozygotes and heterozygotes are more sensitive to ionizing radiation than the cells of presumably ''normal'' homozygotes (35), and breast cancer is associated with ionizing radiation exposure, especially, but not exclusively, when the exposure occurs in the late teens and early 20s (36). If true, this is an example of AD inheritance of cancer susceptibility, since only one copy of the altered gene may increase cancer risk. This may also be an example of gene-environmental interaction, in which the risk of breast cancer is greatly increased if one has both the susceptible genotype (heterozygous for ATM) and sufficient exposure (ionizing radiation exposure).

IV. EVIDENCE OF GENETIC INHERITANCE OF CANCER SUSCEPTIBILITY

A. Standard Epidemiologic Designs

Standard epidemiologic designs are able to detect familial clustering of specific cancers (e.g., case-control studies; cohort studies of relatives, based on family information collected in case-control studies)—an indication of genetic and/or shared environmental influences on cancer risk (33). Unless specific genes are measured, such designs cannot tease out the effects of genetic variation from shared environmental factors.

B. Special Designs

Special designs, such as twin studies, migrant studies, and adoption studies, have been used to disentangle the effects of genes from the effects of environmental factors. Each of these designs ''controls'' for genes or environment under a set of assumptions. In the basic twin design, one compares the similarity of identical (monozygotic [MZ]) twins with the similarity of fraternal (dizygotic [DZ]) twins. MZ twins share all their genes; DZ twins share, on average, 50% of their genes. If one assumes that the twinning process itself does not affect the trait of interest and that MZ twin pairs and DZ twin pairs share relevant environmental factors to the same degree, then greater similarity among MZ twins indicates that genetic variability affects individual variation in the trait under study. Twin studies of diseases like cancer are particularly challenging to carry out in the United States because it's difficult to identify a population-based sample of twin pairs—one that is unbiased with respect to concordance for cancer. (Concordant twin pairs are more likely to come to a researcher's attention or to volunteer for a study.) Areas or countries with population-based birth registries that can be linked to population-based cancer registries have had the most success (38), although it is possible to identify twin pairs in certain cohorts in the United States (39).

In cancer epidemiology, migrant studies have been used to assess the effects of different diets (e.g., a ''Westernized'' diet versus the ''native'' diet) on cancer risk (e.g., effect of fat intake and colon cancer risk [40]). This design assumes that individuals who migrate have a distribution of genes similar to those who don't migrate, that the new diet is adopted after migration, and that the individuals migrate at a time before the environmental factors of interest have their impact on risk (e.g., if childhood exposures are most influential, studying people who migrated as adults will fail to find an environmental effect). Under these assumptions, if cancer rates among migrants are similar to those in the new country, environmental factors have greater impact; if cancer rates among migrants are similar to those in their country of origin, genetic variation has a greater impact.

Some researchers have included second-generation immigrants as well, under the assumption that they are more likely to have adopted the ''new'' diet. This circumvents the assumption about timing of exposure, but does not address the issue of gene distribution.

C. Genetic Segregation Analysis

Genetic segregation analysis is a statistical modeling approach to teasing out the effect of genes and shared environment (41). One compares the distribution of a trait or condition within families with that expected under various models of inheritance, focusing on a series of genetic models. In its simplest form, segregation analysis is how Mendel developed principles of genetic inheritance (Mendelian inheritance)—he examined the phenotypic distribution of offspring from certain types of matings. Well into the 1970s, this simple form of genetic segregation analysis was the only type of segregation analysis available to most researchers. In its current form, genetic segregation analysis (sometimes called complex segregation analysis) includes the impact of genes at one genetic locus (major gene effect), multifactorial or polygenic inheritance, and, in some instances, major effects at more than one genetic locus (e.g., two-locus models), the impact of specific environmental factors, the impact of specific genes, or gene-environment interaction. To conclude genetic inheritance, one compares all models to a general model and must reject nongenetic models (e.g., nontransmitted environmental factors) and not reject a specific genetic model (e.g., autosomal-dominant inheritance). Genetic segregation analysis is typically used to indicate that one genetic locus may have a large impact on the trait, and to estimate the contribution of genetic variation at that locus to interindividual variability in the population (e.g., predisposition to lung cancer [42]). Therefore, it's important to have a well-defined sample of families, ideally, identified from a population-based resource such as a cancer registry.

V. LOCATING CANCER SUSCEPTIBILITY GENES

Often, the first step in locating cancer susceptibility genes is studying families with a high incidence of cancer in which it appears there is genetically inherited cancer susceptibility (high-risk families). The rationale is that, if there is heritable genetic variation in cancer susceptibility, this is where you will find it. Once the gene is discovered and mapped, additional studies can tell us more about the contribution of variation at that particular genetic locus to population variability in cancer risk. This strategy works well for genes that have a large impact on risk and that, in the current environment, do not depend highly on nongenetic factors for their expression (e.g., the effect of variation at BRCA1 and BRCA2 on breast and ovarian cancer risk [43,44]).

In linkage analysis, one examines the cosegregation of the trait of interest with genetic variation at specific loci within high-risk families (45,46) (e.g., collaborative linkage study of breast cancer [43]). Other approaches involve studying sibling pairs (concordant or discordant for cancer) or concordant relative pairs rather than families, but the general approach is the same (46). At the beginning of such a journey, one does not usually have a particular gene to assess but, rather, one wants to assess possible locations throughout the genome. This is typically accomplished by looking at anonymous genetic markers (e.g., variable number of tandem repeats; VNTRs) located throughout the genome, and assessing which ones show evidence of cosegregation with the trait. Once you find evidence for a particular area of a chromosome, one narrows it down to a relatively small region by analyzing genetic markers that are closer and closer together. Then it's time to actually find the gene using laboratory techniques such as positional cloning and DNA sequencing (45,47).

Sometimes one wants to study a particular gene, a candidate gene—it's a gene for which there is evidence that it is related to cancer risk (e.g., a carcinogen metabolism gene). If such genes have a large impact of cancer risk, they can be studied through linkage analysis as described above. More common genes, or genes for which there is evidence of gene-environment interaction, may be better studied using traditional epidemiologic designs (such as case-control studies). For example, the evidence for an association between breast cancer and cigarette smoking is conflicting (48,49). Emerging evidence suggests that breast cancer may be associated with cigarette smoking only with specific genotypes (50,51).

Cancer susceptibility genes have normal functions in the body (e.g., tumor suppressor genes control cell proliferation). Scientists may identify many mutations or variants in a gene. The challenge is to figure out what is ''normal'' variation, and what mutations or variants increase (or decrease) cancer risk. A gene mutation is simply an alteration from the usual or ''wild-type'' gene; polymorphism refers to a mutation that is common in the population (gene frequency of at least 1%) (52). Contrary to how these terms are sometimes used in cancer genetics literature, in population genetics, neither of these terms implies a deleterious or neutral effect. A mutation could be deleterious, advantageous, or neutral; likewise, a polymorphism could have any of these effects.

VI. UNDERSTANDING THE IMPACT OF SPECIFIC GENES

The story does not end with gene discovery; that's just the beginning. To implement genetic testing programs with appropriate prevention and screening interventions, scientists must understand the impact of specific genes on cancer risk, and how their effects may be modified by other genetic or nongenetic factors.

Genes produce proteins; functional studies of these proteins may be particularly useful in developing cancer prevention strategies. They may also help in developing appropriate cancer screening protocols. For example, if functional studies showed that a particular genetic mutation increased a person's sensitivity ionizing radiation, would it be appropriate for such women to have earlier and more frequent mammography to detect breast cancer? In that latter example, we would need to evaluate the risk:benefit ratio of a screening test based on ionizing radiation very carefully.

VII. APPLICATION OF GENETICS TO CANCER SCREENING

Cancer screening is still generally based primarily on age and cancer family history, with benefit estimates based almost exclusively on the population at large. Genetic cancer susceptibility testing is commercially available, and is being performed despite limited knowledge of the impact of specific genes and what we can do to change a person's cancer risk. Enhanced cancer screening, in the form of earlier or more frequent screening or less proven procedures, is often recommended in the literature despite lack of clear evidence that these interventions will have a beneficial impact on cancer mortality. For example, consensus follow-up recommendations include: colonoscopy every 1 to 3 years starting at age 25 years and endometrial cancer screening for individuals who have a genetic predisposition to hereditary nonpolyposis colon cancer; early breast cancer and ovarian cancer screening for individuals who have a high-risk BRCA1 mutation; and early breast cancer screening for those with a high-risk BRCA2 mutation (53,54).

VIII. THE FUTURE

As genetic information is becoming a greater part of health care, we desperately need to answer many questions. Genetic testing has the potential to discriminate individuals at increased cancer risk from those at ''usual'' cancer risk within high-risk families—something cancer family history alone cannot do. We can approach genetic testing like a screening test.

Before genetic testing for cancer risk can be accepted and/or recommended as routine practice, we should know more about cancer risks (gene penetrance):

What are the cancer risks associated with specific mutations or types of mutations in cancer susceptibility genes? How do these estimates differ in various populations? Almost all of the currently available information is based on families who were selected because they have a very high incidence of cancer. For example, breast and ovarian cancer risks associated with BRCA1 mutations have been estimated from selected populations: families selected for having a very high incidence of breast and/or ovarian cancer (43); and a U.S. Ashkenazi-Jewish population (44). The risks associated with specific mutations in very high risk families may differ from those found outside of such families; the estimated cumulative breast cancer risk by age 70 was 82% in the families (43), but was 56% by age 70 in the Ashkenazi-Jewish population (44). Cancer risks may differ because, in addition to sharing genes at a specific cancer susceptibility locus, very high risk families also share other genes and lifestyles that may affect cancer risk, or because cancer risk varies with different mutations in the same gene (55,56),

How is the identified gene's effect modified by other genes and environmental exposures? There is evidence that, even among high-risk mutation carriers, variation in other genes can affect cancer risk (57), and that one mutation can have varied expression (58).

In addition to the primary cancer(s) of interest (e.g., breast and ovarian cancer risk for BRCA1 mutation carriers), what other conditions are associated with specific genes? For example, in addition to increased colon cancer risk, women who carry an altered MSH2 gene are at increased risk of endometrial and ovarian cancer (59).

Which mutations are likely or unlikely to alter an individual's cancer risk? Some mutations will result in structurally altered proteins (e.g., mutations that produce truncated proteins) and will almost certainly increase cancer risk. Some mutations do not alter protein structure at all and will almost certainly not increase cancer risk. However, there is preliminary evidence that some mutations do not alter protein structure, but instead create an unstable region in the chromosome, making acquired mutations more likely to occur (60). There are gene mutations whose functional significance is unknown.

What is the cancer risk associated with a ''normal'' test result? The interpretation will differ according to the individual's family history of cancer and the known presence or absence of specific mutations in his or her family. For example, if a person comes from a very high risk family with no prior evidence of a high-risk mutation in the family, a ''normal'' test result for an individual means very little; he or she has the same estimated cancer risk as before testing.

We also need to understand what interventions may decrease cancer risk and/or mortality:

Does early or more frequent cancer screening improve cancer mortality among populations at increased cancer risk, whether risk status is based on family history of cancer or on specific genetic information? This is a common recommendation for individuals with a family history of breast, colon, or prostate cancer (1,2), and is strongly emphasized for individuals who have high-risk mutations in BRCA1, BRCA2, or a genetic predisposition to hereditary nonpolyposis colon cancer (53,54). However, there is little or no information available regarding evidence of benefit.

Do standard cancer screening approaches improve cancer mortality for high-risk populations? In some instances, the emphasis for populations at increased cancer risk is on ''adherence to'' cancer screening recommendations for the general population. We don't know if they are equally effective in these targeted populations. (There is evidence that the age-specific sensitivity of mammography is lower for women with a family history of breast cancer [61]).

For unproved screening modalities, is the risk:benefit balance likely to improve for high-risk populations? Some cancer screening strategies show little promise in the general population, yet some researchers recommend them for high-risk populations (e.g., ovarian cancer screening for BRCA1 high-risk mutation carriers [54]). There is little or no information about benefit in targeted populations.

Genetic testing for cancer risk has several unusual, but not unique, characteristics that need to be considered as well:

Genetic testing for cancer susceptibility is "predictive testing" in which the lifetime risk of developing cancer is less than 100%.

The test result may have important health and social implications for biologic relatives, including future generations.

The test result is not modifiable (although the cancer risk associated with the test result may be) and therefore cannot be used to measure the impact of preventive strategies on risk.

These characteristics raise a host of social, legal, and ethical issues for the individuals being tested and for their relatives, particularly in our current health care system. These issues, which have not yet been resolved, include difficulty obtaining insurance and restricted educational and employment opportunities.

One final consideration—public opinion:

Genetic testing for cancer susceptibility must be acceptable to the public. Members of very high risk cancer families, as well as the general public, are interested in genetic testing for cancer risk. However, it is clear that clinicians, scientists, and the general public are only beginning to understand the limitations and social, legal, and ethical issues surrounding such testing. Under what conditions/situations would the public consider genetic testing to be acceptable? Current genetic testing strategies are focused on genes that confer greatly increased cancer risk, sometimes referred to as "inherited cancer" or "inherited cancer susceptibility" genes; these genes account for about 5% to 10% of all breast and colon cancers. In the future, genetic testing strategies may be broader, including more common, moderate-risk cancer susceptibility genes whose expression depends more on exposures and other genes.

Genetic testing protocols that are acceptable to scientists, clinicians, and the public must be available. Like genetic testing in general, genetic counseling is an integral part of the process of genetic testing for cancer susceptibility; specialists in oncology and mental health will also need to be a part of the process (62,63). Genetic testing protocols are currently being developed and tested (64,65). Testing minors for cancer risk should be given special consideration because minors are particularly vulnerable. It should be considered only when it is clear that intervention will provide medical benefits for young people.

Genetics applications in health care are only going to increase; new genes are being discovered every day. Unfortunately, knowledge that helps us apply this information to improve the health of individuals and populations is extremely limited. This limited knowledge has not stopped the development of commercial genetic testing for cancer susceptibility, although such uncertainties and concerns about the ethical, legal, and social ramifications of such testing may temper use by the public. We need to make substantial progress in these areas to make genetic testing for cancer susceptibility beneficial to the public.

REFERENCES

1. American Cancer Society's Web page: http://www.cancer.org/bottomcancinfo.html.
2. U.S. Preventive Services Task Force. Guide to Clinical Preventive Services. 2nd ed. Baltimore: Williams and Wilkins, 1996. Web page: http://text.nlm.nih.gov.
3. Sattin R W, Rubin G L, Webster L A, Huezo C M, Wingo P A, Ory H W, Layde P M, and the Cancer and Steroid Hormone Study. Family history and the risk of breast cancer. JAMA 1985; 253:1908–1913.
4. Colditz G A, Willett W C, Hunter D J, Stampfer M J, Manson J E, Hennekens C H, Rosner B A, Speizer F E. Family history, age, and risk of breast cancer. Prospective data from the Nurses' Health Study. JAMA 1993; 270:338–343.
5. Slattery M L, Kerber R A. A comprehensive evaluations of family history and breast cancer risk. The Utah population database. JAMA 1993; 270:1563–1568.
6. Slattery M L, Kerber R A. Family history of cancer and colon cancer risk: the Utah Population Database [published erratum appears in J Natl Cancer Inst 1994; 86: 1802]. J Natl Cancer Inst 1994; 86:1618–1626.
7. Fuchs C S, Giovannucci E L, Colditz G A, Hunter D J, Speizer F E, Willett W C. A prospective study of family history and the risk of colorectal cancer. N Engl J Med 1994; 331:1669–1674.
8. Hayes R B, Liff J M, Pottern L M, Greenberg R S, Schoenberg J B, Schwartz A G, Swanson G M, Silverman D T, Brown L M, Hoover R N et al. Prostate cancer risk in U.S. blacks and whites with a family history of cancer. Int J Cancer 1995; 60: 361–364.
9. Whittemore A S, Wu A H, Kolonel L N, John E M, Gallagher R P, Howe G R, West D W, Teh C Z, Stamey T. Family history and prostate cancer risk in black, white, and Asian men in the United States and Canada. Am J Epidemiol 1995; 141:732–740.
10. Grönberg H, Damber L, Damber J E. Familial prostate cancer in Sweden. A nationwide register cohort study. Cancer 1996; 77:138–143.
11. Le Marchand L, Zhao L P, Quiaoit F, Wilkens L R, Kolonel L N. Family history and risk of colorectal cancer in the multiethnic population of Hawaii. Am J Epidemiol 1996; 144:1122–1128.
12. Lesko S M, Rosenberg L, Shapiro S. Family history and prostate cancer risk. Am J Epidemiol 1996; 144:1041–1047.
13. Lee W H, Bookstein R, Hong F, Young L J, Shew J Y, Lee E Y. Human retinoblastoma susceptibility gene: cloning, identification, and sequence. Science 1987; 235:1394–1399.
14. Schubert E L, Hansen M F, Strong L C. The retinoblastoma gene and its significance. Ann Med 1994; 26:177–184.

15. Malkin D, Li F P, Strong L C, Fraumeni J F Jr, Nelson C E, Kim D H, Kassel J, Gryka M A, Bischoff F Z, Tainsky M A, Friend S H. Germ line p53 mutations in a familial syndrome of breast cancer, sarcomas and other neoplasms. Science 1990; 250:1233–1238.
16. Miki Y, Swensen J, Shattuck-Eidens D, et al. A strong candidate for the breast and ovarian cancer susceptibility gene BRCA1. Science 1994; 266:66–71.
17. Wooster R, Bignell G, Lancaster J, Swift S, Seal S, Mangion J, Collins N, Gregory S, Gumbs C, Micklem G. Identification of the breast cancer susceptibility gene BRCA2 [published erratum appears in Nature 1996; 379:749]. Nature 1995; 378: 789–792.
18. Latif F, Tory K, Gnarra J, Yao M, Duh F M, Orcutt M L, Stackhouse T, Kuzmin I, Modi W, Geil L, et al. Identification of the von Hippel–Lindau disease tumor suppressor gene. Science 1993; 260:1317–1320.
19. Maher E R, Webster A R, Richards F M, Green J S, Crossey P A, Payne S J, Moore A T. Phenotypic expression in von Hippel–Lindau disease: correlations with germline VHL gene mutations. J Med Genet 1996; 33:328–332.
20. Weinstein I, Santella R M, Perera F. Molecular biology and epidemiology of cancer. In: Greenwald P, Kramer B S, Weed D L eds. Cancer Prevention and Control. New York: Marcel Dekker, 1995:83–110.
21. Fishel R, Lescoe M K, Rao M R, Copeland N G, Jenkins N A, Garber J, Kane M, Kolodner R. The human mutator gene homolog MSH2 and its association with hereditary nonpolyposis colon cancer. Cell 1993; 75:1027–1038.
22. Papadopoulos N, Nicolaides N C, Wei Y F, Ruben S M, Carter K C, Rosen C A, Haseltine W A, Fleischmann R D, Fraser C M, Adams M D, et al. Mutation of a mutL homolog in hereditary colon cancer. Science 1994; 263:1625–1629.
23. Kraemer K H, Lee M M, Andrews A D, et al. The role of sunlight and DNA repair in melanoma and nonmelanoma skin cancer. Arch Dermatol 1994; 130:1018–1021.
24. Athma P, Rappaport R, Swift M. Molecular genotyping shows that ataxia-telangiectasia heterozygotes are predisposed to breast cancer. Cancer Genet Cytogenet 1996; 92:130–134.
25. FitzGerald M G, Bean J M, Hegde S R, Unsal H, MacDonald D J, Harkin D P, Finkelstein D M, Isselbacher K J, Haber D A. Heterozygous ATM mutations do not contribute to early onset of breast cancer. Nat Genet 1997; 15:307–310.
26. Helzlsouer K J, Harris E L, Parshad R, Perry H R, Price F M, Sanford K K. DNA repair proficiency: potential susceptibility factor for breast cancer. J Natl Cancer Inst 1996; 88:754–755.
27. McWilliams J E, Sanderson B J, Harris E L, Richert-Boe K E, Burton K L, Henner W D. Glutathione S-transferase M1 (GSTM1) deficiency and lung cancer risk. Cancer Epidemiol Biomarkers Prev 1995; 4:589–594.
28. MacLeod S, Sinha R, Kadlubar F F, Lang N P. Polymorphisms of CYP1A1 and GSTM1 influence the in vivo function of CYP1A2. Mutat Res 1997; 376:135–142.
29. Apple R J, Becker T M, Wheeler C M, Erlich H A. Comparison of human leukocyte antigen DR-DQ disease associations found with cervical dysplasia and invasive cervical carcinoma. J Natl Cancer Inst 1995; 87:427–436.

30. Odunsi K, Terry G, Ho L, Bell J, Cuzick J, Ganesan T S. Susceptibility to human papillomavirus-associated cervical intra-epithelial neoplasia is determined by specific HLA DR-DQ alleles. Int J Cancer 1996; 67:595–602.
31. Kinzler K W, Nilbert M C, Su L K, Vogelstein B, Bryan T M, Levy D B, Smith K J, Preisinger A C, Hedge P, McKechnie D, et al. Identification of FAP locus genes from chromosome 5q21. Science 1991; 253:661–5.
32. Muller H. Recessively inherited deficiencies predisposing to cancer. Anticancer Res 1990; 10:513–518.
33. Kraemer K H, Levy D D, Parris C N, Gozukara E M, Moriwaki S, Adelberg S, Seidman M M. Xeroderma pigmentosum and related disorders: examining the linkage between defective DNA repair and cancer. J Invest Dermatol 1994; 103(suppl): 96S-101S.
34. Gatti R A. Ataxia telangiectasia. Dermatol Clin 1995; 13:1–6.
35. West C M, Elyan S A, Berry P, Cowan R, Scott D. A comparison of the radiosensitivity of lymphocytes from normal donors, cancer patients, individuals with ataxia-telangiectasia (A-T) and A-T heterozygotes. Int J Radiat Biol 1995; 68:197–203.
36. Bier V. Health Effects of Low Levels of Ionizing Radiation. Washington, DC: National Academy Press, 1990.
37. Khoury M J, Beaty T H, Cohen B H. Fundamentals of Genetic Epidemiology. New York: Oxford University Press, 1993:164–199.
38. Ahlbom, A Lichtenstein P, Malmstrom H, Feychting M, Hemminki K, Pedersen N L. Cancer in twins: genetic and nongenetic familial risk factors. J Natl Cancer Inst 1997; 89:287–293.
39. Braun M M, Caporaso N E, Page W F, Hoover R N. A cohort study of twins and cancer. Cancer Epidemiol Biomarkers Prev 1995; 4:469–473.
40. Kolonel J N, Harkin J H, Nomura A M, et al. Multiethnic studies of diet, nutrition and cancer in Hawaii. In: Huyashi, ed. Diet, Nutrition and Cancer. Toyko: Science Society Press, 1986:29–40.
41. Khoury M J, Beaty T H, Cohen B H. Fundamentals of Genetic Epidemiology. New York: Oxford University Press, 1993:233–283.
42. Sellers T A, Bailey-Wilson J E, Elston R C, Wilson A F, Elston G Z, Ooi W L, Rothschild H. Evidence for mendelian inheritance in the pathogenesis of lung cancer. J Natl Cancer Inst 1990; 82:1272–1279.
43. Easton D F, Bishop D T, Ford D, et al. Genetic linkage analysis in familial breast and ovarian cancer: results from 214 families. Am J Hum Genet 1993; 52: 678–701.
44. Struewing J P, Hartge P, Wacholder S, Baker S M, Berlin M, McAdams M, Timmerman M M, Brody L C, Tucker M A. The risk of cancer associated with specific mutations of BRCA1 and BRCA2 among Ashkenazi Jews. N Engl J Med 1997; 336:1401–1408.
45. Thompson M W, McInnes R R, Willard H F. Thompson & Thompson Genetics in Medicine. 5th ed. Philadelphia: W. B. Saunders, 1991:167–199.
46. Khoury M J, Beaty T H, Cohen B H. Fundamentals of Genetic Epidemiology. New York: Oxford University Press, 1993:284–311.

47. Thompson M W, McInnes R R, Willard H F. Thompson & Thompson Genetics in Medicine. 5th ed. Philadelphia: W. B. Saunders, 1991:97–114.
48. Baron J A, Newcomb P A, Longnecker M P, Mittendorf R, Storer B E, Clapp R W, Bogdan G, Yuen J. Cigarette smoking and breast cancer. Cancer Epidemiol Biomarkers Prev 1996; 5:399–403.
49. Morabia A, Bernstein M, Heritier S, Khatchatrian N. Relation of breast cancer with passive and active exposure to tobacco smoke. Am J Epidemiol 1996; 143:918–928.
50. Ambrosone C B, Freudenheim J L, Graham S, Marshall J R, Vena J E, Brasure J R, Michalek A M, Laughlin R, Nemoto T, Gillenwater K A, Shields PG. Cigarette smoking, N-acetyltransferase 2 genetic polymorphisms, and breast cancer risk. JAMA 1996; 276:1494–501.
51. Ishibe N, Hankinson S E, Colditz G A, Spiegelman D, Willett W C, Speizer F E, Kelsey K T, Hunter D J. Cigarette smoking, cytochrome P450 1A1 polymorphisms, and breast cancer risk in the Nurses' Health Study. Cancer Res 1998; 58:667–671.
52. Cavalli-Sforza L L, Bodmer W F. The Genetics of Human Populations. San Francisco: W. H. Freeman, 1971:118.
53. Burke W, Petersen G, Lynch P, Botkin J, Daly M, Garber J, Kahn M J, McTiernan A, Offit K, Thomson E, Varricchio C. Recommendations for follow-up care of individuals with an inherited predisposition to cancer. I. Hereditary nonpolyposis colon cancer. Cancer Genetics Studies Consortium. JAMA 1997; 277:915–919.
54. Burke W, Daly M, Garber J, Botkin J, Kahn M J, Lynch P, McTiernan A, Offit K, Perlman J, Petersen G, Thomson E, Varricchio C. Recommendations for follow-up care of individuals with an inherited predisposition to cancer. II. BRCA1 and BRCA2. Cancer Genetics Studies Consortium. JAMA 1997; 277:997–1003.
55. Gayther S A, Mangion J, Russell P, Seal S, Barfoot R, Ponder B A J, Stratton M R, Easton D. Variation of risks of breast and ovarian cancer associated with different germline mutations of the BRCA2 gene. Nat Genet 1997; 15:103–105.
56. Neuhausen S L, Mazoyer S, Friedman L, Stratton M, Offit K, Caligo A, Tomlinson G, Cannon-Albright L, Bishop T, Kelsell D, et al. Haplotype and phenotype analysis of six recurrent BRCA1 mutations in 61 families: Results of an international study. Am J Hum Genet 1996; 58:271–280.
57. Phelan C M, Rebbeck T R, Weber B L, Devilee P, Ruttledge M H, Lynch H T, Lenoir G M, Stratton M R, Easton D F, Ponder B A J, Cannon-Albright K, Larsson C, Goldgar D E, Narod S A. Ovarian cancer risk in BRCA1 carriers is modified by the HRAS1 variable number of tandem repeat (VNTR) locus. Nat Genet 1996; 12: 309–311.
58. Thorlacius S, Olafdottir G, Tryggvadottir L, Neuhausen S, Jonasson J G, Tavtigian S V, Tulinius H, Ögmundsdottir H M, Eyfjörd J E. A single BRCA2 mutation in male and female breast cancer families from Iceland with varied cancer phenotypes. Nat Genet 1996; 13:117–119.
59. Lynch H T, Smyrk T C, Watson P, et al. Genetics, natural history, tumor spectrum, and pathology of hereditary nonpolyposis colorectal cancer: an updated review. Gastroenterology 1993; 104:1535–1549.

60. Laken S J, Petersen G M, Gruber S B, Oddoux C, Ostrer H, Giardiello F M, Hamilton S R, Hampel H, Markowitz A, Klimstra D, Jhanwar S, Winawer S, Offit K, Luce MC, Kinzler K, Vogelstein B. Familial colorectal cancer in Ashkenazim due to a hypermutable tract in APC. Nat Genet 1997; 17:79–83.
61. Kerlikowske K, Grady D, Barclay J, Sickles E A, Ernster V. Effect of age, breast density, and family history on the sensitivity of first screening mammography. JAMA 1996; 276:33–38.
62. Anonymous: Statement of the American Society of Human Genetics on genetic testing for breast and ovarian cancer predisposition. Am J Hum Genet 1994; 55:i–v.
63. American Society of Clinical Oncology. Statement of the American Society of Clinical Oncology: genetic testing for cancer susceptibility, adopted February 20, 1996. J Clin Oncol 1996; 14:1730–1736.
64. Geller G, Botkin J R, Green M J, Press N, Biesecker B B, Wilfond B, Grana G, Daly M B, Schneider K, Kahn M J E. Genetic testing for susceptibility to adult-onset cancer. The process and content of informed consent. JAMA 1997; 277:1467–1474.
65. Lerman C, Biesecker B, Benkendorf J L, Kerner J, Gomez-Caminero A, Hughes C, Reed M M. Controlled trial of pretest education approaches to enhance informed decision-making for BRCA1 gene testing. J Natl Cancer Inst 1997; 89:148–157.

Index

About the Editors

BARNETT S. KRAMER is Deputy Director, Division of Cancer Prevention, National Cancer Institute, Bethesda, Maryland. Additionally, he is a Clinical Professor in the Department of Medicine, Uniformed Services University for the Health Sciences, Bethesda, Maryland. The Editor-in-Chief of the *Journal of the National Cancer Institute,* Dr. Kramer is the author, coauthor, editor, or coeditor of over 160 research articles, invited papers, book chapters, abstracts, and books, including *Cancer Prevention and Control* (Marcel Dekker, Inc.). A Fellow of the American College of Physicians, he is a member of the American Society of Clinical Oncology and the American Association for Cancer Research. Dr. Kramer received the M.D. degree (1973) with honors from the University of Maryland Medical School, Baltimore, and the M.P.H. degree (1991) from the Johns Hopkins University School of Hygiene and Public Health, Baltimore, Maryland.

JOHN K. GOHAGAN is Chief, Early Detection Branch, Division of Cancer Prevention, National Cancer Institute, Bethesda, Maryland. The author or coauthor of more than 40 related publications, he is a Fellow of the American College of Epidemiology and a member of Sigma Xi, the Scientific Research Society. He received the B.A. degree (1964) in physics from South College, Philadelphia, Pennsylvania, the M.A. degree (1967) in physics from Temple University, Philadelphia, and the Ph.D. degree (1973) from the Massachusetts Institute of Technology, Cambridge.

PHILIP C. PROROK is Chief, Screening Section, and Acting Chief, Biometry Branch, Division of Cancer Prevention, National Cancer Institute, Bethesda, Maryland. A Statistical Editor of the *Journal of the National Cancer Institute,* he is the author or coauthor of over 70 research articles and book chapters. A member of the Biometric Society, the American Statistical Association, and the Society for Clinical Trials, Dr. Prorok received the B.A. degree (1965) in chemistry from the State University of New York at Buffalo, the M.S. degree (1966) in chemistry from the University of Chicago, Illinois, and the Ph.D. degree (1973) in statistics from the State University of New York at Buffalo.